MATHEMATICAL CONNECTIONS: A MODELING APPROACH TO BUSINESS CALCULUS AND FINITE MATHEMATICS

Preliminary Edition

Audrey Fredrick Borchardt &
Bruce Pollack-Johnson

The Villanova Project

PRENTICE HALL, Upper Saddle River, New Jersey 07458

Senior Acquisitions Editor: **SALLY SIMPSON**
Marketing Manager: **PATRICE LUMUMBA JONES**
Editorial Assistant: **APRIL THROWER**
Assistant Vice-President of Production
and Manufacturing: **DAVID W. RICCARDI**
Editorial/Production Supervision: **RICHARD DeLORENZO**
Managing Editor: **LINDA MIHATOV BEHRENS**
Executive Managing Editor: **KATHLEEN SCHIAPARELLI**
Manufacturing Buyer: **ALAN FISCHER**
Manufacturing Manager: **TRUDY PISCIOTTI**
Marketing Assistant: **PATRICK MURPHY**
Creative Director: **PAULA MAYLAHN**
Art Director: **JAYNE CONTE**
Cover Designer: **BRUCE KENSELAAR**

Simon & Schuster / A Viacom Company
Upper Saddle River, NJ 07458

Printed in the United States of America

10 9 8 7 6 5 4 3 2 1

ISBN 0-13-576687-7

Prentice-Hall International (UK) Limited, London
Prentice-Hall of Australia Pty. Limited, Sydney
Prentice-Hall Canada Inc., Toronto
Prentice-Hall Hispanoamericana, S.A., Mexico
Prentice-Hall of India Private Limited, New Delhi
Prentice-Hall of Japan, Inc., Tokyo
Simon & Schuster Asia Pte. Ltd., Singapore
Editora Prentice-Hall do Brasil, Ltda., Rio de Janeiro

CONTENTS

---

* These sections are optional and can be omitted without loss of continuity.

Preface to Volume 2

Introduction to Chapters 5-9

In the first four chapters of this book, we worked with problems that involved a single dependent (output) variable that was a function of a single independent (input) variable. We went over the steps of the problem-solving process: saw how to formulate and fully define models from verbal descriptions and from data, learned how to define and interpret different kinds of rates of change (such as the derivative), and used this knowledge to find optimal solutions to problems involving minimization and maximization. We also saw how *reversing* the derivative process (finding an *antiderivative*) can help to calculate areas under curves, such as in calculating probabilities of continuous random variables given their probability density functions, finding net distance from velocity, or finding total accumulated net profit from a profit per unit time function.

In the last five chapters of the book, we essentially follow the same process for functions in which a single output variable is a function of *two or more input variables*. In Chapters 5 and 6, we see how the notion of a function generalizes to two or more input variables, how to fully define **multivariable models** involving such functions, and how to formulate such models, both from verbal descriptions ("word problems") and from data (using least-squares regression). In Chapter 7, we learn how to use **matrices** (rectangular tables of numbers) in various useful situations, including solving systems of linear equations using the idea of the **inverse** of a matrix. In Chapter 8, we use these matrix skills to help **maximize or minimize** certain kinds of multivariable functions (such as minimizing the SSE to find the best-fit regression model), based on a generalization of the idea of setting the derivative equal to 0 to find local extrema. Finally, in Chapter 9, we discuss optimizing a linear function subject to linear constraints (restrictions), called **linear programming**, utilizing the concept of a partial derivative from Chapter 8 and the matrix theory of Chapter 7.

To get a feel for some of the kinds of problems we will be studying in these four chapters, here are a few examples:

Example 1: You have decided to start a physical fitness / health / diet routine tomorrow. After consulting with nutrition books and your doctor, you are now working out guidelines for yourself on what to eat for breakfast. You decide that you want to get from your breakfast, at least on average, a certain limited amount of fat and calories, at least a fixed amount of fiber and a specific amount of protein You like to have just cereal for breakfast, and you have two favorite cereals. Given the costs of the cereals, what should you eat for breakfast to meet your nutritional goals at the least expense?

Example 2: You have an eccentric friend who owns a newspaper delivery business. He is planning to go on a year-long walkabout in the Australian bush with an Aboriginal guide. He will not be in communication with the States, so he is looking for someone to take temporary ownership of the business (and keep the profits) while he is away. He is asking for an initial investment. You have some money you would like to invest, but you also have another investment opportunity to help another friend who is a musician and wants to make a first CD. This would require some money up front for studio expenses and then some more six months later to pay for production. You and your friend have great confidence in the success of the album, and your friend agrees to give you a guaranteed amount one year from now. If your cash reserves are in a special account earning interest compounded continuously, which investment is better for you? (You don't have enough cash for both.)

Example 3: We all have to learn how to manage our time. This can be very difficult if you are a first-year college student and this had always been done for you before now. Someone was always telling you what to do and when to do it. You have to balance your need for sleep, exercise, studying, social life, extra-curricular activities, etc. How can you get the maximum satisfaction with the way you divide your time? What is the best time of the day to do your math homework to maximize the percentage of correct problems? How many hours of exercise a week is best for you? How much sleep should you get every night to feel your best?

Example 4: You are planning to mail a package and want to be sure that the post office will accept the package. You called and they gave you the information concerning the restrictions on the size of a package that they will accept. What dimension box should you get to stay within the regulations and still hold the greatest amount?

Chapter 5: Multivariable Models from Verbal Descriptions: Interest, NPV, SSE

Introduction

In Volume 1, we focused on solving problems that could be formulated in terms of optimizing and forecasting functions of a *single* decision variable. In this chapter, we focus on the idea of extending the idea of a function from a function of *one* independent variable to a function of *two or more* independent variables, called a **multivariable function**. We discuss formulating and fully defining models based upon multivariable functions, and on graphing such functions *if* we are only working with *two* input variables (so we can use a three-dimensional graph). In Section 5.1, we focus on defining the concept of a multivariable function, verbal definitions of multivariable functions and models, the general format for **multivariable models**, and **3-D graphs** of functions with *two* input variables (including a generalization of the Vertical Line Test). In Section 5.2, we discuss **formulating models from verbal descriptions** ("story problems") and how to translate from words to symbols. In Section 5.3 we discuss a particular category of multivariable functions, involving **simple and compound interest**, including **continuous compounding**. We then apply these concepts in Section 5.4 in defining the concepts of **Future Value** and **Present Value**, as well as **Net Present Value (NPV)** and **Internal Rate of Return (IRR)**, which are essential to financial analyses of business and investment opportunities. In Section 5.5, we discuss errors, the **Sum of the Squared Errors (SSE)**, leading up to a formulation of a **model for SSE in terms of the parameters of a regression model** within a specific category (linear quadratic, etc.). These are necessary to know how to *derive* a best-fit lest-squares regression model from raw data using calculus.

Here are some examples of the kinds of problems the material in this chapter can help you solve:

- The post office has strict regulations about the size of a package that you can mail first class. You want to be able to ship the maximum volume of goods while staying within the size limits. Formulate a mathematical model for this problem, so that you can later find the optimal dimensions for your box.
- You have a small furniture business. Presently you make and sell only two products: tables and chairs (just a single style of each). You know how much raw material each table and chair uses and how much profit you realize from each table and chair. You are trying to plan your production for next week and will not have time to order more raw materials, so you will have to work with what you have. You know the demand for the chairs is limited, but it is not for the tables. Find a model for the profit.

- In comparing two bank accounts, one offers a certain APR compounded monthly, while the other gives its effective annual rate (also called the annual percentage yield, or APY). Which is better?

- Your rich uncle just gave you a gift of money. You want to spend some of it now, but put the rest in a CD to pay for a trip to visit him in California a year from now. Given the interest rate and type of compounding for the CD, how much do you need to put in it now to have the amount of money you need for the trip?

- You have an investment opportunity to help a friend who is a musician and wants to make a first CD. This would require some money up front for studio expenses and then some more money six months later to pay for production. You and your friend have great confidence in the success of the album, and your friend agrees to give you a guaranteed amount one year from now. If your cash reserves are in a special account earning interest compounded continuously, is the CD investment a good idea for you financially?

- You have data from your coffeehouse concert series of past ticket prices and paid attendance for a particular performer. You would like to fit a linear model to represent the demand function. To find the best possible linear model, you want to get an expression for the total error (SSE) in terms of the slope (m) and y-intercept (b) parameters, so that you can minimize it. You may also want to do the same thing for quadratic and exponential models, finding the SSE in terms of their respective parameters.

- You are trying to determine what combination of sleep and exercise will maximize your energy level on a daily basis. You gather data, and now want to fit a multivariate quadratic model. You want to find an expression for the SSE as a function of the model parameters.

In addition to being able to solve problems like those above, after you have studied this chapter, you should also:

- Understand the concept of a multivariable function, and be able to formulate multivariable models from verbal descriptions.

- Know how to interpret 3-D graphs and be able to visually inspect them to get an idea of whether or not they represent a function of 2 variables.

- Understand the basics of simple and compound interest, including continuously compounded interest, and be able to make calculations involving them.

- Understand and be able to apply the concepts of: the time value of money, Future Value, Present Value, Net Present Value, and Internal Rate of Return.

- Be able to formulate the SSE in terms of the parameters of a model category.

Section 5.1: Multivariable Functions, Models, and 3-D Graphs

So far in this book, we have been analyzing problems that involve one independent variable and one dependent variable: How many hours a week should Meg run to optimize her energy level? How much should the coffeehouse charge to realize the most profit? What should you charge for T-shirts to optimize your profit? There are many problems that fall into this category. These problems are modeled using single variable functions. (The "single" here refers to the independent variable, the input that can be changed.) However, there are many situations where the outcome depends on more than one input: Meg might want to consider how much sleep she got over the week as well as how many hours she ran. What if you decided to sell baseball caps as well as T-shirts? The models that describe these situations are called **multivariable functions** because they involve more than one independent variable. Such models, like the single-variable ones, can be derived directly from given information (like traditional "story problems"), or derived from raw data. In this section we will begin our study of such multivariable functions and models.

Here are the kind of problems that the material in this section will help you solve:

- Your group decides to sell both T-shirts and caps as a fund raiser. You want to determine the quantity to order and the price to charge for each item in order to optimize profit. Give the verbal definitions for a model for the profit in terms of the number of caps and T-shirts ordered and sold.
- The post office has strict regulations about the size package that you can mail first class. Give the verbal definitions for a model for the volume of the package as a function of its dimensions that could help you find the dimensions that would maximize the volume.
- You know that your semester grade will depend on your scores on quizzes, tests, homeworks, semester project and final exam. Give the verbal definitions for a model of your semester grade in terms of your scores on these components.
- Your shot percentage in basketball depends not only on your distance from the basket, but also on the angle describing your position with respect to the basket. Give the verbal definitions for a model of your average (expected) shooting percentage as a function of your angle and distance from the basket.

When you have finished this section you should:

- Recognize problems that involve three or more variables, where two or more of the variables could serve as independent variables (inputs, decision variables).
- Know when the relationship between one variable and two or more other variables is a *functional* relationship (that is, know when the one variable is a function of the two or more other variables).
- Know that in such multivariable functions, one variable is the *dependent variable* (output), and the other variables are the *independent variables* (inputs).
- Understand that, throughout this book, *models* for problems having two or more independent variables involve multivariable *functions*.
- Know how to write the verbal definitions of multivariable models, using functional notation.
- Know the format for defining a complete mathematical model involving a multivariable function.
- Know how to *evaluate* multivariable functions.

Multivariable Functions

Sample Problem 1: Four people - Chris, Pat, Tracy, and Fran - have been told their ideal weights by their fitness advisor at a health club. They also each told you their height and frame size (shoulder width divided by height), shown in Table 5.1-1 .

Table 5.1-1

Name	Height	Frame Size	Ideal Weight (lbs)
Chris	5′9″	20%	147
Pat	6′0″	25%	155
Tracy	5′11″	30%	168
Fran	5′9″	20%	136

From this information alone, is a person's ideal weight completely determined by their height and frame size?

Solution: This amounts to answering the question: "Given height and frame size, can you determine ideal weight?" or "Is a person's ideal weight a **function** of their height and frame size?" Recall our basic definition of a function: each input in the domain is paired with exactly one output. So far, we have only dealt with relationships between *two* variables (usually called x and y). Can we generalize the concept to a situation like that above, where we have *three* or more variables?

Our full definition of a (single-variable) function, from Section 1.1, was as follows: A mathematical function is a relationship between two variables (a set of ordered pairs, or relation) in which each value of the first variable from the set of allowable values (called the domain) is paired with *exactly one* value of the second variable. We must now broaden our understanding of this definition. The "first variable" in the examples that we have done before has always consisted of just one thing: one number, one name, one price, one grade range. For multivariable functions the "first variable" will consist of more than just one number or name or price. The "first variable" will consist of a *list* or *sequence* of variables, whose values could be numbers or names or prices or anything else. This list or sequence can be written out in parentheses, with commas between the values. In other words, if there are *two* components to the "first variable" (two inputs), its "value" could be written as an ordered pair.

For example, in Sample Problem 1, if we convert the heights to inches and the frame sizes to decimal form, then the "first variable" value for Chris could be written as the ordered pair (69, 0.20), since 5'9" is $5(12)+9=60+9=69$ inches, and $20\%=0.20$. We could do the same for each of the four people, and obtain the following revision of Table 5.1-1 (Table 5.1-2):

Table 5.1-2

Name	"1st Variable" Value (height in in., frame size as decimal)	Ideal Weight (lbs)
Chris	(69, 0.20)	147
Pat	(72, 0.25)	155
Tracy	(71, 0.30)	168
Fran	(69, 0.20)	136

With this understanding of "first variable," we can see that Chris and Fran have the *same* "value" of the "first variable": both have the "value" given by the ordered pair (69, 0.20). This is because the *pairs* are identical. On the other hand, the pairs (71, 0.30) and (71, 0.25) would be considered *different* "values" of the "first variable" because, even though the heights are the same, the frame sizes are *not*, and so the *pairs* are *different*. In other words, two pairs are the *same* only if *both* corresponding elements are the same (so they must be in the same *positions* as well).

Given this understanding of "first variable value," our original definition of a function, designed for the single-variable case, also works in the multivariable case. So, for Table 5.1-2 to represent a function, each "first variable value" in the domain must be paired with *exactly one* second value. You have probably noticed by now that the "first variable value" (69, 0.20) is paired with *both* 147 *and* 136 . This means that the "first variable value" (69, 0.20) is *not* paired with exactly one second value, so Table 5.1-2 does *not* represent a function. In other words, ideal weight is *not* a function of height and frame size for the given information. □

Based on the above discussion, we can now define a multivariable function: **A multivariable (or multivariate) function is a set of ordered sequences of the same length (at least 3, to be denoted $n+1$) such that every possible different sequence of the first n values allowed (in the *domain*) is paired with *exactly one* last value.**

3-D Graphs and the 3-D Vertical Line Test for Functions of Two Input Variables

You may recall from Section 1.1 that when a relation (set of points) where the values are all real numbers satisfies the Vertical Line Test graphically, the relation is a function. This happens if there is *no* vertical line that hits the graph of the points in two or more places (or, equivalently, if *every* vertical line hits the graph in *no more than one* place). Is there a graphical equivalent of the Vertical Line Test for multi-variable functions?

The first part of the answer to this question is that we can only easily handle graphs in at most three dimensions (in other words, with three variables). So if we are interested in one variable as a function of *three* or more other variables, we simply don't have a good way to graph the situation. On the other hand, it *is* quite possible to graph three variables, such as if we are trying to determine whether one variable is a function of two others, as in Sample Problem 1. Let's first go over a few basics of three-dimensional (3-D) graphs.

In Chapters 1-4 of this book (when we were working with single-variable functions), we had just two variables to work with, which we usually called x and y , and so we drew two perpendicular axes. Usually the horizontal axis was the *x*-axis and the vertical axis was the *y*-axis. When we add a third variable, which is often called z , we need a third perpendicular axis for the *z*-axis. You can *think* of this as an axis coming straight out of your paper toward the ceiling, with the *x*- and *y*-axes staying flat on the paper. But it is usually easier to draw in the three axes differently, as in a perspective painting of the corner of a room with bare walls and floor. In this version, the *z*-axis looks similar to the *y*-axis in a 2-D graph - it is vertical. The *y*-axis is often represented horizontally, in the place where the *x*-axis would usually go in a 2-D graph. The *x*-axis then turns out to be coming out from the origin at an angle, toward the lower left of the page. Many arrangements are possible, especially of the *x*- and *y*-axes, but the one we have described here is depicted in Figure 5.1-1:

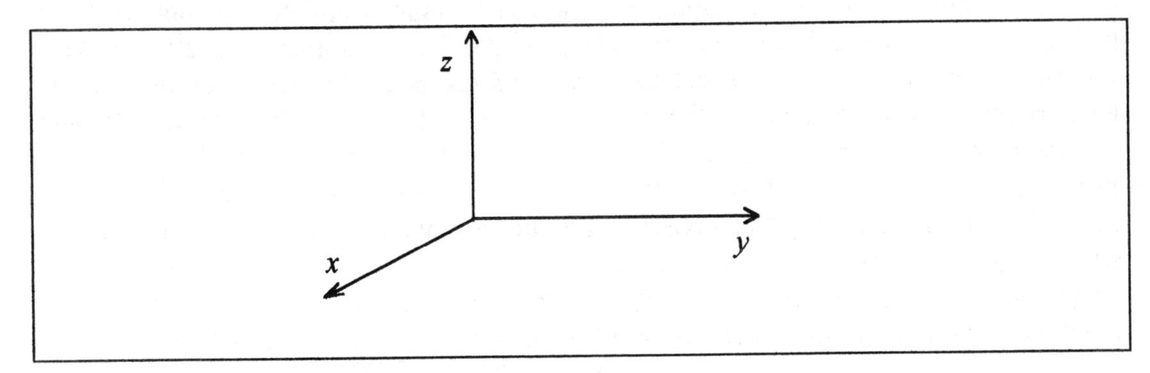

Figure 5.1-1

To graph points on a 3-D graph, you first must mark off units on the three axes. For example, if we wanted to graph the point $(x, y, z) = (1,2,3)$, we would start at the origin, move 1 unit along the x-axis (to the point marked "1" on the x-axis), then *from that point* move 2 units in the direction of the positive y-axis (to the right, parallel to the y-axis), then from *that* point move 3 units up (vertical, parallel to the z-axis in the positive direction). As in 2-D graphs, the convention is to put an arrow on an axis to show the *positive* direction. The graph of the point (1,2,3) is depicted in Figure 5.1-2 .

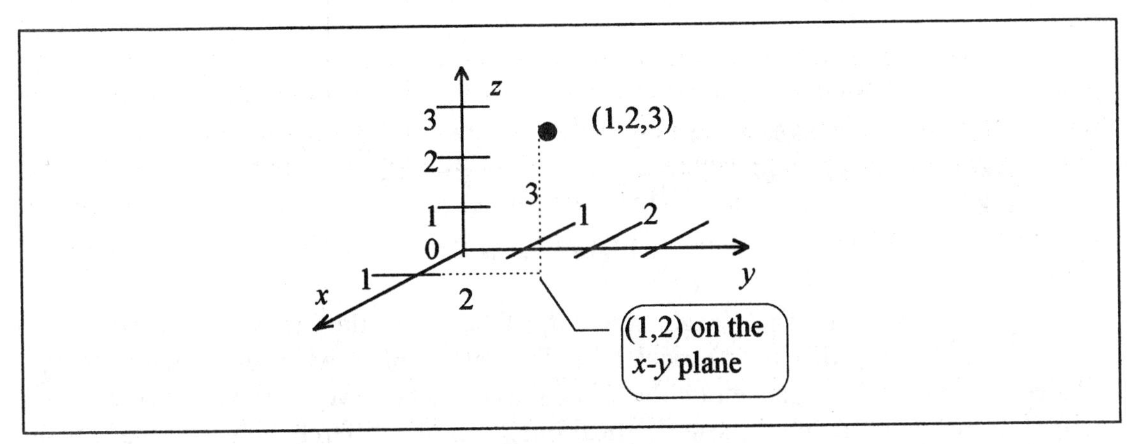

Figure 5.1-2

In the context of our discussion of functions, the point (1,2,3) could be considered as being broken down into a "first variable value" of (1,2) and a "second variable value" of 3. Notationally, we are sort of thinking of it as the point ((1,2) , 3) . Graphically, the "first variable value" (1,2) corresponds to the point on the x-y plane (the plane that is determined by the x- and y-axes) that we got to by going 1 unit in the x direction and 2 units in the y direction from the origin, as indicated in Figure 5.1-2.

For Sample Problem 1, Figure 5.1-3 shows the graphs of the four points involving the four people's height, frame size, and ideal weight. The height and frame size are the x and y variables, respectively, while the ideal weight is the z variable and corresponds to the vertical axis, which is marked off with numerical values. The "first variable value" ordered pairs correspond to the points on the horizontal x-y plane *directly below* each of the points. Since Chris and Fran had the *same* "first variable values", their points are on the *same* vertical line (they line up vertically, and are different), as can be seen on the graph.

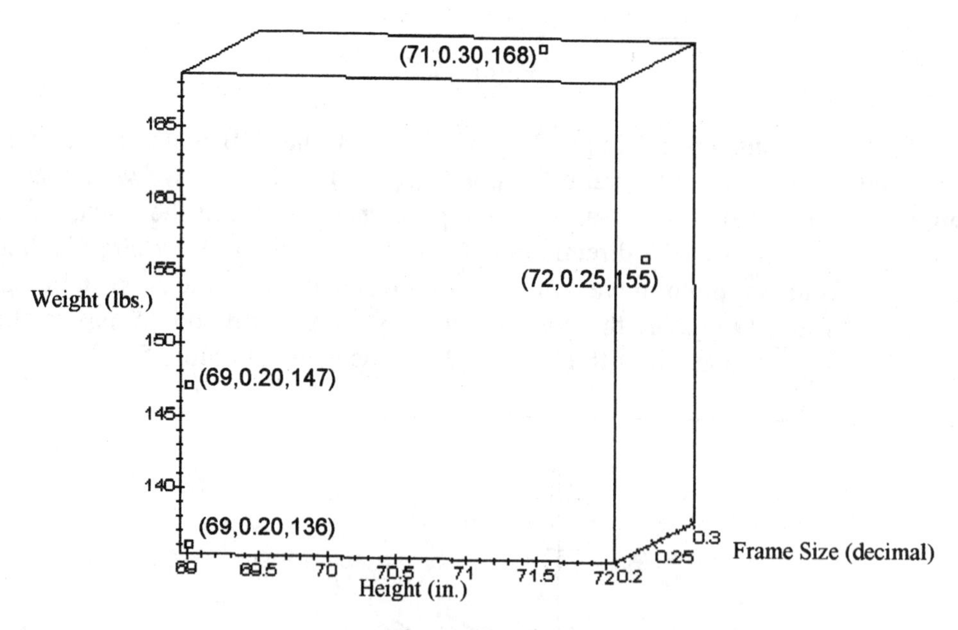

Figure 5.1-3

We saw that the ideal weight was *not* a function of the height and frame size in Problem 1. We see that the Vertical Line Test still applies when the potential output (dependent) variable is graphed on the vertical axis against two other variables in a 3-D graph. If two points lie on the same vertical line, then two different values of the "second variable" are paired with the same values of the "first variable".

The Vertical Line Test for 3-D Graphs: If *no* vertical line touches a 3-D graph in two or more points (or, equivalently, if *every* vertical line hits the graph in *at most one point*), then the vertical axis variable is a function of the other two variables.

We can make one final comment on Sample Problem 1. It turns out that the reason Chris and Fran were told different ideal weights is that Chris is male and Fran is female. Ideal weight charts tend to be broken down, not only by height and frame size, but also by sex. As a result, in the way they are usually given, ideal weight (defined as the midpoint of the range usually given, rounded off) *is not* a function of height and frame size alone, but *is* a function of sex, height, and frame size. Thus, if your population is restricted to all members of the same sex, then the ideal weight *would be* a function of height and frame size.

Multivariable Models, Function Notation, and Function Evaluation

Sample Problem 2: A group you belong to is planning to run a fund raiser that involves selling T-shirts and baseball caps. After comparing costs at different suppliers, you have decided to place the order with a firm that has a fixed cost of $50 for the order, and then a cost of $7 for each shirt and $3 for each cap. Is the total cost of an order a *function* of the number of shirts and caps ordered? If so, write a verbal definition for the function.

Solution: Let's first give verbal definitions for the variables (quantities) involved in this problem. These verbal definitions must be very complete, carefully defining each of the variables, including the units. It can be helpful to assign meaningful letters to the variables, as a reminder of what they stand for. Although this is not necessary, it is often helpful, especially when you are working with multivariable functions, since there are several independent variables to try to keep straight.

s = the number of shirts ordered
c = the number of caps ordered
T = the total cost (in dollars) of an order, including fixed and variable costs

In the above situation, the "first variable" can be thought of as the *combination* of shirts and caps ordered, or the *ordered pair* (s, c). For example, if we ordered 2 shirts and 4 caps, we could think of our "first variable" as the ordered pair $(2,4)$. We can calculate directly that our cost in this case would be

$$\$50 + (\$7)(2) + (\$3)(4) = 50 + 14 + 12 = \$76\ ,$$

so our second variable value would be 76. Thus we could think of this case as the ordered pair $(\,(2,4)\,,\,76\,)$. If you order 2 shirts and 4 caps your cost is 76 dollars. The "first variable" value of (2,4) *must* be paired with the cost value of 76 from the description of the problem, so it is paired with *exactly one* value of the second variable. In general, if we order s shirts and c caps, the cost must be

$$\$50 + (\$7)(s) + (\$3)(c) \quad = \quad 50 + 7s + 3c$$

dollars. The combination (ordered pair) (s,c) is paired with a cost of $50 + 7s + 3c$, which could be thought of as the ordered pair $((s,c), 50 + 7s + 3c)$. Again, the general value of the "first variable," (s,c), is paired with *exactly one* value of the second variable, where that one value is the result you get from substituting s and c into the formula $50 + 7s + 3c$. Since this must be true of *any* possible combination of shirts and caps ordered, this relationship is a function.

With this understanding of the "first variable," our definition of a function is just as valid for multivariable functions as it was for single variable functions. When a multivariable relationship represents a function, the "first variable" consists of a list or sequence of two or more variables, *each* of which is considered an **independent variable** (or **input**). We can write them in parentheses in a specified order, separated by commas. As before, the "second variable" is called the **dependent variable** (or **output**). This is illustrated in two forms in Figure 5.1-4. Figure 5.1-4(a) is analogous to Figure 1.1-9, showing how the ordered pair (x,y) can be thought of as the value of the "first variable" of a single-variable function. Figure 5.1-4(b) shows perhaps the more natural way to think of a multivariable function, as a rule operating on the separate values of the input variables, producing exactly one output value for each different sequence of input values that is in the domain. Figure 1.1-6 still holds equally well in the multivariable context, since a multivariable function is also completely characterized by its domain and rule, just as in the single-variable case.

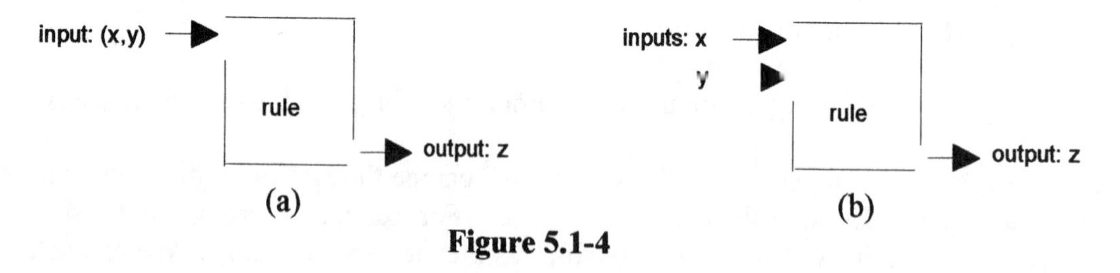

Figure 5.1-4

Function (or functional) notation for multivariable functions is like that for single variable functions: we assign letters to the dependent variable and each of the independent variables. The function name is listed first, followed by parentheses enclosing the independent variables. Where for single variable functions we had $y = f(x)$ or simply $y(x)$, for multivariable functions we have: $z = f(x,y)$ or simply $z(x,y)$. This is sometimes written $y = f(x_1,x_2)$ and can be generalized to include any number (let's call it n) of independent variables: $y = f(x_1,x_2,\ldots,x_n)$. This way, the y variable is the dependent variable, as it was for single-variable functions, and the *single* independent variable (x) in the single-variable case becomes a *list* of independent variables (the x_j's) in the multivariable case.

In this situation, an individual point for this function could be written in the form $((x_1, x_2, \ldots, x_n), y)$ or, for simplicity, $(x_1, x_2, \ldots, x_n, y)$. If we wanted to use the letter f to represent this function, the corresponding **function notation** to what we did in Section 1.2 is:

$$y = f(x_1, x_2, \ldots, x_n)$$

Thus, $f(x_1, x_2, \ldots, x_n)$ can be interpreted as "the value of the dependent variable (y) that is paired with the ordered sequence of independent variable values $(x_1, x_2, \ldots, x_n)$."

A verbal definition for the cost of ordering shirts and caps for the fund raiser is:

$T(s,c)$ = the total cost, in dollars, to order s shirts and c caps . □

With all the work that we have done above for Sample Problem 2, we can in fact *fully define* our cost model. In Section 1.2 of Volume 1 we defined a mathematical model as follows:

A mathematical model is used to represent a real-world problem or situation mathematically. It consists of a specification of a verbal definition of a function and its variables (including units), a symbol definition (algebraic formula) for the function (including the domain), and the assumptions being made.

In Volume 1 (Chapters 1-4), we applied this concept only to functions of one variable. The diagram we used to illustrate this structure, originally in Figure 1.2-1, is repeated here in Figure 5.1-5.

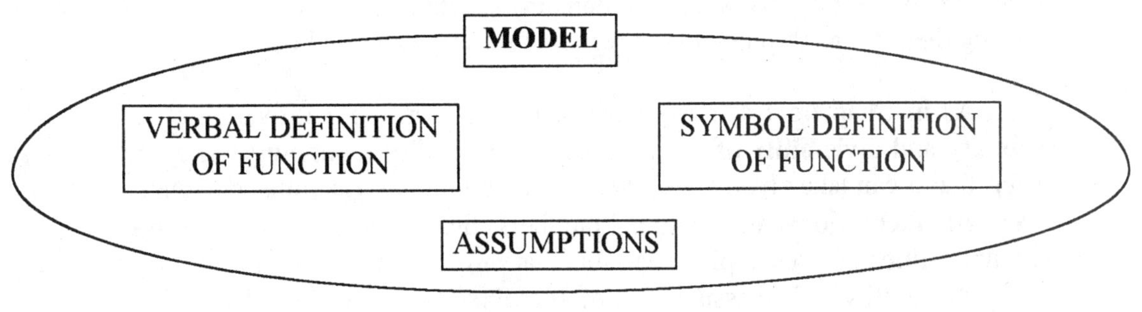

Figure 5.1-5

The *exact same* definition also applies to *multivariable* functions! To be specific and give more detail, we **define a multivariable (multivariate) mathematical model** to be

(1) a **verbal definition** of the function and all of its **independent variables**, as complete, **clear**, and unambiguous as possible, including **units** for *all* of the quantities involved;
(2) a **symbol (mathematical) definition** of the function, including the **domain**; and
(3) a specification of the **assumptions** of the model and any **implications** you can describe.

Figure 1.2-1 applies to multivariable models as well as single-variable models, since both types of models are characterized by the same three components: Verbal Definition, Symbol Definition, and Assumptions.

For multivariable models, these components remain exactly the same. Since the model is multivariate, you have to be sure to define *all* of your independent variables (including their units). The **domain** will often become more complicated for multivariable models, because there can be complex restrictions, or **constraints**, in the real-world situation, that involve complicated combinations of different variables. Now instead of simple intervals, we will often have **systems of equations and inequalities** to define the domain. We will still write the domain in the same place, however: immediately following the symbolic definition of the function rule (which will usually be an equation). Often the domain constraints are preceded by the words "**subject to**" or "**such that**" (usually abbreviated "**s.t.**").

In the multivariable function context, the model function will often be called the **objective function**, and the independent variables the **decision variables**. The set of values satisfying the constraints (the domain) is often called the **feasible region**. We will be using these terms throughout the remaining sections of the book.

As in Chapters 1-4, our major assumptions in our multivariable models will be **certainty** and **divisibility** of the variables. **Divisibility** (continuity) of a variable simply means that it can take on *any real number* (including any fractions, decimals, multiples of π, and all other rational and irrational numbers) that satisfy the given constraint equations and inequalities. For example, a variable that must, by its nature, be an integer would *not* satisfy divisibility. The assumption of **certainty** says that the relationship between the variables is *exactly* as the symbolic definition specifies - with no uncertainty or probabilistic fluctuations. As we saw in the first semester, this rarely holds perfectly in the real world, but can make sense when trying to model the value of the output value *on average*, or its *expected* value. The random fluctuations around such an average is a large part of what is studied in a statistics course. When presenting a model, it can be helpful to point out some of the *implications* of these assumptions if they are not obvious.

Let's now fully define our cost model for the T-shirts and caps:

Verbal Definition: $T(s,c)$ = the total cost, in dollars, to order s shirts and c caps

Symbol Definition: $T(s,c) = 7s + 3c + 50$, such that $s \geq 0$
$c \geq 0$

Assumptions: Certainty and divisibility. In this case, the divisibility assumption means that fractions of T-shirts and caps can be ordered and are charged for proportionately. Certainty means the relationship is exact, so implies that there are no sales, quantity discounts, coupon deals, and so forth.

Clearly, the assumption of divisibility is not realistic in this case. In cases such as this where this assumption is not realistic, it may be necessary to make the assumption to solve certain kinds of problems. After finding a solution, we can always round our answer to the nearest whole number. If the numbers are large, this is usually well within the margin of error of the problem, so is not a serious difficulty. If the magnitudes of the numbers are small, more complicated methods such as *integer programming* may be necessary, which are beyond the scope of this book, but can be studied in courses on *Operations Research (OR)* and *Operations Management (OM)*.

Functional notation can be used to indicate **evaluation of a multivariable function at a point**. In our example, suppose we wanted to find the total cost if we order 2 shirts and 4 caps, as we calculated earlier. The notation for this would be $T(2,4)$, and since we know that (2,4) is in the domain and that the rule is given by

$$T(s,c) = 50 + 7s + 3c ,$$

we can now find the value of the function at (2,4) by substituting 2 for s and 4 for t:

$$T(2,4) = 50 + 7(2) + 3(4) = 50 + 14 + 12 = 76 .$$

Translation: The total cost of ordering 2 shirts and 4 caps would be \$76, as we saw before.

Evaluating multivariable functions is exactly analogous to evaluating single-variable functions: **you simply substitute the values within the parentheses of the functional notation for the corresponding independent variables in the symbol definition of the function, after making sure the sequence of independent variable values is in the domain.**

Verbal Definitions of Multivariable Models

Sample Problem 3: Meg decided to expand her first-semester project. At the time of her original project, her sleep was very consistent, so it was essentially held constant and did not affect her energy level in any way that varied over the time she collected her data. In her second semester, however, her sleep had become less consistent, and so had a more noticeable effect on her energy level. Thus she wanted to find a model for her energy level (still measured on the same 0-100 scale as that given in Section 1.0) as a function of *both* the number of hours' sleep she got each week and the number of minutes of running each week. From this model, she could then try to determine the best values for *both* variables to maximize her expected energy level. Table 5.1-3 gives the results of her data collection:

Table 5.1-3

Run (min)	375	415	375	530	475	500	390	450	510	580
Sleep (hrs)	50	35	42	36	39	32	50	45	36	34
Energy (0-100)	65	70	60	80	85	80	70	85	80	70

From just considering this data, is Meg's energy level a function of her running time alone? Is Meg's energy level a function of her sleep alone? Is Meg's energy level a function of her running *and* her sleep (together)?

Solution: Let's apply our definition of a multivariable function to Meg's data and see if the relationship between her energy level and the combination of her running and her sleep is a function. The "first variable values" consist of the ordered pairs of running minutes and sleep hours recorded (in that order) each week:

(375,50), (415,35), (375,42), (530,36), (475,39), (500,32), (390,50), (450,45), (510,36), (580,34).

The "second variable values" are the energy levels recorded each week: 65, 70, 60, 80, 85, 80, 70, 85, 80, 70, based on the following definitions:

100: Infinite energy
90: *Extremely* high energy level
80: Very high energy level
70: High energy level
60: Pretty high level of energy
50: Medium energy level
40: Moderately low energy level

30: Low energy level
20: Very low energy level
10: *Extremely* low energy level
0: No energy

The ordered "pairs" for the multivariable relation are:

((375,50) , 65), ((415,35) , 70), ((375,42) , 60), ...

and so on. Looking at Table 5.1-3 , the ordered pairs consist of the first two rows themselves forming the inner ordered pairs) paired with the third row. This is shown more explicitly in Table 5.1-4 .

Table 5.1-4

Run (min)	375	415	375	530	475	500	390	450	510	580
Sleep (hrs)	50	35	42	36	39	32	50	45	36	34
Energy (0-100)	65	70	60	80	85	80	70	85	80	70

In the first row of Table 5.1-4 we have two weeks that included 375 minutes of running. For one of these weeks the energy level was 65 and for the other the energy level was 60. However, the hours of sleep for these weeks was not the same: 50 hours for the first week and 42 hours for the second week. The same is true for the sleep: there are two weeks when she got 50 hours of sleep, but the energy levels were not the same. In this case the energy level is not a single-variable function of *either* the running minutes (which would include the ordered pairs (375,65) and (375,60)) *or* the sleeping hours (which would include the ordered pairs (375,50) and (375,42)) *individually*, but it *is* a multivariable function of the running minutes *and* the sleeping hours (together). This is easy to see since all of the input pairs are different, which means each gets paired with exactly one energy level. □

Sample Problem 4: Since Meg's energy level is a function of her running minutes and sleeping hours, we can define a model that expresses this relationship, at least on average. Write the verbal definition of this model.

Solution: Before we solve this problem, let's review the concept of the verbal definition part of a mathematical model. The verbal definition consists of the functional notation for the model *followed* by the definition of the function (dependent variable) in terms of the independent variables used in the functional notation. The independent variables (including units) should either be clearly defined *implicitly* in the function definition, or should be defined separately and explicitly in the same vicinity, or both. This verbal definition must be very complete, carefully defining each of the variables, including the units. It can be helpful to assign meaningful letters to the variables, as a reminder of what they stand for. Although this is not necessary, it is often helpful, especially when you are working with multivariable functions, since there are several independent variables to try to keep straight.

We might choose E for the energy level as before, but instead of using x for the minutes of running each week, we might choose r to represent minutes of running and s for hours of sleep. We could then write the verbal definition for the model:

Verbal Definition: $E(r,s)$ = Meg's average (expected) overall energy level on her 0-100 scale, in a week when she runs r minutes and sleeps s hours that week. □

Sample Problem 5: You are planning to mail a lot of boxes and want to be sure that you get the maximum package volume while staying within the size limits that the Post Office will accept. Give the verbal definition of a model that you could use to help solve this problem.

Solution: We don't have enough information here to have a complete understanding of the problem (we will develop it more in Section 5.2). It does seem that the volume of the box is maximized, within some sort of restrictions. Thus we could use the following verbal definition:

Verbal Definition: $V(l,w,h)$ = the volume of the box in cubic inches when the length is l inches, the width is w inches and the height is h inches. □

Sample Problem 6: Your tennis coach has been helping you map out a strategy for your matches. The following record was kept for the last set that you played: the number of forehand and backhand shots you made in each game and the percentage of points you won in that game. This is shown in Table 5.1-5 .

Table 5.1- 5

Forehand	2	4	3	3	2	3	5	4	4	3
Backhand	3	2	3	2	4	2	6	6	1	4
% Points Won	66	37.5	62.5	0	80	25	66.6	60	37.5	58.3

Just based on this data, is the percentage of points won a function of the numbers of forehands and backhands in a game?

Solution: By our definition, we must determine if, for each pair of independent values (forehand and backhand shots taken), there is exactly one dependent value, percentage of points won. Again, the first two rows are the sets of independent variables, and the third row is the dependent variable. Let's look at the first two rows and see if any of the input pairs are repeated: (3,2) appears twice. In two games you took 3 forehand and 2 backhand shots. Did you win the same percentage of points in those games? No; in the first game you didn't win *any* points, but in the second of those games you won 25% of the points. For each "first variable value" (ordered pair of forehands and backhands) there is *not* exactly one "second variable value" (point-winning percentage), since (3,2) gets paired with both 0 and 25. For this particular study, the percentage of points won is *not* a function of the number of forehand and backhand shots taken in the game. □

We should note here that there are occasions when we might want to *fit* a function (model) to *data* that is not *itself* a function. You may recall that this was also true with single-variable functions. Sometimes the same "first variable values" may get paired with different "second variable values," but it would still be useful to make a model of the *average* or *expected* value for the "second variable" for any given "first variable value" sequence. In this problem, there does seem to be a pattern. To help you analyze your game, it might be useful to fit a model to this data for the *average* (expected) percentage of points won as a function of the numbers of forehands and backhands in a game. In fact, this kind of regression model is very common. Most technologies can fit a function to data even if the data do not themselves represent a function. In Sample Problem 6, the data could be fit exactly as given in the Table 5.1-5 (including the *two* data points, (3,2,0) and (3,2,25)), or the two output values (0 and 25) paired with the input pair (3,2) could be *averaged* (12.5), and the *single* data point (3,2,12.5) used.

There is an old saying: "Everybody complains about the weather, but nobody does anything about it." This is quite true, because there really isn't much we can do about it, but it would be helpful if we knew in advance what the weather was going to be so we could plan accordingly. Is there actually a model for the weather?

One answer is "Yes, in fact there are actually quite a few models for the weather." Right there, that should tell you something. If there were *one* model for which the assumption of certainty held perfectly, there wouldn't *be* any other models. The weather is *really* a multivariable problem! There are not only many, many independent variables, such as upper altitude wind speed and direction, positions of local high and low pressure areas, etc., but also many possible dependent variables: the predicted high and low temperatures for the period, the amount and form of any future precipitation, and the wind speed and direction. The models are constantly being improved, as are the amount and quality of weather data being gathered. Maybe someday we will have a reliable weather model (or a reliable *system* of models), for which the assumption of certainty holds fairly well.

On the other hand, some weather statistics are actually based on multivariable functions, such as the wind-chill factor, which is based on temperature and wind speed:

$C(t,w) =$ the wind chill in degrees Fahrenheit when the temperature is t degrees Fahrenheit and the wind speed is w miles per hour.

Sample Problem 7: You have decided to start a physical fitness / health / diet routine. After consulting with nutrition books and your doctor, you are now working out guidelines on what to eat for breakfast. You have decided that you want to get from your breakfast cereal a certain *maximum* amount of fat and calories, at *least* a fixed amount of fiber, and a *specified* amount of protein. You have two favorite cereals, Special K^{TM} and Raisin Bran. Would the total cost of your cereal (given the current prices at your local store) and the total amounts of fat, calories, fiber, and protein be functions of the amounts of each type of cereal that you eat in a day?

Solution: The cost of a box of each type of cereal is known (since we are given the prices at the local store), and the number of servings in each box is given on the label (the size of each serving is then defined implicitly by the weight of the box), so you can calculate the cost of one serving of each type of cereal. Since each possible combination (which could be fractional) of servings of each cereal would result in a unique cost for that combination, the total cost of your cereal that day will be a function of the number of servings of each type of cereal you eat that day. The verbal definition would be:

$C(K,R) =$ the cost (in dollars) of your breakfast cereal on a day when you eat K servings of Special K and R servings of Raisin Bran.

(As discussed earlier, this definition assumes that the prices are given.)

The same thing will be true for the amounts of fat, calories, fiber and protein consumed from eating the cereals:

$f(K,R) =$ the amount of fat (in grams) consumed when you eat K servings of Special K and R servings of Raisin Bran;
$c(K,R) =$ the number of calories consumed when you eat K servings of Special K and R servings of Raisin Bran;
$b(K,R) =$ the amount of fiber (in grams) consumed when you eat K servings of Special K and R servings of Raisin Bran; and
$p(K,R) =$ the amount of protein (in grams) consumed when you eat K servings of Special K and R servings of Raisin Bran. □

Sample Problem 8: You are planning to sell souvenir caps and T-shirts. You plan to take a survey to determine approximately how many hats and T-shirts you can sell at different prices. You figure that the price of the T-shirts will affect not only the number of T-shirts sold but also the number of caps, and vice versa. Will your profit be a function of the price you charge for the caps and T-shirts? If so, write the verbal definition of the function.

Solution: Recall that the **profit equals the revenue minus the cost** and the **revenue equals the selling price times the quantity sold (price times quantity)**. You have decided that the quantity of each item sold is a function of the price you charge, and will use your survey to find the exact form of the two demand functions. The cost will also depend on the quantities, which are functions of the prices you charge (the demand functions). Therefore, the profit will be a function of the prices that you charge and the verbal definition of the model is:

Verbal Definition: $\pi(s,c)$ = the profit in dollars from the sale of T-shirts at s dollars per shirt and caps at c dollars per cap.

We should point out that the above verbal definition *assumes* that we will order exactly the number of shirts and caps specified by the demand functions at the given prices, and that we will be able to sell *exactly* those numbers at those prices. Even assuming the sales numbers are correct, we *could* of course order more (or less) shirts and caps than the demand functions specify. In such a case, the profit would no longer be *just* a function of the two selling prices; it would have to include the order quantities as independent variables as well to be a function. This is because the same prices (and sales) with different order quantities could have different profits (for example, if you over-order, your profit goes down). □

Sample Problem 9: You make hand-crafted chairs and table. The chairs and tables each use hardware and wood, and you make a specified amount of profit on each chair and each table. Are the quantities of wood and hardware used, and the total profit realized, functions of the numbers of chairs and tables that you make and sell?

Solution: Since each chair and each table requires a known amount of hardware and wood and you can determine how much of these materials you need if you know how many tables and chairs you plan to make, the amounts of wood and hardware needed are functions of the numbers of chairs and tables. The same is true of the profit: if you know how many chairs and tables you make and sell (this assumes demand is sufficient at the current selling price so that you can sell all that you make), and there is a constant profit per chair and table, you will know how much profit you will realize. The relevant verbal definitions would be:

$\pi(c,t)$ = profit, in dollars, from the sale of c chairs and t tables;
$W(c,t)$ = number of board-feet of wood you need to make c chairs and t tables;
$H(c,t)$ = the number of packages of hardware you need to make c chairs and t tables. □

Using 3-D Graphs to Identify When a Relation Is a Function

We have already seen the basic idea behind graphing *points* on 3-D graphs. When we graphed a single-variable *function*, the graph was a *curve*, or a series of lines in the case of a step function. When we graph a function of two variables, the graph will normally be a *surface*. If the function is linear, the surface will be a flat plane. If the function is not linear, the surface will be curved. We will not try to draw graphs of functions with more than two independent variables because there is no known simple system to represent a fourth dimension.

Sample Problem 10: Figure 5.1-5 is the graph (from two different viewpoints) of the cost function for the T-shirts and caps in Problem 2:

$T(s,c)$ = the cost in dollars to order s shirt and c caps
$T(s,c) = 50 + 7s + 3c$, for $s \geq 0$, $c \geq 0$.

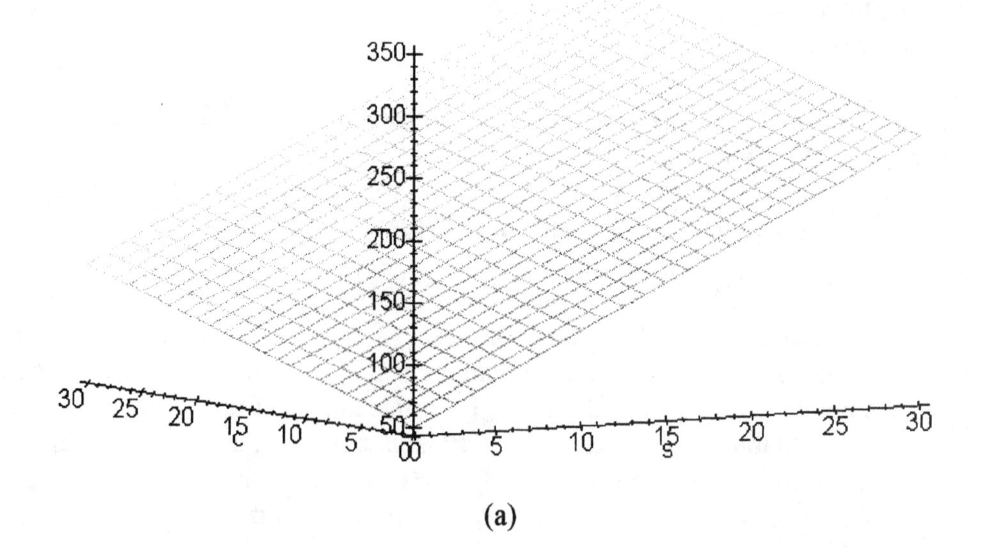

(a)

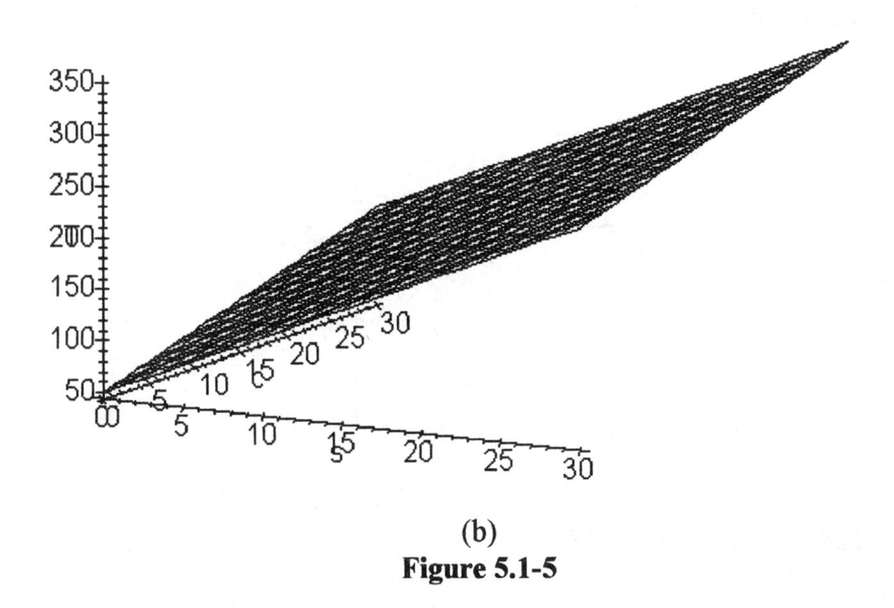

(b)
Figure 5.1-5

For the graph, the domain was: $0 \le s \le 30$, $0 \le c \le 30$. These were arbitrary upper limits to make the graph easier to read. If you imagine the origin representing the corner of a bare room, this graph would be like a flat blanket or tarp, slanting up over the room. Notice that the plane (tarp) is slanted more in the s direction than in the c direction, by looking at the two bottom edges (along the two walls of the room) in Figure 5.1-5(a). We know that the total cost is a function of the number of shirts and caps ordered. Could we determine this just by looking at the graph?

Solution: Recall the Vertical Line Test from the first chapter and our earlier discussion: The graph represents a function if any vertical line drawn anywhere on the graph intersects the graph in at most at one point. The same is true for graphs of relations with two independent variables, since the vertical axis again represents the dependent variable. If a vertical line drawn anywhere on the graph intersects the surface at most at one point, the graph represents a function.

This graph passes the Vertical Line Test because a vertical line (a line parallel to the z axis) drawn anywhere on this graph will cut the plane in exactly one place. Another way of thinking of this is that, for every pair of (s,c) coordinates we can see on the graph, there is exactly one T coordinate. □

Sample Problem 11: Figure 5.1-6 shows the volume of a box with certain restrictions on its dimensions (see Sample Problem 5 and Section 5.2 for more details). We will see later how to express the volume (V on the graph) as a function of just two variables, the width and the height of the box (w and h , respectively), as the graph represents. By looking at the graph *only*, can you determine if the volume is a function of just the height and width of the box over the domain shown?

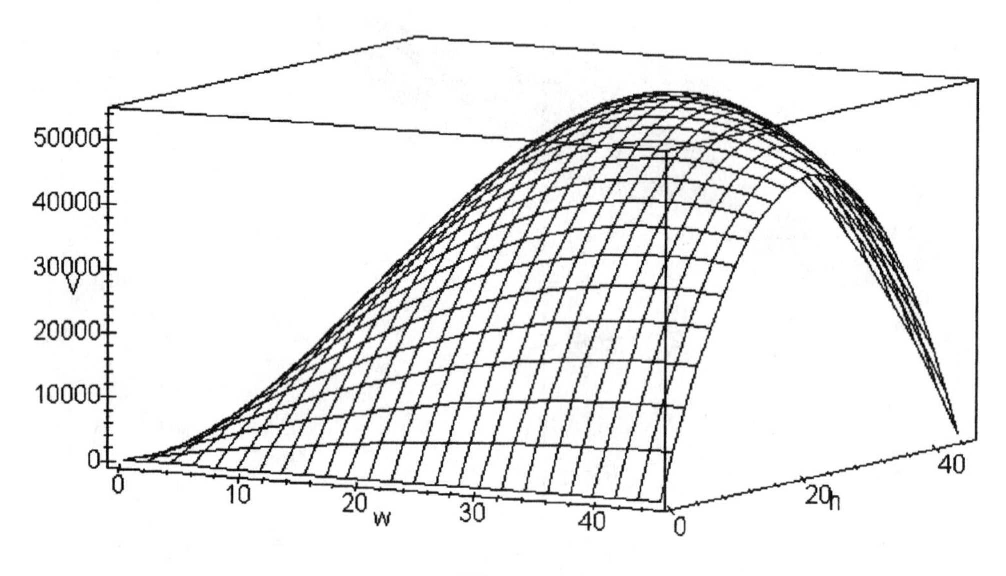

Figure 5.1-6

Solution: You can see that this is a curved surface. Applying the Vertical Line Test to the graph as shown, we see that this graph does represent a function: If we know the w and h coordinates, the width and height of the box, there will be exactly one value for the volume, the V value on the graph. □

Sample Problem 12: Look at the 3-D graph in Figure 5.1-7 . It this the graph of a function?

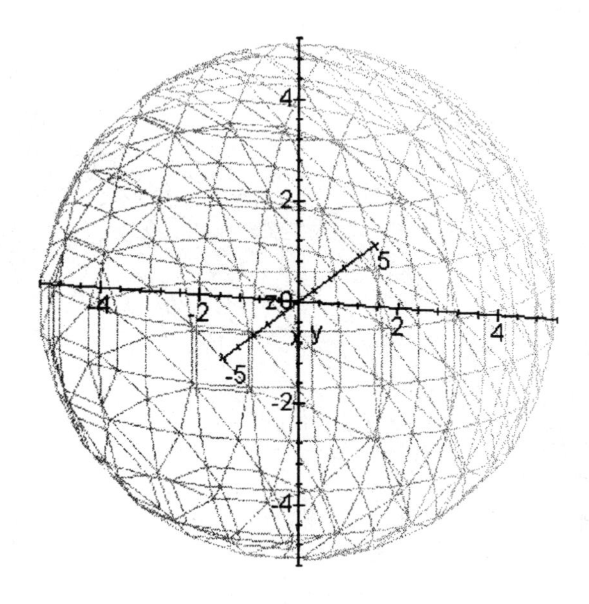

Figure 5.1-7

Solution: This is *not* the graph of a function because it does not pass the vertical line test. For example, a vertical line drawn through the point x=0, y=0 on the x-y plane (through the origin) will pass through the "ball" twice, at (0,0,-5) and (0,0,5). □

Sample Problem 13: The graph in Figure 5.1-8 is an interesting and rather complicated graph, but is it the graph of a function?

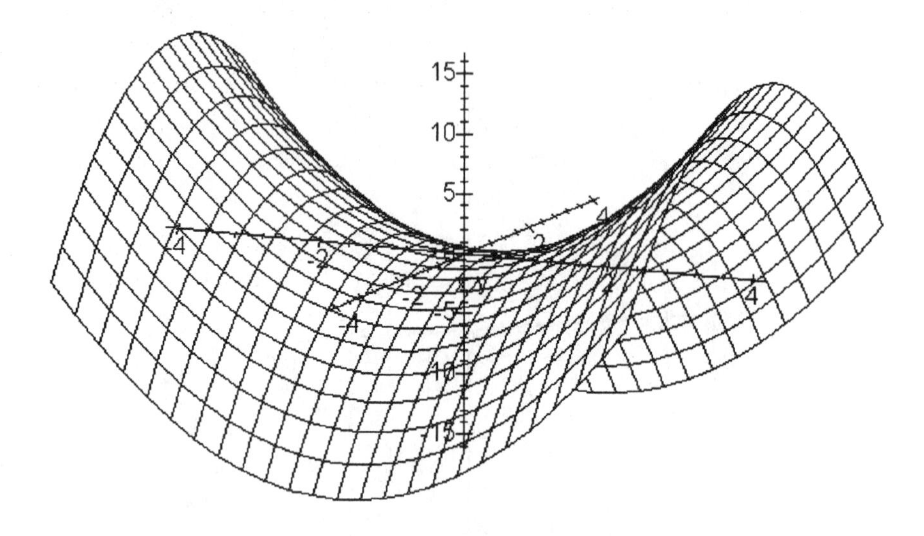

Figure 5.1- 8

Solution: Although this graph is a bit complicated, if we apply the Vertical Line Test, we see that this *is* the graph of a function, at least over the domain shown in the graph, since no vertical line will hit the graph in more than one point. □

Section Summary

Before you begin the exercises, be sure that you:

- Can recognize problems that involve multivariable relationships. In data form, this would normally be expressed in a table that consists of three or more rows (or columns) of variables. If one or more equations can be derived directly from a verbal description, the equation(s) would involve three or more variables, one dependent variable and two or more independent variables.
- Know that a single variable is a **function** of a group of two or more other variables when, each possible *different* ordered combination (sequence, list) of values of the *group* of variables is paired with *exactly one* value of the *single* variable. The variables in the group are called the **independent variables** or the **input variables**, and the single variable is called the **dependent variable** or the **output variable**. The set of possible combinations of values of the input variables is called the **domain**.

- Know that such functions are called **multivariable functions**, and know how to **evaluate** them by substitution.
- Know that the **mathematical models** for problems involving multivariable functions consist of the **verbal definition** of the function and its variables, including all **units**, the **symbol (mathematical) definition** of the function, including the **domain** (constraints or restrictions on the independent variables), and the **assumptions** (normally certainty and divisibility, as before) and their implications.
- Can write clear, unambiguous verbal definitions of multivariable functions using functional notation, including units for the function value and for all variables.

EXERCISES FOR SECTION 5.1:

Warm Up

1. If $T(s,c) = 50 + 7s + 3c$, evaluate the following:
   a) $T(3,8)$
   b) $T(0,10)$
   c) $T(0,0)$

2. If $\pi(c,t) = 140c + 200t$, evaluate the following:
   a) $\pi(10,10)$
   b) $\pi(4,15)$

3. If $P(a,b,c) = 0.3a + 0.5b + 0.15c$, evaluate the following:
   a) $P(2,3.5,4.2)$
   b) $P(4.5,0,3.9)$
   c) $P(2a,b+h,c)$

4. If $C(S,R,G) = 150S + 210R + 240G$, evaluate the following:
   a) $C(0.29,0.35,1.2)$
   b) $C(0,3.5,2.9)$
   c) $C(1.5S,R+h,G+1)$

5. If $A(P,i,n) = P(1+i)^n$, evaluate the following:
   a) $A(100,0.08,5)$
   b) $A(2000,0.045,10)$

6. If $A(P,r,t) = Pe^{rt}$, evaluate the following:
   a) $A(1000,0.045,5)$
   b) $A(5000,0.083,3)$

7. If $R^2(SSR,TSS) = \dfrac{SSR}{TSS}$, evaluate the following:
   a) $R^2(239,310)$
   b) $R^2(139,486)$

8. If $MSE(SSE,n)=\dfrac{SSE}{n}$, evaluate the following:
   a) $MSE(2348,29)$
   b) $MSE(482,31)$

9. If $P(c,t)=4.04c^2+1.39ct+139c-12.4t^2+209t-700$, evaluate the following:
   a) $P(5,10)$
   b) $P(10,15)$

10. If $E(c,x)=-93.1+0.260c+0.146x-0.0000944c^2+0.0000041cx-0.0019x^2$, evaluate the following:
   a) $E(1850,75)$
   b) $E(2000,85)$

Game Time

11. The following table shows the time (in seconds) to cover ¼ mile for several different high performance cars:[1]

Automobile	**Horsepower**	**Torque (ft-lbs)**	**Weight (lbs)**	**sec to ¼ mile**
Chevy Corvette	300	340	3420	14
Mazda RX-7 Turbo	255	217	2840	14
Mitsubishi 3000GT Turbo	300	307	3860	14.1
Nissan 300ZX Turbo	300	283	3540	14.2
Porsche 968	236	225	3540	14.6
Toyota Supra Turbo	320	315	3480	13.8

a) Excluding the name of the automobile, identify natural dependent and independent variables.
b) Is the dependent variable you identified in part (a) a *function* of the independent variables you identified? Why or why not?
c) If you found that the dependent variable you identified *is* a function of the independent variables you identified, write a verbal definition of a model.
d) Based on the data, is the horsepower a function of the automobile? Why or why not?
e) Based on the data, is the automobile a function of the horsepower? Why or why not?
f) Based on the data, is the weight a function of the torque? Why or why not?

---

[1] Data adapted from a student project. The project cited *Car and Driver* April, 1995, pp. 50-58.

12. The following table shows the latitude and longitude of several cities, rounded to the nearest degree. [One degree of latitude = 60 nautical miles, or approximately 52 land miles, and is measured from the equator. One degree of longitude varies from 0 up to 60 nautical miles (0 at the poles, and 60 at the equator), and is measured from Greenwich, England]

City	Latitude	Longitude
Rio Grande, Argentina	54° S	68°W
Sydney, Australia	34° S	151°E
Rome, Italy	42° N	12° E
Tokyo, Japan	36° N	140° E
London, Eng. U.K.	52° N	0° W
London, Tex., U.S.	31° N	100° W
New York, N.Y., U.S.	41° N	74° W
Etah, Greenland	78° N	73° W

a) Based on this table alone, is the city a function of the latitude and longitude? That is, given the latitude and longitude, could you find the city?
b) If so, write a verbal definition of the function.
c) Again based only on the table, are the latitude and longitude functions of the city?
d) If so, write a verbal definition of the respective functions.
e) Do you believe that, considering the *whole world* right *now*, a city is a function of just its latitude or just its longitude? That is, given just the latitude *or* given just the latitude, could you find the city? Why or why not? What if you were given *both* the *precise* latitude *and* the *precise* longitude? Explain.

13. You want to install a fish tank of the traditional (rectangular box) shape. Identify dependent and independent variables related to the volume of the tank and write a verbal definition of a model for the volume.

14. Hot water heaters are generally cylindrical in shape. Identify dependent and independent variables related to the volume of the water heater and write a verbal definition of a model for the volume.

15. An athlete competes in the weight throw event for the Track and Field team. He has been trying to decide how many practice throws he should take and how many warm-up laps he should run before a match to achieve the maximum distance on his throws. He has recorded the following information:[2]

Distance (ft)	Warm-up Throws	Laps	Distance (ft)	Warm-up Throws	Laps
25.2	0	0	27	1	4
25.8	1	0	27.4	1	5
26	2	0	27.1	2	2
26.14	3	0	27.9	3	2
26.3	4	0	28.4	4	2
27	5	0	29.8	5	2
25.5	0	1	27.6	2	3
25.8	0	2	28	2	4
26.1	0	3	28.8	2	5
26.18	0	4	29	3	3
26.35	0	5	33	5	3
26.8	2	1	30.8	3	5
27.2	3	1	32	4	4
28	4	1	30.6	5	4
28.46	5	1	31.2	4	5
26.5	1	2	29.9	5	5

a) As shown by this table, is the distance of the toss a function of the number of warm-up throws?
b) As shown by the table, is the distance of the toss a function of the number of laps run?
c) As shown by the table, is the distance of the toss a function of the number of warm-up throws *and* the number of laps run (together)?
d) Write a verbal definition(s) of the relationships in part (a)-(c) that are functions.

[2] Data adapted from a student project.

16. A student wished to increase her energy level. She decided to split her exercising between cardiovascular workouts and weight lifting. She recorded the time she spent on each type of exercise and her energy level on a scale of 0 to 100 for a number of days. The results are shown in the table below:[3]

Cardio Minutes	Weight Minutes	Total Minutes	Energy Level
40	20	60	70
25	15	40	62
30	15	45	67
30	20	50	72
30	15	45	63
35	20	55	77
35	25	60	79
25	20	45	73
45	20	65	81

a) According to the table, is her energy level a function of the minutes spent doing cardiovascular exercises?
b) According to the table, is her energy level a function of the minutes spent doing weight exercises?
c) According to the table, is her energy level a function of the total minutes spent doing exercises?
d) According to the table, is her energy level a function of the minutes spent doing cardiovascular exercises *and* weight exercises , considered together?
e) Do you think it would be reasonable to use a function to model this data even if the table indicates that the energy level is not a function of the cardiovascular minutes and/or the weight lifting minutes? If so, write a verbal definition of a model.

17. You take your car in to have the oil changed. The bill lists total charges for parts (including the price of a new oil filter), the price per quart of oil, and the hourly charges for labor. *Given* the prices for oil and labor, is the total bill a function of the total cost of the parts, the number of quarts of oil, and the hours of labor? If so, write a verbal definition of a model for the bill.

18. Your utility bill lists the price for gas per Ccf (Hundreds of cubic feet) and the price for electricity per kWh (Kilowatt hours). There is also a given fixed basic charge for each of these services, regardless of the amount of gas or electricity used. *Given* the basic fixed charges, is your total bill a function of the amounts of gas and electricity used? If so, write a verbal definition for a model of your total utility bill.

---

[3] Data adapted from student project. (Lindsey Kridler)

19. Your telephone bill shows a given fixed charge for basic service plus a charge per hour for long distance peak time and a charge per hour for long distance off-peak time. *Given* the basic fixed charge, is your total bill a function of the hours of peak long distance use *and* the hours of off-peak long distance use (together)? If so, write a verbal definition for a model of your phone bill.

20. You received some money as a gift and plan to use it to take a trip when you graduate. You plan to put the money into a savings account until you need it for the trip. The account pays interest compounded continuously. You know the amount you have to deposit, the effective annual interest rate (yield), and how long you can leave it in the account before you have to withdraw it to pay for your trip. Is the amount accumulated in the savings account a function of the amount deposited, the effective annual interest rate (yield), and the amount of time the money is left in the account? If so, write a verbal definition of a model for the amount in your savings account when you take it out.

21. You are looking into buying a microwave oven, but know that you should consider how much it will cost to run, as well as the initial cost. Electricity is normally billed at a given price per kilowatt hour (kWh). The number of kWh's an appliance uses depends on the amps (or on watts, since $\text{watts} = (\text{volts})(\text{amps})$, and volts are constant at 110 for most U.S. electrical outlets) that the appliance draws and how long the appliance is used, which you assume can be expressed as a function of the amps. The higher the amps (or watts) for which the appliance is rated, the less time it will take to heat or cool. Write a verbal definition for a model of the total expected cost over the life of the oven, including both the initial cost and the cost of running the appliance, as a function of the initial cost and the amps it draws.

22. You have to build some containers to ship heavy cargo, such as lead pellets. The specifications call for the containers to be in the shape of rectangular boxes. The containers must hold a fixed number of pellets, thus they have a fixed volume. Because of the weight of the contents, the bases of the containers must be built out of stronger (and more expensive) material than the sides and top. You know the cost in dollars per square foot for the materials for the bottom and for the sides and top. Write a verbal definition for a model of the cost of the containers as a function of two of its dimensions.

Section 5.2: Formulating Models from Verbal Descriptions

As we mentioned before, multivariable models can be derived directly from given information (the traditional word problem, or "story problem") or derived from data. Those derived from data are handled very much the same way as we handled single variable models from data in Chapter 1. We use technology to help us fit a least squares regression model to the data. This process will be discussed in Chapter 6. In this section we discuss the process of deriving a mathematical model of a problem from its verbal description. This process involves clarifying the problem and formulating a model, the first two stages of the 12-step process for problem solving discussed in Chapter 1, analogous to "getting the plane off the ground and up to cruising altitude." We strongly recommend that you look back over the material in the Preface to Volume 1 and Sections 1.0-1.2 as background to this section.

Here are the kinds of problems that the material in this section will help you solve:

- The post office has strict regulations about the size of package that you can mail first class. You want to be able to ship the maximum volume of goods while staying within the size limits. Formulate a mathematical model for this problem, so that you can later find the optimal dimensions for your box.
- You have to build some containers to ship a certain kind of pellets. The containers must hold a fixed number of pellets, so they have a fixed volume. Because of the weight of the contents, the bases of the containers must be built of stronger (and more expensive) material than the sides and top. Write a mathematical model for the cost of the containers, so that you can eventually find the optimal container shape and dimensions that minimize your cost.
- You have a small furniture business. Presently you make and sell only two products: tables and chairs (just a single style of each). You know how much material each table and chair uses and how much profit you realize from each table and chair. You are trying to plan your production for next week and will not have time to order more materials, so you will have to work with what you have. You know the demand for the chairs is limited, but not for the tables. Find a model for the profit.
- You have just started work at a new company. They use a predetermined profit margin, based on the selling price, to set the prices that they charge for their products. The last company that you worked for set their prices using a markup, based on the cost to the company. How do you determine how much to charge for a product using a predetermined profit margin, given the cost?

When you have finished this section you should:

- Be able to identify the dependent variable, sometimes called the objective function, that eventually is to be optimized or forecast in a problem.
- Be able to identify the independent variables, sometimes called the decision variables, in a problem.
- Know how to define all relevant variables clearly and unambiguously, including the units involved.
- Recognize problems that involve known formulas and relate those formulas to the dependent and independent variables of a problem.
- Recognize problems in which the verbal description of the problem defines the relationship between the dependent and independent variables.
- For problems in which there is not a known general formula, know how to assign one or more sets of specific numerical values to the variables, do the calculations, and look for patterns in the calculations, so you can generalize the relationship symbolically using variables and parameters.
- Be able to interpret expressions such as "no more than", "no less than", "at least", "at most", and "exactly", and use the correct mathematical symbols to represent them.

Recall that the first stage (the first two steps) of our 12-step program for problem solving (see Section 1.0) involved clarifying the problem:

Stage A: Clarify the Problem

1) **Identify, bound and define the problem.** Identify the dependent variable to be optimized. Bound the problem: make sure that you have focused on the essentials and specified the scope and scale of the problem you plan to tackle. Define the problem: what exactly do you want to do?
2) **Understand the problem.** This could involve making **sketches** if a physical object is involved. If appropriate, organize the data into **tables or graphs**. Consider any **assumptions** that are being made. Finally, **restate the problem** clearly, concisely, and unambiguously.

The second stage of the problem-solving process involved the formulation of a model:

Stage B: Formulate a Model

3) **Define all variables clearly** and unambiguously, including the **units** involved.
4) **Plan for and gather data as necessary**. (This step will usually not apply here, as you will be formulating the model from the verbal description that is given, so no further data should be needed.)
5) **Formulate a mathematical model for the problem**, including unambiguously **defining** the objective function **verbally and symbolically** (including its **domain** which could involve one or more constraints and **units** for all of the quantities) and specifying all modeling **assumptions**.

If steps 1-3 have been done carefully, step 5, the actual model formulation for the problem, will usually not be too difficult.

As a reminder, recall that, **throughout this book and course, in our terminology, a *model* is assumed to involve *one* function**. If a situation of interest involves several functions, we call *each* one a model.

Model Formulation

Let's start with an example of the simpler type of formulation, where you are given the constants (parameters) and just need to define your variables and put them together appropriately.

Sample Problem 1: You have a small hand-crafted furniture business. You have designed unique chairs and tables (one style for each), which you build and sell. You know that each chair uses 3 board feet of wood and 8 packages of hardware, and that each table uses 6 board feet of wood and 4 packages of hardware. You always try to anticipate your need for wood and hardware, but delivery is sometimes uncertain. You would like to plan a month ahead. Find models for your raw material requirements as a function of the number of chairs and tables that you make.

Solution: You have identified your problem: find models for the raw material requirements. You will limit these models to the requirements for wood and hardware. There are *two dependent* variables: the amount of wood and the amount of hardware, therefore there will be *two functions* and *two models*: one for the wood requirement and one for the hardware requirement.

Let's start with the wood requirement first. Your decision (independent) variables will be the number of chairs and tables that you plan to make that month, since these are what you can control and what determine the raw material requirements. The units for the independent variables will be simply the number of chairs and the number of tables. The dependent variable is the total amount of wood needed to make a given number of chairs and tables, and is in units of board feet of wood.

Let's define:

c = the number of chairs made during the month,
t = the number of tables made during the month, and
$w(c,t)$ = the total amount of wood (in board feet) needed during the month if c chairs and t tables are made.

For example, if you make 2 chairs, you will need (3)(2) board feet of lumber. If you make 3 tables you will need (6)(3) board feet of lumber. If you make 2 chairs and 3 tables you will need (3)(2) + (6)(3) board feet of lumber, or 6+18 = 24 board feet of lumber.

$$w(2,3) = 3(2) + 6(3) = 6 + 18 = 24.$$

If you make 4 chairs and 5 tables, you will need (3)(4) board feet of lumber for the chairs and (6)(5) board feet of lumber for the tables, so

$$w(4,5) = 3(4) + 6(5) = 12 + 30 = 42.$$

By now, the general pattern should be clear: $w(c,t)$ is just 3 times the number of chairs (c) plus 6 times the number of tables (t), so

$$w(c,t) = 3(c) + 6(t) = 3c + 6t.$$

As with functions of one variable, to find the value of a multivariable function at specified values for each of the variables, just plug in the value for each variable wherever it occurs in the algebraic definition. For example,

$$w(5,8) = 3(5) + 6(8) = 15 + 48 = 63 \text{ board feet of wood.}$$

If you have no chairs and an unknown quantity of tables, such as $2a$ tables, you do exactly the same thing: plug in $2a$ where you have t in the function:

$$w(0, 2a) = 3(0) + 6(2a) = 0 + 12a = 12a,$$

since here $c = 0$ and $t = 2a$. In other words, to make no chairs and $2a$ tables, you need $12a$ board feet of wood.

The domain of this function can be assumed to be all nonnegative numbers (for each variable). We could say that these have to be integers, because we are working with chairs and tables, but we could also allow non-integer values, and interpret them as the **average** number of an item to make each month ($c = 3.5$ could mean to make 7 chairs every 2 months, or alternate between making 3 chairs one month and 4 chairs the next). We can now formulate our model for the amount of wood needed:

Verbal Definition: $w(c,t)$ = the amount of wood, in board feet, needed to make c chairs and t tables per month.

Symbol Definition: $w(c,t) = 3c + 6t$, for $c,t \geq 0$

Assumptions: Certainty and divisibility. Certainty means that the relationship is exact. Divisibility means that any fractional values for the wood, chairs, and tables are possible, which could be interpreted as averages per month.

Let's define in words the function for hardware in term of the number of chairs and tables.

c = the number of chairs made during the month,
t = the number of tables made during the month, and
$h(c,t)$ = the total number of hardware packages needed during the month if c chairs and t tables are made.

For example, if you make 2 chairs, you will need (8)(2) packages of hardware. If you make 3 tables you will need (4)(3) packages of hardware. So to make 2 chairs and 3 tables you need (8)(2) + (4)(3) packages of hardware:

$$h(2,3) = 8(2) + 4(3) = 28.$$

Generalizing as we did above, we see that

$$h(c,t) = 8(c) + 4(t) = 8c + 4t.$$

The model for the number of hardware packages needed is then:

Verbal Definition: $h(c,t)$ = the number of hardware packages needed to make c chairs and t tables per month.

Symbol Definition: $h(c,t) = 8c + 4t$, for $c,t \geq 0$

Assumptions: Certainty and divisibility. Certainty means that the relationship is exact. Divisibility means that any fractional values for the wood, chairs and table are possible, which could be interpreted as averages per month. □

Sample Problem 2: An innovative farm experiments with varying diets for its animals. Currently it is working with three basic types of feeds: A, B, and C. Food A contains 30% protein, 2% fat, and 20% carbohydrate; Food B contains 50% protein, 6% fat, and 10% carbohydrate; and Food C contains 15% protein, 5% fat, and 10% carbohydrate. The farm wants to make models of the different nutrients the animals are getting as a function of the amount of each feed in the mix. They plan to use these models to determine the mixture that will provide the desired nutrients with a minimum cost.

Solution: As always, our first step is to be sure that we have correctly identified and defined the problem. The farm is interested in the different nutrients the animals are getting, so the protein, fat, and carbohydrate will be the dependent variables. Again, we have more than one dependent variable, so we will have more than one function and more than one model. The farm has decided to limit the animals' food intake to these three foods (A, B, and C), so we need only consider the contents of these foods. The amounts of each of these foods that the animals consume will be the independent variables (decision variables), since they can be controlled by the farm and determine the values of the dependent variables. The foods for the animals are measured in grams, so we can define:

$a =$ grams of feed A ,
$b =$ grams of feed B,
$c =$ grams of feed C,

The labels on the foods list the different nutrients as percents of the total food. For example, food A has a content of 30% protein, so if you fed an animal 100 grams of food A, the animal would get (0.30)(100) gram or 30 grams of protein. Therefore, it will be convenient to make the units of the different nutrients *grams*, too. Having done this calculation, it should be clear that the protein is, in fact, a *function* of the quantities of each food in the mix:

$P(a,b,c) =$ the number of grams of protein in a mixture of a grams of Food A, b grams of Food B, and c grams of Food C

Suppose we want to calculate $P(50,40,100)$. This means we want to know how much protein there would be in a mixture of 50 grams of Food A, 40 grams of Food B, and 100 grams of Food C. Since we know that Food A is 30% protein, 50 grams of it must have 30% of 50 grams, or

$0.3(50) = 15$ grams of protein.

Similarly, Food B is 50% protein, so 40 grams of Food B must have $0.5(40) = 20$ g of protein, and Food C is 15% protein, so 100 grams of Food C must have $0.15(100) = 15$ g of protein. Thus the total amount of protein is

$$P(50,40,100) = 0.3(50) + 0.5(40) + 0.15(100) = 15 + 20 + 15 = 50$$

(in other words, 50 grams). The pattern would be similar for other combinations of *a*, *b*, and *c*. Thus, the general algebraic form for $P(a,b,c)$ is given by

$$P(a,b,c) = 0.3a + 0.5b + 0.15c$$

The model for the amount of protein received is then:

Verbal Definition: $P(a,b,c)$ = the number of grams of protein in a mixture of a grams of Food A, b grams of Food B, and c grams of Food C.

Symbol Definition: $P(a,b,c) = 0.3a + 0.5b + 0.15c, \quad \text{for } a,b,c \geq 0.$

Assumptions: Certainty and divisibility. Certainty means that the relationship is exact. Divisibility means that any fractional values of grams of foods and protein are possible. □

Challenge: On your own, try to formulate the models for the amounts of fat and carbohydrates the animals would receive.

Demand Functions and Revenue Functions

Sample Problem 3: A small start-up company sells three products: shirts, belts, and slacks. Find the revenue as a function of the selling price and quantity sold of each..

Solution: This problem appears to be well defined. The dependent variable is the revenue received from the sale of shirts, belts, and slacks, which we will define to be in dollars. The independent variables are the prices and the numbers of shirts, belts, and slacks sold. Let's denote the price and quantity of the shirts as p_1 and q_1, respectively, the price and quantity of the belts as p_2 and q_2, respectively, and the price and quantity of the slacks as p_3 and q_3, respectively, assuming all prices are in dollars. We can then call the revenue function $R(p_1,q_1, p_2,q_2, p_3,q_3)$, in dollars. As we saw earlier in the course, for a single product sold at a fixed price, revenue is simply price times quantity, selling price times quantity sold. This will still be true for each item, so we just need to add the results to get the total revenue. For example, if you sold 3 shirts at $20.00 each, 5 belts at $10.00 each and 4 pairs of slacks at $45.00 each, the revenue from the sales would be:

$$\begin{aligned} R &= (\$20)(3) + (\$10)(5) + (\$45)(4) \\ &= \$60 + \$50 + \$180 \\ &= \$290 \end{aligned}$$

If you changed your prices to $18.00 per shirt, $12.00 per belt and $42.00 per pair of slacks and sold the same number of shirts, belts, and slacks as above, the revenue would be:

$$R = (\$18)(3) + (\$12)(5) + (\$42)(4)$$
$$= \$54 + \$60 + \$168$$
$$= \$282$$

If you kept the original prices and sold 5 shirts, 3 belts and 2 pair of slacks the revenue would be:

$$R = (\$20)(5) + (\$10)(3) + (\$45)(2)$$
$$= \$100 + \$30 + \$90$$
$$= \$220$$

Look at the pattern here. We multiply the price of the shirts times the quantity of shirts sold and add to that the price of the belts times the quantity of belts sold and finally, add the price of the slacks times the number of slacks sold. In general, we will have

Verbal Definition: $R(p_1, q_1, p_2, q_2, p_3, q_3)$ = revenue in dollars when q_1 shirts are sold at p_1 dollars per shirt and q_2 belts are sold at p_2 dollars per belt and q_3 pairs of slacks are sold at p_3 dollars per pair of slacks.

Symbol Definition: $R(p_1, q_1, p_2, q_2, p_3, q_3) = p_1q_1 + p_2q_2 + p_3q_3$, for $p_1, q_1, p_2, q_2, p_3, q_3 \geq 0$

Assumptions: Certainty and divisibility. Certainty means that the relationship is exact: that there are no sales, quantity discounts, coupon sales, etc. Divisibility means that any fractional values for shirts, belts, slacks and dollars are possible.

This rather complicated function can usually be simplified by the fact that price and quantity are related, as we saw earlier in the course. For a single commodity, this relationship is called a demand function or demand curve. The quantity of an item that is available can influence the price, as seen when shortages occur. Thus, the *price* can be thought of as a function of the *quantity*:

$$p = D(q)$$

It is also clear that price influences quantity, as shoppers at a bargain store can attest. As a result, it is also true that *quantity* can be thought of as a function of *price*:

$$q = d(p).$$

The two quantities are directly related, so it is just a question of which variable you solve for. Mathematically, the two functions are inverses if the relationship is known exactly.

Revenue = (selling price)(quantity sold) or $R = pq$,

Thus, revenue can be expressed as a function of quantity:

$$R(q) = [D(q)] \cdot (q)$$

(by substituting $D(q)$ for p), or as a function of price:

$$R(p) = (p) \cdot [d(p)]$$

(by substituting $d(p)$ for q).

In some cases, the quantity of an item that is sold depends not only on the price of that item, but also on the price of competing items (such as Pepsi™ and Coke™) as well as the price of complementary items (items that are usually bought together, such as cameras and film, pretzels and soda). We will look at such cases more closely when we study the formulation of multivariable models from data.

Profit Margin and Markup

Sample Problem 4: You just bought a book at a local bookstore for \$19.95 . When you questioned their pricing policy, they told you that they operate on a 20% profit margin. How much did the store pay for the book?

Solution: This seems like a very straightforward problem. You know the selling price of the book (what you paid for the book, \$19.95) and you know the profit margin. You have to determine the cost. Let's be sure that we understand each of these terms. The selling price is the price, in dollars, that you pay for the book. The cost of the book is the amount, in dollars, that the bookstore paid for the book. We are considering the straight cost per book, and are assuming there are *no discounts or fixed costs.*

What exactly is profit margin? **Profit margin** is not the same thing as marginal profit. Recall that the word "marginal" when used as an adjective with cost, revenue or profit, refers to the rate of change of cost, revenue or profit, normally with respect to quantity: for example, the derivative of the cost, revenue or profit function with respect to quantity. **Profit margin**, on the other hand, **is the profit made from the sale of an item expressed as a fraction of the selling price (often in percent form).** If an item cost \$70 to produce and the price charged for that item was \$100, what was the profit margin? The profit made from the item is the selling price (revenue) minus the cost:

profit = \$100 - \$70 = \$30.

Expressing this profit as a fraction of the selling price, we get: $\frac{\$30}{\$100} = 0.3 = 30\%$.

If we wrote out the detailed calculation, it would have been

$$\text{profit margin} = m = \frac{100-70}{100} = \frac{\text{selling price - cost}}{\text{cost}}$$

Profit margin is profit (sales price minus cost) divided by the selling price, so profit margin, *m,* is a function of the cost, *c*, and the selling price, p. The **model for profit margin is:**

Verbal Definition: $m(p,c)$ = the profit margin (expressed as a pure decimal) when the selling price, in dollars, is p, and the cost, in dollars, is c.

Symbol Definition. $m(p,c) = \frac{p-c}{p} = 1 - \frac{c}{p}$ for $p > 0 \quad c > 0 \quad c \le p$

Assumptions: Certainty and divisibility. Certainty means that the relationship is exact. Divisibility means that any fractional values of dollars are possible. Certainty implies that there are no quantity discounts, sales, coupon deals, etc.

This concept is generally used by the seller. As the seller, you typically know the cost of the product and you know what profit margin you want to make and you want to determine the price. That is, often we want to solve the equation given above for price, p. This means we want to manipulate the equation algebraically to get the p by itself. One way to do this is as follows:

$\frac{p-c}{p} = m$ (Switching the two sides.)

$p - c = mp$ (Multiplying both sides by p (OK since $p>0$).)

$p - mp = c$ (Putting terms with p on the left: adding $c\text{-}mp$ to both sides.)

$p(1-m) = c$ (Factoring out the p on the left.)

$p = \frac{c}{1-m}$ (Dividing both sides by $1-m$, which is OK since

$c > 0, p > 0$, and $c \le p$, so $\frac{p-c}{p} = m < 1$, so $1 - m \ne 0$)

Thus, we now have an expression for selling price in terms of cost and profit margin, which we can formalize by defining a model:

Verbal Definition: $p(c,m)$ = the selling price, in dollars, when the cost is c dollars and the profit margin is m [1]

Symbol Definition: $p(c,m) = \frac{c}{1-m}$, *for* $c > 0,\ 0 \le m < 1$.

Assumptions: Certainty and divisibility. Certainty means that the relationship is exact. Divisibility means that any fractional values of dollars and profit margin are possible. Certainty implies that there are no quantity discounts, sales, coupon deals, etc.

In our problem we know the selling price and the profit margin and want to determine the cost. We want to express cost as a function of selling price and profit margin. We have already found this functional relationship in the process of solving the profit margin equation for p (see the third line of that derivation: $p - mp = c$)! This is an example of the value of writing out your steps in solving a problem; sometimes you uncover something that you weren't even looking for, but that will be useful later and save time. From above, we see that we get the following model for c :

Verbal Definition: $c(p,m)$ = the cost in dollars when the selling price is p dollars and the profit margin is m (expressed as a decimal).

Symbol Definition: $c(p,m) = p - mp = p(1-m)$, for $p > 0$, $0 \le m < 1$

Assumptions: Certainty and divisibility. Certainty means that the relationship is exact. Divisibility means that any fractional values of dollars and profit margin are possible. Certainty implies that there are no quantity discounts, sales, coupon deals, etc.

The selling price was \$19.95 and the profit margin was 20%. Plugging these figures into our model for the cost we get:

$$c = 19.95 - (0.2)(19.95)$$
$$c = 19.95(1\text{-}0.2)$$
$$c = 19.95(.8)$$
$$c = 15.96$$

Translation: The cost of the book to the bookstore was \$15.96 .

Notice that this formula makes intuitive sense, since if the profit margin is 20% of the selling price, then the cost must be 80% of the selling price. In general, if the profit is

[1] Remember that the profit margin is defined as the *fraction* of the selling price represented by the profit, so will normally be a number between 0 and 1. If the profit margin is 30%, then m is 0.30 .

a fraction m of the selling price, then the cost must be the fraction $(1\text{-}m)$ of the selling price. □

Sample Problem 5: You heard that there is a store not too far away that advertises that it sells all its books at a discount. In fact, they claim that they mark up their books only 5%. They sell the same book as the one described in Sample Problem 4 for $16.95. If you can determine the cost of the book to the store, can you check to see if their claim of the 5% markup is true or is this false advertising?

Solution: First let's clarify and define the problem. What is it that we really want to know? We want to know if the discount store is actually using a 5% markup. Their claim is that this is the markup for all their books, that they don't vary the markup by specific type of book. We are going to assume that they use the same pricing system for all books and so, if we check the data on one book, our conclusion will be valid for all their books. Now, let's be sure that we are clear on what a 5% markup means. The (fractional) markup means the fraction of the cost (to the *seller*) added onto the cost to get the selling price. Thus the **amount of the markup**[2] is the profit on each unit. If the book costs them $10.00 and they have a 5% markup, the markup amount, or profit, is

$$(0.05)(\$10.00) = \$0.50$$

The book would sell for the cost, $10.00, plus the markup amount, $0.50, or $10.50. If we write out the calculation from scratch, it would be

$$\text{selling price} = 10 + (0.05)(10)$$

If the book cost them $15.00 and the markup is 5%, the markup amount would be

$$(0.05)(\$15.00) = \$0.75$$

and the selling price would be $15.75. This time the detailed calculation from scratch would be

$$\text{selling price} = p = 15 + (0.05)(15) = \text{cost} + (\text{fractional markup})(\text{cost})$$

So we can generalize and say that the selling price of the book, p, is a function of the cost of the book, c, and the fractional markup, k:

---

[2] Sometimes "markup" can be used in two different ways. It can mean the *dollar amount* that was added to the cost to get the selling price (the profit), which we will usually call the **markup amount**, or the *fraction* or the cost represented by the markup amount, which we will call the **fractional markup**, or simply **markup** for short. Usually it is clear from the context which is meant, but we will try to be explicit wherever possible.

$p(c,k)$ = the selling price in dollars when c is the cost in dollars and k is the fractional markup.

$p(c,k) = c + k \cdot c$

$p(c,k) = c \cdot (1+k)$ (Factoring out the c.)

Once again, we can define a full model:

Verbal Definition: $p(c,k)$ = the selling price in dollars when c is the cost in dollars and k is the fractional markup.

Symbol Definition: $p(c,k) = c \cdot (1+k)$, for $c > 0$

Assumptions: Certainty and divisibility. Certainty means that the relationship is exact. Divisibility means that any fractional values of dollars and fractional markup are possible. Certainty implies that there are no quantity discounts, sales, coupon deals, etc.

Notice that we have not put a constraint on k here. Normally, it would be positive (to justify the name mark*up*), but it could quite conceivably be a negative number, up to the magnitude of the cost. This concept is used in pricing "loss leaders", a pricing scheme used to entice shoppers into a store.

This problem, like the preceding one, deals with the relationship between the cost to the seller and the price charged. As expressed above, the selling price is a function of the cost and the fractional markup. In other words, we use this functional definition when the cost and fractional markup are known. In our case, **we assume that the cost of the book was the same to both stores**. So we know the selling price and the cost and we want to determine the fractional markup. We must solve the equation for *k:*

$p = c + kc$

$c + kc = p$ (Reversing the order.)

$kc = p - c$ (Subtracting c from both sides.)

$k = \dfrac{p-c}{c}$ (Dividing both sides by c ; defined since $c > 0$ and not = 0)

Verbal Definition: $k(c,p)$ = the fractional markup when p is the selling price, in dollars, and c is the cost, in dollars.

Symbol Definition: $k = \dfrac{p-c}{c} = \dfrac{p}{c} - 1$ for $p, c > 0$

Assumptions: Certainty and divisibility. Certainty means that the relationship is exact. Divisibility means that any fractional values of dollars and fractional markup are possible. Certainty implies that there are no quantity discounts, sales, coupon deals, etc.

We are assuming that the cost of the book was \$15.96 and the selling price was \$16.95. Plugging these numbers in, we find the fractional markup was

$$k = \frac{p-c}{c} = \frac{p}{c} - 1 = \frac{16.95}{15.96} - 1 \approx 1.062 - 1 = 0.062$$

Translation: The markup on the book is about 6.2% . Even considering possible rounding error, this is more than 1 percentage point above their advertised markup of 5%, or 24% (1.2/5 = 0.24) over their claim. □

Formulating Models with Constraints

Sample Problem 6 : You have to build some containers to ship special pellets. To conform with customs regulations and take advantage of duty pricing, the containers must hold exactly 35 pounds of pellets. In the past, you shipped the pellets in 1 ft by 1 ft by 1 ft containers, which held 5 pounds of pellets, but you are now getting larger orders and the small containers are not economical. Because of the weight of the contents, the bases of the containers must be built of stronger (and more expensive) material than the sides and top. The base of the container costs \$0.84 per square foot and the sides and top cost \$0.62 per square foot. Develop a mathematical model for the cost of the containers so you can determine what size containers to build.

Solution: First, you have to identify, bound and define the problem. What exactly is it that you want? You want to build the container at the least cost. So the cost is the dependent variable to be minimized. You are interested in the cost of the materials, and will not consider the cost of the actual construction of the container. This seems fairly well defined, but is it really? What kind of container should you build? Does it have to be a rectangular box, or could you use a cylindrical canister? You had not really thought about using a cylindrical canister, so should take this possibility into consideration. After consulting your shipper, you have determined that a rectangular box is better, but you would like to price out both types. We have left the problem of finding the cost model for the cylindrical canister as an exercise at the end of the section, and will only work with the rectangular box model here. This is an example of **bounding** a problem.

In the case of the rectangular box, there is a physical object involved here, so a sketch is appropriate. See Figure 5.2 - 1 .

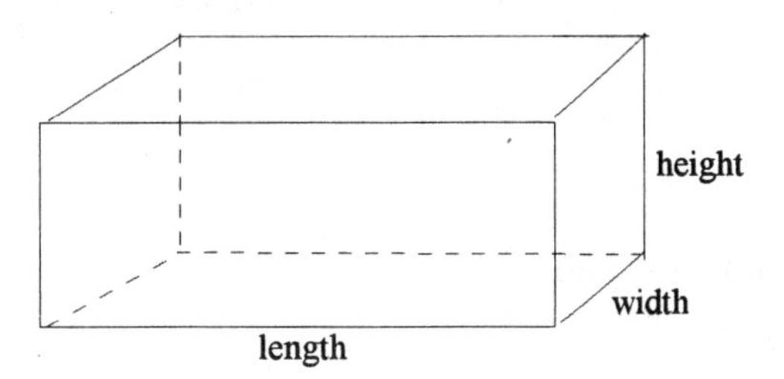

Figure 5.2 - 1

The cost of the box will be determined by the amount of material used in it. The area of the base and the top of the box each would be the length times the width. The front and back sides of the box will each be the length times the height, and the other sides will each be the width times the height. So the independent variables, the decision variables, will be the dimensions of the box - its length, width, and height.

We can now define our variables:

l = the length of the box in feet
w = the width of the box in feet
h = the height of the box in feet
$c(l,w,h)$ = the cost in dollars to build a box of length l feet, width w feet, and height h feet.

The cost, as we have noted, is a function of the length, width, and height: l, w, and h, respectively. The base of the box costs 0.84 dollars per square foot, so the base will cost 0.84 times the length times the width, or $0.84lw$. The front and the back cost 0.62 dollars per square foot, or 0.62 times the length times the height for each. So the cost of the front and back of the box is $2(0.62)lh$, or $1.24lh$. The other two sides also cost 0.62 dollars per square foot, so the cost for each side is 0.62 times the width times the height. So the cost of the other two sides is $2(0.62)wh$, or $1.24wh$. The top of the box also costs 0.62 dollars per square foot, and is the length times the width, or $0.62lw$. The total cost of the box will be the sum of these costs:

$$c(l,w,h) = 0.84lw + 1.24lh + 1.24wh + 0.62lw$$

We can combine like terms and simplify to:

$$c(l,w,h) = 1.46lw + 1.24lh + 1.24wh$$

We are also interested in the volume of the box, which will be the length times the width times the height:

$$V(l,w,h) = lwh.$$

You want a box that will hold 35 pounds of pellets. You used to ship in 1 ft by 1 ft by 1 ft containers, which held 5 pounds of pellets:

$$V(1,1,1) = (1)(1)(1) = 1 \text{ cubic foot.}$$

So a 1-cubic-foot box holds 5 pounds. How many cubic feet will you need to hold 35 pounds of pellets? 35 pounds is 7 times as much as 5 pounds, so you will need 7 times as much space in the box, or 7 times 1 cubic foot. The volume of the new box, length times width times height, must be 7 cubic feet. Using the variables we have defined above:

$$lwh = 7$$

We want to minimize the cost of the box subject to the constraint that $lwh = 7$. How can we express this in our mathematical model? The statement "$lwh = 7$" sets limits on the values that the variables w, h, and l can take. The values that the independent variables can take are the **domain** of the function. The domain of this function will be that w, h, and l must all be greater than zero (if any one of the three were zero, there would be no box) and that $lwh = 7$. We are now ready to write the model for the cost of the box:

Verbal Definition: $c(l,w,h)$ = the cost in dollars to build a box of length l feet, width w feet and height h feet.

Symbol Definition: $c(l,w,h) = 1.46lw + 1.24lh + 1.24wh$, for $lwh = 7$ and $l, w, h, > 0$.

Assumptions: Certainty and divisibility. Certainty means that the relationship is exact. Divisibility means that any fractional values of dollars and feet are possible.

Once again, we leave the cylindrical canister calculation as an exercise.

Example 1: The post office has strict regulations about the size of package that you can mail first class. You want to be able to ship the maximum amount of goods while staying within the size limits. You called the Post Office and they gave you the following information: The total measurement of each package must be no more then 108 inches. What do they mean by this?

What exactly is the problem? You want to be able to ship the maximum amount of goods. This means that you want your shipping boxes to have the maximum volume. So the volume of the box will be the dependent variable. You want to use a standard rectangular box; that is, a base, four sides and a top. You are not interested in shipping tubes, such as those used to ship such things as blueprints or posters. You have identified, bounded and defined the problem.

The next step is to be sure that you understand the problem. Once again we will draw a picture. See Figure 5.2 - 2.

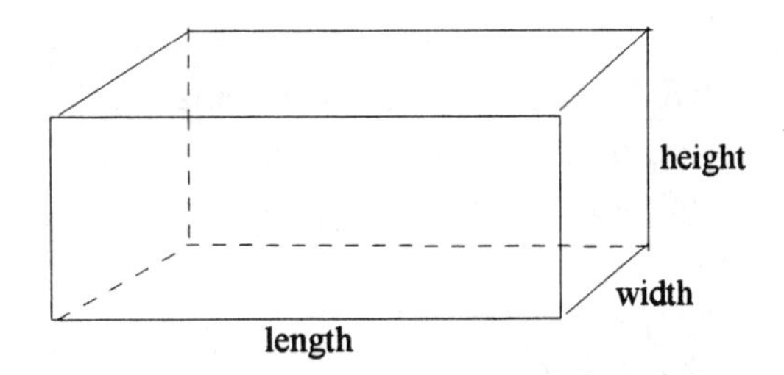

Figure 5.2 - 2

You know that the amount that the box will hold is called the volume of the box, and that the volume of a box equals the length times the width times the height. You really understand the problem pretty well, except for the Post Office requirement. What do they mean by "total measurement"? This requires another call to the Post Office. They explain that the total measurement of the package is the given by the girth (the distance all the way around the package) plus the length of the package, where the girth is measured in the direction in which the package is addressed.[3] You think about this for a while, and decide that you will address the package on the top, and run the address parallel to the width of the package, as shown in Figure 5.2 - 3

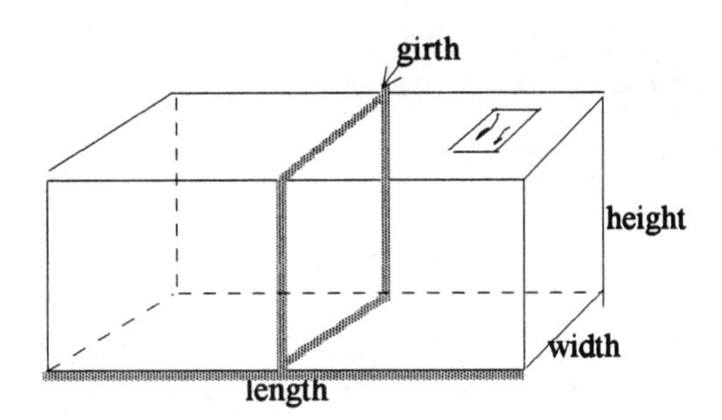

Figure 5.2 - 3

[3] U.S. Postal Service regulations for packages mailed in the United States.

The girth then is the width plus the height, plus the width plus the height again, and the "total measurement" will be this sum plus the length. The total measurement must be no more than 108 inches. Another way to say this is that the total measurement must be less than or equal to 108 inches. Recall that the mathematical symbol for "less than or equal to" is $\leq$. Symbolically,

$$2w+2h+l \leq 108$$ □

A quick review of the translation between words and symbols for inequalities is shown below.

Verbal Expression	Mathematical Symbol
"a is no more than b"	$a \leq b$
"a is at most b"	$a \leq b$
"a is no less than b"	$a \geq b$
"a is at least b"	$a \geq b$
"a is exactly b"	$a = b$

Sample Problem 7: Now that you have a good understanding of the restrictions set by the Post Office, that the sum of the length and the girth can be no more than 108 inches, formulate a model for the package dimensions that will help you find the greatest possible volume.

Solution: We are ready to define the variables. Since the total measurement is given in inches, the length, width, and height will be in inches, and the volume will be in cubic inches.

w = the width of the box in inches
l = the length of the box in inches
h = the height of the box in inches
V = the volume of the box in cubic inches

The volume of the box is given by

$$V = lwh$$

The total measurement of the box is given by

$$w+h+w+h+l = 2w+2h+l$$

We want to maximize

$$V(l,w,h) = lwh$$

subject to the constraint

$$2w+2h+l \leq 108.$$

Exactly what do we mean by "subject to"? How can we express this in our mathematical model? The statement "$2w+2h+l \leq 108$" sets limits on the values that the variables w, h, and l can take. The values that the independent variables can take are the **domain** of the function. The domain of this function will be that w, h, and l must all be greater than zero (if any one of the three were zero, there would be no box) and that $2w+2h+l \leq 108$. We are now ready to write our model for the problem:

Verbal Definition: $V(l,w,h) =$ volume of a box, in cubic inches, of length l inches, width w inches, and height h inches.

Symbol Definition: $V(l,w,h) = lwh$, subject to $2w+2h+l \leq 108$ and $l > 0,\ w > 0,\ h > 0$

Assumptions: Certainty and divisibility Certainty means that the relationship is exact. Divisibility means that any fractional values of inches and cubic inches are possible.

Elimination of Variables

Sample Problem 8: Let's look again at Sample Problem 6. You have to build some containers that must have a volume of 7 cubic feet to hold exactly 35 pounds of lead pellets. The base of the container costs \$0.84 per square foot and the sides and top cost \$0.62 per square foot. Develop a mathematical model for the cost of the containers as a function of only *two* of the dimensions so you can determine what size containers to build.

Solution: We have already stated that the box must have a volume of 7 cubic feet - not at least 7 cubic feet or at most 7 cubic feet, but exactly 7 cubic feet. This means that, if we knew two of the dimensions we could figure out the third, since the product of multiplying the three values must be 7. With this in mind, we can eliminate one variable from our model of the problem. We can do this by solving the equation for the volume for one of the variables, say h:

$lwh = 7$, so

$h = \frac{7}{lw}$ (dividing both sides by lw ; defined since $l,w > 0$, so $lw \neq 0$)

We can now substitute this value for h in our function for the cost and reduce the number of input variables from three to two. The cost function will not be particularly "simplified," as you will see, but the problem will be simplified in some ways - for example, it could now be graphed. Thus we can replace h with $\frac{7}{lw}$ in our cost function:

$$c(l,w) = 1.46lw + 1.24lh + 1.24wh\ , \quad \text{for } l,w,h>0$$

$$c(l,w)=1.46lw+1.24l\left(\frac{7}{lw}\right)+1.24w\left(\frac{7}{lw}\right), \quad \text{for } l,w,\left(\frac{7}{lw}\right)>0\ .$$

Notice that we replaced h in the $h > 0$ constraint as well, but in fact the resulting inequality

$$\left(\frac{7}{lw}\right)>0$$

is redundant. This is because if $l>0$ and $w>0$, then $lw>0$, so $\left(\frac{7}{lw}\right)>0$ also. Thus we do not need to retain this constraint.

Since l and w cannot be equal to 0, as stated in the domain, we can simplify the cost function further by canceling in the last two terms:

$$c(l,w)=1.46lw+1.24l\left(\frac{7}{lw}\right)+1.24w\left(\frac{7}{lw}\right), \quad \text{for } l,w,\left(\frac{7}{lw}\right)>0\ .$$

$$c(l,w)=1.46lw+1.24\left(\frac{7}{w}\right)+1.24\left(\frac{7}{l}\right), \text{ for } w,l>0$$

We can further simplify by multiplying the constants:

$$c(l,w)=1.46lw+\left(\frac{8.68}{w}\right)+\left(\frac{8.68}{l}\right), \text{ for } \ l,w>0$$

Our model is now:

Verbal Definition: $c(l,w)$ = the cost in dollars to build a box of length l feet and width w feet.

Symbol Definition: $c(l,w)= 1.46lw+\left(\frac{8.68}{w}\right)+\left(\frac{8.68}{l}\right)$ for $l,w>0$

Assumptions: Certainty and divisibility. Certainty means that the relationship is exact. Divisibility means that any fractional values of inches and cubic inches are possible. We also assume that the height of the box will be given by $h = \frac{7}{lw}$. □

Sample Problem 9: Let's look again at Sample Problem 7. We want to determine the measurements of a box with the greatest volume that satisfies the condition that the length plus the girth must be less than or equal to 108 inches. Can we eliminate a variable in this problem, as we did in Sample Problem 8?

Solution: Is there any way that we can restate this problem? Having a mathematical statement such as

$$2w+2h+l \leq 108$$

as part of the domain complicates the problem quite a bit. If we *assume* that we will want to use all of the possible girth, we could rewrite that statement as

$$2w+2h+l = 108.$$

This makes intuitive sense: since we are trying to *maximize* the volume, it seems reasonable that we would use *all* of the total measurement that we are allowed. The volume, the dependent variable, is given by

$$V(l,w,h) = lwh.$$

Can we incorporate these two statements into one statement? Look carefully at the condition: $2w+2h+l = 108$. We can easily find the length l as a *function* of w and h:

$$2w + 2h + l = 108$$
$$l = 108 - 2w - 2h \qquad \text{(subtracting } 2w + 2h \text{ from both sides)}$$

To emphasize this functional relationship, we could even write

$$l = g(w,h) = 108 - 2w - 2h.$$

Now that we have solved for one variable (l), we can substitute the expression we got for l into the volume function (including its domain constraints):

$$V(l,w,h) = lwh, \qquad \text{for } l,w,h > 0$$
$$V(w,h) = (108 - 2w - 2h)wh, \qquad \text{for } (108\text{-}2w\text{-}2h),w,h > 0 .$$

Note that the original constraint $l > 0$ has become

$108 - 2w - 2h > 0$
$0 < 108 - 2w - 2h$ (rewriting the inequality by flipping it)
$2w + 2h < 108$ (adding $2w + 2h$ to both sides)
$w + h < 54$ (dividing both sides by $2 > 0$)

With this calculation, and multiplying through in the $V(w,h)$ function, we obtain:

$$V(w,h) = 108wh - 2w^2h - 2wh^2 \quad \text{such that} \quad w,h > 0 \text{ and } w + h < 54$$

We must be sure to *state* that we have made the *assumption* about using the *entire* total measurement possible in developing our model. Our model for this problem is thus:

Verbal Definition: $V(w,h)$ = the volume in cubic inches for a box of width w inches and height h inches.

Symbol Definition: $V(w,h) = 108wh - 2w^2h - 2wh^2$, subject to $w + h < 54$ and w and $h > .0$.

Assumptions: Certainty, divisibility , and the assumption that the box will have *exactly* the maximum allowable total measurement, which is 108 inches. Certainty means that the relationship is exact, divisibility means that any fractional values of inches and cubic inches are possible.

In other words, the answer is: *yes*, we can eliminate a variable. As an exercise, see if you can do the same thing, but eliminate w or h. □

Sample Problem 10: Let's look at the hand-crafted furniture problem again. You have been trying to anticipate your needs for wood and hardware each month and have set up a model to predict this. You have also set up a model for your expected profits from the sales of the chairs and tables: Your expected profit from the sale of each chair is $140 and from each table is $200. You are ready to order your supplies for next month when you get a notice from your supplier: they can send you only 100 packages of hardware and 60 board feet of wood. They promise to be up and running next month and will be able to fill all of your orders, but that is all that is available for this month. Orders for the chairs have been a little slow, and you are sure that, at most, you can sell 11 chairs next month. On the other hand, you are sure that you can sell all the tables that you make. How many chairs and how many tables should you make? Set up the model for this problem.

Solution: Let's assume that you want to make the greatest profit possible, so the profit will be your dependent variable (objective function). You are only going to consider the profit from the sales of chairs and tables, so more specifically, the profit from the sales of chairs and tables will be your dependent variable. Since it is the profit from the sales of chairs and tables, the number of chairs and tables will be your independent variables. Let's define:

c = the number of chairs made and sold during the month
t = the number of tables made and sold during the month
$\pi(c,t)$ = the profit in dollars when c chairs are made and sold and t tables are made and sold.

This is pretty straight forward, and similar to the preceding problems. However, we do have some other things to consider: the limited amount of wood and hardware that we have. At this point, it would probably be a good idea to set up a table of the information that we have.

When you set up a table of your data, *it is usually best to put the independent variables across the top of the table.* This will make it much easier when it comes time to translate your problem into mathematical terms. We have done this for this problem in Table 1 .

Table 1

	Per Chair	Per Table	Constraint
Profit in Dollars	140	200	
Wood in board feet	3	6	at most 60
Hardware, packages	8	4	at most 100

Following the reasoning from Sample Problem 1, if you make and sell 1 chair and 2 tables during the month, your profit will be:

$$\pi(1,2) = (140)(1) + (200)(2) = 140 + 400 = 540$$

dollars. If you make and sell 2 chairs and 3 tables your profit will be:

$$\pi(2,3) = (140)(2) + (200)(3) = 280 + 600 = 880$$

dollars. In general, your profit from the sale of c chairs and t tables will be:

$$\pi(c,t) = 140c + 200t$$

dollars. Since the profit from each chair is \$140, $140c$ is the profit from c chairs and, similarly, $200t$ is the profit from t tables.

In Sample Problem 1 we wrote the amounts of wood and hardware needed as functions of the numbers of chairs and tables made:

$w(c,t) = 3c + 6t$ and $w(c,t) \leq 60$, so $3c + 6t \leq 60$, still requiring $c, t \geq 0$
$h(c,t) = 8c + 4t$ and $h(c,t) \leq 100$, so $8c + 4t \leq 100$, still requiring $c, t \geq 0$.

One other restriction on what you can do was mentioned in the description of the problem: it said you could only sell at most 11 chairs. This means that the number of chairs you make and sell should be at most 11. From the chart at the end of Sample Problem 6, or from your own common sense, this corresponds to the constraint

$c \leq 11$.

Our model will then be:

Verbal Definition: $\pi(c,t) =$ profit in dollars if c chairs and t tables are made and sold.
Symbol Definition: $\pi(c,t) = 140c + 200t$ subject to $3c + 6t \leq 60$
$8c + 4t \leq 100$.
$c \leq 11$
$c, t \geq 0$

Assumptions: Certainty and continuity. Certainty means that the relationship is exact. Divisibility means that any fractional values of chairs, tables, and dollars are possible. Certainty assumes that you can sell all of the tables you make, as stated in the problem.

In this case we cannot eliminate any variables because there are no equations relating just the independent variables to each other. We will learn how to work with this kind of problem in Chapter 9. □

Sample Problem 11: You like two breakfast cereals, Special K™ and Raisin Bran™. You have just set up a dietary plan for your consumption of fat, fiber, protein, and calories. You want to choose your breakfast cereal so that you minimize your costs while meeting your dietary goals. You check the prices and determine that the large economy size of Special K contains 12 ounces or 340 grams and sells for $3.39.[4] [Be an intelligent consumer. Check the unit prices. We've seen examples where the large "economy " size actually costs more per unit than a smaller container.] As we become more health conscious in the U.S., food packages now also have a detailed list of nutritional information. One serving of Special K (1 cup or 31 grams or 1.1 ounces) contains 0 grams of fat, 1 gram of dietary fiber, 6 grams of protein and 150 calories. (This

[4] Prices as of 19 October, 1996 at a local grocery store.

includes ½ cup of skim milk; eating dry cereal is not very appealing.) You also like Raisin Bran and like to have a change so you check it out: the large economy size contains 20 ounces or 567 grams and sells for \$3.69. One serving of Raisin Bran (also 1 cup, but weighing 61 grams or 2.2 ounces) contains 1.5 grams of fat, 8 grams of dietary fiber, 6 grams of protein, and 240 calories (again, including ½ cup of skim milk). You have decided that you want to get at most 2 grams of fat, at least 5 grams of fiber, at most 250 calories, and exactly 9 grams of protein at breakfast. (In other words, you have *bounded* and *clarified* your problem already, which is often the hardest part of a problem like this.) Find a model for the cost.

Solution: First we must identify and define the problem. What you want to do is minimize your cost, so the cost is the dependent variable. Next bound the problem: You have decided to concentrate on breakfast; specifically, the cereal that you have for breakfast. You have also decided to consider just two cereals, Special K and Raisin Bran. You have checked out the information on the two cereals and have created the table shown in Table 2 (each serving includes ½ cup of skim milk):

Table 2

Cereal	**Cost/ Box**	**Amount in Box**	**Serving Size**	**Fat/ Serving**	**Fiber/ Serving**	**Protein/ Serving**	**Calories/ Serving**
Special K	\$3.39	18 ounces 510 grams	1.1 ounces 31 grams	0 grams	1 grams	6 grams	150
Raisin Bran	\$3.69	20 ounces 569 grams	2.2 ounces 61 grams	1 grams	7 grams	4 grams	210

Take a minute here and look carefully at the table. Could you make a guess as to the most economical combination of cereals that meet your requirements? The cost per box is not too different. What is your answer?

Your next step is to define all relevant variables, including units. What actually are the independent variables, what can you change? You can change the amount of each type of cereal that you eat every day. If you restrict yourself to using whole "servings" as listed on the boxes, you really don't have a lot of choices. One serving of either cereal does not satisfy your protein requirement of exactly 9 grams. Two servings of Special K does not meet your fiber requirement. Two servings of Raisin Bran does meet your fiber requirement but exceeds your calorie restriction. One serving of each also exceeds your calorie restriction. You will have to have a combination of partial "servings." Since you could eat a "partial serving" and the nutritional contents are given in grams per serving, you could decide to use "servings" as the units for your independent variables. You are now ready to define your variables:

c = the cost in dollars for cereal (ignoring the cost of milk)

K = the number of servings of Special K (including 4 oz skim milk)
R = the number of servings of Raisin Bran (including 4 oz skim milk)

Notice that we are ignoring the cost of milk here. This will probably not affect our decision significantly, so is a good starting model, but this cost could be incorporated for a more complete model of the real problem. Also, K and R could be interpreted in the obvious way as the number of servings every day, or they could also represent the *average* number of servings of each cereal per day. For example, $K = .5$ could mean "half a serving every day" or "one serving every other day." The model would be the same either way.

The verbal definition of your model is:

$c(K,R)$ = the cost in dollars for K servings of Special K and R servings of Raisin Bran (ignoring milk costs).

The boxes don't give the cost per serving, so you have to calculate that. Special K says there are "about 16" servings per box. You want to check that out. The box contains 18 ounces and 1 serving is 1.1 ounce, so there are

(18 ounces per box)/(1.1 ounces per serving) = 16.3636...servings per box.

If you use the gram measure, there are

(510 grams per box)/(31 grams per serving) = 16.4516...servings per box.

Before you decide how to handle these decimals, let's look at the Raisin Bran. One box contains 20 ounces and each serving is 2.2 ounces, so there are 20 ounces/2.2 ounces per serving or 9.09 servings per box. Checking the gram measurements for Raisin Bran, 569 grams per box/61 grams per serving is 9.37 servings per box. You decide to round to 16.5 and 9.4, since the gram measurements seem the most precise.

The cost per serving for Special K would then be \$3.39 per box/16.5 servings per box = \$0.2054 per serving. The cost per serving for the Raisin Bran is \$3.69 per box/9.4 servings per box = \$0.39255 per serving. Again you could decide to round to \$0.21 for Special K and \$0.39 for Raisin Bran. (You could also check the cost using the average of the two *servings per box* values rather than rounded figures and see how the costs change.) Exploring some of these possibilities is a perfect example of sensitivity analysis, since it involves uncertainty about your data.

Let's look at the new table (Table 3).

Table 3

Cereal	**Cost/ Serving**	**Fat/ Serving**	**Fiber/ Serving**	**Protein/ Serving**	**Calories/ Serving**
Special K	\$0.21	0 grams	1 grams	6 grams	150
Raisin Bran	\$0.39	1 grams	7 grams	4 grams	210

As we mentioned earlier, a good practice when you are working with this type of data is to set up your table with the independent variables across the top row. In fact, it would have been easier if we had just done that from the very beginning! Let's switch our table now and add our dietary constraints (Table 4).

Table 4

	Special K	**Raisin Bran**	**Constraint**
Cost/Serving	\$0.21	\$0.39	
Fat Grams/Serving	0	1	at most 2
Fiber Grams/Serving	1	7	at least 5
Protein Grams/Serving	6	4	exactly 9
Calories/Serving	150	210	at most 250

The cost of K servings of Special K, since each serving costs \$0.21, would be $0.21K$ dollars. Similarly, the cost of R servings of Raisin Bran, at \$0.39/serving, would be $0.39R$ dollars. Thus our total cereal cost would be

$$c(K,R) = 0.21K + 0.39R \ .$$

In a similar way, since Special K has 0 grams of fat per serving, K servings would have $(0)(K) = 0$ grams of fat. And since Raisin Bran has 1 gram of fat per serving, R servings would have $(1)(R) = R$ grams of fat. Thus the total amount of fat would be $0 + R$, and this total is supposed to be no more than 2 grams. Thus, the fat constraint will be

$$0 + R \leq 2$$

Notice that the coefficients and the right-hand side (RHS) of this constraint are exactly the numbers in the "Fat" row of the table (Table 4). You may have already guessed that the pattern is exactly the same for the other constraints. Only the relation symbol (= or an inequality) will differ, based on the verbal conditions. As a result, reading from Table 4 and substituting K for Special K and R for Raisin Bran, and mathematical symbols for the verbal conditions, we see that our nutritional constraints are:

$0K + 1R \leq 2$
$1K + 7R \geq 5$
$6K + 4R = 9$
$150K + 210R \leq 250$

The only other constraints are **non-negativity constraints**: $K, R \geq 0$.

We are now ready to set up our model:

Verbal Definition: $c(K, R)$ = the cost in dollars for K servings (31 grams each) of Special K and R servings (61 grams each) of Raisin Bran, ignoring the cost of milk.

Symbol Definition: $c(K, R) = 0.21K + 0.39R$ s.t.

$1R \leq 2$
$1K + 7R \geq 5$
$6K + 4R = 9$
$150K + 210R \leq 250$
$K, R, \geq 0$

Assumptions: Certainty and divisibility Certainty means that the relationship is exact. Divisibility means that any fraction values of servings, nutrient amounts, and dollars are possible. The constraints are assuming that each serving of cereal is eaten with 4 oz. of skim milk, although the milk cost is not included in the objective function. □

Section Summary

In this section we have emphasized some important points in problem solving. We carefully defined and bounded our problems. We defined the variables. For some of the problems we studied several concrete examples and looked for a pattern to help us develop the mathematical model. These are the skills that you must bring to formulating models from verbal descriptions.

Before you begin the exercises, be sure that you:

- Can identify the dependent variable, sometimes called the objective function, that is to be optimized or analyzed.
- Can identify the independent variables, sometimes called the decision variables.
- Know how to define all relevant variables clearly and unambiguously, including the units involved.

- Recognize problems that involve known formulas and relate those formulas to your dependent and independent variables.
- Recognize problems in which the verbal description defines the relationship between the dependent and independent variables, and the domain constraints on the independent variables.
- If necessary, know how to assign one or more sets of specific numerical values to the variables, do the calculations, and look for patterns in the calculations, so you can generalize symbolically with variables and parameters for problems in which there is not a known general formula.
- Are able to interpret expressions such as "no more than", "no less than", "at least", "at most", "exactly", and use the correct mathematical symbols:

Verbal Expression	Mathematical Symbol
"a is no more than b"	$a \leq b$
"a is at most b"	$a \leq b$
"a is no less than b"	$a \geq b$
"a is at least b"	$a \geq b$
"a is exactly b"	$a = b$

EXERCISES FOR SECTION 5.2

Warm Up

1. Calculate the selling price for the following profit margins:
   a) Cost = $98, profit margin = 45%
   b) Cost = $98, profit margin = 85%
   c) Cost = $12.95, profit margin = 80%

2. Calculate the selling price for the following profit margins:
   a) Cost = $98, profit margin = 20%
   b) Cost = $12.95, profit margin = 50%
   c) Cost = $12.95, profit margin = 75%

3. Calculate the profit margins for the following:
   a) Cost = $29.76, sales price = $49.95.
   b) Cost = $29.76, sales price = $89.95.
   c) Cost = $1.35, sales price = $12.95.
   d) Cost = $1.35, sales price = $4.29.

4. Calculate the profit margins for the following:
   a) Cost = $103.56, sales price = $249.35
   b) Cost = $103.56, sales price = $356.
   c) Cost = $0.95, sales price = $1.35.
   d) Cost = $0.95, sales price = $4.00

5. Calculate the percent markup for the following:
   a) Cost = $29.76, sales price = $49.95.
   b) Cost = $29.76, sales price = $89.95.
   c) Cost = $1.35, sales price = $12.95.
   d) Cost = $1.35, sales price = $4.29.

6. Calculate the percent markup for the following:
   a) Cost = $103.56, sales price = $249.35
   b) Cost = $103.56, sales price = $356.
   c) Cost = $0.95, sales price = $1.35.
   d) Cost = $0.95, sales price = $4.00

Game Time

7. You have been calculating your selling prices based on a profit margin of 20%.
   a) What would the selling price be for an item that costs you $10?
   b) What percent markup is implied by this profit margin? Formulate a mathematical model that gives fractional markup as a function of profit margin.

8. You have been calculating your selling prices based on a markup of 20%.
   a) What would the selling price and profit margin be for an item that costs you $10?
   b) Formulate a mathematical model that gives the profit margin as a function of the fractional markup.

9. You plan to sell caps and T-shirts as part of a fund raising project. Your initial plan is just to price the caps at $5 each and the T-shirts at $12 each. Formulate the mathematical model for the revenue from the sale of the caps and T-shirts as a function of the quantities sold of each.

10. You are planning a small bake sale to raise money for your club. The members have promised to bring in 15 cakes and 18 pies to sell. Formulate the mathematical model for the revenue from the sale of the cakes and pies as a function of the prices charged for each. (Assume that all cakes are the same price and all pies are the same price, but the prices for the cakes and pies can be different.)

11. You have to build some containers to ship special pellets. To conform with customs regulations and take advantage of duty pricing, the containers must hold 35 pounds of pellets. You used to ship the pellets in 1 ft by 1 ft by 1 ft containers, which held 5 pounds of pellets, but you are now getting larger orders and the small containers are not economical. Because of the weight of the contents, the bases of the containers must be built of stronger (and more expensive) material than the sides and top. The base of the container costs $0.84 per square foot and the sides and top cost $0.62 per square foot. You have speculated that cylindrical canisters might be better than boxes. Develop a mathematical model for the cost of the containers so you can determine what kind of containers to build. [The area of a circle is πr^2, where r is the radius, and the volume of a cylinder is $\pi r^2 h$, where r is the radius and h is the height (or altitude).]

12. You have decided to add another possible cereal, Grape Nuts™, to your breakfast menu. The nutritional information for Grape Nuts is:

Cereal	Cost/ Box	Amount in Box	Serving Size	Fat/ Serving	Fiber/ Serving	Protein/ Serving	Calories/ Serving
Grape-Nuts	$3.39	24 ounces 680 grams	2.0 ounces 58 grams	0 grams	5 grams	6 grams	240

The rest of the information from Sample Problem 12 remains the same:

Cereal	Cost/ Box	Amount in Box	Serving Size	Fat/ Serving	Fiber/ Serving	Protein/ Serving	Calories/ Serving
Special K	\$3.39	18 ounces 510 grams	1.1 ounces 31 grams	0 grams	1 grams	6 grams	150
Raisin Bran	\$3.69	20 ounces 569 grams	2.2 ounces 61 grams	1 grams	7 grams	4 grams	210

Formulate a new model for the cereal cost to help you determine how you can satisfy your requirements at the lowest cost.

13. You have to manufacture boxes similar to those in Sample Problem 6, that is, the sum of the length and girth can be no more than 108 inches. Assume that you will use the entire allowance: the sum of the length and girth will be exactly 108 inches. The cost for the top and sides is \$2 per square foot and the cost for the bottom of the box is \$3 per square foot.
    (a) Fully define a mathematical model for the cost of the box, using three input variables.
    (b) Fully define a mathematical model for the cost of the box, using two input variables.

14. You have decided to expand your hand-crafted furniture business. A friend in another area wants to work with you on this project. As before, the chairs require 3 board feet of lumber and 8 packages of hardware and the table require 6 board feet of lumber and 4 packages of hardware. You will now be making the furniture in two different locations. Suppose the first location has orders for c_1 chairs and orders for t_1 tables in a week, and the second location has c_2 orders for chairs and t_2 orders for tables in a week.
    a) Set up 4 mathematical models that shows how much lumber and how many packages of hardware should be sent *to each location* in a week, given the above orders.
    b) Set up 2 mathematical models that shows how much lumber and how many packages of hardware are needed in a week at both locations *combined*, given the above orders.

15. The following table shows the nutritional values per serving for three different snack foods:

	Chocolate Cookie	Hard Pretzel	Dried Apricots
Fat (grams)	7	0	0
Saturated Fat (grams)	5	0	0
Cholesterol (mg)	0	0	0
Sodium (mg)	70	655	10
Total Carbohydrates (g)	20	22	22
Fiber (grams)	0.5	1	3
Sugars (grams)	12	0.5	19
Protein (grams)	1	3	1
Calories	150	110	90
Calories from Fat	60	0	0

You keep a record of the amount of each snack you eat during a week.

a) Develop 2 mathematical models that shows the amounts of saturated fat and calories that you would get each week from various amounts of these snack foods.
b) Develop 2 mathematical models that shows the amounts of protein and fiber that you would get each week from various amounts of these snack foods

16. Six Star Refinery blends four petroleum components into three grades of gasoline: regular, premium and ultra-premium. The component specifications are given[5]:

GRADE	COMPONENT SPECIFICATIONS
REGULAR	40% component A 20% component B 10% component C remainder component D
PREMIUM	50% component C remainder component D
ULTRA-PREMIUM	10% component A 60% component C remainder component D

a) Express the amount of each component needed as a function of the specified quantities of each grade of gasoline if you produce r barrels of regular, p barrels of premium, and u barrels of ultra-premium. Be sure to define your variables and your functions unambiguously (including units for everything and the domains of the four functions).

b) If component A costs \$9 per barrel, B costs \$7 per barrel, C costs \$12 per barrel and D costs \$6 per barrel, write a function expressing the total cost of the components if you produce r barrels of regular, p barrels of premium, and u barrels of ultra-premium.

c) If regular sells for \$12 per barrel, premium sells for \$15 per barrel, and ultra-premium sells for \$18 per barrel, develop a *model* for the total profit as a function of r, p, and u, assuming that there are no other costs than those of the components (or that the fixed costs are averaged into the variable costs) given above.

17 You are trying to raise money for your sports team by selling T-shirts and hats. You can order both items from the same company, which will charge you \$4 for each hat and \$6 for each shirt, with a fixed cost of \$30 for any order. Your team has \$100 to spend for everything, and has decided to sell the shirts for \$12 and the hats for \$7, at which price you think you can sell no more than 15 hats and 10 shirts. Assume that both items are "one size fits all" and that the demands for the two products are independent for this exercise. *Formulate* **(do not solve!)** a mathematical model for this problem.

---

[5] Problem adapted from notes: Introduction to Operations and Production Management, Nydick et al.

18. You run a small candle-making business, making regular and deluxe candles. You can sell the regular candles for a profit of $4 each, and the deluxe candles for a profit of $7 each. Regular candles use 6 oz. of wax and 5 inches of wick material, while deluxe candles use 9 oz. of wax and 6 inches of wick material. You have 120 oz. of wax and 110 inches of wick material in your studio at the moment, and will not get any more supplies until tomorrow. You need to determine how many candles of each type you should make today. *Formulate* (**do not solve!**) the mathematical model for this problem.

19. You are trying to decide between two dog foods for your Golden Retriever, Ralph. Purina™ costs $32 for a 10-pound bag, and each ounce of food has 4 grams of fat, 8 grams of protein, and 200 calories, while Iams™ costs $20 for a 5-pound bag, and each ounce of food has 1 gram of fat, 6 grams of protein, and 220 calories. You want Ralph to get at least 1000 calories per day, but not more than 1400 calories. You also want him to get no more than 10 grams of fat, and at least 40 grams of protein. You have to decide what combination of the two dog foods you should give Ralph every day. *Formulate* (**do not solve!**) the mathematical model for this problem.

20. XYZ Toys Inc. wants to maximize its profit from the production of five different toys, which use six different materials. The quantities of each material used in each toy, along with the amount of the materials available and the unit profit from each toy, are shown in the spreadsheet below:[6]

Material	Toy A	Toy B	Toy C	Toy D	Toy E		Amount Available	
Red Paint	0	1	0	1	3		625	
Blue Paint	3	1	0	1	0		640	
White Paint	2	1	2	0	2		1100	
Plastic	1	5	2	2	1		875	
Wood	3	0	3	5	5		2200	
Glue	1	2	3	2	3		1500	
Unit Profit ($)	15	30	20	25	25			

Formulate (**do not solve!**) a complete model for the profit realized from the sale of all the toys combined, including the domain (constraints). (Note: The source did not give any units for the materials, so just use "units".)

[6] Walkenbach, John, Excel for Windows 95 Bible, p 623, (Foster City, CA, IDG Books Worldwide, Inc.,1995)

21. A city government has begun a service named Dirt Cheap, Inc. The service uses many existing resources to produce a positive income stream for the city. In addition Dirt Cheap reduces and recycles garbage and landscape trimmings. Dirt Cheap has a collection program for organic garbage, park trimmings, Christmas trees, and so on. The service mulches or composts these items, and combines them in different blends with soil and mineral additives to produce high-quality mulch, soil and growing mixtures. Some of the labor is volunteer, and material costs, except for collection costs, are low. The spreadsheet below shows the amounts of each component in each product, as well as the selling prices, inventory, and unit costs: [7]

Components:	**Soil**	**Compost**	**Minerals**	**Mulch**			Selling Price
	Ounces of each Component per Bag of each Product:						
Products:	*(oz./bag)*	*(oz./bag)*	*(oz./bag)*	*(oz./bag)*			*($/bag)*
Garden	55	54	76	23			105.00
Yard	64	32	45	20			84.00
Top Soil	43	32	98	44			105.00
Veggie Gro	18	45	23	18			57.00
Inventory *(oz.)*	4100	3200	3500	1600			
Unit Cost *(per ounce)*	$ 0.20	$ 0.15	$ 0.10	$ 0.23			

a) Identify, bound, and define the basic problem facing Dirt Cheap, and define decision variables.
b) Find the cost of *one* bag of each product.
c) Calculate the profits from the sale of one bag of each product.
d) Formulate a model for the total amount of soil used.
e) *Fully define* a model for the total profit from the combined sales, including the full domain (all constraints).

[7] Adapted from Person, Ron, Special Edition Using Excel for Windows 95, p841, Copyright 1995 by Que® Corporation

Section 5.3: Interest and Investments

In the first two sections of this chapter, we have talked about functions and models of two or more variables, called multivariable functions and models, and how to formulate models based on verbal descriptions (as opposed to raw data), as in traditional word problems. In this section, we examine one particular category of this kind of problem: situations involving interest. We develop models for dealing with simple and compound interest, including both discrete and continuous compound interest. We also discuss how interest rates are presented by banks and other lending institutions. This can help you make calculations related to interest on investments and loans.

Here are some examples of the kind of problems you should be able to solve after studying this section:

- You are interested in investing a gift of money from your grandparents, to save up for a trip in two years. You are considering a certificate of deposit (the *other* kind of CD) that compounds interest at a certain rate. How much money (total) would you have after the two years?
- In comparing two credit card offers, one offers a certain (nominal) annual percentage rate (APR) compounded monthly, while the other gives its effective annual rate (also called the annual percentage yield, or APY). Which is better?
- You just won the lottery! You want to save a certain portion of your winnings for the future. One option pays a certain (nominal) annual percentage rate (APR) compounded quarterly, and another pays a slightly lower APR, but is compounded continuously. Which should you go for?

After studying this section, you should be able to solve problems like those above and also:

- Calculate your total monetary accumulation (as well as just the interest portion by itself) for any investment using simple interest, given the principal, interest rate per period, and the number of periods (or the frequency of interest calculations and the amount of time).
- Calculate your total monetary accumulation (as well as just the interest portion by itself) for any investment using discrete compound interest, given the principal, annual percentage rate (APR), number of compounding periods per year, and the number of years (or given the principal, interest rate per period, and the number of periods).

- Calculate your total monetary accumulation (as well as just the interest portion by itself) for any investment using continuous compound interest, given the principal, annual percentage rate (APR), and the number of years.
- Calculate any one of the *other* quantities (other than accumulated money) listed for the above cases, given the rest of the quantities.
- Calculate the annual percentage yield (APY, also called the yield or the effective annual percentage rate) for an investment given the APR and the compound interest structure, and vice-versa.
- Understand what Euler's number (e) corresponds to mathematically and in everyday terms.

Personal and business finances revolve heavily around the idea of *interest.* There are two major categories of interest: *simple interest* and *compound interest.* Simple interest does not occur very frequently - mainly in certain kinds of loans and bonds. Generally, when most of us have heard of interest, it has probably involved compound interest (like the sample problems above).

Simple Interest

Let's start with some basic terminology. **Principal** is the initial amount of money borrowed or invested. **Interest** is the amount paid for using borrowed money. If you borrow money, you are paying interest for having the use of the money. If you deposit money in a bank, the bank pays you interest, because it is using your money. **Simple interest** involves accumulating interest always **based on the *original principal* (amount invested or borrowed), even for multiple time periods.**

Sample Problem 1: You have invested \$100 and it accumulated *simple* interest at a rate of 10% per year for 4 years. How much accumulated interest would you have after the 4 years? If you did not take the interest, but allowed it to accumulate, how much would the accumulated total be after 4 years?

Solution: The accumulated interest at the end of the first year would be \$10 (10% of \$100). The interest accumulated during the second year would *also* be \$10, because it is *still* computed from the *original* principal of \$100. The third and fourth years would again be the same, so the total accumulation of interest would be $(4)(\$10) = \$40$. The total amount (principal and interest) of your investment at the end of the 4 years would thus be the original principal plus the interest, or $\$100 + \$40 = \$140$.□

In this calculation we have used the basic relationship

total accumulation = principal + interest ,

which holds for *both* simple and compound interest.

Our **model for simple interest** is:

Verbal Definition: $S(P,i,n)$ = the accumulated money when P is the original principal, i is the simple interest rate per period expressed as a pure decimal, and n is the number of periods, where the accumulated money is in the same money units as the principal.

Symbol Definition: $S(P,i,n) = P(1+in)$ for $P,i,n \geq 0$

Assumptions: Certainty and divisibility. Certainty implies that the relationship is exact. Divisibility implies that any fractional value of money units, interest rate and periods is possible

If the interest is only accrued at the end of each period, then this model is only fully valid when the number of periods, n , is a whole number. It is also possible to think of the interest as being accrued continuously, in which case the model would be fully valid for *all* nonnegative values of n , but in practice simple interest is usually discrete.

Note: **For this and the other models we define in Sections 5.3 and 5.4 , we have not specified money units for the principal, the total accumulated value, and the amount of interest**. We could specify dollars as the units for each of these, but the result is much more general than that (it could be cents, cruzieros, pesos, pounds, marks, yen, etc.). The important thing to remember is that, **within a given model, the money units for all of the monetary quantities should be the same**. For example, in the model for $S(P,i,n)$ above, this means that the accumulated money is in the same money units as the principal. All of the formulas in these two sections are based on this assumption. Also, **we will always assume in our formulas that interest rates are expressed in pure decimal form** (so 10% would be written as 0.10).

Basic Compound Interest

Compound interest involves accumulating interest **based upon the current running *total accumulation* at the time when the interest is compounded**. Thus in our example of investing $100, suppose this time it accumulates interest at a percentage rate of 10%, compounded annually. Our calculation of interest for the first year is the same as above, since the running total accumulation during the first year is the *same* as the principal, the original $100. So the interest for the first year is again $10, making a total accumulation of $110 after the interest is added in.

For the second year, the interest added will now be 10% of the $110 (which is the running total accumulation during the second year):

$$(0.1)(\$110) = \$11 \;.$$

After adding in this second-year interest, the new total accumulation will be

$$\$110 + \$11 \;=\; \$121 \;.$$

Continuing the same process into the third year, the interest added will be:

$$(0.1)(\$121) \;=\; \$12.10 \;,$$

and so the new accumulation will be

$$\$121 + \$12.10 \;=\; \$133.10 \;.$$

Combining these operations into a single calculation, the total accumulation *after* adding the interest for the fourth year will be:

$$\$133.10 + (0.1)(\$133.10) \;=\; \$133.10 + \$13.31 \;=\; \$146.41$$

Let's generalize what we just did. Suppose our principal is P, i is the interest rate applied each compounding period (which was one year in our example), n compounding periods is the length of time we are accumulating. Thus we have

P = principal (initial amount of money)
i = interest rate (expressed as a pure decimal) *per compounding period*
n = the number of compounding periods

Let's also use the notation

A_n = the total accumulation (principal and interest) after n compounding periods.

Recall that in our example the accumulation at the end of the first compounding period (year) was

$$100 + (0.1)(100) \;=\; 100 + 10 \;=\; 110$$

(ignoring the units for the time being, to focus on the numerical calculations). In symbols, this would be written:

$$\begin{aligned} A_1 &= P + (i)(P) \\ &= P + Pi \\ &= P(1+i) \ . \qquad \text{(factoring out } P \text{ from both terms)} \end{aligned}$$

Notice that this means we can get the total accumulation after one period by taking the total accumulation at the beginning of the period and multiplying it by the factor $(1+i)$. In our example, that means we could have gotten the $110 by multiplying the original $100 by the factor 1.1 (because $1 + 10\% = 1 + 0.1 = 1.1$):

$$(100)(1.1) = 110 \ .$$

Remember that $\boldsymbol{i}$ **should be expressed in *pure decimal* form when doing calculations!**

What about the second year? In our example, we calculated

$$110 + (0.1)(110) = 110 + 11 = 121 \ .$$

In symbols, this would be:

$$A_2 = A_1 + (i)(A_1) = A_1 + iA_1 = A_1(1+i)$$

However, recall that we already found an expression for A_1 in terms of P and i:

$$A_1 = P(1+i) \ ,$$

which we can now substitute into the above expression for A_2 to get:

$$\begin{aligned} A_2 &= A_1(1+i) \\ &= [P(1+i)](1+i) \\ &= P(1+i)^2 \ . \end{aligned}$$

Notice that the effect of compounding the interest in the second year was to multiply by the factor $(1+i)$, just as it was for the first year. As you can probably see by now, the same pattern holds for *every* year, so, for example,

$$\begin{aligned} A_3 &= A_2(1+i) \\ &= [P(1+i)^2](1+i) \\ &= P(1+i)^3 \ . \end{aligned}$$

In summary, we see that

$$A_1 = P(1+i)$$
$$A_2 = P(1+i)^2$$
$$A_3 = P(1+i)^3$$

The pattern here is quite simple: for example A_4 will be $P(1+i)^4$, and in general, we will have

$$A_n = P(1+i)^n$$,where n is the number of compounding periods.

This is the **basic compound interest formula**, upon which all other compound interest results are based. The **model for the total money accumulated using compound interest and interest rate per period is:**

Verbal Definition: $A(P,i,n)$ = the accumulated money when the original principal is P, i is the interest rate per period expressed as a decimal and n is the number of compounding periods, where the accumulated money is in the same money units as the principal.

Symbol Definition: $A(P,i,n) = P(1+i)^n$ for $P,i,n \geq 0$

Assumptions: Certainty and divisibility. Certainty implies that the relationship is exact. Divisibility implies that any fractional value of money units, interest rate, and periods is possible

Sample Problem 2: Use the basic compound interest formula to find the total accumulation if $100 is invested at an interest rate of 10% compounded annually for 4 years.

Solution: We have already solved this numerically the long way, so all we are doing here is to **verify** (double-check) the result, using our formula. In this example,

$$P = 100,\ i = 10\% = 0.10\ ,\ \text{and}\ n = 4\ ,$$

and we are looking for $A(P,i,n) = P(1+i)^n$, so we see that

$$\begin{aligned} A_4 = A(100,0.10,4) &= 100(1+.10)^4 \\ &= 100(1.1)^4 \\ &= 100(1.4641) \\ &= 146.41\ . \end{aligned}$$

Translation: The total accumulation from one hundred dollars invested at 10% compounded annually for four years is $146.41. □

Notice that we do indeed get the same answer as before. [Authors' Note: When we first wrote this, we made an arithmetic mistake in the numerical calculations. It wasn't until we verified the answer with the formula that we realized this. That is *exactly* why verification is so important!]

Sample Problem 3: Seven years ago, your parents bought you a savings bond for $50 that compounded interest at a rate of 1/2 percent every month. How much is it worth now?

Solution: In this case the principal is $50, the compounding period is one month, and the interest rate per compounding period is 1/2 percent. The total time is 7 years, but we need to figure out how many *compounding periods* this corresponds to. Since a compounding period is one month, there are 12 compounding periods in a year, so in 7 years we see that we get

$$n = (12)(7) = 84$$

(since there are 84 months in 7 years). The other values in the problem are

$$P = 50 \text{ and } i = (1/2)\% = 0.5\% = 0.005 .$$

We are looking for $A(P,i,n)$, so we calculate

$$\begin{aligned} A(P,i,n) &= P(1+i)^n \\ A(50,0.005,84) &= 50(1+0.005)^{84} \\ &= 50(1.005)^{84} \\ &\approx 50(1.52037) \\ &\approx 76.02 . \end{aligned}$$

Translation: A $50 savings bond that compounded interest at a rate of 1/2 percent every month will be worth about $76.02 after 7 years. □

Discrete Compound Interest with a Given APR and Compounding Structure

You may be aware that most banks, credit card companies, and loan companies do not usually express compounding in the format described just above. Instead of giving the *actual* interest rate *per compounding period,* they usually give some kind of an *annual* interest rate. This helps to standardize things to some extent, but one common form in which the rate is often expressed *still* does not make it easy to compare investment opportunities.

The common form referred to above is usually called the **APR**, or **annual percentage rate**, or the **nominal annual interest rate**. These terms all mean the same thing. For example, a car loan might advertise an APR of 6%. Like most loans, this is usually compounded monthly, which is the same as saying it is compounded 12 times per year. What "6% compounded monthly" *really* means is an *actual* interest rate of

$$(6\%)/12 = (6/12)\% = .5\% = 0.005$$

compounded every month. Notice that this is the exact *same* interest structure as the savings bond in Sample Problem 3. In other words, the APR and the number of times interest is compounded per year are the components of a formula for finding the actual interest rate per compounding period:

(Actual interest rate per period) = (APR) / (# of compounding periods per year)

(just dividing up the APR into equal pieces). To make the discussion simpler, let's define some new notation:

r = the APR, or nominal annual interest rate
m = the number of compounding periods per year

In our example, r was 6% ($r = 0.06$) and m was 12 (12 months in a year). We said that the actual interest rate per compounding period (which we call i) was equal to $6\%/12$; in general the calculation for i is given by

$$i = r/m \ .$$

The number of compounding periods per year (m) is normally assumed to be a positive integer, usually 1, 2, 4, 12, 52, or 365 (compounding annually, semi-annually, quarterly, monthly, weekly, or daily, respectively). Notice that some of these values are only approximate (for example 52 weeks is only 364 days). Also, financial institutions have special standard rules about how to round fractions when computing interest. However, the formulas we are giving you should give answers that are very close to exactly what would happen in real life, give or take a penny or two here and there.

We saw in our example that the number of compounding periods (months) in 7 years was $(12)(7) = 84$. To generalize this, let's define

t = the number of *years* over which interest is accumulated

(as opposed to the number of *compounding periods*, which we called n). If we are accumulating for t years, and there are m compounding periods per year, then the total number of *compounding periods* over which we are accumulating is given by

$$n = mt$$

(as in 84 = (12)(7) in our example).

Since we've added three new definitions, let's recap all of our definitions:

P = principal (initial amount of money)
i= interest rate, expressed as a decimal, per compounding period
n = the number of compounding periods over which interest is accumulated
r = APR, or nominal annual interest rate
m = the number of compounding periods per year
t = the number of years over which interest is accumulated

At this point, we should note that, for now, **when *working* with discrete compounding, we will assume that we only focus on values of n and t that correspond to *whole numbers* (integer values) of compounding periods (so mt must be a whole number, or t a multiple of $(1/m)$)**. This violates our usual assumption of divisibility (or, more precisely, it means that our assumption of divisibility does not match reality perfectly here, and could yield misleading results for other values of n and t). Thus, our models will be defined more generally, but will only be *valid* for values satisfying this condition.

In this more complicated situation (given the APR and compounding structure), we can still reduce the problem to our basic compound interest formula by using the above equations to calculate i and n :

$$A_n = A(P,i,n) = P(1+i)^n \quad , where\ i = r/m\ and\ n = mt\ .$$

To write this in a different form, let's use a different notation for the total accumulation:

$A(t)$ = the total accumulation after t years.

This looks quite similar to A_n , but uses *function* notation instead. This means that **when we use a *subscript*, we interpret the subscript to mean the number of *compounding periods*, but when we use *function notation*, we interpret the time variable to mean**

the number of *years*. Since we saw before that $n = mt$, we see that these forms are related by the equation

$$A(t) = A_{mt} \ .$$

Putting all of the pieces together, we find that

$$\begin{aligned} A(t) &= A_{mt} = A_n = P(1+i)^n \\ &= P(1+\frac{r}{m})^{mt} \ . \end{aligned}$$

Seeing how this applies to Sample Problem 3, we would have

$$\begin{aligned} A(t) &= P(1+\frac{r}{m})^{mt} \quad \text{, so} \\ A(7) &= 50\left(1+\frac{0.06}{12}\right)^{(12)(7)} \\ &= 50(1.005)^{84} \\ &\approx 50(1.52037) \approx 76.02 \ . \end{aligned}$$

Once again, you can think of this as simply reducing to the basic compound interest formula by calculating i and n , or you can apply the above formula directly. From a multivariable function perspective, we could think of this new formula as a function of four variables. Thus we have the following **model for accumulated total money using compound interest, annual percentage rate, and compounding periods per year:**

Verbal Definition: $A(P,r,m,t)$ = the total accumulation when the principal is P , the annual percentage rate is r , the number of compounding periods per year is m , and the number of years over which the interest is accumulated is t , where the accumulated money is in the same money units as the principal

Symbol Definition: $A(P,r,m,t) = P(1+\frac{r}{m})^{mt}$ for $P,r,t \geq 0 \ ,m > 0$

Assumptions: Certainty and divisibility. Certainty implies that the relationship is exact. Divisibility implies that any fractional values of dollars, percentages, compounding periods and years are possible.

Remember that, as we discussed earlier, **in practice we will only focus on values of t that correspond to *whole numbers* (integer values) of compounding periods, which is the same as saying that t must be a multiple of $(1/m)$, or that mt must be a whole number.** This means that **our assumption of divisibility (the general formula)** does not match reality perfectly here, and **could yield misleading results for other values of m and t.** The results could have a meaningful or useful interpretation, but would not correspond to the actual monetary accumulation in an investment with the given structure. **In order to make sure this model is fully *valid* for a given real situation, be sure to check this condition!**

You may have noticed that we have defined two different functions, both called A.

$$A(P,i,n) = P(1+i)^n \text{ for } P,i,n \geq 0$$
$$A(P,r,m,t) = P(1+\tfrac{r}{m})^{mt} \text{ for } P,r,t \geq 0,\ m > 0$$

This is not a real problem, since each has a different number of independent variables. As long as we use function notation, which will show the number of independent variables in the parentheses, we will know which version we are using. Since $i = r/m$ and $n = mt$ they are actually variations of the same formula.

We also have two functions with the same independent variables, but the functions have different names, S for simple interest and A for discrete compound interest.

$$S(P,i,n) = P(1+in) \text{ for } P,i,n \geq 0 \quad \text{for simple interest}$$
$$A(P,i,n) = P(1+i)^n \text{ for } P,i,n \geq 0 \quad \text{for discrete compound interest}$$

Sample Problem 4: You are considering two investment opportunities. One (investment A) offers 6% interest compounded daily, and the other (investment B) offers 6.05% compounded semi-annually.

(a) For comparison purposes, if you invested \$1 in each investment for a year, which would end with a larger total accumulation of money? Carry your calculations out to 4 decimal places.

(b) What would be your total accumulation in each investment after a year if each started with \$100?

(c) What would be your total accumulation in each investment after 4 years if each started with \$100?

Solution: In all parts of this question, for investment A:

$m = 365$ and $r = 0.06$, so $i = r/m = 0.06/365 \approx 0.00016438$.

(Notice that for simplicity we are assuming 365 days in a year.)

For investment B:

$m = 2$ and $r = 0.0605$, so $i = 0.0605/2 = 0.03025$.

(a) Here, $P = 1$ and $t = 1$. Let us define

$A(t)$ = total accumulation (in dollars) in investment A after t years
$B(t)$ = total accumulation (in dollars) in investment B after t years

So in this case we have

$$A(1) = 1(1+0.00016438)^{(365)(1)} = (1.00016438)^{365} \approx 1.0618$$
$$B(1) = 1(1+0.03025)^{(2)(1)} = (1.03025)^{2} \approx 1.0614 \ .$$

So, when rounded off, the investments would both pay \$1.06 . However, for larger amounts of principal, investment A, *even though it has the smaller APR*, would accumulate slightly more money. Figure 5.3-1 shows how the interest is accumulated using the different compounding periods. The daily compounding appears to be almost a continuous line, although it is actually a tiny step function, while the semi-annual compounding is clearly a step function. The accumulated amount does not change for the semi-annual compounding until six months have passed. It is difficult to determine from this graph which investment is better.

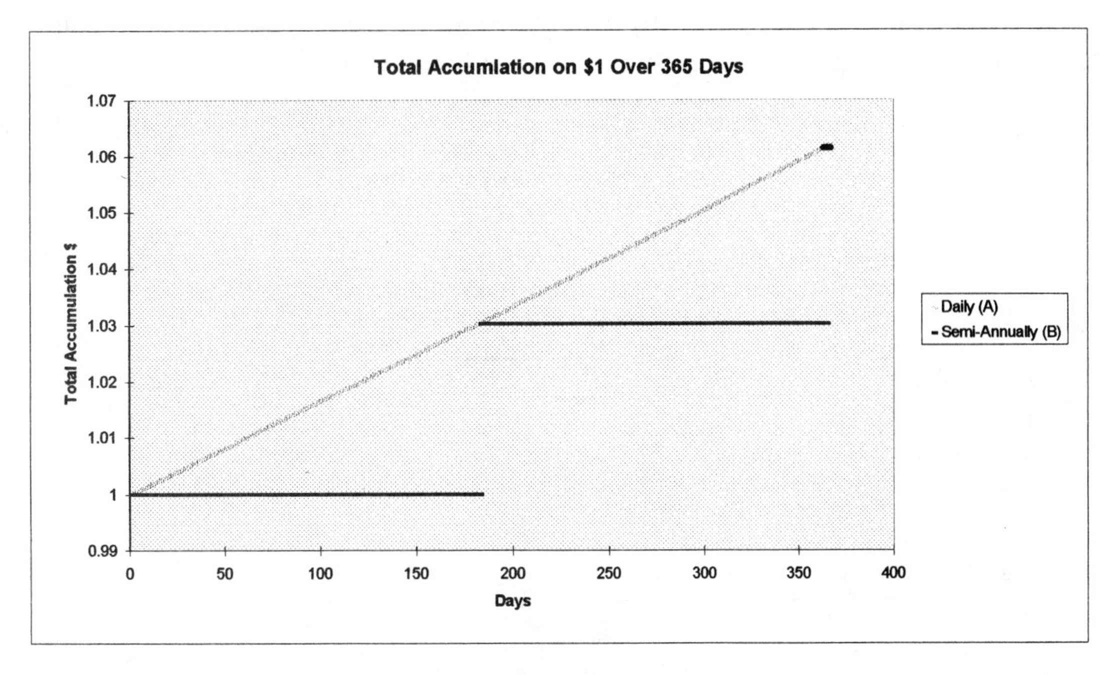

Figure 5.3-1

(b) This time, the only difference is the principal, $P = 100$ instead of $P = 1$. So the answers are given by

$$A(1) = 100(1+0.00016438)^{(365)(1)} = 100(1.00016438)^{365} \approx 106.18$$
$$B(1) = 100(1+0.03025)^{(2)(1)} = 100(1.03025)^{2} \approx 106.14 \ .$$

Notice that these answers could have been obtained by simply moving the decimal point from the answers in (a) two places to the right. Investment A accumulates $106.18 , while investment B accumulates $106.14 , so investment A is slightly better. The comparison graph for (b) would then be the same as Figure 5.3-1, except the y-axis values would be considered to be in units of $100, or rewritten by moving the decimal points two places to the right.

(c) This is the same situation as (b), except that now $t = 4$ instead of $t = 1$. So now we find that

$$A(4) = 100(1+0.00016438)^{(365)(4)} = 100(1.00016438)^{1460} \approx 127.12$$
$$B(4) = 100(1+0.03025)^{(2)(4)} = 100(1.03025)^{8} \approx 126.92 \ .$$

Once again, investment A is slightly better than investment B ($127.12 vs. $126.92). □

Figure 5.3-2 shows the total accumulation on $100 over 4 years. Again, it is difficult to determine from the graph which investment is better because of the scale.

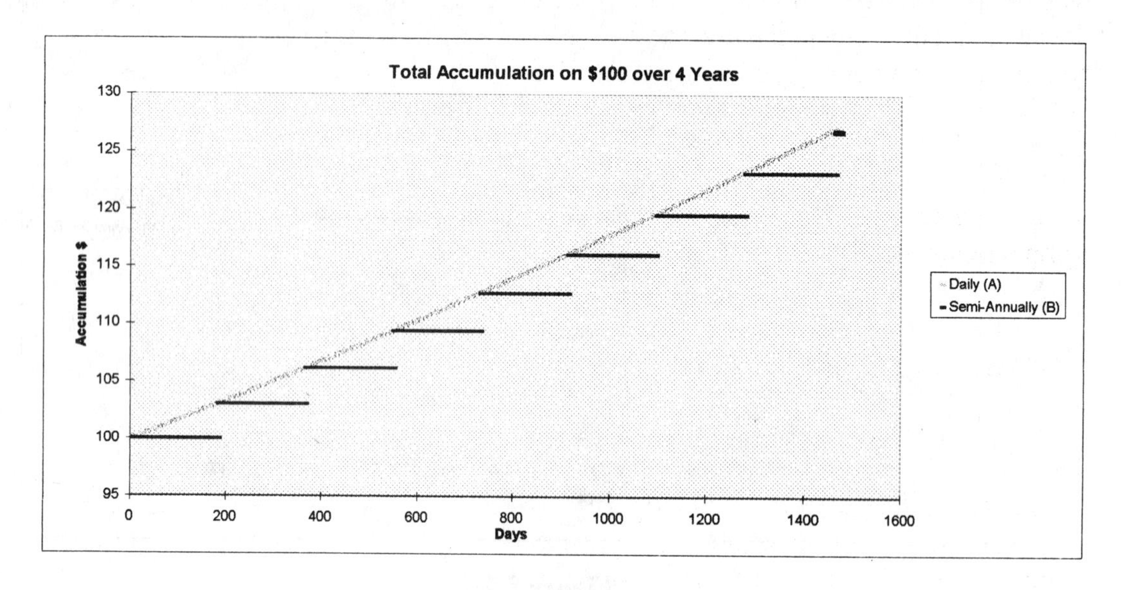

Figure 5.3-2

Figure 5.3-3 shows just the last five days of the four year period. In this graph we can see that the total accumulated in investment A is more than that for investment B. We can also see the discrete nature of the data.

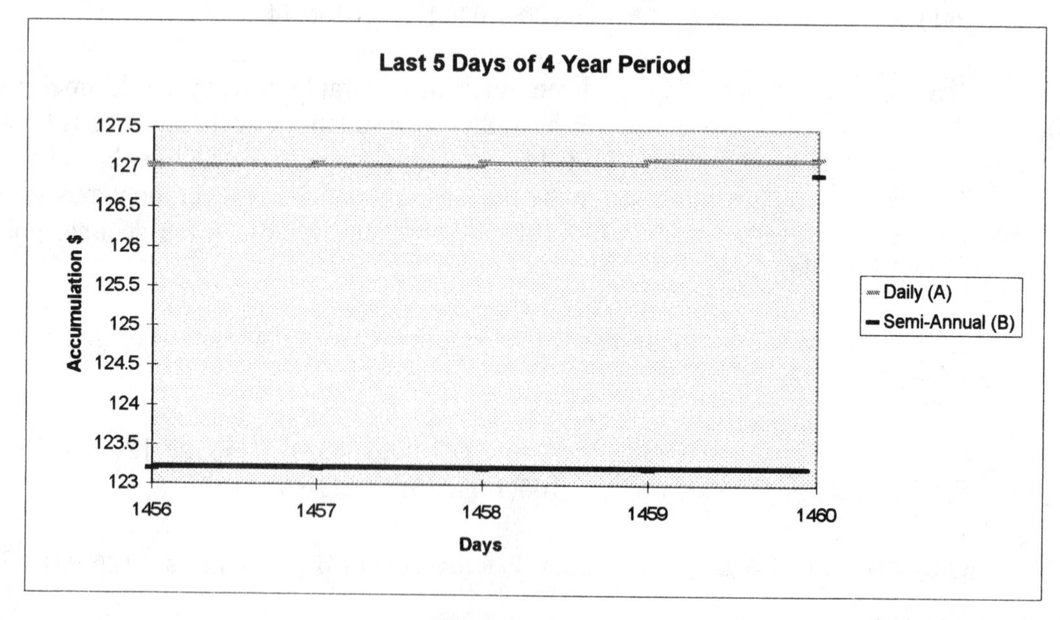

Figure 5.3-3

Note that we have used 1460 days for four years (365 times 4), when in fact, there will almost always be 1461 days in four years (because of leap years).[1] Adding one more day would only widen the difference, since the daily accumulation would have one additional compounding, unless the actual daily interest rate were also adjusted (m = 365.25), in which case the result would be almost identical.

Also, notice that for all of Sample Problem 4, in all of the cases studied, mt was a whole number (in fact, both m and t were whole numbers), so the discrete compounding model was valid.

Annual Percentage Yield (APY)

The answers to part (a) of Sample Problem 4 give us very useful information about the two investments. We can see from them that our *actual* annual interest rate (technically called the **effective annual interest rate**, or the **annual percentage yield**, often abbreviated **APY** or called the **yield** for short), as opposed to the APR (nominal annual interest rate), is 6.18% for A and 6.14% for B. Numerically, we get this by

1. Calculating the amount of *interest* in the total accumulation after a year (just subtract the total accumulation *minus* the principal, which was 1 in this case).
2. Dividing by the principal (which was 1 in this case, so does not change the value).
3. Converting the result to a percent by moving the decimal place two places to the right.

For investment A, the total accumulation was 1.0618 . Since the principal was 1, the amount of interest was

$$\text{interest} = (\text{total accumulation}) - (\text{principal}) = 1.0618 - 1 = 0.0618 \ .$$

To find what fraction of the original principal this represents, we simply divide the amount of interest by the principal

$$\begin{aligned}(\text{fraction of principal represented by interest}) &= (\text{interest}) / (\text{principal}) \\ &= (0.0618)/(1) = 0.0618\end{aligned}$$

Converting this to a percent (moving the decimal point 2 places to the right and adding the % sign, which is equivalent to multiplying by 100%, or 1), we get

$$.0618 = (.0618)(100\%) = (.0618)(100)\% = 6.18\% \ .$$

[1] There is no leap year in "century years," such as the years 1900 and 2000.

For investment B, if we streamline the above process, we get

(1.0614 - 1)/(1) = .0614 = 6.14%

Notice that, by starting with principal of 1, we really don't need to do the division, so the APY (in decimal form) is given simply by the total accumulation minus 1 . Thus we have the following **model for APY** :

Verbal definition: $APY(r,m)$ = the total accumulation, in dollars, after one year if the principal is \$1, r is the Annual Percentage Rate (APR), and m is the number of compounding periods per year.

Symbol Definition: $APY(r,m) = (1+\frac{r}{m})^m - 1 \quad \text{for } r,m > 0$

Assumptions: Certainty and divisibility. Certainty implies that the relationship is exact. Divisibility implies that any fractional values of the Annual Percentage Rate, and the number of compounding periods per year are possible.

In this model, we have specified the money units (dollars). The answer for the APY is actually independent of the money units chosen; we have chosen a familiar unit just for simplicity, to help understand the concept more easily. Technically, the APY corresponds to the *numerical value* (ignoring the money units), which is the pure decimal form of the APY, and can easily be converted to percent form, as above. We could just as well have *not* specified the money units, as in the other models of this section (the function would be exactly the same), or think of the APY as the interest after one year in *money units of interest per money unit of principal*. Since the money units cancel, it is really a pure number.

As long as you are comparing investments for equal periods of time that are multiples of all of the compounding periods involved, the best way to make the comparison is with the APY, or effective interest rate. As we saw above, the APR does not necessarily indicate which investment is better, but the APY does, even when the principal is not 1 and the time period is not 1 year (as long as the time is an exact number of compounding periods for all interest structures being compared).

In everyday language, **the APY (in pure decimal form) can be thought of as the amount of interest (in dollars) after 1 year for a principal of \$1.** Again, technically, the APY corresponds to the *numerical value* (ignoring the dollar units), which corresponds to the pure decimal form of the APY, and can easily be converted to percent form, as above.

As before, this model and result is only fully valid for a given situation if mt is a whole number. In this case, we know that t is 1, so full validity is assured if m is a whole number, as it was for both investments in the above example.

Continuous Compounding

What is the effect of changing the number of compounding periods per year? Intuitively, the more times you compound in the year, the more interest you can get on your interest, so the APY seems as if it should increase. Let's try it, and see what happens!

Sample Problem 5: In Brazil in the early 1980's, inflation was so high that you could actually find banks where the APR for savings accounts was as high as 100%, or even more! If you started with a principal of 1 cruziero (the Brazilian monetary unit) and look at investing it at an APR of 100% for 1 year,

(a) Find the total accumulation for compounding that is annual, semi-annual, quarterly, monthly, weekly, and daily.

(b) What would be the logical conclusion of this process?

Solution: Once again, $P = 1$ and $t = 1$, and this time we have $r = 1$ (since 100% = 1.00). All money units will be assumed to be in cruzieros. Since $r = 1$, our calculation of the total accumulation will be

$$(1+1/m)^m \quad ,$$

(a) Annual means $m = 1$, semi-annual means $m = 2$ (twice a year, or every 6 months), quarterly means $m = 4$ (4 times a year or every quarter, which means every 3 months), and monthly means $m = 12$. We will assume weekly means $m = 52$, and we will assume that daily means $m = 365$. Using technology, we get Table 1.

Table 1

m (compoundings/yr)	Total Accumulation $(P=1, r=1, t=1)$: $A(1)=(1+1/m)^m$
1	$(1+1/1)^1 = 2$
2	$(1+1/2)^2 = (3/2)^2 = 9/4 = 2.25$
4	$(1+1/4)^4 = (5/4)^4 = 625/256 \approx 2.441406$
12	$(1+1/12)^{12} \approx 2.613035$
52	$(1+1/52)^{52} \approx 2.692597$
365	$(1+1/365)^{365} \approx 2.714567$

(b) Notice that the total accumulation increases a lot at first, but then the improvement gets less and less. It looks as though the total accumulation might be approaching a *limit*, just as the average rates of change did as we reduced the size of the interval being considered to estimate the slope of the tangent line at a point, and as the rectangle sums did as we increased the number of subintervals to find an area. So let's use the same idea: let's keep putting in larger and larger values of m and see if the values in the table *converge* to a limit. To get there reasonably quickly, let's use powers of 10, starting with 1,000 (Table 2).

Table 2

m (compoundings/yr.)	Total Accumulation in 1 year $(P=1, r=1, t=1)$: $A(1)=(1+1/m)^m$ (cruzieros)
1,000	2.7169
10,000	2.7181
100,000	2.7183
1,000,000	2.7183

From the table we see that, to 5 significant figures, the total accumulation seems to be stabilizing at around 2.7183 cruzieros. Mathematically, we could say we were finding

$$\lim_{m\to\infty}\left(1+\frac{1}{m}\right)^m$$

This is in fact the definition of e (as in e^x), or *Euler's number*:

$$e = \lim_{m\to\infty}\left(1+\frac{1}{m}\right)^m \approx 2.7183$$

For reasons that we hope are clear to you, this situation is commonly called **continuous compounding**. You could also call it compound interest with $m = \infty$. Thus, in everyday terms, you can say that ***e* is the total accumulation (in dollars) of $1 after 1 year at an APR of 100%, compounded continuously**. □

How can we find the total accumulation under continuous compounding for different values of P, r, and t? Intuitively, we would say

$$A(t) = \lim_{m\to\infty}\left[P\left(1+\frac{r}{m}\right)^{mt}\right]$$

How can this be simplified? Without getting into all of the mathematical details, let's look at how this can be understood, at least intuitively:

$$\lim_{m\to\infty}\left[P\left(1+\frac{r}{m}\right)^{mt}\right] = P\left\{\lim_{m\to\infty}\left[\left(1+\frac{r}{m}\right)^{mt}\right]\right\} \text{ (since } P \text{ does not depend on } m\text{, it can be factored out)}$$

$$= P\left\{\lim_{m\to\infty}\left[\left(1+\frac{1}{\frac{m}{r}}\right)^{mt}\right]\right\} \qquad \left(\text{since } \frac{1}{\frac{m}{r}} = \frac{r}{m}\right)$$

$$= P\left\{\lim_{m\to\infty}\left[\left(1+\frac{1}{\frac{m}{r}}\right)^{\frac{m}{r}(rt)}\right]\right\} \qquad \left(\text{since } \tfrac{m}{r}(rt) = mt\right)$$

$$= P\left\{\lim_{m\to\infty}\left[\left(1+\frac{1}{\frac{m}{r}}\right)^{\frac{m}{r}}\right]^{rt}\right\} \qquad \left(\text{since } \left(x^a\right)^b = x^{ab}\right)$$

$$= P\left\{\left[\lim_{m\to\infty}\left(1+\frac{1}{\frac{m}{r}}\right)^{\frac{m}{r}}\right]^{rt}\right\} \qquad (\text{since } rt \text{ does not depend on } m)$$

$$= P\left\{\left[\lim_{\frac{m}{r}\to\infty}\left(1+\frac{1}{\frac{m}{r}}\right)^{\frac{m}{r}}\right]^{rt}\right\} \qquad \left(\text{since, for fixed } r>0,\ m\to\infty \text{ means } \tfrac{m}{r}\to\infty\right)$$

$= Pe^{rt}$ (think of $\frac{m}{r}$ as another letter, like n - you get the definition of e)

In other words, when compounding continuously,

$A(t) = Pe^{rt}$.

As a multivariable **model for continuous compounding** we write:

Verbal Definition: $F(P,r,t)$ = total monetary accumulation when the principal is P, the annual percentage rate is r, compounded continuously, and the number of years is t, where the accumulated money is in the same money units as the principal.

Symbol Definition: $F(P,r,t) = Pe^{rt}$ for $P,r,t \geq 0$

Assumptions: Certainty and divisibility. Certainty implies that the relationship is exact. Divisibility implies that any fractional values of dollars, percents and years are possible.

Since we have already defined a function of three variables in this section called A, we have called this function F, which you can think of as standing for the Future Value of your money after t years (we will in fact define this formally in Section 5.4). In this case, the total accumulation of \$1 after 1 year will be e^r, so the effective rate (APY) will be just that value minus 1 **for continuous compounding with an APR of r:**

APY $= y = e^r - 1$.

Notice again that, whenever we used the discrete compounding formula above, the condition that mt be a whole number was satisfied, so the results are fully valid.

Sample Problem 6: Suppose investment C pays 5.9% compounded continuously.

(a) What would be the total accumulation for \$1 after 1 year?
(b) What is the APY?
(c) What would be the total accumulation for \$100 after 1 year?
(d) What would be the total accumulation for \$100 after 4 years?
(e) How does investment C compare to investments A and B from Sample Problem 3?

Solution: This time, $r = 0.059$.

(a) $C(1) = (1)e^{(.059)(1)} = e^{.0.059} \approx 1.0608$.

(b) APY $= e^r - 1 = e^{0.059} - 1 \approx 0.0608$ (or 6.08%)

(c) $C(1) = (100)e^{(0.059)(1)} = 100(1.0608) = 106.08$.

(d) $C(4) = (100)e^{(0.059)(4)} = 100e^{0.236} \approx 100(1.2662) = 126.62$.

(e) Since $A(4)$ was 127.12, $B(4)$ was 126.82, and $C(4)$ was 126.62, investment C is worse than both A and B. The yield for A was 0.0618 . The yield for B was 1.0614 - 1 = 0.0614 . The yield for C is 1.0608 - 1 = 0.0608 . Figure 5.3-4 shows the accumulated amounts for the last five days for the three investments. (The continuous compounding is a continuos curve, as shown in Figure 5.3-5.)

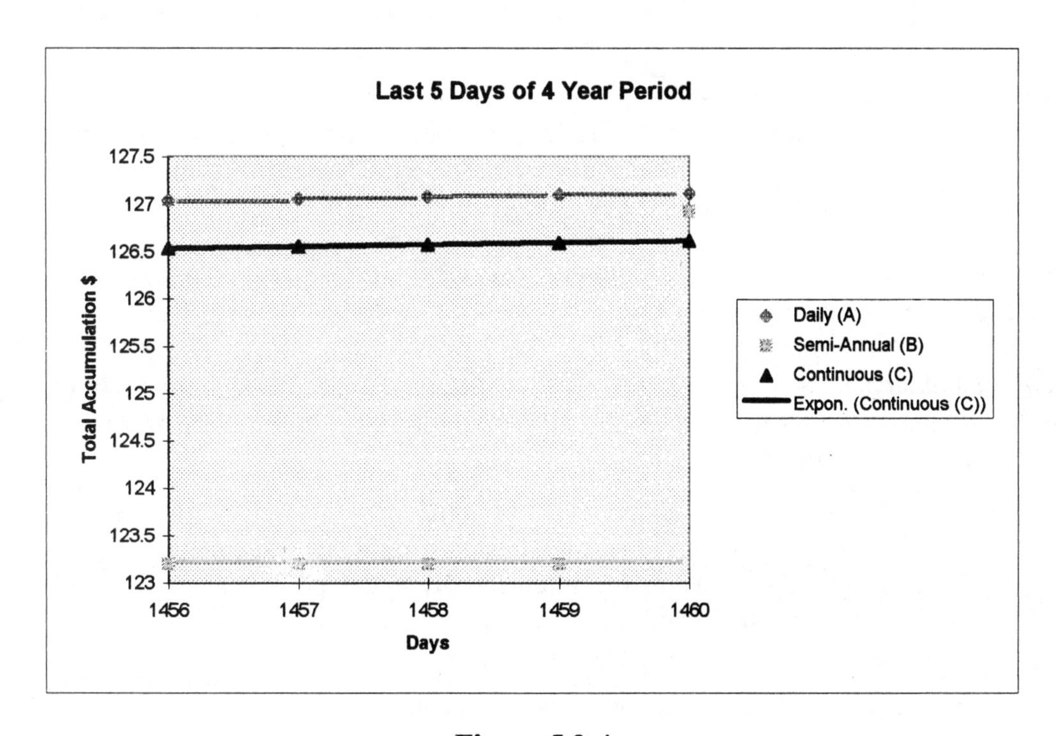

Figure 5.3-4

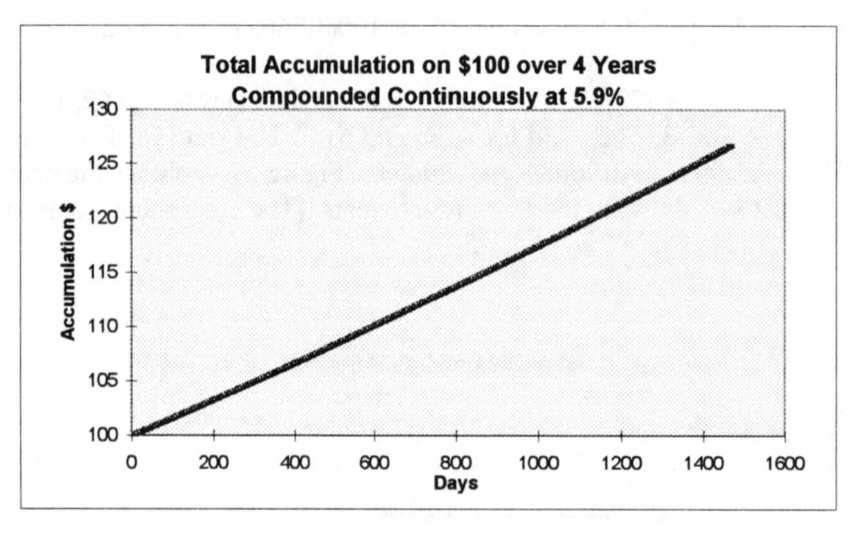

Figure 5.3-5

Section Summary

Before trying the exercises, be sure that you

- Can calculate your total monetary accumulation (as well as just the interest portion by itself) for any investment using **simple interest** (interest only on the original principal, not on the interest), given the principal, interest rate per period, and the number of periods (or the frequency of interest calculations and the amount of time), using the formula

  $S(P,i,n) = P(1+in)$ where S is the accumulated money, P is the original principal, i is the simple interest rate per period, expressed as a pure decimal, and n is the number of periods (a whole number) , where the accumulated money is in the same money units as the principal.

  To be fully valid, the number of periods, n , must be a whole number, if the interest is accrued only at the end of each period.

- Understand that **compound interest** involves getting interest on *accumulating interest* as well as on the original principal.

- Can calculate your total monetary accumulation (as well as just the interest portion by itself) for any investment using **discrete compound interest**, given the principal, **actual interest rate per period**, and the **number of periods**, using the following formula for accumulated money:

$A(P,i,n) = P(1+i)^n$, where P is the principal, i is the actual interest rate per period, n is the number of periods, and the accumulated money is in the same money units as the principal.

To be fully valid, the number of periods, n, must be a whole number.

- Can calculate your total monetary accumulation (as well as just the interest portion by itself) for any investment using **discrete compound interest**, given the principal, annual percentage rate (**APR**), **number of compounding periods per year**, and the **number of years**, using the following formula for accumulated money:

  $A(P,r,m,t) = P\left(1+\frac{r}{m}\right)^{mt}$, where P is the principal, r is the APR, m is the number of compounding periods per year, and t is the number of years, where the accumulated money is in the same money units as the principal.

  To be fully valid, mt must be a whole number (t must be an integer multiple of $(1/m)$, corresponding to a whole number of compounding periods).

- Understand that Euler's number (e) is defined mathematically by $e = \lim_{m\to\infty}\left(1+\frac{1}{m}\right)^m \approx 2.7183$, and can be thought of as the precise accumulated money value (in dollars) of \$1 after exactly 1 year, with interest compounded continuously (letting the number of compoundings per year get arbitrarily large, so m approaches infinity), at an APR of 100%

- Can calculate your total monetary accumulation (as well as just the interest portion by itself) for any investment using ***continuously* compounded interest**, given the principal, annual percentage rate (APR), and the number of years, using the following formula:

  $F(P,r,t) = Pe^{rt}$, where P is the principal, r is the APR and t is the number of years, where the accumulated money (future value) is in the same money units as the principal.

- Can calculate each of the *other* quantities (other than total monetary accumulation) in each of the cases listed above, given the *rest* of the quantities.

- Understand that the **Annual Percentage Yield (APY)**, also called the **yield** or the **effective annual interest rate**, is the actual percentage return on your money after 1 year, or, in everyday terms, the numerical value of the interest (in dollars) on $1 after 1 year.

- Understand that the **APR** (the **Annual Percentage Rate** or **nominal annual interest rate**), designated r, is an artificial number used to find the actual interest rate per compounding period (i), by dividing the APR by the number of compoundings per year ($i = \text{APR}/m = r/m$).
- Can **calculate the APY**, or y, for an investment, given the APR and compounding structure, using one of the following two formulas:

Discrete Compounding:

$$y = \left(1 + \frac{r}{m}\right)^m - 1$$, where y is the yield (APY), r is the APR, and m is the number of compounding periods per year.

This is fully valid only if m is a whole number.

Continuous Compounding:

$$y = e^r - 1$$, where y is the yield (APY) and r is the APR.

and understand how the idea that the APY is the interest (in dollars) on $1 after 1 year leads to these formulas.

EXERCISES FOR SECTION 5.3:

Warm Up

1. If $P = 2000$, $i = 0.08$, and $n = 6$, find S_n (the total accumulation using simple interest).

2. If $P = 600$, $i = 0.05$, and $n = 10$, find S_n (the total accumulation using simple interest).

3. If $P = 2000$, $i = 0.08$, and $n = 6$, find A_n (the total accumulation using compound interest). How does this compare to your answer in Exercise 1?

4. If $P = 600$, $i = 0.05$, and $n = 10$, find A_n (the total accumulation using compound interest). How does this compare to your answer in Exercise 2?

5. If $P = 1500$, $r = 0.07$, $m = 4$, and $t = 5$, find $A(t)$ and the APY.

6. If $P = 500$, $r = 0.09$, $m = 52$, and $t = 7$, find $A(t)$ and the APY.

7. If $P = 1500$, $r = 0.07$, $m = \infty$, and $t = 5$, find $A(t)$ and the APY. How does your answer compare to your answer in Exercise 5?

8. If $P = 500$, $r = 0.09$, $m = \infty$, and $t = 7$, find $A(t)$ and the APY. How does your answer compare to your answer in Exercise 6?

Game Time

9. You have just moved to a new city, and are choosing a bank. You've narrowed your choice down to two banks. Their fee structures are identical, except for their interest structure. One gives 3% interest, compounded quarterly, and the other gives 2.9%, compounded continuously. Which is better? How did you decide?

10. You pay your car insurance on an installment plan, with a nominal annual interest rate (APR) of 15%, compounded quarterly, while your credit card charges you 14.7% interest, compounded monthly. Which has a lower effective annual interest rate (APY)? How did you do your calculations?

11. Eight years ago, your grandparents put $3000 in a special account for you that accumulates interest at a rate of 8%, compounded monthly.
    (a) How much money is in the account now?
    (b) What is the yield of the account?
    (c) When will the money have doubled (approximately)?

12. Four years ago, you received a gift of $1000, and put it in an account that earns 5%, compounded continuously.
    (a) How much money is in the account now?
    (b) What is the yield of the account?
    (c) If you want to use the money to put a $1500 down payment on a car, when will the account have accumulated that much money?

13. You have just won some money in a lottery, and you want to put a portion of it, $2500, into an investment that pays 9%, compounded continuously.
    (a) Fully define a model for your total monetary accumulation as a function of time.
    (b) Find the yield for this investment, and use it to express the function for your model in (a) in the form $y = ab^x$.
    (c) *Using the function you derived in part (b),* find an expression for the rate of change of your money at any time in the future.
    (d) At what rate is your money growing after 3 years? To what does this correspond graphically?
    (e) Express your numerical answer to (d) as a percentage of your accumulation at that time (this will be the **percent rate of change at that time**).
    (f) Comment on the assumptions of your model in (a). How would your comments change if the compounding had been monthly instead of continuously?

14. You just received a gift from a favorite uncle, and you want to put a portion of it, $800, into an investment that pays 5%, compounded continuously.
    (a) Fully define a model for your total monetary accumulation as a function of time.
    (b) Comment on the assumptions in your model definition. How would your comments change if the compounding had been quarterly instead of continuously?
    (c) Find an expression for the rate of change of your money at any time in the future.
    (d) At what rate is your money growing after 3 years? To what does your answer correspond to graphically?
    (e) Express your numerical answer to (d) as a percentage of your accumulation at that time (this will be the **percent rate of change at that time**).
    (f) What is the yield of this investment? How could you use your answer to write your function in a different form?

15. Find $A(600, 0.075, 365, 4)$, as defined in this section, explain what it means in words, and make up a situation where this calculation might apply.

16. Find $F(1200, 0.04, 7)$, as defined in this section, explain what it means in words, and make up a situation where this calculation might apply.

17. Explain in your own words the idea behind the effective yield of an investment.

18. Explain in your own words what e (Euler's number) corresponds to in a real-world context.

Section 5.4: The Time Value of Money (Present Value and Future Value) and Loans

In the last section, we talked about how compound interest works. In this section, we discuss **the time value of money,** the fact that as long as it is possible to earn interest, the value of money depends upon when it is received: the same face value of money is worth more if received earlier rather than later. For example, if the money is received earlier, it can then be invested in some virtually certain investment, such as a bank account or CD, and worth *more* by the time the later money would have been received. The value of a certain amount of money (principal) after a given amount of time at a given compound interest rate and structure is called its **future value** at that later time. Sometimes this process is reversed, such as when you are saving up for something you want at some point in the future. You want to know how much you need to invest *now* to achieve the desired amount of money at the given point in time in the future, at a given compound interest rate and structure. The amount you need now is the **present value** of the future amount. When you are looking at a business opportunity that involves revenues and expenses at different points of time in the future, it is helpful to get a measure of what that opportunity would be *equivalent to* in dollars *right now*, at the present time. By taking the present value of each positive and negative cash transaction and adding them, you can get such a measure, called the **net present value** of the cash flow at a given interest rate and structure. You can think of this as how much the opportunity would be worth to you (the most you would pay for it) right now. As a result, this idea can also be used to *compare* different opportunities. Another way to compare different investment opportunities is to compare the effective rate of return (the *annual interest rate* that the return on the investment is *equivalent to*), called the **internal rate of return**, for each.

Here are some of the kinds of problems that the contents of this section will help you be able to solve:

- You are interested in investing a gift from your grandparents, to save up to buy a car in two years. You are considering a certificate of deposit that compounds interest at a certain rate. How much money (total) would you have after the two years?
- Your rich uncle just gave you a gift of money. You want to spend some of it now, but put the rest in a CD to pay for a trip to visit him in California a year from now. Given the interest rate and type of compounding for the CD, how much do you need to put in it now to have the amount of money you need for the trip?
- You have an investment opportunity to help a friend who is a musician and wants to make a first CD. This would require some money up front for studio expenses

and then some more money six months later to pay for production. You and your friend have great confidence in the success of the album, and your friend agrees to give you a guaranteed amount one year from now. If your cash reserves are in a special account earning interest compounded continuously, is the CD investment a good idea for you financially?

- You want to help finance a community literacy program. You are considering an investment with a known single initial payment (investment cost) and a guaranteed single cash receipt (revenue) at a given time in the future. What interest rate (APY) would make you indifferent between investing in this opportunity and not investing in it? That is, what APY on an account would make the return from this investment exactly equal to the return you would get from investing your cash in that account for the same period of time?

The purpose of this section is to help you solve problems like those above and:

- To understand the concept of **the time value of money**: why the same amount of money received or spent at different times is worth different amounts now (or at any other given time).
- To understand what is meant by the **future value** at a given time of a cash transaction at another given point in time, given a particular interest rate and compounding structure, as well as how to calculate this future value.
- To understand what is meant by a **cash flow**.
- To understand what is meant by the **present value** of a cash transaction at a given point in time, given a particular interest rate and compounding structure, as well as how to calculate this present value.
- To understand how to calculate the future value at a given time or the present value (usually called the **net present value**) of a cash flow, and understand what each of these quantities means in real-world terms.
- To understand what is meant by the **internal rate of return (IRR)** of a cash flow or investment opportunity, and how to calculate it.
- To understand how the amount of each **payment on a loan** can be calculated, given the amount of the loan, interest rate and structure, and the payment structure.
- To understand how these concepts are useful in business, in policy-making, and in our personal lives.

Future Value

Sample Problem 1: Monty Corridor gives you a choice between a prize of \$1000 now (Door #1), or \$1200 to be received three years from now (Door #2). The best sure investment you know for your available cash over this time period is at a guaranteed rate of 6% compounded continuously. Which prize should you choose, to have the most cash available from the prize three years from now?

Solution: The cash available from the Door #2 prize three years from now will be \$1200 . What about the Door #1 prize? From the last section, we know that

$$F(P,r,t) = Pe^{rt}$$

for continuous compounding. In this case

P = principal = \$1000
r = nominal annual interest rate (APR) = 6% = 0.06
t = years over which interest is accumulated = 3

Thus, we get

$$F(1000,0.06,3) = (1000)e^{(0.06)(3)} = 1000e^{.18} \approx 1000(1.19722) = 1197.22 \ .$$

Translation: The prize from Door #1 will be worth \$1197.22 three years from now, so the prize from Door #2 is slightly better. □

The quantity we calculated for the first cash transaction (the prize behind Door #1), was the accumulated cash after three years if the principal were invested now at the given interest rate and compounding structure. This is what is meant by the **future value** of a transaction at a given point in the future, given an interest rate and a compounding structure. In this case, we would say, "the future value 3 years from now of \$1000 received now, at 6% compounded continuously, is (will be) \$1197.22 ."

If compounding is continuous, we can redefine the model $F(P,r,t)$ we defined in Section 5.3 to reflect the fact that it also gives us the future value:

Verbal Definition: $F(P,r,t)$ = the future value (or the total accumulated money, including interest), in the same money units as the principal, after t years, when P is the principal (amount of the cash transaction), and r is the nominal annual interest rate (APR).

Symbol Definition: $F(P,r,t) = Pe^{rt}$ for $P,r,t \geq 0$

Assumptions: Certainty and divisibility. Certainty implies that the relationship is exact. Divisibility implies that any fractional values of money, APR, and years are possible.

Sample Problem 2: You are about to buy a stereo on sale, and are given the choice between paying $500 now, or laying it away for 6 months and paying $510 at that time. You currently have $600 in your savings account, and it earns 4% compounded quarterly.

(a) Which payment method will leave you with the most cash in your savings account 6 months from now?

(b) What is the future value 6 months from now of the $500 cost (now) for the first option?

Solution: (a) *For Option 1 ($500 now):* You start with $600, so after paying the $500 immediately, you would have $100, which would get compounded for 6 months. Recall that when interest is compounded m times per year at an APR of r for t years, the total accumulation is

$$A(P,r,m,t) = P(1 + r/m)^{mt} \quad .$$

In this case,

P = principal = $100 ,
r = APR = 4% = 0.04 ,
m = number of compoundings per year = 4 , and
t = number of years = 0.5

Thus we get that

$$A(100,0.04,4,0.5) = (100)\left(1+\frac{0.04}{4}\right)^{(4)(.5)} = (100)(1+0.01)^2 = (100)(1.01)^2 = (100)(1.0201)$$
$$= 102.01$$

Translation:, Six months from now, the $100 would have grown to $102.01 .

Notice that, since we have used the discrete compounding formula, we should check to make sure that it is fully valid by checking to see that mt (the total number of compounding periods) is a whole number. In this case,

$$mt = (4)(0.5) = 2 ,$$

so our model is fully valid.

For Option 2 ($510 in 6 months): In this case, we would get interest on our $600 for the 6 months, so we would use the same formula as above, but with

$P = \$600$
$r = 0.04$
$m = 4$
$t = 0.5$.

So for this case, we would get

$$A(600,0.04,4,0.5) = (600)\left(1+\frac{0.04}{4}\right)^{(4)(0.5)} = (600)(1+0.01)^2 = (600)(1.01)^2 = (600)(1.0201)$$
$$= 612.06$$

Translation: After the six months, our savings would have accumulated to $612.06 .

At this point, we would then have to pay the $510 cost, so we would be left with

$612.06 - $510 = $102.06 .

Thus, Option 2 ($510 in 6 months) is a tiny bit (about 1 piece of bubble gum) better, since $102.06 is better than $102.01 .

Once again, to check the validity of this calculation, we check the value of mt :

$mt = (4)(0.5) = 2$,

which is a whole number (of periods), so our results are fully valid.

(b) We want to determine the future value 6 months from now of the $500 cost that occurs now. We calculated that, if we made the $500 payment now, we would end up with a balance of $102.01 in the bank 6 months from now. We also know that if we had been able to keep our $600 in the savings account for the 6 months, it would have grown to $612.06 . Thus, if the payment were to occur 6 months from now, we would have $612.06 in the savings account (*before* making the payment). To *end up* with the *same* balance of $102.01 (as in the "pay now" scheme) *after* making the payment (in the "pay later" scheme), the payment 6 months from now would then have to be

$612.06 - $102.01 = $510.05 .

We can thus think of this value as what the \$500 payment now is ***equivalent to financially*** 6 months from now. This is what we mean by the **future value** of an expenditure, or negative cash transaction. Out of curiosity, let's see what the future value calculation for a *positive* \$500 after 6 months would be:

$$F(500,0.04,4,0..5) = (500)\left(1+\frac{.04}{4}\right)^{(4)(.5)} = (500)(1+.01)^2 = (500)(1.01)^2 = (500)(1.0201)$$
$$= 510.05$$

We get the same numerical value! In other words, we can use the formula for future value given above for *both* positive and negative cash transactions. Thus the future value 6 months from now of the \$500 payment now is \$510.05 . □

To repeat, **if P is *any* cash transaction (a receipt *or* a payment), its future value t years later is given by**

$F = F(P,r,t) = Pe^{rt}$ (for continuous compounding) or
$F = F(P,r,m,t) = P(1 + r/m)^{mt}$ (for discrete compounding m times per year).

In simple terms, **the future value of a transaction (whenever it occurs) at a given time in the future for a given interest rate and structure is the transaction's *financially equivalent dollar value* at that *future* time.** Technically, the future value can be interpreted as the *net difference* between the *balance* you would have in a savings account at the specified time in the future, at the given interest rate and compounding structure, *including* the transaction (putting a receipt into, or taking an expense out of, the account), and the balance you would have in the account at the same specified future time *without* the transaction, assuming that there is no other activity in the account over that period except the accumulation of interest. Specifically, the calculation takes the balance at the future time *including* the transaction *minus* the balance at the future time *without* the transaction. In practice, the **future value is really just a compound interest accumulated money calculation**.

If the transaction occurs at time t_1 and the future value is to be calculated at time t_2 (assumed to be greater than or equal to t_1), both in years, then the value of t in the formula (the time span over which the transaction amount is to be compounded) will be

$t = t_2 - t_1$.

Let's look at Sample Problem 2 a little bit more. In the continuous compounding case, the effective annual interest rate (annual percentage yield, or APY), as we saw in the last section, is given by

$$y = \text{APY} = (1 + r/m)^m - 1$$
$$= \left(1+\frac{0.04}{4}\right)^4 - 1 = (1+0.01)^4 - 1 = (1.01)^4 - 1 \approx 1.0406 - 1$$
$$= 0.0406$$

Translation The effective annual interest rate (yield, or APY) of 4% interest compounded quarterly is approximately 4.06%.

In an effort to simplify the formula for the future value using discrete compounding we make the following substitutions:.

The expression for the future value could be written in the form

$$F = A(P,r,m,t) = P\left(1+\frac{r}{m}\right)^{mt} = P\left[\left(1+\frac{r}{m}\right)^m\right]^t$$

Since APY = $y = \left(1+\frac{r}{m}\right)^m - 1$, this means that $1+y = \left(1+\frac{r}{m}\right)^m$, so if we replace the $\left(1+\frac{r}{m}\right)^m$ by $(1+y)$ in our expression for F above, we get

$$F = A(P,y,t) = P(1+y)^t .$$

If we let $b = 1+y$, we can further simplify the formula to read:

$$F = A(P,b,t) = Pb^t .$$

This puts the expression for the future value in the form of an exponential function, $y = f(x) = ab^x$. For example, in Sample Problem 2, we would have

$$b = \left(1+\frac{r}{m}\right)^m = 1+y \approx 1+0.0406 = 1.0406 \quad .$$

As defined above, b equals the yield plus one, $1+y$: b (1.0406 in Sample Problem 2) is the APY (0.0406, or 4.06%) plus 1.

To see how this could be used, to calculate the future value of the $500 after 6 months, we could have calculated

$$F = 500(1.0406)^{0.5} \approx 510.05 \quad .$$

Referring back to Section 1.4, the value we are calling b here is the *constant ratio* of the exponential function, while the value of y is the value of the *constant percent increase* (as usual, expressed in pure decimal form). In fact, **the constant ratio will always be equal to 1 plus the constant percent increase**. If we generated data from

$$F(t) = 500(1.0406)^t$$

using $t = 0,1,2,3,\ldots$, we would get

Table 1

t	$F(t)$	b=Ratio ($F(t)/F(t\text{-}1)$)	y = % Increase from $F(t\text{-}1)$ to $F(t)$
0	500	N/A	N/A
1	520.3	520.3/500 ≈ 1.0406	(520.3-500)/500 ≈ .0406 = 4.06%
2	541.42	541.42/520.3 ≈ 1.0406	(541.42-520.3)/520.3 ≈ .0406 = 4.06%
3	563.41	563.41/541.42 ≈ 1.0406	(563.41-541.42)/541.42 ≈ .0406 = 4.06%
4	586.28	586.28/563.41 ≈ 1.0406	(586.28-563.41)/563.41 ≈ .0406 = 4.06%

Figure 5.4-1 shows the relationship between $F(t)$ and t.

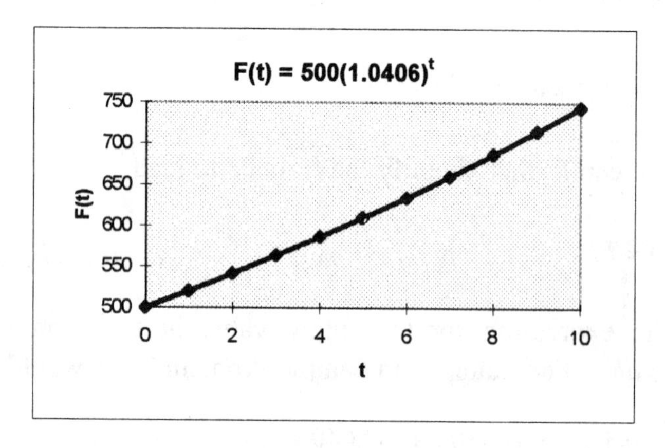

Figure 5.4-1

A General Model for Future Value

Now consider Sample Problem 1 again, when we needed to find the future value three years from now of $1000 received now, at 6% compounded continuously. In the last section, we saw that for continuous compounding, the yield is given by

$$y = \text{APY} = e^r - 1$$

which in the example would be

$$y = e^r - 1 = e^{0.06} - 1 \approx 1.061837 - 1 = 0.061837 \ .$$

Notice that the formula for future value in the case of *continuous* compounding can also be written in the same form as that derived above for *discrete* compounding m times per year:

$$F = F(P,r,t) = Pe^{rt} = P(e^r)^t = Pb^t = P(1+y)^t \ .$$

In the example, we would have

$$b = 1 + y \approx 1 + 0.061837 = 1.061837 \ ,$$

so the future value of \$1000 after three years would be

$$F = Pb^t \approx (1000)(1.061837)^3 \approx 1197.22 \ .$$

This result is the same as before, verifying that the formulas are equivalent to each other. If we used the formula

$$F = 1000e^{0.06t} \approx 1000(1.061837)^t \ ,$$

and created a table like Table 1 above, the constant ratio would be 1.061837 and the constant percent increase would be $0.061837 = 6.1837\%$.

In other words, the idea of using the APY (y) can give us a single formula for future value:

$$F = P(1+y)^t = Pb^t$$

With this in mind, we can write a **general model for the future value**:

Verbal Definition: $FV(P,y,t) =$ the future value, in the same money units as the principal, when P is the principal, or the amount of the cash transaction, y is the APY, and t is the number of years

Symbol Definition: $FV(P,y,t) = P(1+y)^t$ for $P,y,t \geq 0$

Assumptions: Certainty and divisibility. Certainty implies that the relationship is exact. Divisibility implies that any fractional values of P, y, and t are possible.

This is clearly the best way to do the calculation, especially if you are *given* the yield, but there may be times when it makes sense even without being given it. And for general discussions about future value, it is convenient to have a single expression for the calculation. Remember, however, that in the case of discrete compounding, this model is only fully valid if t represents a whole number of compounding periods (only if mt is a whole number).

As a final note, you can think of y as the *actual* interest rate (what we called i at the beginning of our compound interest discussion) for *one year* and t as the number of periods (what we called n before), where in this case we would be thinking of the periods as *years*. The only difference is that the i and n notation is completely general (for any size compounding period), while the y and t notation is working specifically with *years* (it acts *as if* one year is the compounding period, even if it is not). Then the above formula is just a substituted version of

$$A_n = P(1+i)^n \quad ,$$

where we have substituted y for i and t for n. In fact, if a particular problem makes units other than years most natural, we could use the formula involving i and n for the future value.

From all of the above discussion, the bottom line is that future value corresponds to the idea of the total accumulation (initial amount plus compounded interest) of a cash amount after a certain period of time.

Present Value

Sample Problem 3: Your rich uncle just gave you a gift of \$2000. You want to spend some of it now, but put the rest in a CD to pay for a trip to visit him in California two years from now, the cost of which you estimate will be \$1200 at the time. Given that the CD compounds interest continuously at 7%, how much do you need to put in it now to have the amount of money you need for the trip two years from now?

Solution: The basic question is, how much do you need to invest *now*, so that after two years the total accumulation (principal plus interest) will be \$1200? We know that the appropriate formula for this situation is

$$F(P,r,t) = Pe^{rt} \quad ,$$

where, in this problem,

$r = 7\% = 0.07$ and
$t = 2$.

What makes the problem different is that this time we know $F(P,r,t)$ but not P . Specifically, we know that

$$F(P,r,t) = F(P,0.07,2) = 1200 \ .$$

Thus, our equation becomes

$$1200 = Pe^{(0.07)(2)}$$
$$1200 = Pe^{0.14}$$
$$1200 \approx (P)(1.150274) \ .$$

Dividing both sides by 1.150274, we get

$$P \approx 1200/1.150274 \approx 1043.23 \ .$$

Translation: We would need to invest \$1043.23 in the CD now to accumulate \$1200 after two years. This would leave us with

$$\$2000 - \$1043.23 = \$956.77$$

to do with as we like right now. □

The answer we got of **the amount we'd have to invest *now* to accumulate a given total at a given time in the future** is the idea of **present value**. In this case, we would say "the present value of \$1200 two years from now, at 7% compounded continuously, is \$1043.23 ." Analogous to the future value, **the present value of a transaction that occurs at a given time in the future for a given interest rate and structure is the transaction's *financially equivalent dollar value* at the *present* time**

As we said before, the total accumulation, $F(P,r,t)$, is the same thing as the future value, so we could write this as

$$F = Pe^{rt} \ ,$$

and want to solve this for P . We can do this by simply dividing both sides by e^{rt} :

$$F = Pe^{rt}$$
$$Pe^{rt} = F \qquad \text{(switching sides)}$$

$$P = \frac{F}{e^{rt}} = Fe^{-rt}$$ (dividing both sides by $e^{rt} > 0$)

(Recall that $x^{-n} = \frac{1}{x^n}$, so $\frac{1}{e^{rt}} = e^{-rt}$.)

Let's try this formula on Sample Problem 3. Here,

$F = 1200$, $r = 0.07$, and $t = 2$, so we get

$$P = Fe^{-rt} = (1200)e^{-(0.07)(2)} = 1200e^{-014} \approx 1200(0.869358) \approx \$1043.23 \; .$$

Translation: The present value of \$1200 two years from now, at 7% compounded continuously, is \$1043.23. The same answer - it works!

A General Model for Present Value

When deriving the general model for the future value we derived the relationship

$$(1+y) = e^r$$

If we substitute this in the expression for the present value we have

$$P = \frac{F}{e^{rt}} = \frac{F}{\left(e^r\right)^t} = \frac{F}{(1+y)^t} = F(1+y)^{-t}$$

You have probably figured out already that, for discrete compounding, depending on the formula for compound interest that is chosen:

$$F = P(1+i)^n \qquad F = P\left(1+\frac{r}{m}\right)^{mt} \qquad F = P(1+y)^t$$

the Present Value formula could also be found by solving these formulas for P, and written as

$$P = \frac{F}{(1+i)^n} = F(1+i)^{-n} = \frac{F}{\left(1+\frac{r}{m}\right)^{mt}} = F\left(1+\frac{r}{m}\right)^{-mt} = \frac{F}{(1+y)^t} = F(1+y)^{-t}$$

as long as values of t are restricted to whole number multiples of the compounding period. The most **general model for present value is** thus:

Verbal Definition: $P(F,y,t)$ = the present value of a future transaction F, t years from the present when the annual yield (APY) is y, where the money units for the present value are the same as for F.

Symbol Definition: $P = \dfrac{F}{(1+y)^t} = F(1+y)^{-t}$, for $F,y,t \geq 0$

Assumption: Certainty and divisibility. Certainty implies that the relationship is exact. Divisibility implies that any fractional values of F, y and t are possible.

Again, for discrete compounding, this model is only fully valid if t is an integer multiple of the length of one compounding period (only if mt is a whole number).

For continuous compounding, then,

$$1+y = e^r \ ,$$

and for discrete compounding,

$$1+y = \left(1+\frac{r}{m}\right)^m$$

For Sample Problem 3, we would have

$$1+y = e^{0.07} \approx 1.072508$$

so

$$y \approx 1.072508 - 1 = 0.072508 \ .$$

Thus the present value could be calculated as:

$$P(1200,0.072508,2) = \frac{F}{(1+y)^t} = \frac{1200}{(1+0.072508)^2} \approx \frac{1200}{1.150273} \approx 1043.23 \ .$$

Once again, we get the same answer.

Discount Factors

You may have heard the term "discount factor" relating to the selling of bonds or borrowing cash. The symbol α is commonly used to represent the **annual discount factor**, which can be thought of numerically as **the Present Value of $1 one year from now.** Since it is actually that present value divided by the $1, it is a pure decimal (has no units). From our general formula for Present Value, this tells us that $\alpha = \frac{1}{1+y}$.

We can thus define

$$\alpha = \frac{1}{1+y} = \frac{1}{\left(1+\frac{r}{m}\right)^m} = \left(1+\frac{r}{m}\right)^{-m}$$

for discrete compounding and

$$\alpha = \frac{1}{1+y} = \frac{1}{e^r} = e^{-r}$$

for continuous compounding. Since α equals $\frac{1}{1+y}$, α will always be smaller than 1.

Since we already know that $P = F\left(\frac{1}{1+y}\right)^t$ in general, this leads us to the general formula

$$P = F\alpha^t$$

For example, in Sample Problem 3 we would have an annual discount factor of

$$\alpha = e^{-r} = e^{-.07} \approx .932394 \text{ .}$$

This means that a dollar a year from now is worth about 93 cents today for that problem. Since we want to know the present value of $1200 two years from now, we can do the calculation as

$$P = F\alpha^t \approx 1200(.932394)^2 \approx \$1043.23 \text{ ,}$$

Translation: The present value of $1200 received two years from now is $1043.23, as before.

Once again, for discrete compounding, this **discounting** (present value) formula is only fully valid if t is a whole number multiple of the compounding period (only if mt is a whole number). Discount factors for varying interest rates and years are frequently presented in tables such as Table 2. The values for 1 year are annual discount factors, so correspond to what we have called α. We have not defined a symbol for the others, but for t years, they correspond to α^t. Logically, as the interest rate and time increase, the discount factor decreases.

Table 2

Discount Factors (on $1) for Different Numbers of Years and Interest Rates

	Interest (Discount) Rate (Yield/APY Form)			
Years	5%	6%	7%	8%
1	0.952	0.943	0.935	0.926
2	0.907	0.890	0.873	0.857
3	0.864	0.840	0.816	0.794
4	0.823	0.792	0.763	0.735
5	0.784	0.747	0.713	0.681

To show how these values connect to what we have already done, consider an APY of 7% ($y = 0.07$). The annual discount factor would then be

$$\alpha = \frac{1}{1+y} = \frac{1}{1+0.07} = \frac{1}{1.07} \approx 0.934579 \ .$$

If you look in Table 2 for the entry corresponding to 1 year and 7%, you see the entry is 0.935, which is the same answer, rounded to 3 decimal places. Notice that in our earlier discussion for Sample Problem 3, we got a discount factor of 0.932 , which is not the same as the value in Table 2. That is because the discount factor of 0.932 corresponds to an *APR* (*nominal* annual interest rate) of 7% rather than an *APY* (annual *yield*, or *effective* annual interest rate). When using a table such as Table 2, **make sure that the way you are expressing your interest rate corresponds to the assumptions of the table!** Similarly, the entry corresponding to 2 years and 7% is the rounded value of

$$\alpha^2 = \left(\frac{1}{1+y}\right)^2 = \left(\frac{1}{1.07}\right)^2 \approx 0.873439 \ .$$

The Net Present Value of a Cash Flow

Sample Problem 4. You have $5000 you would like to invest, and you have an investment opportunity to help another friend who is a musician and wants to make a first CD. This would require $3000 now, for studio expenses, and $2000 in 6 months, to pay for physical production of the CD's. You and your friend have great confidence in the success of the album, and your friend agrees to give you $6000, guaranteed, one year from now. If your cash reserves are in a special account earning 5% compounded continuously and you consider this your cost of capital, is the CD investment a good idea for you?

Solution: This problem involves a situation in which there is a *combination* of *receipts* (income) and *expenditures* (costs), or *positive* and *negative* cash transactions, respectively. Such a *mixture* of cash transactions is often called a **cash flow**. If we use the subscript i to represent the i'th transaction in a cash flow, the investment opportunity in this problem could be represented with a table of times (t_i) and cash transactions (F_i) as follows:

Table 3

i	Time, t_i, *(years from now)*	Cash Transaction Face Value, F_i, at time t_i *($)*
1	0	-3000
2	0.5	-2000
3	1	6000

Thus, in this example, $t_1 = 0$, $F_1 = -3000$, $t_2 = 0.5$, $F_2 = -2000$, $t_3 = 1$, and $F_3 = 6000$. We could also represent the cash flow schematically with the diagram in Figure 5.4-2. This gives a nice visual way of picturing what is going on.

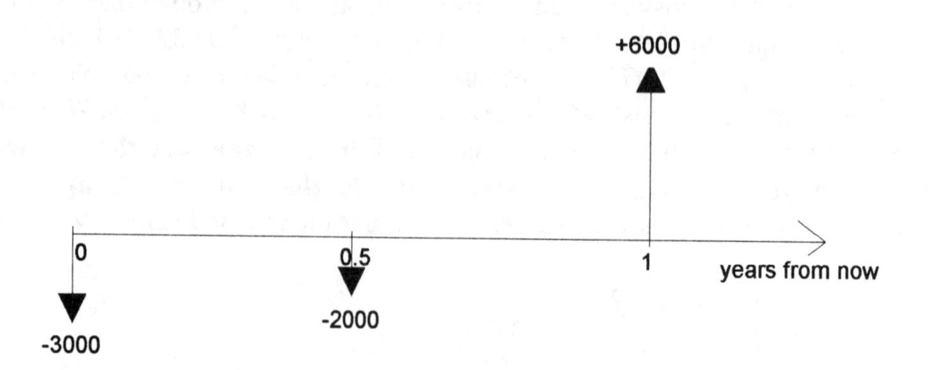

Figure 5.4-2

Based on our present value calculations earlier, we can find the present value of each of these individual transactions, and then combine them by adding (carrying the *signs* as well) to see what **the *net* effect of all the transactions** would be, **expressed in terms of**

what they are financially equivalent to, at the present time. The "net" part means that we take the sum of the present values of the positive transactions *minus* the sum of the *magnitudes* of the present values of the negative transactions (analogous to profit being revenue *minus* cost). Mathematically, this just means we *add* all of the individual present values, *including* the sign of each. Thus this is called the **Net Present Value (NPV)** of the cash flow. If we use the present value formula for continuous compounding

$$P_i = F_i e^{-rt_i} \quad ,$$

then, since $r = 0.05$, we get

$$\begin{aligned} \text{NPV} &= -3000e^{-0.05(0)} + (-2000)e^{-0.05(.5)} + 6000e^{-0.05(1)} \\ &\approx -3000 - 1950.62 + 5707.38 \\ &\approx -4950.62 + 5707.38 = 756.76 \end{aligned}$$

Translation: Paying the \$3000 now is (obviously) like paying \$3000 now, paying \$2000 in 6 months is like paying \$1950.62 now, and receiving \$6000 in a year is equivalent to receiving \$5707.38 now, based on the 5% interest possible in your special account. Notice how we have retained the signs of the original transactions. The net result of the entire CD investment would thus be equivalent to receiving \$756.76 right now, so this is the NPV of that investment (cash flow).

Is that a good idea? Would you like someone to give you \$756.76 right now? *You'd better believe it!!!!!* □

Using the general form for present value, we could say that, if there are n transactions in the cash flow, then a **general model for Net Present Value is given by**

Verbal Definition: $NPV(F_1, t_1, F_2, t_2, ..., F_n, t_n, y)$ = The net present value, in dollars, of the cash flow defined by cash transactions of F_i, in dollars, at time t_i years from now (for $i = 1, 2, \ldots, n$), at an annual percentage yield of y (in pure decimal form).

Symbol Definition: $NPV = \dfrac{F_1}{(1+y)^{t_1}} + \dfrac{F_2}{(1+y)^{t_2}} + \cdots + \dfrac{F_n}{(1+y)^{t_n}} = \sum_{i=1}^{n} \dfrac{F_i}{(1+y)^{t_i}}$ for $t_i, y \geq 0$

Assumption: Certainty and divisibility. Certainty implies that the relationship is exact. Divisibility implies that any fractional values of *F*, *y*, and *t* are possible.

As before, *for discrete compounding, this formula is only fully valid if all of the* t_i *values are whole number multiples of the compounding period* (that is, if mt_i is a whole number for all of the t_i values).

The idea of Net Present Value could also be applied to the future value, and could logically be called "Net Future Value" (NFV) of a cash flow at a specified point in time, say, t, although this term is not commonly used. An easy way to get this answer, however, if you already have the NPV of the cash flow, is to simply calculate the future value of the NPV at the specified time. In Sample Problem 4, this would mean compounding the NPV of $756.76 from the present (time 0) to one year from now (time 1):

$$NFV(1) = (NPV)e^{rt} = (756.76)e^{(0.05)(1)} \approx \$795.56 .$$

In the general case, we could say that

$$NFV(t) = (NPV)(1+y)^t .$$

Either way, we still need to remember that, *for discrete compounding, this is only fully valid if t is an integer multiple of one compounding period.*

The Internal Rate of Return

Sample Problem 5: Consider again the CD investment in Sample Problem 4. How high would the interest rate on your special account have to be so that you would be *just indifferent* to the choice between making the investment or leaving your money in the account? That is, at what interest rate on the special account would you have exactly the same amount of money at the end of the period, whichever choice you made (keeping your money in the special account, or investing in the investment and keeping the *rest* of your money in the special account)?

Solution: We saw that when $r = 0.05$ (an APR of 5%, compounded continuously), the NPV was

$$\text{NPV} = -3000e^{-0.05(0)} + (-2000)e^{-0.05(.5)} + 6000e^{-0.05(1)} \approx \$756.76 .$$

For other r values, we could do the analogous calculation. Let's try a few:

Table 4

APR (r)	NPV *(in $)*
0.05	756.76
0.10	526.57
0.15	308.76
0.20	102.71
0.25	-92.19

You can see that the CD investment would be a good idea, even if your account were earning 20% !!! (Let us know if you find such an account!) But it wouldn't be a good idea at 25% . This suggests that our answer is somewhere between 20% and 25% . The precise answer would be the value of r at which the NPV was exactly zero.

To get a precise answer to the question, we need to formulate a *function* to express the NPV in terms of r . In a sense, we have already done this; it would be

$$\begin{aligned} \text{NPV}(r) &= -3000e^{(-r)(0)} - 2000e^{(-r)(0.5)} + 6000e^{(-r)(1)} \\ &= -3000 - 2000e^{-0.5r} + 6000e^{-r} \end{aligned}$$

Using your equation-solving operation on your favorite technology, you just want to know when this function will be equal to 0. Doing so should give an answer of

$$r \approx 0.2260 \ .$$

Translation: At an APR of about 22.6% on your special account, the NPV of the CD investment would be about $0, so making the investment or not would make no difference to you. If the interest rate on the account were a little lower than 22.6%, the investment in the CD project would be a good idea (the NPV would be positive). If the APR were a little higher on your account, you'd be better off just leaving your money in the account (the NPV of the CD investment would be negative). □

In effect, this is saying that the **equivalent rate of return** (expressed as an equivalent continuous APR) for this investment is about 22.6% . Converting this to a yield value, we can calculate that

$$e^{0.2260} \approx 1.2536 \ .$$

This means that the yield (effective annual interest rate, or APY) is about 25.36% . This number is called the **internal rate of return (IRR)** of the CD investment, commonly expressed as an APY. In simple terms, then, **the internal rate of return of an investment cash flow is the equivalent rate of return** (annual interest rate, usually expressed as an APY) **of that investment**. More precisely, you can think of it as the *interest rate* on a special account for which you would end up with the *same final balance* after the last transaction of the investment cash flow, whether you just *left your money in the special account the whole time*, or *used the account for all the receipts and expenditures of the cash flow*. As before, this assumes that there are no other transactions on the account other than the accumulation of interest, which is considered to be compounded continuously.

If we wanted to solve this problem for the APY from the beginning, we could use our general present value formula and solve the equation

$$\frac{-3000}{(1+y)^0} - \frac{2000}{(1+y)^{0.5}} + \frac{6000}{(1+y)^1} = 0$$

which simplifies to

$$-3000 - \frac{2000}{\sqrt{1+y}} + \frac{6000}{1+y} = 0$$

This could be solved by hand (we don't necessarily recommend it!), but can be solved much faster using technology, which yields once again a solution of

$$y \approx 0.2536$$

(a yield of about 25.36%).

Sample Problem 6: Suppose you are offered a simple investment opportunity: you can invest \$2000 now, and are guaranteed to receive \$2500 2 years from now. What is the IRR of this investment?

Solution: Based on what we said above, if we define y to be the APY, then the expression for the NPV of this cash flow is

$$\text{NPV}(y) = \frac{-2000}{(1+y)^0} + \frac{2500}{(1+y)^2} = -2000 + \frac{2500}{(1+y)^2}$$

As we said before, to find the IRR, we set this equal to 0 and solve for y. Notice what happens if we set it equal to 0 and rearrange the equation as follows:

$$-2000 + \frac{2500}{(1+y)^2} = 0$$

$$\frac{2500}{(1+y)^2} = 2000 \qquad \text{(adding 2000 to both sides)}$$

$$2000(1+y)^2 = 2500 \qquad \text{(multiplying both sides by } (1+y)^2 > 0 \text{ and switching sides)}$$

In this form, we can clearly see that y is indeed simply the interest rate (APY) at which \$2000 will accumulate to \$2500 in 2 years. Thus, putting \$2000 into an account with that interest rate will yield \$2500, so the result of the investment or putting your money into the account would be the same.

Notice that we can solve this equation fairly easily by hand:

$$2000(1+y)^2 = 2500$$

$$(1+y)^2 = \tfrac{2500}{2000} = 1.25 \qquad \text{(dividing both sides by 2000)}$$

$$1+y = \sqrt{1.25} \qquad \text{(taking the square root of both sides } (y \geq 0)\text{)}$$

$$y = \sqrt{1.25} - 1 \qquad \text{(subtracting 1 from both sides)}$$

$$\approx 1.1180 - 1 = 0.1180 = 11.80\% \qquad \text{(simplifying)}$$

Thus the IRR of the investment is about 11.80% (in APY form). □

Loan Structure

Sample Problem 7: Suppose you borrow \$1000 from your parents to get a stereo for your dorm room, and agree to pay it back plus 10% after a year. From your perspective, what is the NPV of the cash flow, at 10% compounded annually?

Solution: As we did in the previous problem, we can make a table of the individual cash transactions:

Table 5

i	**Time, t_i** *(years from now)*	**Cash Transaction Face Value, F_i** *(\$)*
1	0	1000
2	1	-1100

In this case, the discount factor is

$$\alpha = \frac{1}{\left(1+\frac{r}{m}\right)^m} = \frac{1}{\left(1+\frac{0.1}{1}\right)^1} = \frac{1}{1+0.1} = \frac{1}{1.1} \approx 0.090909...$$

so the NPV is given by

$$\text{NPV} = 1000\left(\frac{1}{1.1}\right)^0 - 1100\left(\frac{1}{1.1}\right)^1 = 1000 - 1000 = 0 \ .$$

So the NPV is 0! □

This answer sounds a bit odd, but it makes sense if you think about it, since the interest rate is essentially the IRR (internal rate of return) for the person lending you the money (since for them, that is an investment). In fact, this is the way all loans are structured, including car loans, home mortgages, loans for computers and other consumer goods, etc.

Sample Problem 8: Suppose you wanted to pay off the same $1000 loan from your parents, still at 10% compounded annually, but in two payments over two years. How much should each payment be?

Solution: If we call the payment PMT, the expression for the NPV would be

$$NPV(PMT) = 1000 - (PMT)\left(\frac{1}{1.1}\right)^1 - (PMT)\left(\frac{1}{1.1}\right)^2$$

From what we saw in the previous problem, **the payment** ***PMT*** **should be fixed at the value that would make the NPV of the entire loan cash flow equal to 0, using the interest rate and structure of the loan**. In other words, we simply set the expression for *NPV(PMT)* equal to 0:

$$1000 - (PMT)\left(\frac{1}{1.1}\right) - (PMT)\left(\frac{1}{1.1}\right)^2 = 0 \quad \text{(set the equation equal to 0)}$$

$$1000 = (PMT)\left(\frac{1}{1.1}\right) + (PMT)\left(\frac{1}{1.21}\right) \quad \text{(square } \tfrac{1}{1.1}\text{, and add the PMT terms to both sides)}$$

$$PMT\left[\frac{1}{1.1} + \frac{1}{1.21}\right] = 1000 \quad \text{(switch the sides and combine like terms)}$$

$$PMT = \frac{1000}{\left[\frac{1}{1.1} + \frac{1}{1.21}\right]} \approx 576.19 \quad \text{(divide both sides by } \left[\frac{1}{1.1} + \frac{1}{1.21}\right]\text{)}$$

Translation: Our calculations indicate that, to pay off a loan of \$1000 at 10% compounded annually over two years in two payments (at the end of the first and second years), each of the two payments should be \$576.19 . □

In practice, banks have strict conventions about rounding off values as you pay off the principal and interest on a loan, so the mathematical solution may be off by a penny or two, but it will always be very close.

A situation like a loan in which you make regular periodic payments of equal amounts is often referred to as an **annuity**. This can also refer to a retirement fund with regular equal income payments after you retire. When the payments occur at the *end* of the period (like Sample Problem 7), the annuity is called an **ordinary annuity**. When the payments occur at the *beginning* of the period, the annuity is called an **annuity due**. The basic concepts presented here are the foundation of most methods of financing any type of enterprise.

Many graphing calculators and spreadsheet packages will do many of the calculations from this section and Section 5.3 automatically. See your technology supplement for details, and see if you can reproduce the results from the sample problems and examples with your technology.

Before closing this section, we should clarify the difference between **price inflation** versus the **time value of money**. Price inflation is a macroeconomic phenomenon related to the unemployment rate, business inventories, political instability, and many other complex and interrelated factors. It is simply how the "cost of living" goes up with time, usually expressed as the **inflation rate** of the Consumer Price Index (CPI), based on a standard market basket of goods. The idea of time value of money, upon which the concepts of Future Value and Present Value are based, involves the best so-called **risk-free interest rate** for investments. In other words, it's the highest relatively safe and sure rate at which your savings and investments can grow. This could depend on the amount of your assets that you can invest and other factors. In general, this rate should be significantly more than the inflation rate, so that saving money actually increases your effective buying power over time. Thus an NPV calculation is not the same as an "adjustment for inflation." On the other hand, the risk-free rate used in the NPV calculation will be affected by the inflation rate (as inflation increases, the risk-free rate will also tend to increase). At a time of very high inflation, the inflation rate could even exceed the risk-free rate, which can have dramatic and dangerous consequences in an economy.

You may also hear or read about a *net* interest rate in economics. This refers to the risk-free interest rate *over and above* the rate of inflation, so is defined to be the pure risk-free interest rate *minus* the inflation rate.

Section Summary

Before trying the homework exercises, look back over this section to be sure that you

- Understand the concept of the time value of money: the same amount of money at different times is worth different amounts now (or at any other given time), since money at an earlier time can be invested to accumulate to a higher amount at a later time.
- Understand that the **future value** at a given time in the future of a cash transaction at another given (earlier) point in time, given a particular interest rate and compounding method, is how much *money* the earlier transaction is *worth* (financially equivalent to) at the *later* time. Put differently, it is how much the *transaction* amount would *accumulate to* (with compound interest, starting from the time of the transaction) by the *later* point in time.
- Know how to calculate the future value, F, t years (or n periods) into the future from the occurrence of a transaction P :

Discrete Compounding: $F = P\left(1+\frac{r}{m}\right)^{mt} = P(1+y)^t = Pb^t = P(1+i)^n$

(fully valid only if mt is a whole number)

Continuous Compounding: $F = Pe^{rt} = P(1+y)^t = Pb^t$,

where:

r is the APR,

m is the number of compounding periods per year,

P is the transaction amount (face value, at the time it occurs),

t is the number of years into the future the transaction is being compounded (the time point at which the future value is desired minus the time point of the transaction),

y is the yield,

b is the compounding factor (ratio), equal to $(1+y)$,

i is the actual interest rate per compounding period (equal to r/m), and

n is the number of compounding periods over which the transaction is being compounded (equal to mt),

and where the money units of the future value are the same as the money units of P.

- Understand that the **present value** of a cash transaction that occurs at a given point in time in the future, given a particular interest rate and compounding structure, is how much *money* the future transaction is *worth* (financially equivalent to) at the *present* time. Put differently, it is the amount which, if invested at the given interest rate and compounding structure *right now*, would *accumulate to* the amount of the *future* transaction at the future time when it occurs.

- Know how to calculate the present value, P, of a cash transaction of F which occurs t years from the present:

Discrete Compounding: $P = F\left(1+\dfrac{r}{m}\right)^{-mt} = F(1+y)^{-t} = \dfrac{F}{(1+y)^t}$

$= Fb^{-t} = F\alpha^t = \dfrac{F}{(1+i)^n} = F(1+i)^{-n}$

(fully valid only if mt is a whole number)

Continuous Compounding: $F = Fe^{-rt} = F(1+y)^{-t} = Fb^{-t} = F\alpha^t$,

where:

r is the APR,

m is the number of compounding periods per year,

t is the number of years from the present that the transaction occurs,

F is the transaction amount (face value, at the time it occurs),

y is the yield,

b is the compounding factor (ratio), equal to $(1+y)$,

α is the discount factor (the reciprocal of b, the compounding factor/ratio),

i is the actual interest rate per compounding period (equal to r/m), and

n is the number of compounding periods over which the transaction is being compounded (equal to mt),

and where the money units of the present value are the same as the money units of F.

- Understand that a **cash flow** is a sequence of positive and negative cash transactions (revenues and expenses, or profits and losses), and can be represented schematically using a timeline.
- Understand that the **Net Present Value** of a cash flow corresponds to the amount of cash at the *present* time that is *financially equivalent to* (worth the same as) the cash flow. This means that *doing the transactions out of* an account with the given interest rate and compounding structure would yield the *same balance* after the cash flow was over as *just keeping your money in* that account.
- Understand how to calculate the present value of a cash flow (usually called the **Net Present Value**, or **NPV**) at a given interest rate and compounding structure by calculating the sum of the present values of the individual transactions (including negative signs for expenditures or losses).

$$NPV = \frac{F_1}{(1+y)^{t_1}} + \frac{F_2}{(1+y)^{t_2}} + \cdots + \frac{F_n}{(1+y)^{t_n}} = \sum_{i=1}^{n} \frac{F_i}{(1+y)^{t_i}}$$

$$NFV(t) = F_1(1+y)^{t-t_1} + F_2(1+y)^{t-t_2} + \cdots + F_n(1+y)^{t-t_n} = \sum_{i=1}^{n} F_i(1+y)^{t-t_i}$$

where the i'th transaction of the cash flow is F_i and occurs t_i years from the present. With discrete compounding, the formulas are only fully valid when the exponents are whole number multiples of the compounding period.

- Understand that the idea of **Net Future Value** is analogous to NPV, but is calculated at some *future* point in time, t (rather than at the *present* time), and corresponds to the future value at that future time of the NPV.

  $NFV(t) = (NPV)(1+y)^t$

- Understand that the **internal rate of return (IRR)** of an investment opportunity is the *effective* annual *interest rate* (APY) or *equivalent rate of return* of the investment. In other words, it is the interest rate that would leave you with the *same* total money balance at the end of the time period, whether you just left your money in an account at that interest rate, or used the account for the transactions of the investment. Put still differently, the IRR is the interest rate that makes the NPV of the entire cash flow equal to 0. To calculate it, first get an expression for the NPV in terms of the APY, then set it equal to 0, and finally solve for the APY (usually using technology).

- Understand that a loan is structured so that the NPV of the entire cash flow of the loan is 0, using the interest rate and structure of the loan. The interest rate is thus also the IRR of the loan.

- Understand how these concepts are useful in business, in policy-making, and in our personal lives.

EXERCISES FOR SECTION 5.4:

Warm Up:

For Problems 1-6, find the Future Value of the given amount of money (P) after t years, at the given interest rate and compounding structure:

1. $P = 400$, $t = 3$ years, APR = 5%, compounded monthly.
2. $P = 2000$, $t = 7.5$ years, APR = 6.5%, compounded quarterly.
3. $P = 1000$, $t = 4.25$ years, APR = 7%, compounded continuously.
4. $P = 1500$, $t = \pi$ years, APR = 7.13%, compounded continuously.
5. $P = 2000$, $t = 7.5$ years, APY = 6.5%, compounded quarterly (be careful!).
6. $P = 1500$, $t = \pi$ years, APY = 7.13%, compounded continuously.

For Problems 7-12, find the Present Value, based on the given interest rate and compounding structure, of the given future amount of money (F), if the transaction is to occur t years from the present:

7. $F = 400$, $t = 3$ years, APR = 5%, compounded daily.
8. $F = 2000$, $t = 7.5$ years, APR = 6.5%, compounded weekly.
9. $F = -1000$, $t = 4.25$ years, APR = 7%, compounded continuously.
10. $F = 1500$, $t = e$ years, APR = 7.13%, compounded continuously.
11. $F = 2000$, $t = 7.5$ years, APY = 6.5%, compounded weekly (be careful!).
12. $F = 1500$, $t = e$ years, APY = 7.13%, compounded continuously.

For Problems 13-16, a cash flow is given, where F_i is the ith transaction and t_i is how many years into the future it occurs. Find the NPV for each cash flow, at the given interest rate and compounding rule:

13. $F_1 = -2000$, $t_1 = 0$; $F_2 = 2500$, $t_2 = 2$; APR = 8%, compounded semiannually.
14. $F_1 = -1000$, $t_1 = 0$; $F_2 = 2500$, $t_2 = 5$; APY = 10%, compounded monthly.
15. $F_1 = -1000$, $t_1 = 0$; $F_2 = -1000$, $t_2 = 1$; $F_3 = 1500$, $t_3 = 3$; $F_4 = 1500$, $t_4 = 5$; APY = 5%, compounded continuously.
16. $F_1 = -1000$, $t_1 = 0$; $F_2 = -1000$, $t_2 = 1$; $F_3 = 1500$, $t_3 = 3$; $F_4 = 2000$, $t_4 = 5$; APY = 5%, compounded continuously.

For Problems 17-18, find the IRR for each cash flow:

17. $F_1 = -2000$, $t_1 = 0$; $F_2 = 2500$, $t_2 = 2$.
18. $F_1 = -1000$, $t_1 = 0$; $F_2 = 2500$, $t_2 = 5$.

For Problems 19-20, find the Future Value (as of the last t_i value) for each cash flow by finding the Future Value of each transaction and adding:

19. $F_1 = -1000$, $t_1 = 0$; $F_2 = 2500$, $t_2 = 5$; APY = 10%, compounded monthly.
20. $F_1 = -1000$, $t_1 = 0$; $F_2 = -1000$, $t_2 = 1$; $F_3 = 1500$, $t_3 = 3$; $F_4 = 1500$, $t_4 = 5$; APY = 5%, compounded continuously.

Game Time:

21. You just received $200 for your birthday. If you put it in your savings account, which earns 4.25% compounded quarterly, how much will it be worth after a year? What is the economic term for your answer?

22. You want to put away some cash in a bank CD to buy a stereo that costs $1400 6 months from now. The CD earns 6.14%, compounded weekly.
    (a) How much would you need to put away now to have the money you need in 6 months?
    (b) What is the economic term for your answer to (a)?
    (c) What is the discount factor (α) in this situation?

23. For Problem 22, suppose you do not have the cash to put in now (all at once), but you have a Money Market account that earns 5.7%, compounded monthly. How much money would you have to put into the account at the beginning of each of the next 6 months to save the total you need for the stereo?

24. You want to save for a down payment on a car. If you put away $100 per month (at the end of each month), in an account that earns 4.75% compounded continuously, how much will you have saved after 8 months? What is the economic term for your answer?

25. You have a musician friend who wants you to invest in a recording project for a new album. She needs $2000 now for recording studio expenses, and then $1500 in 6 months to pay for physical production of the CD's. She has agreed to pay you back $4000 one year from now, after going on tour and having time to sell albums. If you don't make this investment, you will invest in a special account, which is guaranteed to earn a 6.5% yield (compounded continuously).

(a) What is the Net Present Value of this investment?

(b) Is the investment a good idea financially (assuming you have total confidence in both investments, and consider the risks to be equivalently negligible)? Explain.

(c) What is the Future Value of the cash flow, as of 1 year from now?

(d) Suppose you have $5000 available for this investment in the special account right now. How much would you have in the account a year from now if you left it all there for the year? How much would you have in the account if you invested in your friend out of it, and put the returns back in, each at the appropriate times?

e) What is the annual discount factor (α) in this situation? Show how you could use it to calculate the answer to part (a).

Section 5.5: Formulating SSE in Terms of Model Parameters

In the first four sections of this chapter, we talked about formulating multivariable functions and models of various kinds. One particular multivariable function is so important that it is the primary focus of this section. In Chapter 1, we introduced the concept of the errors associated with a particular model as related to a particular data set. Graphically, the error associated with each data point was the vertical difference between the actual value and the model value (the actual value *minus* the model value). To get a measure of the fit of the function to the data (actually, a measure of the *lack* of fit), we squared the individual errors and added them, to get a measure of the total error. We squared the errors so that positive and negative errors would not cancel, which would make it seem like there was less total error than was really the case, and to put a larger penalty on large deviations, to try to avoid them. This led to the concept of the Sum of the Squared Errors, abbreviated SSE. In Chapter 1 we then jumped to fitting models from data by using technology and least squares regression. We briefly discussed the least squares concept and saw how the sum of the squares of the errors was calculated, but never explained how the least squares regression line was calculated. How is this done? In this section, we show how to formulate a model of the SSE as a function of the parameters of the type of function being considered. Can we set up a model for calculating the best fit line using least squares regression? In *this* section, we will show how to *express* the SSE as a *function* of the parameters of a model category, for a given set of data. The idea is that *later*, in Chapter 8, we will show how to *find* the combination of *values* of all of the parameters that *minimizes* this SSE function, and these will be the best-fit least-squares regression model parameters.

Here are some examples of the kinds of problem the material in this section will help you to solve:

- You have data from your coffeehouse concert series of past ticket prices and paid attendance for a particular performer. You would like to fit a linear model to represent the demand function. To find the best possible linear model, you want to get an expression for the total error (SSE) in terms of the slope (m) and y-intercept (b) parameters, so that you can minimize it. You may also want to do the same thing for quadratic and exponential models, finding the SSE in terms of their respective parameters.

- You have information about your total costs and the numbers of chairs and tables you have made in recent months in your furniture business. You want to find a linear model for the cost as a function of the numbers of chairs and tables made, to estimate your fixed cost and variable costs for each type of furniture. To find the best linear model, you want to express the SSE as a function of the parameters of your model: the constant (fixed cost) and the coefficients for each of the independent variables (variable costs).

- You are trying to determine what amount of exercise will maximize your energy level on a daily basis. You gather data, and now want to fit a multivariate quadratic model. You want to find an expression for the SSE as a function of the model parameters.

In addition to being able to solve problems like those above, after studying this section, you should:

- Understand the concept of the **error** between a specific data point and a specific model and what it corresponds to graphically.
- Understand the reasoning behind using squared errors, and the concept of the SSE, to measure the total error between a specific set of data and a specific model.
- Be able to calculate the SSE for a given model and set of data
- Be able to formulate an expression for SSE as a function of the parameters of a model category which you want to fit to a set of data

Calculating the Errors and the Sum of the Squared Errors (SSE) for a Particular Data Set

Sample Problem 1: You kept a record of how many three-point basketball shots you made out of 15 tries after an hour of practice. After 1 hour of practice you could only make 3 shots out of 15 tries. After two hours, you made 8. One more hour of practice and you made 11 of the shots, but after the fourth hour of practice you were down to 10 shots. You are considering modeling the relationship between hours of practice and shots made out of 15 as a linear function, $f(x) = mx+b$, where $f(x)$ is the number of shots made out of 15 after practicing for x hours. Find an expression for the SSE as a function of the parameters of the model (m and b), $\text{SSE}(m,b)$.

Solution: In this case, the problem is fairly well defined: we want to generalize the calculations discussed in Section 1.3 to formulate a model for the SSE in terms of the parameters of a general linear model and the data given in the question. We bound our problem by saying that we will consider only single-variable linear regression. We want to eventually minimize the sum of the squares of the errors, so this sum will be our dependent variable. We are fitting a line, which has the general form: $y = mx + b$, where m is the slope of the line and b is the vertical intercept. Our independent variables will be the m and the b (the parameters) of the fit line. Now we must formulate a mathematical model for the SSE. Let's start by taking another look at the data in tabular form (Table 1).

Table 1

Hours	Shots made (out of 15)
1	3
2	8
3	11
4	10

For this example we will let x = the number of hours practiced and y = the number of shots made out of 15. Let's take a look at a plot of this data. (Figure 5.5-1)

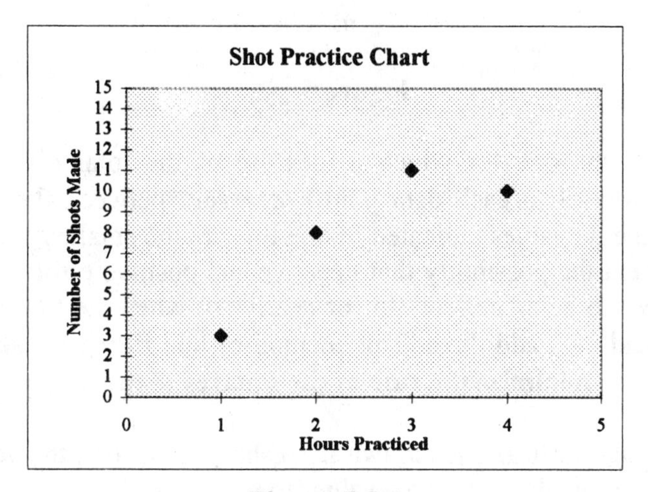

Figure 5.5-1

Suppose, for example, we had guessed that the best model to describe this data was $y = f(x) = 2x+3$, where x = the number hours practiced and y = the number of shots made out of 15 tries. From the graph, (Figure 5.5-2) it looks like this line, with a vertical intercept of 3 and a slope of 2, would fit pretty well.

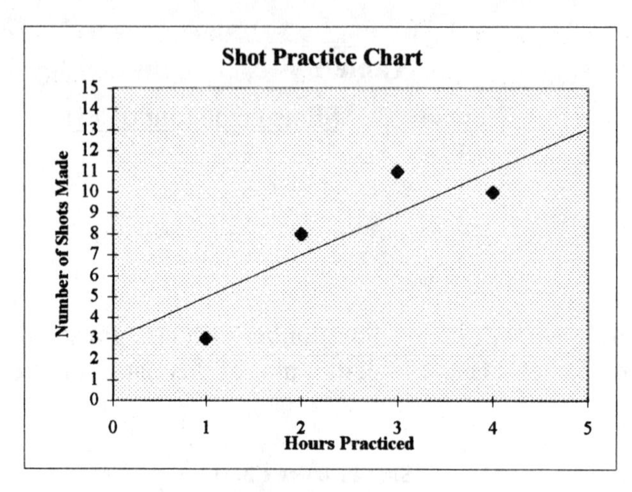

Figure 5.5-2

The best fit will be the one for which a measure of the total differences (measured vertically from the model to each data point) or **errors** between the y values of the plotted points and the y values indicated by the line are the smallest. As we discussed in Section 1.3, to prevent the possibility that negative and positive errors would cancel each other out and *appear* to be a "perfect" fit (or nearly so) when the fit was *not* perfect, we **square** the errors and then add them. This method of line fitting is called **least squares regression** because we **minimize the sum of the squares of the errors**.

Figure 5.5-3 has vertical lines drawn in to show the errors, the vertical differences between the y value of the data and the y value from the model.

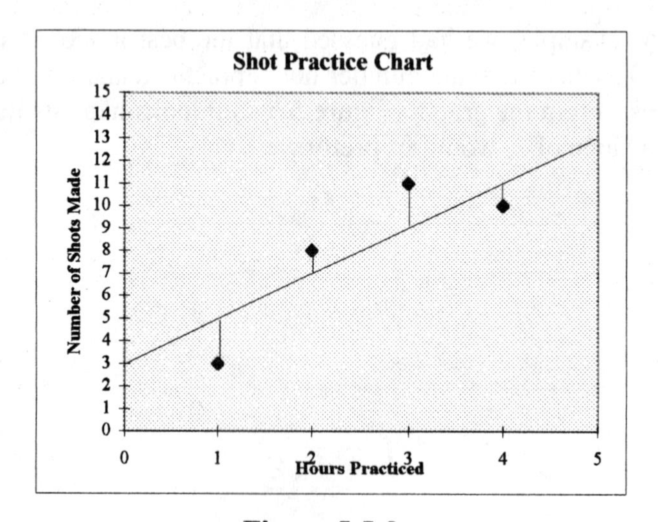

Figure 5.5-3

Our table for the SSE is shown in Table 2 .(In this table we have used the notation $\hat{y}$ to denote the y-value predicted by the model. We will continue to use this notation throughout the book.)

Table 2

Actual x	**Actual y**	**Predicted y**	**Error**	**Error Squared**
x_{data}	y_{data}	$y_{model} = f(x_{data}) = 2x_{data}+3$	$y_{data} - y_{model}$	$(y_{data}-y_{model})^2$
x	y	$\hat{y} = f(x) = 2x + 3$	$y - \hat{y}$	$(y - \hat{y})^2$
1	3	$2(1) + 3 = 5$	$3 - 5 = -2$	$(-2)^2 = 4$
2	8	$2(2) + 3 = 7$	$8 - 7 = +1$	$(1)^2 = 1$
3	11	$2(3) + 3 = 9$	$11 - 9 = +2$	$(2)^2 = 4$
4	10	$2(4) + 3 = 11$	$10 - 11 = -1$	$(-1)^2 = 1$
		Sums ➔	$-2 + 1 + 2 + -1 = 0$	$4 + 1 + 4 + 1 = 10$

The sum of the errors (differences) is zero! This does not present an accurate picture of how bad the fit is, because we can see that *none* of the points actually lie *on* the line. The sum of the *errors* is zero, but the sum of the *squares* of the errors is 10. From this example it is rather easy to see the importance of **squaring the differences (errors)**. **The smaller the Sum of the Squared Errors (SSE) is, the better the fit.** Also, as we discussed in Section 1.3, by trying to minimize the sum of the squares of the errors, larger errors get weighted (penalized) more heavily, so the final best-fit model tends to avoid larger errors if possible. In our example, the fit of our guessed model is good, but perhaps not as good as it could be.

Using technology, we found the least squares regression line for this data to be: $y = 2.4x + 2$. The calculations for the SSE for this model are shown in Table 3.

Table 3

Actual x	**Actual y**	**Predicted y**	**Error**	**Error Squared**
x_{data}	y_{data}	$y_{model} = f(x_{data}) = 2.4x_{data}+2$	$y_{data} - y_{model}$	$(y_{data}-y_{model})^2$
x	y	$\hat{y} = 2.4x + 2$	$y - \hat{y}$	$(y - \hat{y})^2$
1	3	$2.4(1)+2 = 4.4$	$3 - 4.4 = -1.4$	$(-1.4)^2 = 1.96$
2	8	$2.4(2)+2 = 6.8$	$8 - 6.8 = +1.2$	$1.2^2 = 1.44$
3	11	$2.4(3)+2 = 9.2$	$11 - 9.2 = +1.8$	$1.8^2 = 3.24$
4	10	$2.4(4)+2 = 11.6$	$10 - 11.6 = -1.6$	$(-1.6)^2 = 2.56$
		Sums ➔	0	9.20

The sum of the squared errors went from 10 to 9.2. This is definitely a better fit. In fact, it is the *best*-fitting line because the technology found the model for which the SSE is minimized.

Can we generalize from the two examples above to get a model for the SSE as a function of the parameters of our model, m and b? Let's look at Tables 2 and 3 again. In each case we made a table of the actual data and the value that the model predicted by plugging the x value from the data into the formula for the model. Then we calculated the difference (**error**) between the y value of the actual data and the y value predicted from the model (**the actual data value *minus* the predicted model value**). Finally, that error was **squared**, and all of these squared errors were then **added** together. This is fairly easy to do when we have numbers to plug in. We'll always have the actual x and y values, but we don't know the m and b of the linear model $y = mx + b$; this is what we are trying to find. Since we don't have this, let's replace the given specific values for m and b with the independent variables "m" and "b".

If we do this for *either* of the two tables, we get Table 4 .

Table 4

Actual x	**Actual y**	**Predicted y**	**Error**	**Error^2**
x_{data}	y_{data}	$y_{model} = f(x_{data}) = mx_{data}+b$	$y_{data} - y_{model}$	$(y_{data}-y_{model})^2$
x	y	$\hat{y} = f(x) = mx + b$	$y - \hat{y}$	$(y - \hat{y})^2$
1	3	$m(1) + b$	$3\text{-}(1m + b)$	$(3\text{-}1m \;\; b)^2$
2	8	$m(2) + b$	$8\text{-}(2m + b)$	$(8\text{-}2m - b)^2$
3	11	$m(3) + b$	$11\text{-}(3m + b)$	$(11\text{-}3m - b)^2$
4	10	$m(4) + b$	$10\text{-}(4m + b)$	$(10\text{-}4m - b)^2$

Once we have replaced the numerical values for m and the b with the variable letters, the tables are the same. So we now have a model for the SSE for this example:

Verbal Definition: $SSE(m,b)$ = the Sum of the Squared Errors (SSE), for a linear model with slope m and y-intercept b and the data points: (1,3), (2,8), (3,11), and (4,10)

Symbol Definition: $SSE(m,b) = (3 - m - b)^2 + (8 - 2m - b)^2 + (11 - 3m - b)^2 + (10 - 4m - b)^2$, for m and b any real numbers.

Assumptions: Certainty and divisibility. Certainty implies that the relationship is exact. Divisibility implies that any fractional values of the variables are possible.

The formula could be multiplied out and simplified by combining like terms, but this does not suit our purpose right now. We will study this further in Chapter 8. □

SSE(m,b) is a quadratic function of two variables. A quadratic function is one in which the exponents are all whole numbers and the sums of the exponents of the variables in any one term is never more than 2, and equals 2 for at least one nonzero term. The graph of *this* quadratic function is shaped sort of like a bowl (or more like a hammock), as shown in Figure 5.5-4.

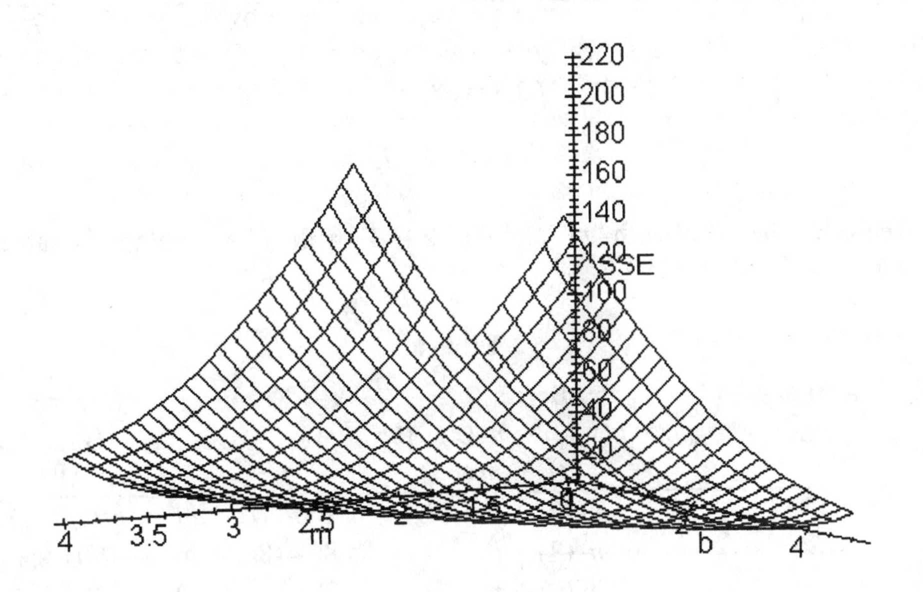

Figure 5.5-4

Our two model examples provide us with two points of this function:

SSE(2,3) = 10
SSE(2.4,2) = 9.2

General Model for the SSE for Single Variable Linear Functions

Sample Problem 2: Find a general model for SSE as a function of m and b for a linear function of the form $y = f(x) = m\,x + b$, based on a given *general* set of data points.

Solution: Let's look at another table with a *different* line fit to a *different* set of data, (ordered pairs of x and y values), to see if we can find the general patterns. We'll use the coffeehouse data from Chapter 1, using the model $y = f(x) = -15x + 203.3$, shown in Table 5 .

Table 5

Actual x	Actual y	Predicted y	Error	Error Squared
x_{data}	y_{data}	$y_{model} = f(x_{data}) = -15x_{data} + 203.3$	$y_{data} - y_{model}$	$(y_{data} - y_{model})^2$
x	y	$\hat{y} = f(x) = -15x + 203.3$	$y - \hat{y}$	$(y - \hat{y})^2$
7	100	$-15(7) + 203.3 = 98.3$	$100-98.3 = 1.7$	$1.7^2 = 2.89$
8	80	$-15(8) + 203.3 = 83.3$	$80-83.3 = -3.3$	$(-3.3)^2 = 10.89$
9	70	$-15(9) + 203.3 = 68.3$	$70-68.3 = 1.7$	$1.7^2 = 2.89$
		Sums ➔	0.1	16.67

Replacing the "-15" with "*m*" and the "203.3" with "*b*" we obtain the table shown in Table 6 .

Table 6

Actual x	Actual y	Predicted y	Error	Error Squared
x_{data}	y_{data}	$y_{model} = f(x_{data}) = m \cdot (x_{data}) + b$	$y_{data} - y_{model}$	$(y_{data} - y_{model})^2$
x	y	$\hat{y} = f(x) = mx + b$	$y - \hat{y}$	$(y - \hat{y})^2$
7	100	$m \cdot (7) + b$	$100-(7m + b)$	$(100 - 7m - b)^2$
8	80	$m \cdot (8) + b$	$80 - (8m + b)$	$(80 - 8m - b)^2$
9	70	$m \cdot (9) + b$	$70 - (9m + b)$	$(70 - 9m - b)^2$

This looks like our previous table except for the different values for the actual x and y data and the fact that there were only three ordered pairs of data points. We are *part* of the way to a general model for the SSE. Now we have to find a way to generalize the x and y data values. We'll use the same kind of notation we used in Sample Problem 3 of Section 5.2, where we had several different products: we'll call the first x data value x_1, the second x data value x_2, the third x data value x_3, and so on. We'll do the same things with the y values: the first y data value is y_1, the second y data value is y_2, and so on. The ordered pairs of data thus are: (x_1, y_1), (x_2, y_2), (x_3, y_3) ...(x_n, y_n). Let's replace the values in the Table 4 from the first example above using this notation, shown in Table 7 .

Table 5.5-7

Actual x	**Actual y**	**Predicted y**	**Error**	**Error Squared**
x_{data}	y_{data}	$y_{model} = m \cdot (x_{data}) + b$	$y_{data} - y_{model}$	$(y_{data} - y_{model})^2$
x	y	$\hat{y} = f(x) = mx + b$	$y - \hat{y}$	$(y - \hat{y})^2$
x_1	y_1	$m \cdot (x_1) + b$	$y_1 - (mx_1 + b)$	$(y_1 - mx_1 - b)^2$
x_2	y_2	$m \cdot (x_2) + b$	$y_2 - (mx_2 + b)$	$(y_2 - mx_2 - b)^2$
x_3	y_3	$m \cdot (x_3) + b$	$y_3 - (mx_3 + b)$	$(y_3 - mx_3 - b)^2$
x_4	y_4	$m \cdot (x_4) + b$	$y_4 - (mx_4 + b)$	$(y_4 - mx_4 - b)^2$

A quick look at Tables 2 and 3 should convince you that replacing the actual data figures with the x_1, y_1, x_2, y_2, etc. and the actual slope and intercept with "m" and "b" will result in exactly the table we have above. In other words, the table is a *general form* for the least squares calculations for a linear model with ordered pairs of data. To make this table truly general we just have to indicate that we are not limited to 3 or 4 data points: we want to have any number of data points. If we define n to be the number of data points, where n is any positive integer, then we can denote the n data points (x_1, y_1), (x_2, y_2), …, (x_n, y_n) . Our general table would now look like Table 8 .

Table 5.5-8

Actual x	**Actual y**	**Predicted y**	**Error**	**Error Squared**
x_{data}	y_{data}	$y_{model} = m \cdot (x_{data}) + b$	$y_{data} - y_{model}$	$(y_{data} - y_{model})^2$
x	y	$\hat{y} = f(x) = mx + b$	$y - \hat{y}$	$(y - \hat{y})^2$
x_1	y_1	$m \cdot (x_1) + b$	$y_1 - (mx_1 + b)$	$(y_1 - mx_1 - b)^2$
x_2	y_2	$m \cdot (x_2) + b$	$y_2 - (mx_2 + b)$	$(y_2 - mx_2 - b)^2$
$\vdots$	$\vdots$	$\vdots$	$\vdots$	$\vdots$
x_n	y_n	$m \cdot (x_n) + b$	$y_n - (mx_n + b)$	$(y_n - mx_n - b)^2$

Note: the "$\vdots$ " in the third row of the table indicates that the pattern continues and some of the entries have been omitted. We can now give a general model for the sum of the squares of the errors for a linear model and n data points:

Verbal Definition: $S(m,b)$ = the sum of the squares of the errors between the y data values and the y model values for a linear regression model $y = mx + b$, based on the data points $(x_1, y_1), (x_2, y_2), \ldots, (x_n, y_n)$.

Symbol Definition: $S(m,b) = (y_1 - mx_1 - b)^2 + (y_2 - mx_2 - b)^2 + \ldots + (y_n - mx_n - b)^2$, for m and b any real numbers.

Assumptions: Certainty and divisibility. Certainty implies that the relationship is exact. Divisibility implies that the sum and the parameters can take on any fractional values.

In Section 4.2 we learned how to use **sigma notation** to indicate a sum. We can greatly simplify our functional definition above by using this notation. We will first replace the actual subscripts of the x and y values with the usual i for the index variable. We want to add the squares of the differences as the index i varies from 1 to n:

$$\sum_{i=1}^{n} (y_i - mx_i - b)^2 = (y_1 - mx_1 - b)^2 + (y_2 - mx_2 -)^2 + \ldots + (y_n - mx_n - b)^2$$

We can now write our model:

Verbal Definition: $SSE(m,b)$ = the sum of the squares of the errors between the y data values and the y model values for a linear model $y = f(x) = mx + b$ based on the data points $(x_1, y_1), (x_2, y_2), \ldots, (x_n, y_n)$.

Symbol Definition: $SSE(m,b) = \sum_{i=1}^{n} (y_i - mx_i - b)^2$ for m and b any real numbers.

Assumptions: Certainty and divisibility. Certainty implies that the relationship is exact. Divisibility implies that the sum and the parameters can take on any fractional values. □

A General Model for the SSE for Single-Variable Quadratic Functions

Sample Problem 3: Find an expression for the SSE as a function of the parameters of a single-variable *quadratic* model $y = ax^2 + bx + c$ based on the data points $(x_1, y_1), (x_2, y_2), \ldots, (x_n, y_n)$.

Solution: The only difference from the linear model case is that now the predicted y value is given by $ax_i^2 + bx_i + c$ instead of $mx_i + b$. Thus our model is given by:

Verbal Definition: SSE(a,b,c) = the sum of the squares of the errors between the y data values and the y model values for a quadratic model ($y = ax^2 + bx + c$) based on the data points $(x_1,y_1),(x_2,y_2),\ \dots\ ,(x_n,y_n)$

Symbol Definition: SSE(a,b,c) $=\sum_{i=1}^{n}(y_i - ax_i^{\,2} - bx_i - c)^2$ for a,b, and c any real numbers

Assumptions: Certainty and divisibility. Certainty implies that the relationship is exact. Divisibility implies that the sum and the parameters can take on any fractional values. □

For the basketball practice example in Sample Problem 1, the function would be

$$SSE(a\ ,b,c) = (3\text{-}a(1)^2\text{-}b(1)\text{-}c)^2\text{+}(8\text{-}a(2)^2\text{-}b(2)\text{-}c)^2\text{+}(10\text{-}a(3)^2\text{-}b(3)\text{-}c)^2\text{+}(11\text{-}a(4)^2\text{-}b(4)\text{-}c)^2$$
$$= (3 - a - b - c)^2 + (8 - 4a - 2b - c)^2 + (10 - 9a - 3b - c)^2 + (11 - 16a - 4b\text{-}\ c)^2$$

General Model for SSE for Any Single Variable Function

We can easily generalize this to *any* single-variable model $y = f(x)$, in which case we get

Verbal Definition: SSE(model parameters) = the sum of the squares of the differences between the y data values and the y model values for a general single-variable model of the form $y = f(x)$ based on the data points $(x_1,y_1),\ (x_2,y_2),\ \dots,(x_n,y_n)$

Symbol Definition: *SSE*(model parameters) = $\sum_{i=1}^{n}\left[y_i - f(x_i)\right]^2$.

Assumptions: Certainty and divisibility. Certainty implies that the relationship is exact. Divisibility implies that the sum and the parameters can take on any fractional values.

We can further abbreviate this model by using the notation $\hat{y}$ for the value predicated by the model. The actual data value is indicated by a plain y. If we wish to refer to a general value, we use y_i for the actual data value and $\hat{y}_i$ for the predicted value.

Thus, our general symbol definition could be written: $SSE = \sum_{i=1}^{n}(y_i - \hat{y}_i)^2$

We might have chosen an exponential function of the form $y = f(x) = ab^x$ for the coffeehouse data of Sample Problem 2. In this case the SSE function would be

$$SSE(a,b) = (100 - ab^{(7)})^2 + (80 - ab^{(8)})^2 + (70 - ab^{(9)})^2$$
$$= (100 - ab^7)^2 + (80 - ab^8)^2 + (70 - ab^9)^2$$

The exact same idea can be generalized to functions of several variables.[1] We will talk about this in Chapter 6 when we look at regression in even more detail. For example, if we wanted to fit a simple linear function of two variables of the form

$$z = f(x,y) = ax + by + c$$

and our data points were given by $(x_1, y_1, z_1), (x_2, y_2, z_2), \ldots, (x_n, y_n, z_n)$, then the SSE expression would be given by

$$SSE(a,b,c) = \sum_{i=1}^{n}\{z_i - [f(x_i, y_i)]\}^2 = \sum_{i=1}^{n}\{z_i - [a(x_i) + b(y_i) + c]\}^2$$
$$= \sum_{i=1}^{n}(z_i - ax_i - by_i - c)^2$$

If we wanted to fit the unique plane to fit the three points (1,2,3), (4,5,6), and (7,8,9), for example, the expression for the SSE would be given by

$$SSE(a,b,c) = (3-a(1)-b(2))^2 + (6-a(4)-b(5))^2 + (9-a(7)-b(8))^2$$
$$= (3 - a - 2b - c)^2 + (6 - 4a - 5b - c)^2 + (9 - 7a - 8b - c)^2$$

Section Summary

Before you proceed to the exercises, you should

- Understand the concept of an **error** associated with a data point in relation to a model. This is the **vertical distance from the model to the data point**, or the difference that results from calculating the **actual** data value of the dependent variable **minus** the dependent variable value **predicted** by the model for the given value(s) of the independent variable(s)

---

[1] For multivariable functions $y = f(x)$ with m independent variables, if the input values of the n data points are denoted $x^{(1)}, x^{(2)}, \ldots, x^{(n)}$, where each $x^{(i)}$ is an ordered list of m values, the general expression for SSE is

$$SSE(\text{model parameters}) = \sum_{i=1}^{n}(y_i - \hat{y}_i)^2 = \sum_{i=1}^{n}\left[y_i - f\left(x^{(i)}\right)\right]^2 .$$

- Know how to calculate the errors and the Sum of the Squared Errors (SSE) for a given specific model and a given specific set of data points, using

  $SSE = \sum_{i=1}^{n}(y_i - \hat{y}_i)^2$, where y_i is the actual value and $\hat{y}_i$ is the predicted value

- Know that the SSE for a single-variable *linear* model, $y = f(x) = mx+b$, as a function of its parameters, given the general data values (x_1,y_1), (x_2,y_2), ...,(x_n,y_n), is $SSE(m,b) = \sum_{i=1}^{n} (y_i - mx_i - b)^2$.

- Know that the SSE for a single-variable *quadratic* model, $y = f(x) = ax^2+bx+c$, as a function of its parameters, given the general data values (x_1,y_1), (x_2,y_2), ...,(x_n,y_n), is $SSE(a,b,c) = \sum_{i=1}^{n} (y_i - ax_i^2 - bx_i - c)^2$.

- Understand that, in general, the SSE for a single-variable *general* model, $f(x)$ as a function of its parameters, given the general data values (x_1,y_1), (x_2,y_2), ...,(x_n,y_n), is given by $SSE = \sum_{i=1}^{n}[y_i - f(x_i)]^2$.

- Understand that, in general, the SSE for a two-variable model, $z = f(x,y)$ as a function of its parameters, given the general data values (x_1,y_1,z_1), (x_2,y_2,z_2), ...,(x_n,y_n,z_n), is given by $SSE = \sum_{i=1}^{n}[z_i - f(x_i,y_i)]^2$.

EXERCISES FOR SECTION 5.5:

Warm Up

1. Consider the following data:

x	y
0	2
1	5
2	6
3	9
4	8

a) Plot the data points and *visually* estimate a linear model with whole number coefficients that would fit it reasonably well.
b) *By hand*, use your model in (a) to calculate the *predicted* y value for each of the x data values given in the table (show your work).
c) *By hand*, calculate the error and the squared error for each data point with respect to your linear model in (a), and then find the SSE (show your work).
d) Write out an expression for the SSE for these data points and a linear model of the form $y = f(x) = mx + b$, $SSE(m,b)$.
e) Simplify your expression for $SSE(m,b)$ in (d), and plug in the coefficients (the m and b values) from your answer to (a) to verify your answer for the SSE in (c).
f) Write an expression for $SSE(a,b,c)$ for a quadratic model in the form $y = f(x) = ax^2 + bx + c$ for the above data.

2. Consider the following data:

x	y
2	9
4	6
6	4
8	5

a) Plot the data points and *visually* estimate a linear model with whole number coefficients that would fit it reasonably well.
b) *By hand*, use your model in (a) to calculate the *predicted* y value for each of the x data values given in the table (show your work).
c) *By hand*, calculate the error and the squared error for each data point with respect to your linear model in (a), and then find the SSE (show your work).
d) Write out an expression for the SSE for these data points and a linear model in the form $y = f(x) = mx + b$, $SSE(m,b)$.

e) Simplify your expression for *SSE(m,b)* in (d), and plug in the coefficients (the m and b values) from your answer to (a) to verify your answer for the SSE in (c).

f) Write an expression for *SSE(a,b,c)* for a quadratic model in the form $y = f(x) = ax^2 + bx + c$ for the above data.

3. Consider the following data:

x	y
0	2
1	5
2	6
3	9
4	8

a) Use technology to find the least-squares best-fit linear model for the above data.

b) Use technology to create the table of calculations of the predicted y value, error, and squared error for each data point using the least-squares linear model in (a), and then calculate the SSE. Copy your table and final answer.

c) If you did Exercise 1, plug in the coefficients of your answer to 3(a) into your SSE function from 1(d) or 1(e) to verify your SSE value in 3(b).

d) If you did Exercise 1, which answer is lower: 1(c) or 3(b)? (If you did not do Exercise 1, check the sample solution in the back of the book.) Can you find *any* combination of values of m and b that achieves a lower SSE than the better of these two?

e) Suppose you also have data for a third variable, z, and the values are: 10,12,13,16,18 (in the same order as in the table above). For the linear model of the form $z = f(x,y) = ax + by + c$, write out the expression for *SSE(a,b,c)*.

4. Consider the following data:

x	y
2	9
4	6
6	4
8	5

a) Use technology to find the least-squares best-fit linear model for the above data.
b) Use technology to create the table of calculations of the predicted y value, error, and squared error for each data point using the least-squares linear model in (a), and then calculate the SSE. Copy your table and final answer.
c) If you did Exercise 2, plug in the coefficients of your answer to 4(a) into your SSE function from 2(d) or 2(e) to verify your SSE value in 4(b).
d) If you did Exercise 2, which answer is lower: 2(c) or 4(b)? Can you find *any* combination of values of m and b that achieves a lower SSE than the better of these two?
e) Suppose you also have data for a third variable, z, and the values are: 2,3,6,8 (in the same order as in the table above). For the linear model of the form $z = f(x,y) = ax + by + c$, write out the expression for $SSE(a,b,c)$.

Game Time

5. In Chapter 1, you worked with the following demand data for T shirt sales, based on a market research poll:

Price (in $)	8	10	12	15
Number Sold	302	231	175	123

a) Plot the data points and *visually* estimate a linear model with whole number coefficients that would fit it reasonably well.
b) *By hand*, use your model in (a) to calculate the *predicted* y value for each of the x data values given in the table (show your work).
c) *By hand*, calculate the error and the squared error for each data point with respect to your linear model in (a), and then find the SSE (show your work).
d) Write out an expression for the SSE for these data points and a linear model $y = f(x) = mx + b$, $SSE(m,b)$.
e) Simplify your expression for $SSE(m,b)$ in (f), and plug in the coefficients (the m and b values) from your answer to (a) to verify your answer for the SSE in (c).
f) Write an expression for $SSE(a,b,c)$ for a quadratic model for the above data in the form $y = f(x) = ax^2 + bx + c$

6. You are trying to determine the best amount of time to spend on studying in one day, with an average workload, to balance your desire to learn and get good grades with the rest of your life (sleeping, eating, social life, extracurricular activities, etc.). You devise a scale to measure the balance for a particular day, from 0 (the worst possible balance between studying and everything else) to 100 (perfect balance). For the first five school nights, you obtain the following data:

Study Hours	2	5	4	6	3
Balance (0-100)	45	75	70	70	55

a) Plot the data points and *visually* estimate a linear model with whole number coefficients that would fit it reasonably well.
b) *By hand*, use your model in (a) to calculate the *predicted* y value for each of the x data values given in the table (show your work).
c) *By hand*, calculate the error and the squared error for each data point with respect to your linear model in (a), and then find the SSE (show your work).
d) Write out an expression for the SSE for these data points and a linear model of the form $y = f(x) = mx + b$, $SSE(m,b)$.
e) Simplify your expression for $SSE(m,b)$ in (d), and plug in the coefficients (the m and b values) from your answer to (a) to verify your answer for the SSE in (c).
f) Write an expression for $SSE(a,b,c)$ for a quadratic model for the above data in the form $y = f(x) = ax^2 + bx + c$

7. In Chapter 1, you worked with the following demand data for T shirt sales, based on a market research poll:

Price (in $)	8	10	12	15
Number Sold	302	231	175	123

a) Use technology to find the least-squares best-fit linear model for the above data.
b) Use technology to create the table of calculations of the predicted y value, error, and squared error for each data point using the least-squares linear model in (a), and then calculate the SSE. Copy your table and final answer.
c) If you did Exercise 5, plug in the coefficients of your answer to 7(a) into your SSE function from 5(d) or 1(e) to verify your SSE value in 7(b).
d) If you did Exercise 5, which answer is lower: 5(c) or 7(b)? (If you did not do Exercise 5, check the sample solution in the back of the book.) Can you find *any* combination of values of m and b that achieves a lower SSE than the better of these two?
e) Write an expression for $SSE(a,b)$ for an exponential model for the above data in the form $y = f(x) = ab^x$.

8. You are trying to determine the best amount of time to spend on studying in one day, with an average workload, to balance your desire to learn and get good grades with the rest of your life (sleeping, eating, social life, extracurricular activities, etc.). You devise a scale to measure the balance for a particular day, from 0 (the worst possible balance between studying and everything else) to 100 (perfect balance). For the first five school nights, you obtain the following data:

Study Hours	2	5	4	6	3
Balance (0-100)	45	75	70	70	55

a) Use technology to find the least-squares best-fit linear model for the above data.

b) Use technology to create the table of calculations of the predicted y value, error, and squared error for each data point using the least-squares linear model in (a), and then calculate the SSE. Copy your table and final answer.

c) If you did Exercise 6, plug in the coefficients of your answer to 8(a) into your SSE function from 6(d) or 6(e) to verify your SSE value in 8(b).

d) If you did Exercise 6, which answer is lower: 6(c) or 8(b)? Can you find *any* combination of values of m and b that achieves a lower SSE than the better of these two? If you didn't do Exercise 6, which answer do you think *should* be lower?

e) Write an expression for $SSE(a,b)$ for an exponential model for the above data in the form $y = f(x) = ab^x$.

Chapter 5 Summary

Multivariable Functions and Models

A **multivariable function** is a set of ordered lists of the form $(x_1, x_2, \ldots, x_n, y)$, where n is 2 or more (a **multivariable relation**), for which each *different* list of the form $(x_1, x_2, \ldots, x_n)$ is paired with *exactly one* value of y. If this is the case, the x_i variables are called the **independent variables (inputs** or **decision variables)** and y is called the **dependent variable (output)**. The set of all possible sequences of values for the independent variables is called the **domain** of the function, and is often specified by equality and inequality **constraints**. If $n = 2$, and the values of the variables are numbers, a relation can be plotted using a **3-D graph**. The **Vertical Line Test** also holds **for 3-D graphs**: if *every* vertical line hits the graph in *no more than 1* point, then the graph *does* represent a function; if there exists at least one vertical line that hits the graph in *2 or more* points, then the graph *does not* represent a function.

Analogous to single-variable models, a **multivariable model** involves a **verbal definition** of a function (clearly stated and **unambiguous**, including **units** for the function value and all of the input values), a **symbol definition** of the function (including its **domain**), and a statement of the **assumptions** and their implications. As before, our standard assumptions are **divisibility** of all of the variables (they can take on *any* real number value satisfying the constraint equations and inequalities) and **certainty** of the relationship (the symbol definition defines the relationship *exactly*). Multivariable functions in *general* do not *have* to assume divisibility of the variables (they could have restrictions to integer values, for example), but such exceptions are not our focus in this text.

Formulation of Multivariable Models from Verbal Descriptions

When trying to formulate a multivariate model from a verbal description, first identify, bound, and clarify your problem. Next, use the given information to see if you can **define the decision variables**, and then try to formulate a model for the **objective function** first by directly translating verbal statements such as:

Verbal Expression	Mathematical Symbol
"*a* is no more than *b*"	$a \le b$
"*a* is at most *b*"	$a \le b$
"*a* is no less than *b*"	$a \ge b$
"*a* is at least *b*"	$a \ge b$
"*a* is exactly *b*"	$a = b$

Sometimes formulations also involve standard formulas (such as areas and volumes, the Pythagorean Theorem, revenue equals selling price times quantity sold, profit equals revenue minus cost, etc.), and you need to be able to recognize these situations.

The number of decision variables in a formulation can sometimes be reduced by solving an *equation* that involves only decision variables: solving for one of them in terms of the others, and then substituting the resulting expression in for that one variable everywhere else in the formulation (including in the constraints).

If you are having difficulty getting a full formulation, try **plugging in one or more sets of specific numerical values for the decision variables** and calculating the results. It is often easier to work with concrete numbers than abstractions. Write out the calculations without simplifying and try to generalize the patterns into formulas involving the variables to get the objective function and/or the constraints. Use values that are different from each other and from the parameters in the problem to make the patterns most obvious.

Simple and Compound Interest

Simple interest accrues interest only on the *principal*, *not* on the accumulated *interest*. Calculations involve the formula

$$S(P,i,n) = P(1+in)$$

where S is the total accumulated money, P is the original principal, i is the simple interest rate per period, expressed as a pure decimal, and n is the number of periods (a whole number), where the accumulated money is in the same money units as the principal. **To be fully valid, the number of periods, n, must be a whole number, if the interest is accrued only at the end of each period.**

Compound interest involves getting interest on the *accumulating interest as well as* on the original *principal*. You can calculate your total monetary accumulation (as well as just the interest portion by itself) for any investment involving **discrete compound interest** by using the following **basic formula** for accumulated money:

$$A(P,i,n) = P(1+i)^n,$$

where P is the principal, i is the actual interest rate per period, n is the number of periods, and the accumulated money is in the same money units as the principal. **To be fully valid, the number of periods, n, must be a whole number.**

Often, **discrete compound interest** is expressed using the **Annual Percentage Rate (APR)**, or **nominal annual interest rate**, and the number of compounding periods per year, instead of directly giving the interest rate per compounding period. The APR, designated r, is an artificial number used to find the actual interest rate per compounding period, by dividing the APR by the number of compounding periods per year, m (i = APR/m = r/m). Calculations are then based on the following formula for accumulated money:

$$A(P,r,m,t) = P\left(1+\frac{r}{m}\right)^{mt},$$

where P is the principal, r is the APR, m is the number of compounding periods per year, and t is the number of years, where the accumulated money is in the same money units as the principal. (Note: to get this formula, substitute $i = \frac{r}{m}$ and $n = mt$ into the basic formula above.) **To be fully valid, $n = mt$ must be a whole number**, corresponding to a whole number of compounding periods).

The idea of ***continuously*** **compounded interest** is to let the number of compounding periods per year get arbitrarily large (take the limit as m goes to infinity), which leads to the following formula for the total accumulated money after t years:

$$F(P,r,t) = \lim_{m\to\infty} P\left(1+\frac{r}{m}\right)^{mt} = Pe^{rt},$$

where P is the principal, r is the APR, and t is the number of years, and where the accumulated money (future value) is in the same money units as the principal. Euler's number (e) is defined mathematically by $e = \lim_{m\to\infty}\left(1+\frac{1}{m}\right)^{m} \approx 2.7183$, and can be thought of as the precise accumulated money value (in dollars) of \$1 after exactly 1 year, with interest compounded continuously at an APR of 100%

The **Annual Percentage Yield** (**APY**), also called the **yield** or the **effective annual interest rate**, is the *actual* percentage return on your money after 1 year (in pure decimal form), or, in everyday terms, the numerical value of the interest (in dollars) on \$1 after 1 year. The APY can be calculated, given the APR and compounding structure, using one of the following two formulas:

Discrete Compounding:

$$y = \left(1 + \frac{r}{m}\right)^m - 1 \ ,$$

where y is the yield (APY), r is the APR, and m is the number of compounding periods per year. **This is fully valid only if m is a whole number.**

Continuous Compounding:

$$y = e^r - 1 \ ,$$

where y is the yield (APY) and r is the APR.

The Time Value of Money: Future Value and Present Value

The **time value of money** refers to the concept which says that the same amount of money at different points in time is worth different amounts *now* (or at any other given time). This is because money at an earlier time can be *invested* to accumulate to a higher amount at a later time. The interest rate used is usually your best-known safe way to invest the money under consideration, *not* the inflation rate. The **Future Value** at a given time of a cash transaction at *another* given (earlier) point in time, with a particular interest rate and compounding structure, is the financially equivalent dollar value of the earlier transaction (how much it is worth) at the later time. Put differently, it is the *value* to which the transaction amount at the *earlier* time would *accumulate* as of the *later* time, at that interest rate and structure. Thus, such a calculation is just a direct application of the compound interest formulas, where the time variable is the *difference* between the earlier time point and the later time point (we could express this in symbols as $t = t_2 - t_1$). Thus the Future Value, F, can be calculated as follows:

Discrete Compounding: $F = P\left(1 + \frac{r}{m}\right)^{mt} = P(1+y)^t = Pb^t = P(1+i)^n$

(fully valid only if mt is a whole number)

Continuous Compounding: $F = Pe^{rt} = P(1+y)^t = Pb^t$

where:

P is the transaction amount (face value, at the time it occurs),

t is the number of years into the future the transaction is being compounded (the time point at which the future value is desired minus the time point of the transaction),

r is the APR,

m is the number of compounding periods per year,

y is the yield,

b is the compounding factor (ratio), equal to $(1+y)$,

i is the actual interest rate per compounding period (equal to r/m), and

n is the number of compounding periods over which the transaction is being compounded (equal to mt),

and where the money units of the future value are the same as the money units of P.

The **Present Value** of a cash transaction that occurs at a given point in time, given a particular interest rate and compounding structure, is the financially equivalent amount of cash (how much it is worth) at the *present* time which, if invested at the given interest rate and compounding structure, would *accumulate* to the amount of the transaction at the time it occurs (later). This can be calculated simply by solving the Future Value formulas for the principal P (the transaction amount at the earlier time). Thus, the Present Value, PV, of a cash transaction of F which occurs t years from the present is given by:

Discrete Compounding: $$PV = P = F\left(1+\frac{r}{m}\right)^{-mt} = F(1+y)^{-t} = \frac{F}{(1+y)^t}$$

$$= F\,b^{-t} = F\alpha^{t} = \frac{F}{(1+i)^n} = F(1+i)^{-n}$$

(fully valid only if *mt* is a whole number)

Continuous Compounding: $PV = P = Fe^{-rt} = F(1+y)^{-t} = F\,b^{-t} = F\alpha^{t}$

where:

F is the transaction amount (face value, at the time it occurs in the future),

t is the number of years from the present that the transaction occurs,

r is the APR,

m is the number of compounding periods per year,

y is the yield,

b is the compounding factor (ratio), equal to $(1+y)$,

α is the discount factor (the reciprocal of b, the compounding factor/ratio),

i is the actual interest rate per compounding period (equal to r/m), and

n is the number of compounding periods over which the transaction is being compounded (equal to mt),

and where the money units of the present value are the same as the money units of F.

Cash Flows, Net Present Value (NPV), and Internal Rate of Return (IRR)

A **cash flow** is a sequence of positive and negative cash transactions (revenues and expenses, or profits and losses), and can be represented schematically using a timeline. The **Net Present Value**, or **NPV**, of a cash flow at a given interest rate and compounding structure is the sum of the present values of the individual transactions (including negative signs for expenditures or losses). The NPV corresponds to the amount of cash at the *present* time that is financially *equivalent to* the cash flow (how much the cash flow is *worth now*). This means that *doing* the transactions *out of* an account with the given interest rate and compounding structure would yield the *same* balance after the cash flow was over as a single transaction *now* in the amount of the NPV using the account would end up with at that same later time. The idea of Net Future Value (NFV) is analogous (though much less commonly used), but is calculated at some future point in time, t, and corresponds to the future value at that future time of the NPV. It can be thought of as how much the cash flow is *worth* (what it is financially *equivalent to*) at that future time.

$$NPV = \frac{F_1}{(1+y)^{t_1}} + \frac{F_2}{(1+y)^{t_2}} + \cdots + \frac{F_n}{(1+y)^{t_n}} = \sum_{i=1}^{n} \frac{F_i}{(1+y)^{t_i}}$$

$$NFV(t) = (NPV)(1+y)^t$$

where the amount of the i'th transaction of the cash flow is F_i and occurs t_i years from the present. Similar formulas could also be obtained using the other forms of the formulas for Present Value. **With discrete compounding, the formulas are only fully valid when the exponents are whole number multiples of the compounding period.**

The **Internal Rate of Return (IRR)** of a cash flow or investment opportunity is its *effective* annual interest rate (APY) or *equivalent* rate of return (the interest rate on a special account that would leave you with the same total money at the end of the time period whether you just left your money in the account or used it for all of the cash flow transactions, and so would make the NPV = 0). To calculate the IRR, formulate an expression for the NPV in terms of the APY (or possibly the APR, but the APY is more general), set it equal to 0, and solve for the APY (usually using technology).

An **annuity** is a financial structure involving payments at equally-spaced time intervals. If the payments are at the *end* of each period, the annuity is called an **ordinary annuity**. If the payments are at the *beginning* of each period, the annuity is called an **annuity due**. Standard car loans and house mortgages are common examples. Payment amounts can be calculated if you know the number and timing of the payments and the interest rate and compounding structure, by setting the NPV (using the interest rate of the loan) equal to 0 and solving for the payment amount.

Expressing SSE in Terms of Model Parameters

To find a best-fit model of a certain type for a certain set of data, we measure the error for each data point as the vertical distance from the model to the data point. This error corresponds to the *actual* output (y) minus the *predicted* output ($\hat{y}$,obtained by plugging the input value into the *general* formula) for that data point. To get an overall measure of the error for a particular model and the data set, we square these errors and add them to get the Sum of the Squared Errors (SSE). For single-variable functions, if the data points are $(x_1, y_1), (x_2, y_2), \ldots, (x_n, y_n)$, then the expressions for the SSE in terms of the parameters for different model categories are

Linear: $SSE(m,b) = \sum_{i=1}^{n} (y_i - mx_i - b)^2$

Quadratic: $\text{SSE}(a,b,c) = \sum_{i=1}^{n} (y_i - ax_i^2 - bx_i - c)^2$

General ($y = f(x)$): $SSE(\text{model parameters}) = \sum_{i=1}^{n} (y_i - \hat{y}_i)^2 = \sum_{i=1}^{n} [y_i - f(x_i)]^2$.

For a multivariable linear function of the form $z = f(x,y) = ax + by + c$, the expression for SSE would be

$$SSE(a,b,c) = \sum_{i=1}^{n} \{z_i - [f(x_i, y_i)]\}^2 = \sum_{i=1}^{n} \{z_i - [a(x_i) + b(y_i) + c]\}^2$$
$$= \sum_{i=1}^{n} (z_i - ax_i - by_i - c)^2$$

Chapter 6: Multivariable Models from Data - Regression and Statistics

In Chapter 5, we discussed formulating multivariable models from verbal descriptions, including special cases involving interest, the time value of money (such as Net Present Value), and expressing the Sum of the Squared Errors between a model from a particular category and a set of data as a function of the parameters of that category of model. The purpose of formulating such a model of the SSE in terms of its parameters is to then find the set of parameters that will *minimize* the SSE; this is what we mean by a **best-fit least-squares regression model**. This is part of the process of **formulating a model from raw data**. In Chapter 8, we discuss the calculus involved in performing this minimization of the SSE, so you could do it by hand yourself for simple examples if you wanted. However, in practice, since the calculations are very complicated, such model-fitting is normally done using technology (as we did for single-variable functions in Volume 1). For multivariable models, at the present time the technology of choice for everyday multivariable regression is usually a **spreadsheet**, so we focus on that technology. In Section 6.1 we discuss **using spreadsheets to find best-fit linear and nonlinear multivariable regression models**, including discussions of linear and quadratic models and pointers on how to select a model. Since some analysis using regression models involves statistics, we then continue our discussion from Chapter 4 about probability distributions and talk about how to find the **mean** (average) of a set of numbers or a continuous probability distribution in Section 6.2 . We also discuss how the mean differs from the median and the mode as a measure of central tendency, and how the choice of which statistic to use can be manipulated for different purposes. We follow this immediately by a discussion of the **variance** (a measure of variability, and one way of measuring the *riskiness* of an investment) of a set of numbers or a continuous distribution and the related concept of the **standard deviation**, each of which can be computed for a **population** as a whole or for a **sample** from a larger population. As part of this development, we discuss the concept of **Mean Squared Error (MSE)**, related to the SSE, and how it can be used to evaluate forecasting methods. In Section 6.3, we then relate these statistical concepts to some of the output information normally obtained from a regression analysis, including the **coefficient of determination (R^2)** and the **standard error** of a model parameter, to help interpret this summary output and use it in model formulation. We also discuss the basic **assumptions of regression analysis** and some inherent **dangers and potential misuses of regression analysis**, to help you when performing analyses yourself and when trying to understand other people's analyses. Finally, in optional Section 6.4, we talk about how the concepts of mean and variance are used in analyzing **investment portfolios**, and define **dominance** of investments and **efficient sets** as part of analyzing **risk-return tradeoffs**.

Here are some examples of the kinds of problems that this chapter will help you solve:

- You run a small furniture manufacturing business. You have been keeping records about the costs associated with making different combinations of chairs and tables. You need to be able to calculate what your costs will be in the future. Find a model for your cost as a function of the number of chairs and tables that you make.
- You believe that your energy level depends not only on the amount of exercise that you get, but also on the amount of sleep that you get. Find a model that expresses your energy level as a function of both exercise and sleep.
- A business competitor is about to come out with a new product that is likely to have a significant impact on your business. You have estimated the probability density function for when the new product will be released. What is the expected (average, mean) time of its release? What is the standard deviation of the release time, to help get a feel for the risks involved?
- Your marketing department has collected data on a health product, and is about to analyze it. Should they use the mean or the median or the mode in presenting the results about the health benefits of the product?
- You are trying to determine the effect of sleep and exercise on your energy level. You gather data and fit a model. What is the margin of error in using this model to predict your energy level for a specific combination of sleep and exercise, both on average and for a particular day?
- Someone has collected data on sales of handguns, the number of death sentences, and the murder rate, on a state-by-state basis, in the U.S. over a certain period of time. When doing a regression of the murder rate on the other two variables, the coefficients on the two variables had a positive sign. The researcher claims that this means that allowing more handgun sales and increasing death sentences in a particular state will increase the murder rate, and therefore both practices should be curtailed. Is this claim justified?
- You are analyzing a number of potential investments. You want to estimate the average (mean) rates of return and the risks (standard deviations) to do a risk/return tradeoff analysis.

By the time you have finished studying this chapter, you should:

- Know the basics of how to use a spreadsheet program, including how to set up and perform multivariable regression analyses.
- Know how to select and use one or more multivariable regression analyses to formulate a multivariable model to help solve a problem, including factors to consider in choosing between potential reasonable models.
- Understand the concepts of the mean, variance, and standard deviation (both for populations and samples) of a set of numbers or a continuous distribution, and how to calculate them by hand and using technology.
- Understand the differences between the mean, the median, and the mode, and how the selection of which to use can be manipulated to serve different purposes.
- Understand the idea behind the **Total Sum of Squares**, the **Sum of the Squares due to the Regression**, and the SSE, including how they are related to each other and how they help define the coefficient of determination (R^2), and how to interpret and use R^2 in selecting a model when formulating a problem.
- Understand how to compare alternative forecasting methods using the idea of the Mean Squared Error, including the idea of **withholding data** for validation of a forecasting model or method.
- Understand what is meant by the **degrees of freedom** of a regression analysis and by the **standard error** of each parameter estimated in a regression, to further help in selecting a model when formulating a problem.
- Understand potential dangers and misuses of regression analysis, know how to avoid them yourself, and know how to detect them in others' analyses.
- Understand the concepts of **dominance**, **efficient sets**, and **Pareto superiority**, and how they can be used as part of **risk-return tradeoff analysis** of potential investments.

Section 6.1: Multivariable Models from Data - Spreadsheets and Regression

In this section we discuss how to formulate multivariable functions, both linear and nonlinear, from raw data. Just as we did with single variable models, we use technology to help us fit functions to the data using least squares regression. Most popular spreadsheet programs can do regression analysis with functions of more than one independent variable. Here are the types of problems that the material in this section will help you solve:

- You run a small furniture manufacturing business. You have been keeping records about the costs associated with making different combinations of chairs and tables. You need to be able to calculate what your costs will be in the future. Find a model for your cost as a function of the number of chairs and tables that you make.
- In the above situation, you have also been keeping records of the relationships between the prices that you charged for the chairs and tables on different occasions and the number of chairs and tables you were able to sell at these prices. In order to price your furniture so that you can maximize your profit you need to know how many chairs and tables will be sold at different prices. Find models for the demands for the chairs and tables and the total revenue and profit from *both* as functions of the prices charged for the chairs and tables.
- You believe that your energy level depends not only on the amount of exercise that you get, but also on the amount of sleep that you get. Find a model that expresses your energy level as a function of both exercise and sleep.

By the time you have finished this section you should:

- Know the basics of how a spreadsheet works, including the setup of individual cells and blocks of cells and how they are named, how to enter data and formulas (using both relative and absolute addresses), and how to copy data and formulas from one place to another.
- Know how to enter data in a spreadsheet to set it up for a multivariable (or single-variable) regression analysis.
- Know how to use a spreadsheet program to perform a multivariable (or single-variable) regression analysis.
- Understand how to find the equation for a model from the output given from the regression.
- Be able to select a model for a particular problem from among several regressions.

- Write out the full model determined by the regression.

The General Form of Multivariable Linear Functions

Sample Problem 1: You run the small furniture manufacturing business discussed in Section 5.2. You have been keeping records about the costs associated with making different combinations of chairs and tables, shown in Table 1. You need to be able to calculate what your costs will be in the future. Find a model for your cost as a function of the number of chairs and tables that you make.

Table 1

	A	B	C	D
1	DATA FOR CHAIRS AND TABLES			
2	COST RECORDS			
3	MONTH	TOTAL	NUMBER	NUMBER
4		COSTS	OF CHAIRS	OF TABLES
5	JAN	$1350	5	3
6	FEB	$1660	8	4
7	MAR	$1480	4	5
8	APR	$1450	5	4
9	MAY	$1460	8	2
10	JUN	$1210	3	3
11	JUL	$1600	10	2
12	AUG	$1940	12	4
13	SEPT	$1750	5	7
14	OCT	$1920	6	8
15	NOV	$1740	12	2
16	DEC	$1080	4	1

Solution: The data in Table 1 appears as it would appear in a spreadsheet. A spreadsheet consists of a grid boxes or **cells**, organized in horizontal rows and vertical columns. The columns are named using letters, and the rows are named using numbers. To identify a particular box or **cell**, the convention is to give the column first, then the row. For example, the entry "$1350" is in cell B5. **To enter text and data on a spreadsheet** like this, simply use the arrow keys to move to the cell you want, and type in the text or number you want in that location. For details about spreadsheets and formatting (such as to make the $ appear in the cost data values, which is not *required* here), see your technology supplement.

We want to find a model for the total costs as a function of the number of chairs and tables that are made. Clearly, the cost is not constant, so the simplest possible model, as in the single-variable case, would be a linear function. The wood and hardware models defined in Section 5.2 were all **multivariable linear functions**:

Verbal Definition: $h(c,t)$ = the number of hardware packages needed to make c chairs and t tables in a month.
Symbol Definition: $h(c,t) = 8c + 4t$, for $c,t \geq 0$
Assumptions: Certainty and divisibility. Certainty implies that the relationship is exact. Divisibility implies that any fractional numbers of chairs, tables, and hardware packages are possible.

Verbal Definition: $w(c,t)$ = the amount of wood, in board feet, needed to make c chairs and t tables in a month.
Symbol Definition: $w(c,t) = 3c + 6t$, for $c,t \geq 0$
Assumptions: Certainty and divisibility. Certainty implies that the relationship is exact. Divisibility implies that any fractional numbers of chairs, tables, and board feet of wood are possible.

Note that in both of these models, each variable only occurred by itself, and with an exponent of 1, just as in single-variable linear functions. But, just as in the single-variable case, it is also possible to have a pure constant term, what we usually think of as the *y*-intercept, the b in $y = mx + b$. For a multivariate linear function we can designate the input variables $x_1, x_2, \ldots x_n$.

The **general form for a multivariate linear function** with input variables $x_1, x_2, \ldots x_n$, is:

$$f(x_1, x_2, \ldots, x_n) = m_1 x_1 + m_2 x_2 + \ldots + m_n x_n + b.$$

You may have seen some multivariable functions expressed in the form $z = f(x,y)$, paralleling the common usage for single variable functions $y = f(x)$. Using the subscript notation for different independent variables avoids confusion about whether y is an input or the output, and allows greater flexibility for generalizing patterns in formulas. We will use both formats to help you get used to them both.

Notice that in our wood and hardware examples above, the constant (b) term was 0, so was not visible. This constant tells you the value of the function when all of the independent variables ($x_1, x_2, \ldots, x_n$) are set to 0. In a function of two variables, expressed as

$$y = f(x_1, x_2) = m_1 x_1 + m_2 x_2 + b,$$

the constant can still be thought of as the *y*-intercept. In this situation, the graph is done in three dimensions. You can think of the x_1 and x_2 axes as sitting flat on your desk, perpendicular to each other, and the y axis then points up vertically. The graph of a linear function is then a flat **plane** extending in all directions, usually slanted at an angle, and the place where this plane intersects the vertical (y) axis is b. Figure 6.1-1 shows an example for the function

$$y = 2x_1 + 5x_2 + 9 .$$

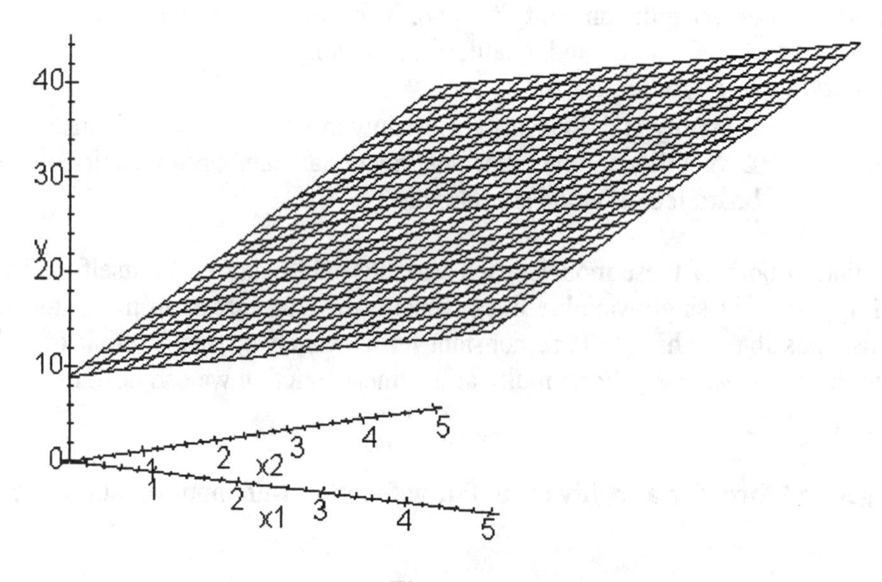

Figure 6.1-1

Notice that the vertical intercept is 9, and that the plane slants more steeply over the x_2 axis than the x_1 axis, since the corresponding slopes are $m_2 = 5$ vs. $m_1 = 2$.

If your problem has only two independent variables and you have graphing software, you can plot the data to see if a flat plane surface would be a reasonable model over your domain of interest. If not, you can try plotting the dependent variable against each of the independent variables individually, and see if each relationship is reasonably close to being linear, at least for a first-cut model.

Fitting Linear Model of Functions of Several Variables

To fit models of functions of several variables, you usually need to use a spreadsheet or statistical computer program, since the current generation of inexpensive graphing calculators cannot yet do this operation easily. The input is quite similar, however. Usually the best format is to set up a spreadsheet of the data, with a column for each variable and each row corresponding to a data point. This is similar to using lists on graphing calculators. It is a good idea to make the first (leftmost) column of this matrix be your dependent variable (such as y above). **In this course, we will only be working with a single dependent variable at a time.** The other columns can go in any order, and it is best to put them alongside the dependent variable column. All columns must be of the same length, and should all have numerical entries everywhere. It is a good idea to label the columns, and possibly the rows as well, if that makes sense. In our example (Table 1) the rows are labeled by the month they correspond to.

Our problem is to find a model for the total cost as a function of the number of chairs and tables made. The problem is well defined. Notice that the data for the costs of chairs and tables given above is already in the form we have just outlined. The dependent variable here is the cost, and that column goes first. It is followed by the columns for the independent variables, one for the number of chairs c, as we defined earlier in place of x_1, and one for the number of tables t, as we defined earlier in place of x_2. If we define

$C(c, t)$ = the expected (average) total cost, in dollars, in a month if c chairs and t tables are made,

then the symbol definition of our linear model will have the form

$$C(c,t) = m_1 c + m_2 t + b\ .$$

Since the data is already in a spreadsheet, to find the coefficients of the variables and the constant term (the parameters) for this model, you simply follow the directions for doing **least-squares multiple linear regression** (see your technology supplement for details). Typically this involves specifying rectangular **blocks (ranges) of cells**. Blocks of cells are named by specifying the cells of two opposite corners (usually the upper left and the lower right), with a symbol such as ".." (two periods) or ":" (a colon) in between. For example, in Table 1, the block of cells corresponding to all of the numerical data (costs, number of chairs, and number of tables) together could be specified as "B5..D16", since the upper left corner is cell B5 and the lower right corner is D16.

In the regression function of your spreadsheet you will normally be prompted to specify the block of cells (which should be just one column) for the dependent variable values (B5..B16 from the given spreadsheet) and then the block corresponding to the independent variables (C5..D16 from the given spreadsheet in Table 1). You can also identify a location for the output. The output you get may look something like Table 2.

Table 2

	A	B	C	D	E	F	G
1	DATA FOR CHAIRS AND TABLES						
2	COST RECORDS						
3	MONTH	TOTAL	NUMBER	NUMBER			
4		COSTS	OF CHAIRS	OF TABLES			
5	JAN	$1350	5	3			
6	FEB	$1660	8	4			
7	MAR	$1480	4	5			
8	APR	$1450	5	4			
9	MAY	$1460	8	2			
10	JUN	$1210	3	3			
11	JUL	$1600	10	2			
12	AUG	$1940	12	4			
13	SEPT	$1750	5	7			
14	OCT	$1920	6	8			
15	NOV	$1740	12	2			
16	DEC	$1080	4	1			
17							
18	SUMMARY OUTPUT						
19							
20	*Regression Statistics*						
21	Multiple R	1					
22	R Square	1					
23	Adjusted R Square	1					
24	Standard Error	2.83585E-13					
25	Observations	12					
26							
27	ANOVA						
28		*df*	*SS*	*MS*	*F*	*Significance F*	
29	Regression	2	779066.6667	389533	4.8E+30	7.1805E-136	
30	Residual	9	7.23783E-25	8E-26			
31	Total	11	779066.6667				
32							
33		*Coefficients*	*Standard Error*	*t Stat*	*P-value*	*Lower 95%*	*Upper 95%*
34	Intercept	700	2.86458E-13	2.4E+15	2E-135	700	700
35	X Variable 1	70	2.80758E-14	2.5E+15	1E-135	70	70
36	X Variable 2	100	4.19395E-14	2.4E+15	2E-135	100	100

To interpret this output, some of the numbers given are more important than others. **The parameters for the model are given under the "Coefficients" label** in the bottom section. In this example the **constant** or **intercept** value (b) is given beside the label "**Intercept**", so b is 700. In essence this is the fixed cost, such as rent, utilities, etc., for just being in business, no matter how many chairs and tables you make. The **coefficients** for the chair and table variables (c and t) are given beside the "**X Variable**" labels, and are listed ***in the same order*** **as they were given in the data**, so the first coefficient here (70) corresponds to chairs (c) and the second (100) corresponds to tables (t), since the chair data was to the left of the table data. Thus our best-fit least-squares multiple linear regression model for the cost is:

Verbal Definition $C(c,t)$ = the expected (average) total cost, in dollars, in a month if c chairs and t tables are made,

Symbol Definition: $C(c,t) = 70c + 100t + 700$ for $c,t \geq 0$

Assumptions: Certainty and divisibility. Certainty implies that the relationship is exact. Divisibility implies that any fractional values of dollars, chairs, and tables are possible.

The domain of this function is the same as we discussed for the wood function (both variables must be non-negative, and could be restricted to be integers or not, but here we will assume they do not have to be integers, using the "average/expected value per month" interpretation). We could also restrict the domain to the limits of the data, such as $3 \leq c \leq 12$ and $1 \leq t \leq 8$, but it makes sense to start both at 0, and as long as there are no possibilities of volume discounts for the major cost components, it makes sense to leave the upper limits open-ended. □

Using Regression Analysis to Evaluate and Compare Models

How good of a fit is this model? And what do all those other numbers in the output mean? Look at the section of the summary output report for the regression in Table 2 labeled "Regression Statistics". The most important single value of these numbers is the value beside the label "**R Square**". This value tells you the **percentage of all the variation in the dependent variable** (cost) from all of your data points **that is explained by your model**. This value can vary from 0 to 1 (0% to 100%). In our case, the value is 1, or **100%, which means that our model fits the data perfectly**! If your R-squared value is below 0.6 (60%), you need to think carefully about how useful your model is. It suggests that you should probably consider using probability theory and statistics to understand how to best use the model, which is beyond the scope of this course. We will discuss the meaning of R^2 in more detail in Section 6.3 .

WARNING: Never use the R-squared value as the *only* criterion for choosing a model! Adding more parameters, for example using a cubic instead of a quadratic model, will always yield a better R-squared value, but is not necessarily a better model. In general for modeling, **always try to choose the simplest model possible** that reasonably represents the real situation. Remember KISS: Keep It Simple, Stupid! If a more complicated model does not add that much or is not needed to capture a certain property of the data, use the simpler model instead. One reasonable approach is to look at a graph of your data (if possible) or think of the nature of it to get a feel for the rough shape, especially whether it seems approximately linear or not. Then if there are a couple of models that would represent that shape reasonably well, use the R-squared value as one factor to help choose between them.

For the purpose of selecting a model to solve a problem, the next most important part of the regression output are the columns labeled "Lower 95%" and "Upper 95%" for each of the parameters. These give what in statistics is called a **confidence interval** (margin of error) for the parameter estimate. For instance, in Table 2 the 95% confidence interval for the intercept in our cost model is given to be the interval [700,700]. This means we are *very highly* confident (have a confidence level of 95%, to be specific) that the true intercept in this situation, given the data and **given that certain statistical assumptions are valid**, is between a lower limit of 700 and an upper limit of 700. In this example, we get the lower and upper values the same (note that this occurs for both *coefficients* as well) because the linear function fit our data perfectly: remember that the R squared value was 100%. More commonly, the values would be different. We will see an example of this in the next sample problem. We are usually most interested in these confidence intervals to determine whether we are reasonably confident of the ***sign* of each parameter** estimate; if not, we may decide that the **term** associated with that parameter does not **belong in the model**. Most packages also give the option of using 99% confidence intervals, for a more conservative criteria (this will tend to eliminate more terms from the model).

Some spreadsheet packages do not calculate confidence intervals automatically for a regression analysis. If not, they usually *do* calculate what is called the **standard error** of that parameter, which is a measure of the error, or variability, in the estimate for that parameter. Notice that for each of the parameters in the "Coefficients" column, there is a value under the label "Standard Error". This can used to *calculate* a **confidence interval** (margin of error) for the parameter estimate. In Section 6.3 , we will explain more about the details of standard errors and confidence intervals, and the statistical assumptions being made when using them the way we are using them in this section, but you really need to take a statistics course to fully understand all that they involve and to be truly confident of using regression analysis properly. If your spreadsheet program does not give confidence intervals, our discussion in Sections 6.2 and 6.3 will help you interpret the standard error in an approximate way to be able to estimate a rough confidence interval in your head. This may be sufficient for the purpose of choosing a model in many cases.

The "Observations" value in the summary output report just **counts the number of data points**. In the section of the report labeled "ANOVA", the column labeled "df" stands for "degrees of freedom." The main number in that column which we will highlight here is the one in the "Residual" row, so it could be called the **degrees of freedom of the residuals**, or sometimes the **degrees of freedom of the regression analysis**. In statistics, **residual** is just a different term for what we have been calling the **error** between a data point and a model, so the "Residual" row could just as well be labeled "Error". **The degrees of freedom of the residuals** (or errors, or regression analysis) **is simply the number of data points** (shown beside "Observations") **minus the number of parameters being estimated** (the number of columns in the last section of the report):

degrees of freedom of residuals = (# of data points) - (number of model parameters)

For example, in the single-variable case, when fitting a line (which has two parameters, m and b) with two data points, you have **no** (2-2=0) degrees of freedom to allow for errors (residuals), because the model is completely determined: a line is completely determined by two points. More generally, the degrees of freedom of the residuals tells you how many data points *more* than the *minimum* necessary are being used to find the kind of model being considered.

In our sample problem, we have 12 data points and 3 parameters (the intercept and the coefficients of the 2 independent variables), so the number of degrees of freedom of the residuals is $12 - 3 = 9$. Since more data points mean more information, and since the KISS! principle of simplicity argues for fewer parameters, in general, **the larger the degrees of freedom of the residuals, the better the model**. Unfortunately, **trying to *improve* the R squared value normally *reduces* the number of degrees of freedom of the residuals** (such as changing from a linear model to a quadratic model, which will always fit at least as well, but will add at least one parameter, and so decrease the degrees of freedom of the residuals).[1] As a result, **picking a model often involves judging trade-offs between the fit, simplicity, and amount of information in different models, based on the context of the problem and other factors.**

Sample Problem 2: In your furniture manufacturing business, you have data for the price you have charged each month for chairs and tables, as well as the number of each sold each month (Table 3). Find demand functions to model the number of each type of furniture sold as a function of the prices.

---

[1] The "Adjusted R Square" entry on the summary output is a way of adjusting the R squared value to adjust for the degrees of freedom of the residuals, to reflect the fact that using more parameters will always yield a better fit (so it lowers the R squared value for lower degrees of freedom). See Section 6.3 and a statistics text for more details.

Table 3

SALES RECORDS				
	CHAIR PRICE	TABLE PRICE	CHAIRS SOLD	TABLES SOLD
JAN	$190	$270	5	3
FEB	$180	$260	8	4
MAR	$200	$245	4	5
APR	$195	$260	5	4
MAY	$180	$285	8	2
JUN	$200	$270	3	3
JUL	$170	$285	10	2
AUG	$165	$265	12	4
SEPT	$200	$220	5	7
OCT	$195	$210	6	8
NOV	$165	$290	12	2
DEC	$195	$290	4	1

Solution: The demand functions for the two products, tables and chairs, will be modeled separately. To determine if the price of chairs depends only on the quantity of chairs available or if it also depends upon the quantity of tables, two regression analyses were done (See Table 4):

Table 4

	A	B	C	D	E	F	G
16	SUMMARY OUTPUT						
17							
18	*Regression Statistics*						
19	Multiple R	0.972327361					
20	R Square	0.945420497					
21	Adjusted R Square	0.939962547					
22	Standard Error	3.350061226					
23	Observations	12					
24							
25	ANOVA						
26		*df*	*SS*	*MS*	*F*	*Significance F*	
27	Regression	1	1944.020898	1944.020898	173.2189655	1.22002E-07	
28	Residual	10	112.2291022	11.22291022			
29	Total	11	2056.25				
30							
31		*Coefficients*	*Standard Error*	*t Stat*	*P-value*	*Lower 95%*	*Upper 95%*
32	Intercept	215.2863777	2.408849787	89.37310201	7.52561E-16	209.919125	220.65363
33	Chairs Sold	-4.249226006	0.322858415	-13.16126763	1.22002E-07	-4.968599509	-3.5298525
34							
35	SUMMARY OUTPUT						
36							
37	*Regression Statistics*						
38	Multiple R	0.994231585					
39	R Square	0.988496445					
40	Adjusted R Square	0.9859401					
41	Standard Error	1.621185599					
42	Observations	12					
43							
44	ANOVA						
45		*df*	*SS*	*MS*	*F*	*Significance F*	
46	Regression	2	2032.595815	1016.297908	386.6834252	1.87821E-09	
47	Residual	9	23.65418472	2.628242747			
48	Total	11	2056.25				
49							
50		*Coefficients*	*Standard Error*	*t Stat*	*P-value*	*Lower 95%*	*Upper 95%*
51	Intercept	208.6093242	1.637610104	127.3864418	5.75144E-16	204.90479	212.31386
52	Chairs Sold	-4.035922264	0.160502271	-25.14557732	1.19565E-09	-4.399003903	-3.6728406
53	Tables Sold	1.39186078	0.239758027	5.805272907	0.000257766	0.849490028	1.9342315
54							

The first Summary Output shows the price of chairs as a function of the quantity of chairs (here rounded to three significant digits):

$$PC(c) = -4.25c + 215.$$

The second Summary Output shows the price of chairs as a function of the quantity of chairs and the quantity of tables (rounded to three significant digits):

$$PC(c,t) = -4.04c + 1.39t + 209 .$$

To run the first regression (the number of chairs the only independent variable) on the spreadsheet using the data in Table 3, the dependent variable block used was B2..B14, the independent variable block used was D2..D14, and the solution report was placed in the block with A18 in the upper left corner. Note that in this case, the label for the columns were *included* in the blocks, so the labels appear on the summary output. This is very helpful, because it identifies the independent variables, but may not be an option on every spreadsheet. When it *is* possible, you should *indicate* whether or not you are including the labels when you run the regression (the default is usually to assume your are *not*). For the second regression (using *both* variables as inputs), the dependent variable block would be the same B2..B14, the independent variable block would be D2..E14 , and the solution report would be placed in the block with A35 in the upper left corner. Create a spreadsheet and try running these regressions yourself, to make sure you understand how to do it. See your technology supplement for more details.

Which of these models should we use? Notice that this time, the upper and lower limits of the confidence intervals for the coefficients are not identical, so they do give true intervals with positive lengths. In the first regression, the 95% confidence interval for the coefficient of c is [-4.97,-3.53]. You can think of this as meaning that we are 95% certain that the true coefficient of c lies within this interval, based on the data (again, this assumes certain statistical assumptions are valid, which you should study statistics to fully understand; we will discuss them briefly in Section 6.3). This suggests that we at least know the *sign* of the coefficient with a high level of confidence, so it makes sense to keep that term in our model. This would also occur even at a 99% level of confidence. In this case, the sign is negative, which makes good economic sense (if you charge more for chairs, you'll sell fewer of them). In general, **if the confidence interval for the coefficient of a variable includes *both* negative *and* positive values, it suggests we are not even highly confident of the *sign* of that coefficient, so should at least *consider* dropping that term from the model and the regression**. We will see an example of this in Sample Problem 5.

Looking at the confidence intervals for the 2 coefficients of the second model, we see that again we are quite confident of their *signs* (this would also be true at the 99% confidence level), and the signs make sense (as before, we expect the coefficient of the chair price to be negative, and a positive coefficient on the table price suggests that if you charge more for tables, people may buy more chairs, since they will be relatively cheaper). These coefficients are really rates of change, and we will discuss them in more detail in Section 8.1 . *If* we had *not* been confident of the sign of the table variable, that would have been a strong factor favoring the first (single-variable) model. As it is, we need to consider other factors to help us decide between the two.

The R squared value for the two-variable model is 0.988, which is better than the 0.945 of the single-variable model, but not that much. On the other hand, the degrees of freedom of the single-variable model is 10, versus 9 for the two-variable model. This decision is a close call, and could hinge largely on whether your perception as owner of the business is that the price of tables influences the sales of chairs or not. The analysis could be tried both ways to see how much difference the choice of the model makes, in the Sensitivity Analysis step of solving your problem.

We will now run regressions to determine the demand function for the tables. The process will be exactly the same, except that the dependent variable is now in C2..C14 and the single variable independent variable is in E2..E14 and the locations for the Summary Outputs have been changed (Table 5) Note for the multivariable regression that the Chairs Sold is again the first independent variable. The variables will appear in the order in which they are entered in the spreadsheet, not necessarily the order that seems most logical to you.

Table 5

	A	B	C	D	E	F	(
55	SUMMARY OUTPUT						
56							
57	*Regression Statistics*						
58	Multiple R	0.991598814					
59	R Square	0.983268208					
60	Adjusted R Square	0.981595029					
61	Standard Error	3.548333915					
62	Observations	12					
63							
64	ANOVA						
65		*df*	*SS*	*MS*	*F*	*Significance F*	
66	Regression	1	7399.093264	7399.093264	587.6646091	3.24985E-10	
67	Residual	10	125.9067358	12.59067358			
68	Total	11	7525				
69							
70		*Coefficients*	*Standard Error*	*t Stat*	*P-value*	*Lower 95%*	*Uppe*
71	Intercept	308.9378238	2.172276317	142.2184744	7.25365E-18	304.0976897	313.
72	Tables Sold	-12.38341969	0.510829328	-24.24179467	3.24985E-10	-13.52161856	-11.2
73							
74	SUMMARY OUTPUT						
75							
76	*Regression Statistics*						
77	Multiple R	0.998233264					
78	R Square	0.99646965					
79	Adjusted R Square	0.995685127					
80	Standard Error	1.718070161					
81	Observations	12					
82							
83	ANOVA						
84		*df*	*SS*	*MS*	*F*	*Significance F*	
85	Regression	2	7498.434114	3749.217057	1270.161059	9.22954E-12	
86	Residual	9	26.5658857	2.951765078			
87	Total	11	7525				
88							
89		*Coefficients*	*Standard Error*	*t Stat*	*P-value*	*Lower 95%*	*Uppe*
90	Intercept	300.9295436	1.73547622	173.3988286	3.58859E-17	297.0036207	304.
91	Chairs Sold	0.986761918	0.170094135	5.801269494	0.000259062	0.601981958	1.37
92	Tables Sold	-12.04597779	0.254086338	-47.40899441	4.14023E-12	-12.62076146	-11.4

The first Summary Output that appears in Table 5 shows the price of tables as a function of the quantity of tables:

$$PT(t) = -12.4t + 309.$$

The second Summary Output shows the price of tables as a function of the quantity of both the tables and chairs:

$PT(c,t) = 0.987c - 12t + 301.$

This time, we are again very confident of the signs of all of the coefficients, but the improvement in the R squared is even less this time from adding the second variable (from 0.983 to 0.996). This difference is probably less than the margin of error in the data. The residual degrees of freedom (df) are the same as for the chair price models, favoring the single-variable model. The combination of the *small* addition to the R squared from the second variable and the resulting *decrease* in residual df suggest that the single-variable model is about as good as the two-variable model, so let's work with that. Of course, if you had strong feelings that the price of the chairs *does* significantly affect the sales of tables, you could include it for completeness. In any case, trying it during Sensitivity Analysis is a good idea.

Our final demand functions, then, are:

Verbal Definition: $PC(c,t)$ = the price you would have to charge for a chair in dollars in order to sell *exactly* c chairs and t tables (on average).

Symbol Definition $PC(c,t) = -4.04c + 1.39t + 209$ for $165 \le c \le 200,\ 210 \le t \le 290$

Assumptions: Certainty and divisibility. Certainty implies that the relationship is exact. Divisibility implies that any fractional values for dollars, chairs, and tables are possible.

Verbal Definition: $PT(t)$ = the price you would have to charge for each table in dollars in order to sell exactly t tables (on average).

Symbolic Definition $PT(t) = -12.4t + 309$ for $210 \le t \le 290$

Assumptions: Certainty and divisibility. Certainty implies that the relationship is exact. Divisibility implies that any fractional values for dollars, chairs, and tables are possible. □

Notice that we have used **the intervals defined by the data values as the default for the domains of both functions**. For each variable, we have simply used the minimum and maximum data values as the extremes of the interval defining the domain for that variable. It is always possible to try extrapolating beyond the data, but it should be done very cautiously, so we will adopt this convention of using the intervals defined by the data wherever possible. Going beyond that should be justified in some way.

Formulating Models from Raw Data *and* Verbal Descriptions

Sample Problem 3: Find a model for the revenue from the sales of the tables and chairs discussed in the Sample Problems 1 and 2.

Solution: As discussed earlier, our revenue function, the revenue in dollars if you make c chairs and t tables and set the prices to have exactly these quantities demanded, will be given by:

$$R = (\text{price chairs})(\text{quantity chairs}) + (\text{price tables})(\text{quantity tables})$$
$$R = [P(c,t)](c) + [P(t)](t).$$

The revenue is thus:

$$R(c,t) = (-4.04c + 1.39t + 209)(c) + (-12.4t + 309)(t)$$
$$R(c,t) = 4.04c^2 + 1.39ct + 209c - 12.4t^2 + 309t \quad \text{(Multiplying out)}$$

The model for the revenue is:

Verbal Definition: $R(c,t)$ = the total revenue in dollars when c chairs and t tables are made.
Symbol Definition: $R(c,t) = 4.04c^2 + 1.39ct + 209c - 12.4t^2 + 309t$ for $165 \le c \le 200$, $210 \le t \le 290$
Assumptions: Certainty and divisibility. Certainty implies that the relationship is exact, which is assuming that the prices charged will be those determined by the demand functions to result in sales of exactly c chairs and t tables. Divisibility implies that any fractional values of dollars, chairs, and tables are possible. □

Sample Problem 4: Find a model for the profit realized from the sales of the tables and chairs in Sample Problems 1 - 3.

Solution: Profit is equal to the revenue minus the cost, or $P = R - C$. Since we already calculated our cost function for this problem to be

$$C(c,t) = 70c + 100t + 700,$$

and our revenue function to be:

$$R(c,t) = 4.04c^2 + 1.39ct + 209c - 12.4t^2 + 309t$$

we can now find the profit function.

$$P(c,t) = R(c,t) - C(c,t)$$
$$P(c,t) = (4.04c^2 + 1.39ct + 209c - 12.4t^2 + 309t) - (70c + 100t + 700)$$
$$P(c,t) = 4.04c^2 + 1.39ct + 139c - 12.4t^2 + 209t - 700 \quad \text{(Combining like terms.)}$$

Thus the model for the profit is:

Verbal Definition: $P(c,t)$ = the total profit in dollars when c chairs and t tables are made and prices are set to force the demand to equal c chairs and t tables

Symbol Definition: $P(c,t) = 4.04c^2 + 1.39ct + 139c - 12.4t^2 + 209t - 700$

for $165 \le c \le 200, 210 \le t \le 290$

Assumptions: Certainty and divisibility . Certainty implies that the relationship is exact, and that the prices are set according to the demand functions defined earlier. Divisibility implies that any fractional values of dollars, chairs, and tables are possible. □

In Chapter 8 you will learn how to maximize this profit function using calculus.

Notice that Sample Problems 3 and 4 involve formulation *both* from raw data *and* from verbal descriptions. All kinds of combinations like this are possible.

General Form of a Multivariate Quadratic Function

The profit function above is an example of a **multivariable quadratic function**. This means that each variable occurs with exponents of only 0, 1, or 2, that the *sum* of the exponents in any individual term is also 0, 1, or 2, and that *at least one* term has exponents that sum to 2 *and* a nonzero coefficient. For a two-variable quadratic function with variables x_1 and x_2, the general form is

$$f(x_1, x_2) = m_1x_1 + m_2x_2 + a_{11}x_1^2 + a_{12}x_1x_2 + a_{22}x_2^2 + b \ .$$

We have seen numerous graphs of two-variable quadratic functions in Section 5.1 . Figures 6.1-2 and 6.1-3 give two of the most common shapes: a bowl (or hammock), and a saddle, respectively.

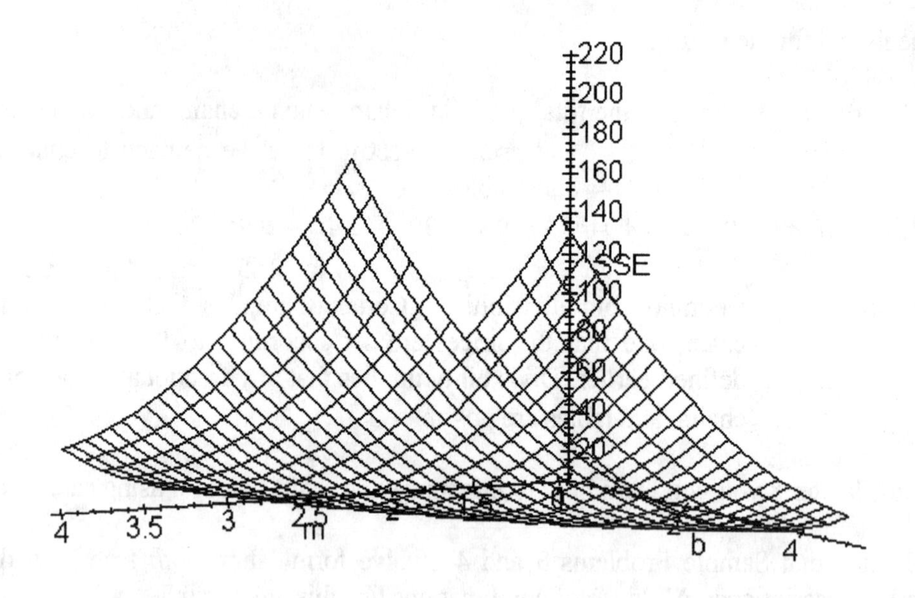

Figure 6.1-2

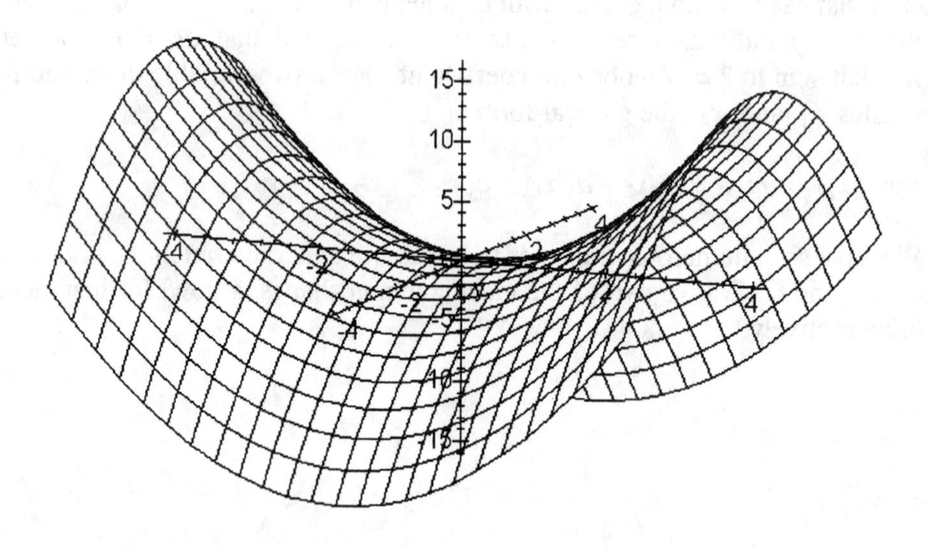

Figure 6.1-3

A dome shape is a third common possibility, similar to Figure 6.1-2 flipped upside down. In Section 8.4 we will discuss the relationship between the formula of a quadratic function and its shape.

For functions of more than two variables, the pattern of the function is the same, and could be written in the form

$$f(x_1, x_2, \ldots, x_n) = \sum_{i=1}^{n} m_i x_i + \sum_{i=1}^{n} \sum_{j=i}^{n} a_{ij} x_i x_j + b \quad .$$

The first summation is just the linear terms, and the second summation just is a way of describing all of the squared terms (when $j = i$) and all of the mixed product terms (when $j > i$). The sum for j starts at i rather than 1 so that each mixed product term occurs only *once* (so that there are not separate terms for x_1x_2 and x_2x_1 , for example). The b , of course, is the intercept or constant term, as usual.

For example, for $n = 2$, this becomes

$$\begin{aligned} f(x_1, x_2) &= \sum_{i=1}^{2} m_i x_i + \sum_{i=1}^{2} \sum_{j=i}^{2} a_{ij} x_i x_j + b \\ &= \left(\sum_{i=1}^{2} m_i x_i \right) + \left\{ \sum_{i=1}^{2} \left(\sum_{j=i}^{2} a_{ij} x_i x_j \right) \right\} + b \\ &= (m_1 x_1 + m_2 x_2) + \left\{ \left(\sum_{j=1}^{2} a_{1j} x_1 x_j \right) + \left(\sum_{j=2}^{2} a_{2j} x_2 x_j \right) \right\} + b \\ &= m_1 x_1 + m_2 x_2 + \{ (a_{11} x_1 x_1 + a_{12} x_1 x_2) + (a_{22} x_2 x_2) \} + b \\ &= m_1 x_1 + m_2 x_2 + a_{11} x_1^2 + a_{12} x_1 x_2 + a_{22} x_2^2 + b \end{aligned}$$

as we stated earlier.

Nonlinear Regression Using Spreadsheets

Example 1: We know that computer spreadsheet programs will fit multivariable linear models, as in our first sample problems. What about multivariable non-linear models? Let's go back and start with a simpler problem. In Chapter 1 we discussed the problem relating to a ball (or egg) tossed into the air. The data for the height of the ball, in feet, for time, in seconds, after it was thrown is shown in Table 6:

Table 6

Time(Sec)	Height(Ft)
0	5
0.25	16
0.5	25
0.75	32
1	37
1.25	40
1.5	41
1.75	40
2	37

We used technology to fit a quadratic model to the data:

Verbal Definition $H(t)$ = the height of the ball in feet t seconds after the toss

Symbol Definition: $H(t) = -16t^2 + 48t + 5$ for $0 \leq t \leq 3$

Assumptions: Certainty and divisibility. Certainty implies that the relationship is exact. Divisibility implies that any fractional values of height and time are possible.

So far, we have only seen *linear* regressions using a spreadsheet package, and in fact regression packages technically really *only* do linear regressions. Could we still have done the above quadratic regression on a spreadsheet using the regression function? The answer is: Yes, we *can* do a non-linear regression on a spreadsheet, as long as the model is a *linear* function of its parameters. For example, quadratic models have the form

$$ax^2 + bx + c \ .$$

If you temporarily think of the variable here (x) as if it were a number or a constant, and think of the parameters as variables (somewhat like we did in Section 5.5), this becomes

$$(x^2)a + (x)b + c \ .$$

Thought of like this, the model is a *linear* function of a, b, and c, since each occurs in a separate term by itself, with an exponent of 1.

Thus, **we simply have to "fool" the spreadsheet by treating the nonlinear variable terms as if they were new independent variables**. In the quadratic example, we think of x^2 (or t^2 in the ball example) as a second independent variable (which we could think of as being named w or whatever). The reason this is a trick is that, of course, this "new" variable is *not at all independent*, since it totally depends on other variables in the model.

To do this, then, we simply add in columns of "data" for the additional "independent" variables, which can easily done by simple calculations. Spreadsheets are especially convenient for doing this kind of calculation. In our ball example, we now want columns for both t^2 and t. Table 7 shows the data setup work needed to perform the regression. As we have said before, the independent variables need to be all together in one block of cells. Since the initial data was entered in a way that does not allow this (there is not room to put a column for t^2 beside the column for t, although it *is* possible to insert a column in a spreadsheet), the simplest way to do this is to *copy* the column for t (Time) *after* the column for Height, and then create the column for t^2 (Time^2) beside it.

To **copy a block of cells on a spreadsheet**, use the mouse to *highlight* the block by clicking on one corner and dragging to the opposite corner, then click on the Copy icon (or click on Edit and then Copy), then click on the upper left corner of the destination block, and click on the Paste icon (or click on Edit and then Paste). In Table 8, we have copied the block in A1..A10 and pasted it to the block with upper left cell C1.

To *square* the column of Time values, we can use a spreadsheet **formula**. One way to do this is to write the formula in one cell, then copy to the rest of the column. In our example, we want to square the value in C1, and put the answer in D1. To do this, in cell D1, we enter the formula "+C1^2" (the quotation marks don't get entered). The formula starts with "+" so that the spreadsheet knows that the C at the beginning is a formula and not text. Other symbols are possible (like "="), but the plus sign works for most spreadsheets (it may get changed to a "=", but can still be entered using "+"). If the natural form of the formula starts with a "-", then no "+" is needed. After typing in the formula, if you hit <Enter> or click on another cell, you will see that the spreadsheet has performed the calculation and that there is a 0 in cell D1.

Now, to **copy the formula** to the rest of the column, you can highlight the cell with the formula you want to copy (D2 in the example), click on the Copy icon, then highlight the entire block of cells to which you want to copy the formula (D3..D10, and click on the Paste icon. Try it, and you will see that you get exactly what you want: the squares of the Time values in column D (D2..D10). Now you can just type in "Time^2" in cell D1 to make the data look just like Table 8.

Now, try clicking on cell D3. If you look near the top of the screen, you will see that the contents of cell D3 is the formula "+C3^2"! How did the spreadsheet know that you wanted to square cell C3 there? (Remember that the formula you copied was "+C2^2".) The answer is that what we just did is so commonly needed, it is what the spreadsheet assumes by default, and uses what is called **relative addressing**. When you copy a formula from a cell (which we will call the **source** cell) to another cell (which we will call the **destination** cell), the address in the source cell is used in its *relative* position to the source, and the *corresponding* address in that *same* relative position to the destination cell is used in the new formula. In our example, the source cell (D2) formula had the address C2, which is the cell in the *same row* (2) and *one column to the left* (C relative to D). When this is copied to the destination D3, the address in the same relative position (in the *same row* and *one column to the left*) is C3, so C3 is what gets put in for the address in the formula.

If you *don't* want this relative addressing to happen (if you want to copy the *exact* row and/or column, not the relative position), you can put a "$" before the column letter or the row number (or both), which is called **absolute addressing**. If our formula in D2 had used "+C$2" it would mean "the address in *Row 2* and *one column to the left*", so if we copied it to D3, the new formula would have the address "+C$2" in it (relative addressing for the column, and absolute addressing for the row). We will see examples of this in Section 8.4 .

Now we are ready to run the regression, which we do using B1..B10 as the dependent variable block and C1..D10 as the independent variable block (being sure to indicate that we are *including* the labels), and putting the summary output in the block with A12 in the upper left corner, as shown in Table 7.

Table 7

Time(Sec)	Height(Ft)	Time(Sec)	Time^2			
0	5	0	0			
0.25	16	0.25	0.0625			
0.5	25	0.5	0.25			
0.75	32	0.75	0.5625			
1	37	1	1			
1.25	40	1.25	1.5625			
1.5	41	1.5	2.25			
1.75	40	1.75	3.0625			
2	37	2	4			
SUMMARY OUTPUT						
Regression Statistics						
Multiple R	1					
R Square	1					
Adjusted R Square	1					
Standard Error	1.3987E-14					
Observations	9					
ANOVA						
	df	*SS*	*MS*	*F*	*Significance F*	
Regression	2	1268	634	3E+30	7.93327E-91	
Residual	6	1.17383E-27	1.96E-28			
Total	8	1268				
	Coefficients	*Standard Error*	*t Stat*	*P-value*	*Lower 95%*	*Upper 95%*
Intercept	5	1.13683E-14	4.4E+14	9E-87	5	5
Time(Sec)	48	2.65066E-14	1.81E+15	2E-90	48	48
Time^2	-16	1.27518E-14	-1.3E+15	2E-89	-16	-16

The best-fit regression function is exactly the same as the one from the quadratic model obtained in Chapter 2. The intercept is 5, the coefficient of t is 48, and the coefficient of t^2 is -16, so the regression function is $-16t^2 + 48t + 5$. □

Sample Problem 5: A young gymnast was interested in the relationship between her overall energy level and the amount of exercise she had during that day plus the amount of calories she consumed during that day. Table 8 shows a record she kept for 20 days of the minutes of exercise and the amount of calories consumed. She did not feel that a linear model would be appropriate because too much or too little exercise would result in a lowering of her energy level. The same thing would be true about the caloric intake: when she ate too much she felt lethargic, and when she had too little to eat she just couldn't get started. She really felt that a quadratic model would be the best fit, and that this should include a term for the cross-product of the exercise and calories. How can she fit a non-linear model to her data?

Table 8

Daily Calorie and Exercise Record			
Energy	Calories	Exercise (min)	
y	c	x	
69	1800	0	
77	1800	30	
74	1800	45	
74	1800	75	
69	1800	0	
79	1750	60	
88	1300	60	
91	1300	45	
54	2000	90	
91	1300	75	
85	1200	60	
91	1500	60	
66	1000	120	
87	1250	60	
88	1200	90	
93	1400	60	
82	1400	120	
78	1000	30	
86	1300	105	
57	2000	60	

Solution: We will use exactly the same idea for the exercise/calorie/energy multivariable nonlinear model that we did for the ball-toss single-variable nonlinear model in Example 1. Table 9 shows the original data entered (with the dependent column first), then the calories squared; then the cross product of the calories and the exercise; and then the exercise squared, all calculated by entering and copying formulas. Try to reproduce these calculations on your own.

Table 9

Energy, Calories, and Exercise						
	Energy	Calories	Exercise (min)			
	y	c	x	c^2	cx	x^2
	69	1800	0	3240000	0	0
	77	1800	30	3240000	54000	900
	74	1800	45	3240000	81000	2025
	74	1800	75	3240000	135000	5625
	69	1800	0	3240000	0	0
	79	1750	60	3062500	105000	3600
	88	1300	60	1690000	78000	3600
	91	1300	45	1690000	58500	2025
	54	2000	90	4000000	180000	8100
	91	1300	75	1690000	97500	5625
	85	1200	60	1440000	72000	3600
	91	1500	60	2250000	90000	3600
	66	1000	120	1000000	120000	14400
	87	1250	60	1562500	75000	3600
	88	1200	90	1440000	108000	8100
	93	1400	60	1960000	84000	3600
	82	1400	120	1960000	168000	14400
	78	1000	30	1000000	30000	900
	86	1300	105	1690000	136500	11025
	57	2000	60	4000000	120000	3600
SUMMARY OUTPUT						
Regression Statistics						
Multiple R	0.9896467					
R Square	0.9794005					
Adjusted R Square	0.9720436					
Standard Error	1.910614					
Observations	20					
ANOVA						
	df	*SS*	*MS*	*F*	*Significance F*	
Regression	5	2429.843756	485.9688	133.1259	2.72703E-11	
Residual	14	51.10624446	3.650446			
Total	19	2480.95				
	Coefficients	*Standard Error*	*t Stat*	*P-value*	*Lower 95%*	*Upper 95%*
Intercept	-93.1423	12.64336946	-7.36689	3.53E-06	-120.2596579	-66.0249
c	0.2602388	0.016248958	16.01572	2.13E-10	0.225388226	0.295089
x	0.1459598	0.135347154	1.07841	0.299081	-0.144331272	0.436251
c^2	-9.44E-05	5.48592E-06	-17.206	8.18E-11	-0.000106157	-8.3E-05
cx	4.098E-05	6.00381E-05	0.682533	0.506039	-8.7791E-05	0.00017
x^2	-0.001958	0.000450741	-4.34474	0.000673	-0.002925098	-0.00099

This suggests that a complete model for the gymnast's problem is:

Verbal Definition $E(c,x)$ = The energy, on average, on a scale of 0-100, for a day when c calories are consumed and x minutes of exercise are taken

Symbol Definition: $E(c,x) = -93.1 + 0.260c + 0.146x - 0.0000944c^2 + 0.0000410cx - 0.00196x^2$

for $1000 \le c \le 2000,\ 0 \le x \le 120$

Assumptions: Certainty and divisibility. Certainty implies that the relationship is exact, and that other factors (sleep, health, etc.) are held constant. Divisibility implies that any fractional values of energy, calories and minutes are possible.

Note that for this regression the R squared is quite high, but two of the standard errors are high enough that the 95% confidence level involves a change of sign for the coefficients (includes 0): the coefficient for the x term and for the cx term change from negative to positive, and this would be even more true at the 99% confidence level. The confidence interval for the cx term is very close to being all negative numbers, but the sign of the x term seems quite uncertain. In such a situation, it makes sense to try running the regression without either or both terms, at least as part of the Sensitivity Analysis. Let's see what happens when we remove the x term (see the summary output in Table 10).

Table 10

SUMMARY OUTPUT						
Regression Statistics						
Multiple R	0.9887818					
R Square	0.9776894					
Adjusted R Square	0.9717398					
Standard Error	1.920965					
Observations	20					
ANOVA						
	df	*SS*	*MS*	*F*	*Significance F*	
Regression	4	2425.598401	606.3996	164.3312	3.42472E-12	
Residual	15	55.35159896	3.690107			
Total	19	2480.95				
	Coefficients	*Standard Error*	*t Stat*	*P-value*	*Lower 95%*	*Upper 95%*
Intercept	-89.18148	12.16368686	-7.33178	2.48E-06	-115.1077824	-63.2552
c	0.2609434	0.016323776	15.98548	7.88E-11	0.226150093	0.295737
c^2	-9.59E-05	5.3287E-06	-18.0002	1.44E-11	-0.000107276	-8.5E-05
cx	0.0001027	1.82076E-05	5.640924	4.69E-05	6.38993E-05	0.000142
x^2	-0.001531	0.0002162	-7.08211	3.73E-06	-0.001991971	-0.00107

This time, we can be reasonably confident of *all* of the signs of the coefficients (including the cx coefficient, which was *not* true for the first model), the R^2 value is essentially identical, and we have added 1 degree of freedom of the residuals, so this model is clearly better. It would have the form:

Verbal Definition $E(c,x)$ = The energy, on average, on a scale of 0-100, for a day when c calories are consumed and x minutes of exercise are taken

Symbol Definition: $E(c,x) = -89.2 + 0.261c - 0.0000959c^2 + 0.0000103cx - 0.00153x^2$, for $1000 \le c \le 2000,\ 0 \le x \le 120$

Assumptions: Certainty and divisibility. Certainty implies that the relationship is exact, and that other factors (sleep, health, etc.) are held constant. Divisibility implies that any fractional values of energy, calories and minutes are possible.

Notice that some of the coefficients in the two models are quite similar, and some are somewhat different. Exploring the optimal solutions for both would give useful information in the Sensitivity Analysis for this example. □

Section Summary

Before you begin the exercises be sure that you:

- Understand the basic organization and naming conventions for spreadsheets: cells, blocks of cells, text, data, formulas, and how to enter and copy them.
- Know how to enter data in a spreadsheet program in order to run a regression: put the column for your dependent variable first, then use adjacent columns for each independent variable (or nonlinear term, acting *like* an independent variable) to the right of the dependent variable column; all columns should be of the same length.
- Know that in most spreadsheet programs you can include column labels (assumed to be in the first row) when you identify the independent and dependent variables, but that you must indicate that you have done so. The labels for the independent variables will then appear in the summary output.
- Know how to use the regression function on your spreadsheet program: identify the dependent and independent variable blocks.
- Understand the output given from a regression analysis: the R squared gives the percentage of all the variations in the dependent variable explained by your model; the upper and lower limits of the 95% confidence interval give a margin of error for the parameter estimates, to help determine if the sign of each

parameter is relatively clear or not, and whether or not that term should be included in the model.

- Understand that the degrees of freedom of the residuals (errors) is the number of data points minus the number of parameters estimated, and should be as large as possible, since more data (more information) and simpler models (fewer parameters) are preferable.
- Be able to select the best model from among regressions: compare the R squared values (the higher the better) and the degrees of freedom (the higher the better), and check the confidence intervals for the coefficients (the fewer that include 0 the better). Remember KISS!
- Know how to write models determined by regression analyses: select appropriate letter designations for the variables; remember that the *coefficients* of the independent variables appear (top to bottom) in the order in which the corresponding independent variable *columns* appear (left to right) in the spreadsheet. As a default, use the intervals defined by the data as your initial domain for each independent variable, unless you have good reasons to enlarge the domain.
- Know how to enter data on a spreadsheet to run a non-linear multivariable regression: enter the dependent variable in one column and the independent variable data in adjacent columns, such as $x_1, x_2, x_1^2, x_1x_2, x_2^2$ for a two-variable quadratic model.
- Know how to enter and copy formulas on a spreadsheet, and how to use relative and absolute addressing.

EXERCISES FOR SECTION 6.1

Warm Up:

For Exercises 1-4, run a regression model for z as a function of x and y. Do you think that this function is a good description of the relationship between the variables? Explain.

1.

x	y	z
1	5	14
2	8	27
3	5	22
4	7	32
5	2	24
6	9	45

2.

x	y	z
2	7	16
5	5	11
6	8	15
8	4	7
4	1	0
3	6	14

3.

x	y	z
-2	7	4
10	2	5
11	5	8
-12	8	1
2	3	3
-3	9	5
7	6	7
-14	10	2

4.

x	y	z
6	5	13
8	3	28
7	7	10
2	1	7
9	9	16
4	4	8
5	3	15
1	8	-15

For Exercises 5-8, run a regression model for x as a function of y and z. Do you think that this function is a good description of the relationship between the variables? Explain your answer.

5.

x	y	z
1	5	14
2	8	27
3	5	22
4	7	32
5	2	24
6	9	45

6.

x	y	z
2	7	16
5	5	11
6	8	15
8	4	7
4	1	0
3	6	14

7.

x	y	z
-2	7	4
10	2	5
11	5	8
-12	8	1
2	3	3
-3	9	5
7	6	7
-14	10	2

8.

x	y	z
6	5	13
8	3	28
7	7	10
2	1	7
9	9	16
4	4	8
5	3	15
1	8	-15

Game Time

9. A small cottage industry involves the making of candy. At the present time they are making only two kinds of candy, fudge and divinity (similar to fudge, but a lighter color). The average costs of producing these two candies are 60 cents per piece for the divinity and 70 cents per piece for the fudge. Since their products compete with each other, they expect that the demand for each product will be related both to its own cost and the cost of the other. They have kept records of prices and sales over the last three months to see if they can determine this relationship. The data is shown below.

LARGE PIECES FUDGE	LARGE PIECES DIVINITY	PRICE FUDGE	PRICE DIVINITY
181	174	$ 1.50	$ 1.25
177	176	$ 1.75	$ 1.50
181	172	$ 1.50	$ 1.35
173	178	$ 1.95	$ 1.75
182	173	$ 1.45	$ 1.25
179	172	$ 1.65	$ 1.55
176	173	$ 1.79	$ 1.69
172	178	$ 1.99	$ 1.79
180	171	$ 1.56	$ 1.47
185	169	$ 1.25	$ 1.10
180	181	$ 1.50	$ 1.00
172	185	$ 1.89	$ 1.38

a) For each product, determine quantity sold as a function of price (or prices) charged. Write the models for these demands.
b) Find the total revenue as a function of the price of each product. Write the model for the revenue as a function of the price of each product.
c) Find the total profit as a function of the price of each product. Write the model for the profit as a function of the price of each product.
d) Determine the price of each product as a function of the quantity(quantities). Write the models for the prices as functions of the quantities.
e) Find the total revenue as a function of the quantity of each product. Write the model for the revenue as a function of the quantity of each product.
f) Find the total profit as a function of the quantity of each product. Write the model for the profit as a function of the quantity of each product.

10. You are running a fund-raiser for your organization and plan to sell T shirts and baseball caps at all sporting events again this spring. You know from past experience that if you price the T shirts too high, people will not buy them, but they may buy a cap instead. You are not sure if the reverse is true. You decided to experiment last year and kept the following records. Determine quantity as a function of price and price as a function of quantity for T shirts and caps.

CAPS SOLD	SHIRTS SOLD	PRICE CAPS	PRICE SHIRTS
15	13	$5.00	$10.00
16	10	$5.50	$12.00
19	10	$4.95	$11.95
16	9	$5.75	$13.00
20	9	$4.95	$12.50
23	6	$5.25	$15.00
21	10	$4.75	$11.95
21	9	$5.00	$12.75
24	11	$3.95	$11.50
25	9	$4.25	$12.95
23	9	$4.50	$12.50

11. The Afton Corporation makes asphalt shingle squares for buildings. Below is a table of information from the 1980s giving the total volume of squares they and their entire region (including them) sold, in thousands:[2]

	Volume of Squares (000s)		Price per Shingle Square	
Year	Region Total	Afton Corp.	Competitors	Afton Corp.
1979	830	500	$17	$17
1980	977	586	$17	$17
1981	1085	651	$15	$15
1982	1205	723	$15	$15
1983	1339	803	$17	$17
1984	1488	893	$18	$18
1985	1600	960	$18	$18
1986	1500	750	$18	$20
1987	1250	500	$18	$20

a) Set up a spreadsheet for this data. Run a regression analysis to determine if Afton Corp.'s volume was a function of their price. Write Afton Corp.'s market volume as a

[2] Problem adapted from Strategic Marketing Problems, by Kerin and Peterson (Needham Heights, MA: Allyn and Bacon, 1990). Volume and Price Behavior for Asphalt Shingle Squares: 1979-1987

function of their price as found by the regression analysis. Would you have confidence using this function to predict future volume for Afton Corp.? Why or why not?

b) Set up a column to show Afton Corp.'s market share (the volume of squares sold by Afton divided by the total volume sold) from 1979 to 1987.

c) Run a regression analysis to determine if Afton Corp.'s market share was a function of their price. Write Afton Corp.'s market share as a function of their price as found by the regression analysis. Would you have confidence using this function to predict future market share? Explain your answer.

d) Run a regression analysis to determine if the volume of the region was a function of the competitors' price. Would you have confidence using this function to predict future regional volume? Explain your answer.

12. The table below shows the average points per game, the number of field goal attempts and free throw attempts and the percentage of field goals and free throws made by some NBA players.

AVG PTS GAME	FIELD GOAL ATTEMPTS	FREE THROW ATTEMPTS	FIELD GOAL %	FREE THROW %
22.6	399	167	0.401	0.653
18.8	417	164	0.384	0.787
18	361	166	0.449	0.753
11	223	84	0.493	0.786
8.2	178	60	0.444	0.633
7	52	13	0.404	0.538
6.9	99	20	0.485	0.65

a) Develop a model for the average number of points per game as a function of the field goal and free throw attempts.
b) Develop a model for the average number of points per game as a function of the percentage of field goal attempts made and the percentage of free throw attempts made.
c) Which is the better model? Explain why you chose this one.

13. You have been making homemade candies to sell at fairs and flea markets and other such events. So far you have restricted your candies to just two varieties, Caramel Cream Fudge and Chocolate Cream Caramels. The fudge costs $3.89 per pound and the caramels costs $3.09 per pound. You have been experimenting with pricing in order to increase your profit. The following spreadsheet shows the prices that you charged and the amounts that you sold:

Price Fudge	Price Caramels	Quantity Lbs.Fudge	Quantity Lbs. Caramel
$ 6.00	$ 5.00	17	13
$ 6.50	$ 4.50	16.75	14.75
$ 5.95	$ 4.25	17	14.5
$ 6.25	$ 6.00	16.75	11.5
$ 5.50	$ 5.50	17.25	11.25
$ 6.75	$ 5.00	16.75	14
$ 5.75	$ 4.95	17	12.75
$ 5.49	$ 4.75	17.25	12.75
$ 6.00	$ 4.50	17	14
$ 6.50	$ 5.00	16.75	13.75
$ 5.95	$ 4.95	17	13
$ 6.25	$ 4.95	16.75	13.5
$ 5.50	$ 4.50	17.25	13.25
$ 6.75	$ 4.50	16.75	15
$ 5.75	$ 4.25	17	14
$ 5.49	$ 5.00	17.25	12.25

a) For each product determine quantity sold as a function of price (or prices) charged. Write the models for these demands.

b) Find the total revenue as a function of the price of each product. Write the model for the revenue as a function of the price of each product.

c) Find the total profit as a function of the price of each product. Write the model for the profit as a function of the price of each product.

d) Determine the price of each product as a function of the quantity(quantities). Write the models for the prices as functions of the quantities.

e) Find the total revenue as a function of the quantity of each product. Write the model for the revenue as a function of the quantity of each product.

f) Find the total profit as a function of the quantity of each product. Write the model for the profit as a function of the quantity of each product.

14. A small sporting goods store is trying to cash in on the popularity of the local NHL hockey team. They sell two hockey jerseys, the official jersey of the NHL and a replica. In order to improve their profits they need to determine the demand functions for the two shirts. They have collected the data shown below:

PRICE OFFICIAL	PRICE REPLICA	QUANTITY OFFICIAL	QUANTITY REPLICA
$ 64.99	$ 48.00	8	10
$ 94.99	$ 30.00	5	15
$ 67.00	$ 49.99	7	9
$ 83.99	$ 49.00	3	8
$ 69.99	$ 59.00	4	5
$ 64.99	$ 48.00	8	10
$ 67.00	$ 49.99	7	9
$ 70.00	$ 54.00	5	7
$ 82.00	$ 47.99	4	9
$ 64.00	$ 53.00	7	8
$ 81.00	$ 39.99	6	12

The officially sanctioned jerseys cost the retailer $19.23 each and the replicas cost $12.38 each, plus there is a one-time order charge of $50.00.

a) For each product, determine quantity sold as a function of price (or prices) charged. Write the models for these demands.
b) Find the total revenue as a function of the price of each product. Write the model for the revenue as a function of the price of each product.
c) Find the total profit as a function of the price of each product. Write the model for the profit as a function of the price of each product.
d) Determine the price of each product as a function of the quantity(quantities). Write the models for the prices as functions of the quantities.
e) Find the total profit as a function of the quantity each product. Write the model for the profit as a function of the quantity of each product.

15. An athlete competes in the weight throw event for the Track and Field team. He has been trying to decide how many practice throws he should take and how many warm-up laps he should run before competing to achieve maximum distance on his throws. He recorded the following information:

Distance Feet	Warm-Up Throws	Laps
27	1	4
27.4	1	5
27.1	2	2
27.9	3	2
28.4	4	2
29.8	5	2
27.6	2	3
28	2	4
28.8	2	5
29	3	3
33	5	3
30.8	3	5
32	4	4
30.6	5	4
31.2	4	5
29.2	5	5

If D equals the distance in feet and t equals the number of warm-up throws and l equals the number of laps, then $D(t,l) = m_1 t + m_2 l + a_{11}t^2 + a_{12}tl + a_{22}l^2 + b$.

a) Use multivariable regression to model $D(t,l)$.

b) Do you think this is a good model? Explain your answer.

16. A student wants to increase her energy level. She decided to split her exercising between cardiovascular workouts and weight lifting. She recorded the time she spent on each type of exercise and her energy level on a scale of 0 to 100. The results are shown below:

Energy Level	Cardio Minutes	Weight Minutes
70	40	20
62	25	15
67	30	15
72	30	20
63	30	15
77	35	20
79	35	25
73	25	20
81	45	20

If E equals the energy level and c equals the minutes spent on cardiovascular exercises and l equals the minutes spent on weight lifting, , then

$E(c,l) = m_1c + m_2l + m_3c^2 + m_4cl + m_5l^2 + b$.

a) Use multivariable regression to model $E(c,l)$.

b) Do you think this is a good model? Explain your answer.

[illegible]

A student wants to [illegible] level. She decided to [illegible] exercising [illegible] [illegible] and [illegible]. She recorded the time she spent on each type [illegible] on a scale of 0 to 10. The results are shown below.

[illegible]

If [illegible] equals the minutes spent [illegible] exercise [illegible] and [illegible]

[illegible]

Section 6.2: Mean, Variance, Standard Deviation, MSE; Misuse of Statistics

After studying Chapter 4 of Volume 1, you should now know a good amount about evaluating definite integrals and how they relate to finding probabilities and medians of continuous random variables. In that chapter we talked about the mode (most likely value) and the median (50th percentile), which are both measures of "central tendency" (two ways of describing the "middle" or "center" of a probability distribution). But the most common measure of the center of a distribution is the **mean**, a generalization of the simple average of a set of numbers. And often, in addition to having representations of the center, we also want a measure of how spread out the values of the distribution are. If you are investing in a stock, and expect the average annual rate of return to be 10%, you probably would like to know if the range is between -30% and 50% rather than between 8% and 12% -- especially if you are retired and are depending on the investment for income! The **range** (the interval of values of the random variable where the relative probability function is not 0) is a crude measure of variability. A more informative measure is called the **standard deviation**, which is related to what is called the **variance**. These are related to accumulating the squares of the differences of the possible values from the mean, analogous to what we do when fitting curves using SSE. In this section, we will define these concepts and show how to calculate them, and then discuss how they can be manipulated to emphasize different properties of a set of data. We will also discuss the idea of finding the average (or mean) of the squared errors, called the **Mean Squared Error** (MSE). This is closely related to the concept of variance in statistics, and is commonly used to evaluate forecasting methods.

After studying this chapter, you should be able to solve problems like the following:

- A business competitor is about to come out with a new product that is likely to have a significant impact on your business. You have estimated the probability density function for when the new product will be released. What is the expected time of its release? What is the standard deviation of the release time, to help get a feel for the risks involved?
- You have found a Web site that carries quotes of stock prices daily going back in time for months. You have a few favorite stocks in which you are thinking of investing, and want to compute their average daily rate of return and their average risk.
- You want to compare two forecasting techniques to project the price of a favorite stock over the next few months. You have data from the past few years, so you want to see how the two techniques would have performed in the past.

- Your marketing department has collected data on a health product, and is about to analyze it. Should they use the mean or the median or the mode in presenting the results about the health benefits of the product?
- Two politicians have given what sound like very different statistics about income in this country. Is it possible that they are both telling the truth? How can you reconcile the two to get a complete picture?

In addition to being able to solve the above kinds of problems, after studying this section, you should also:

- When working with a finite set of numbers, understand the difference between considering that set as your entire **population** for your analysis, and considering the set as a **sample** from some larger population and want to make inferences about the larger population from the sample.
- Be able to find the **mean, (population) variance, and (population) standard deviation** of a *finite set of numbers*.
- Be able to find the **sample variance** and **sample standard deviation** for a finite set of numbers you want to consider as a *sample*.
- Understand how to *interpret* the mean and standard deviation in the real world.
- Understand the *concepts* of mean, error, squared error, SSE, MSE, and variance, and what they correspond to *graphically*.
- Be able to calculate the SSE and MSE for a given model (or forecasting method) and set of data.
- Be able to calculate the **mean, (population) variance, and (population) standard deviation** for a *continuous random variable* with a given probability density function.
- Understand the differences between the mean, the median, and the mode, and how the choice of which to use can be *manipulative*.

The Mean (Average) of a Finite Set of Numbers

Example 1: On a ten-point quiz, consider the set of scores: {9,9,8,5,4}. The mode (most frequent score) is 9. The median (middle score, when they are listed in order) is 8. What is the average? As we all know, for a finite set of numbers like this, the average is given by:

$$\text{average} = \frac{\text{sum of the scores}}{\text{number of scores}}$$

If we call the scores $x_1, x_2, x_3, \ldots, x_n$ in general (so the number of scores is n), and we call the average (mean) $\bar{x}$, then the formula for this can be written:

$$\text{mean} = \bar{x} = \frac{x_1 + x_2 + \ldots + x_n}{n}$$

Notice the sum in the numerator is very repetitive: all of the terms are the same, except for the subscript of the x. As we saw in Section 4.2 of Volume 1 and in Section 5.5, we can use **sigma notation** here as follows:

$$\sum_{i=1}^{n} x_i = x_1 + x_2 + \ldots + x_n \quad \text{, so}$$

$$\bar{x} = \frac{x_1 + x_2 + \ldots + x_n}{n} = \frac{\sum_{i=1}^{n} x_i}{n}$$

In our quiz score example, then, the average (mean) would be

$$\bar{x} = \frac{\sum_{i=1}^{5} x_i}{5} = \frac{x_1 + x_2 + x_3 + x_4 + x_5}{5} = \frac{9+9+8+5+4}{5} = \frac{35}{5} = 7$$

<u>The Mean (Expected Value) of a Continuous Random Variable</u>

Going back to the calculation of the mean above, notice that the calculation could have been written

$$\begin{aligned} \text{average} &= 9\left(\frac{1}{5}\right) + 9\left(\frac{1}{5}\right) + 8\left(\frac{1}{5}\right) + 5\left(\frac{1}{5}\right) + 4\left(\frac{1}{5}\right) \\ &= 9\left(\frac{2}{5}\right) + 8\left(\frac{1}{5}\right) + 5\left(\frac{1}{5}\right) + 4\left(\frac{1}{5}\right) \end{aligned}$$

Graphically, this situation could be represented with a histogram pdf, as in Figure 6.2-1 .

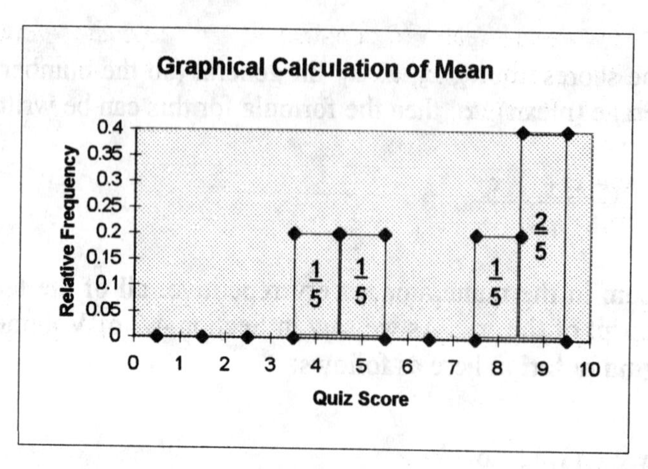

Figure 6.2-1

The rectangles in the graph are all of width 1. You can think of this as reflecting the fact that a score of "4" could have actually been anything from a 3.5 to a 4.499, rounded off. The height of each rectangle reflects the relative frequency of each score, similar to what we did with histograms in Section 4.1 of Volume 1.

Symbolically, more generally, this calculation could be written

$$\bar{x} = \sum_{i=1}^{4} x_i (f(x_i))$$

where $f(x_i)$ is the relative frequency of the score x_i. Notice that we are actually using a different definition of x_i this time, by not counting repeated scores as different values, but instead counting them as *one* value and reflecting the *number* of them in the relative frequency.

Sample Problem 1: Your competitor is about to come out with a new product that will have a large effect on your business (remember him from Chapter 4 of Volume 1?). As before, you estimate that the probability distribution for the number of months before it is released (X) to be

$$f(x) = \begin{cases} 0.2e^{-0.2x} & \text{for } x \geq 0 \\ 0 & \text{otherwise} \end{cases}$$

find the average (mean) number of months before the product is released.

Solution: Now we are ready to consider how to find the average, or **mean**, of a *continuous* probability distribution such as the one in this problem. Since the formula given just before Sample Problem 1,

$$\text{mean} = \bar{x} = \sum_{i=1}^{4} x_i(f(x_i))$$

looks like the sum for one of the rectangle methods (where the Δx is 1) for finding the area under a curve, we can define the mean for a continuous random variable in an analogous way:

$$\text{mean} = \mu = E(X) = \int_{-\infty}^{\infty} x f(x)dx$$

μ is "mu", the lower-case Greek letter corresponding to "*m*" (the first letter in "mean") and is pronounced "mew." It is a common symbol for the mean of a continuous random variable. $E(X)$ is read "ee of ex" and stands for what is called the **expected value** of X (another name for the mean). The $\bar{x}$ notation is usually used just for discrete cases (such as for a finite number of values). The integral is over all real numbers, but remember that many random variables have restricted domains, and $f(x) = 0$ everywhere else, so you can always use the endpoints of the domain over which $f(x)$ is *strictly* positive ($f(x) > 0$) to simplify your calculations.

For example, in our problem, the domain over which the pdf is strictly positive is $[0,\infty)$, so the mean will be

$$\mu = E(X) = \int_{-\infty}^{\infty} x f(x)dx \qquad \text{(definition of the mean)}$$

$$= \int_0^{\infty} x f(x)dx \qquad \text{(since the domain over which } f(x) > 0 \text{ is } [0,\infty)\text{)}$$

$$= \int_0^{\infty} x\left[0.2e^{-0.2x}\right]dx \qquad \text{(substituting in for } f(x)\text{)}$$

$$= \int_0^{\infty} 0.2xe^{-0.2x}\,dx \qquad \text{(simplifying)}$$

Notice that, since $f(x)$ is 0 for all values of x in the interval $(-\infty,0)$, then $xf(x)$ is also 0 over that interval, so the integral of $xf(x)$ over that interval is 0, and so we can in fact just work with the integral from 0 to ∞.

We have not learned enough calculus to find an antiderivative and use the Fundamental Theorem of Calculus to evaluate this integral, so we must find the solution with technology. Since the integral is an improper integral (one of the limits of integration is some kind of infinity), what we really want to find is

$$\lim_{b \to \infty} \int_0^b 0.2xe^{-0.2x}dx$$

If we set up a table to evaluate this integral numerically (for example, using Simpson's Rule and $n = 1000$) for $b = 10, 20, 40$, etc., the results are as follows:[1]

b	Integral on [0,b]
10	2.97
20	4.54
40	4.98
80	5.00
160	5.00

Our conclusion, then, is that the expected (mean, average) number of months before the new product will be released is 5 months. □

A Graphical Interpretation of the Mean

Sample Problem 2: Find the mean of the pdf given by

$$f(x) = \begin{cases} \frac{1}{6} - \frac{1}{72}x & \text{for } 0 \le x \le 12 \\ 0 & \text{otherwise} \end{cases}$$

and explain what the mean corresponds to graphically.

Solution: If we plug into the formula for the mean of a continuous distribution, we get

$$\mu = \int_{-\infty}^{\infty} xf(x)dx = \int_0^{12} x\left(\tfrac{1}{6} - \tfrac{1}{72}x\right)dx$$
$$= \int_0^{12} \left(\tfrac{1}{6}x - \tfrac{1}{72}x^2\right)dx \qquad \text{(multiplying out)}$$

Notice that we can solve this integral easily and exactly using the Fundamental Theorem of Calculus, so let's refresh your memory and do it here:

---

[1] Note: If you are using a spreadsheet for improper integration, be sure to double-check that your final integral (in the above example, the integral from 0 to 160) converges itself. For example, evaluate it using Simpson's Rule for $n = 250$, 500, and 1000, and make sure that the values converge. See your technology supplement.

$$\mu = \int_0^{12} \left(\frac{1}{6}x - \frac{1}{72}x^2\right)dx$$

$$= \left[\frac{1}{6}\frac{x^2}{2} - \frac{1}{72}\frac{x^3}{3}\right]_0^{12}$$ (taking antiderivatives and evaluating)

$$= \left[\frac{x^2}{12} - \frac{x^3}{216}\right]_0^{12}$$ (multiplying out)

$$= \left[\frac{(12)^2}{12} - \frac{(12)^3}{216}\right] - \left[\frac{(0)^2}{12} - \frac{(0)^3}{216}\right]$$ (plugging in the endpoints and subtracting)

$$= [12 - 8] - [0 - 0] = 4 - 0$$ (simplifying)

$$= 4$$

Thus the mean of this distribution is 4.

Graphically, the mean corresponds to the balance point of the density function. In this way of thinking, picture the area under the density curve as a flat sheet of metal or plastic of *uniform* thickness, and imagine you have a stick (dowel) that you want to put under the sheet, with the stick lined up parallel to the vertical axis. The point along the horizontal axis where you could put the stick so that the sheet in the shape of the area under the density curve just balances (like a see-saw) would be the mean.

In our problem, this would look something like Figure 6.2-2 .

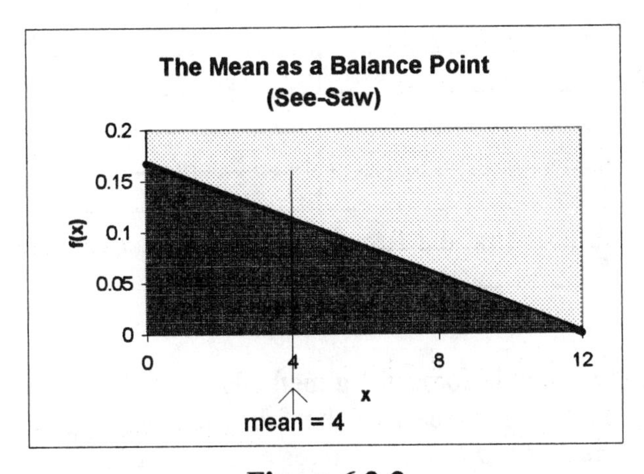

Figure 6.2-2

The vertical line at 4 is the stick (like the frame of the see-saw, which we are looking *down* on from above), and the dark triangle region is the flat sheet of metal (like the moving part of the see-saw, again seen from above), which should balance perfectly when the stick is placed under the sheet parallel to the *y*-axis at the mean ($x = 4$ here). As with a see-saw, the *heavier* kid (wider part of the sheet) must be *closer* to the balance point to balance with the *lighter* kid (narrower part of the sheet), who must be *further* from the balance point. □

Mean Squared Error (MSE) and Forecasting

In Section 1.3 of Volume 1 and in Section 5.5, we have discussed the basic idea of errors in the case of a single-variable model, and the concept of looking at the squares of these errors, or the **squared errors**, as we might call them. We can also talk about other operations we might do on these squared errors. For example, if we are given a particular set of data and just want to compare the fit of two possible models, the SSE (sum of the squared errors) is a perfectly good measure. But what if we wanted to compare the fit for two *different* models associated with *different* data sets? If the data sets are different sizes, the one with more data points is likely to have a higher SSE, even though visually we all might agree that the fit looks better. How could we make an adjustment for such different numbers of data points? Let's look at an example.

Sample Problem 3: You are interested in investing in the stock of your favorite company, Wiz Biz. You have tracked its price over the last few months, and want to try to forecast what it will be into the future. Here is the data:

Table 1

Month	Price
0	28
1	30
2	29
3	32
4	31
5	35
6	34

You are considering two simple forecasting methods: Last Value (just use the price from the *most recent* month as the forecast of the following month) and Last Two Average (take the *average* of the *last two* months' prices as the forecast of the next month). Both methods can be applied sequentially to this data. The Last Value could start by forecasting the value for month 1, but the Last Two Average method could only start by forecasting the value for month 2, since it requires *two* months of data (months 0 and 1) to be averaged.

Thus we can get predicted and actual prices for both methods over a number of different years each, but the number of predictions, and so the number of errors, will be different. How can we get a measure of the error to make a comparison between the methods? Which seems better in this case?

Solution: Let us first apply the methods and calculate the errors:

Table 2

Month	Price	Last Value	Error	Squared Error	Last 2 Avg.	Error	Squared Error
0	28						
1	30	28	30-28=2	$2^2 = 4$			
2	29	30	29-30=-1	$(-1)^2 = 1$	29	29-29=0	$0^2 = 0$
3	32	29	32-29=3	$3^2 = 9$	29.5	32-29.5=2.5	$2.5^2 = 6.25$
4	31	32	31-32=-1	$(-1)^2 = 1$	30.5	31-30.5=.5	$.5^2 = .25$
5	35	31	35-31=4	$4^2 = 16$	31.5	35-31.5=3.5	$3.5^2 = 12.25$
6	34	35	34-35=-1	$(-1)^2 = 1$	33	34-33=1	$1^2 = 1$

For the two methods, the Sum of the Squared Errors (SSE) is

$SSE_{last} = 4+1+9+1+16+1 = 32$
$SSE_{last2avg} = 0+6.25+.25+12.25+1 = 19.75$

Based on these alone, the Average of the Last Two Values method looks better. But the Last Value method made more predictions, so had more data points, which could have been the reason its SSE was larger. How could we adjust for these different numbers of data points?

Perhaps you already thought yourself of the idea of finding the **average** of the squared errors; in other words, add them up and divide by the number of data points (let us as usual call this n). As we have just seen, such an average could be called the **mean** value of the squared errors, or the **Mean Squared Error (MSE)**. This measure is often used when comparing forecasting methods for the exact reasons we have discussed above. As a formula, we could say

$$MSE = \frac{SSE}{n} = \frac{\sum_{i=1}^{n}(y_i - \hat{y}_i)^2}{n}$$

$$= \frac{\sum_{i=1}^{n}(\textit{actual}\text{ output }\textit{minus}\text{ the }\textit{predicted}\text{ output for the }\textit{input}\text{ value of the }i\text{'th data point})^2}{\text{the total number of data points}}$$

In the example, then,

$$\text{MSE}_{\text{last}} = \frac{1+4+1+9+1+16}{6} = \frac{SSE_{last}}{n} = \frac{32}{6} = 5\frac{1}{3} \approx 5.333$$

$$\text{MSE}_{\text{last2avg}} = \frac{0+6.25+.25+12.25+1}{5} = \frac{SSE_{last2avg}}{n} = \frac{19.75}{5} = 3.95$$

This suggests that, even after adjusting for the different numbers of data points, the Average of the Last Two Values method still seems better for predicting Wiz Biz's stock price over this period. □

There are a number of other simple techniques for **forecasting** simple **time series** like this. (A time series is simply a set of values over time, like stock prices, inflation rates, etc.) For instance, instead of a *simple* average of the last 2 values, you could take a *weighted* average of the last 2 values, such as 80% of the last value plus 20% of the value before that. Or you could take a weighted average of the most recent *observation* with the previous *estimate*, called **exponential smoothing**. This turns out to be equivalent to taking a weighted average of all of the previous values, where the weights decrease as you go further back in time from the present. All of the methods mentioned so far are appropriate in a situation where you expect the values to be fairly **stable** over time, with only some possible drift up or down occasionally.

If you believe that there is a significant **trend**, the simplest method would be to fit a line to the last 2 values, which you could call a **trendline**, and use that line to project into the future. Or, you could use 3 or more of the most recent points to fit the trendline, especially if you expect the *trend* to be fairly stable. There are also versions of the concept of exponential smoothing that can be applied in the trend situation, which allows for updating estimates of the trend as well as the value at different times.

For details about these different techniques of forecasting and others, see a book on forecasting.

Measuring Variability: Population and Sample Variance of a Discrete Set of Numbers

Example 2: Suppose you are trying to figure out the best angle from which to shoot a 3-point shot in basketball, to achieve your highest possible shooting percentage on average. You decide to gather data at a number of different angles by shooting and calculating your shooting percentages. Let's suppose you decide to take 50 shots at *each* of 5 angles: 0, 45, 90, 135, and 180 degrees from the right baseline (as viewed from the foul line), but you take your 50 shots in groups of 10. The data are given in Table 3.

Table 3

Angle *(°)*	**1st 10** *(%)*	**2nd 10** *(%)*	**3rd 10** *(%)*	**4th 10** *(%)*	**5th 10** *(%)*	**Overall** *(%)*
0	10	20	20	10	30	18
45	70	50	60	60	70	62
90	50	40	60	70	50	54
135	70	80	50	60	60	64
180	0	20	30	40	10	20

Now, the most obvious way to find a model for this relationship is to plot the data points for the overall shooting percentage versus the angle, as shown in Figure 6.2-3 .

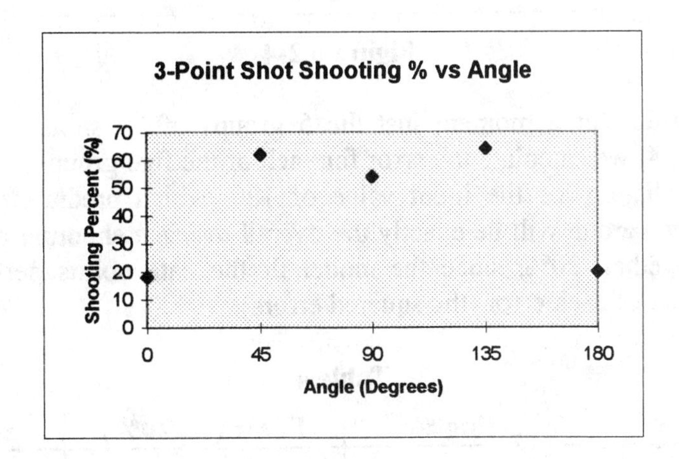

Figure 6.2-3

To capture the wiggles in this relationship, since it would seem to require at least two points of inflection, we would need at least a quartic model. But since we only have 5 data points, we could not fit anything more than a quartic, and a quartic will in fact fit perfectly. This is not necessarily ideal, since it might suggest our model is more accurate than it really is, but for the current discussion it is convenient, so we will work with it. The resulting quartic model (rounded to 4 significant figures) is:

Verbal Definition: $S(a)$ = shooting percent, on average, from an angle of a degrees from the right baseline (viewed from the foul line)

Symbol Definition: $S(a) = -0.000001443a^4 + 0.0005176a^3 + 0.06226a^2 + 2.863a + 18.00$

$0 \leq a \leq 180$

Assumptions: Certainty and divisibility.

The graph of this function is shown in Figure 6.2-4 .

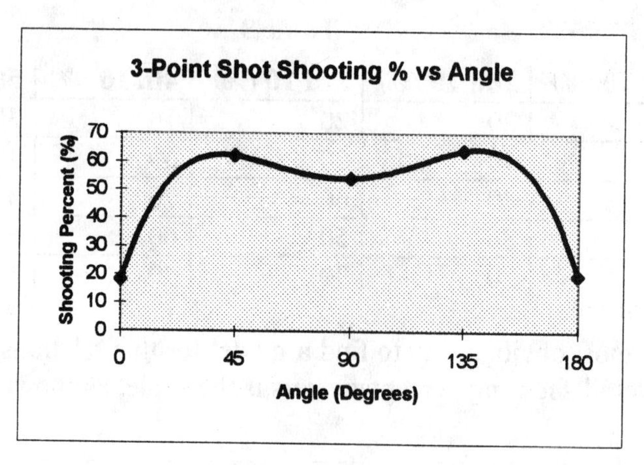

Figure 6.2-4

Let's consider for a moment just the 5 groups of 10 shots each taken at 180 degrees. In Table 4, we calculate the error for each of the five groups of 10, as compared to the model prediction for the input value of 180, which predicts that the shooting percentage for *each* group will be exactly the overall *average* shooting percentage for all of the groups together, 20%, since the model fit the data points perfectly. We also calculate the square of each error (the squared errors).

Table 4

Group (at 180°)	Shooting %	Error (vs. 20%)	Squared Error
1	0	-20	400
2	20	0	0
3	30	10	100
4	40	20	400
5	10	-10	100

The Sum of the Squared Errors will be

$$\text{SSE} = 400 + 0 + 100 + 400 + 100 = 1000$$

Since there are 5 data values, the average squared error (MSE) will be the total (SSE) divided by 5, or

$$\text{MSE} = \frac{SSE}{n} = \frac{1000}{5} = 200 .$$

In this situation, this value of the MSE, 200, also corresponds to what is called the **variance**, or the **population variance**, of the 5 data values. Intuitively, the variance gives us a measure of how spread out data values are. Let's define the concept more generally for finite sets of data values now:

The **(population) variance** of the *n* values $x_1, x_2, ..., x_n$ (denoted by σ^2) is given by

$$\text{(population) variance} = \sigma^2 = \frac{\sum_{i=1}^{n}(x_i - \bar{x})^2}{n},$$

$$\text{where } \bar{x} = \frac{x_1 + x_2 + ... + x_n}{n} \text{ is the mean of the values.}$$

As we have said, this is related to the idea of squared errors, if we think of $\bar{x}$ as the *predicted* value for the set of numbers. Put differently, it is as if we were using a model which was just a *constant* (horizontal line), and the constant we use as the prediction is the *mean.* Since the model is constant, the results would not be affected by different input values (they could be all the same, as in this example, or all different; it wouldn't matter), so all we really have is a set of output values. In this interpretation, it might make more sense to use the letter y for the variable instead of x, so the actual output values would be $y_i = x_i$, and the predicted output values would be $\hat{y}_i = \bar{x}$. In that case, the variance is exactly the Mean Squared Error, MSE, since

$$MSE = \frac{SSE}{n} = \frac{\sum_{i=1}^{n}(y_i - \hat{y}_i)^2}{n}.$$

You can see that replacing y_i with x_i and $\hat{y}_i$ with $\bar{x}$ in the above MSE formula gives exactly the population variance formula.

The symbol for the population variance is σ^2. The symbol "σ" is the *lower*-case Greek letter "sigma". Recall that the summation symbol, $\sum$, is the *upper*-case sigma.

There is a subtle variation on this variance formula. The above formula for the **population variance** is valid **if the x_i values are *all* that you care about,** your entire **population** of concern. But **if the x_i values are considered to be a *sample* of values *from a larger population*,** and we want to *use* the sample to make *inferences* about the larger population, then we will often **use** what is called the **sample variance,** denoted S^2, which is calculated as

$$\text{sample variance} = S^2 = \frac{\sum_{i=1}^{n}(x_i - \bar{x})^2}{n-1}$$

The details of the theoretical difference between these two kinds of variance is beyond the scope of this course, but would normally be covered in an intermediate statistics course. In essence **S^2 is considered an *estimate* of the true variance for the entire larger population** that the sample comes from, and **a better (unbiased) estimate is obtained by using the (n-1) in the denominator instead of n** .

In our basketball example, it makes sense to think of our data values as a *sample* from a larger population, which we could think of as the almost limitless set of possible groups of 10 shots we *could* make at the given angle of 180 degrees on the 3-point line. Our sample variance would then be (since the mean is (0+20+30+40+10)/5 = 20)

$$S^2 = \frac{(0-20)^2 + (20-20)^2 + (30-20)^2 + (40-20)^2 + (10-20)^2}{5-1} = \frac{400+0+100+400+100}{4} = \frac{1000}{4}$$
$$= 250$$

Notice that **the *sample* variance gives us a *larger* estimate of the variability, or margin of error, in our data values, than does the population variance**. This makes sense, since if our values are only a sample, it is possible that a different sample could vary even more (**the uncertainty adds to our margin of error**).

In our example, we could calculate the sample variance for each of the angles in our data table. Try doing at least one of these calculations yourself, to make sure you have the idea clear. You should get the results shown in Table 14.

Table 14

Angle (°)	1st 10	2nd 10	3rd 10	4th 10	5th 10	Mean	Sample Variance
0	10	20	20	10	30	18	70
45	70	50	60	60	70	62	55
90	50	40	60	70	50	54	105
135	70	80	50	60	60	64	130
180	0	20	30	40	10	20	250

Notice that the sample variances are quite different in the different groups. Some of the pieces of a **standard statistical analysis of a least-squares regression model,** such as our use of confidence intervals for the parameter estimates in Section 6.1 , ***assumes* that these variances are all *equal*** (this has the intimidating name of the assumption of **homoscedasticity**). Once again, the details are beyond the scope of this course, but we will approach this topic again in Section 6.3 . More advanced coverage would come in a statistics course.

Example 3: In measuring variability for a discrete set of values like our quiz scores in Example 1, {4,5,8,9,9}, let's look at a detailed breakdown of the population variance calculation for our quiz scores (recall that the mean was 7):

Index	Score	Deviation (Error)	Squared Deviation/Squared Error
i	x_i	$(x_i - \bar{x})$	$(x_i - \bar{x})^2$
1	4	$(4 - 7) = -3$	$(-3)^2 = 9$
2	5	$(5 - 7) = -2$	$(-2)^2 = 4$
3	8	$(8 - 7) = 1$	$(1)^2 = 1$
4	9	$(9 - 7) = 2$	$(2)^2 = 4$
5	9	$(9 - 7) = 2$	$(2)^2 = 4$

Thus we get

$$\text{population variance} = \sigma^2 = \frac{\sum_{i=1}^{5}(x_i - \bar{x})^2}{5}$$

$$= \frac{(4-7)^2 + (5-7)^2 + (8-7)^2 + (9-7)^2 + (9-7)^2}{5}$$

$$= \frac{(-3)^2 + (-2)^2 + (1)^2 + (2)^2 + (2)^2}{5}$$

$$= \frac{9+4+1+4+4}{5} = \frac{22}{5} = 4.4$$

The *population* variance would be the appropriate variance to use if the five quiz scores were all we cared about - for instance, if they corresponded to the only five students in a unique one-time upper-level seminar course. However, if the five scores were just a fraction of one class (maybe the first five quizzes graded, selected at random from a class of 30), or if the class were one out of 15 sections of the same required course into which all freshmen are registered at random, and you wanted to make inferences about the larger group (such as to estimate its population variance), then the *sample* variance would be more appropriate. There could even be a situation where the *same* group of five quiz scores (say for a class of only five students that was one out of 15 sections of a large course) for which *either* type of variance would be appropriate, *depending on the desired analysis*. For example, to just *report* the results to the class of five, the *population* variance would be more appropriate, but to *estimate* the *population* variance of all the students in the larger course the *sample* variance would be more appropriate.

For whatever reason it is wanted, the sample variance in this case would be

$$\text{sample variance} = S^2 = \frac{\sum_{i=1}^{n}(x_i - \bar{x})^2}{n-1}$$
$$= \frac{(4-7)^2 + (5-7)^2 + (8-7)^2 + (9-7)^2 + (9-7)^2}{5-1} = \frac{22}{4} = 5.5$$

(notice that, as will always be the case, the numerator is the *same* as for the population variance). This would then mean that the *better* estimate of the *population* variance of the *larger* group (say, the students in all of the 15 sections of the larger course) is 5.5 (the sample variance), not 4.4 (the population variance), since the five scores are being considered a *sample* from the larger population.

In case you were wondering, we do not make a distinction between the population mean and the sample mean; or, more accurately, the formulas are the same for both the population and sample mean of a finite set of values. Notationally, when you are thinking of the data as a *sample*, it is common to use the $\bar{x}$ notation for the sample mean (and then μ for the population mean of the larger population), and if you consider the values to be your entire population, it is common to use μ for the population mean, but the distinction is not important for the purposes of this text, so we will tend to stick to the $\bar{x}$ notation for the mean of a set of finite values, however it is being treated.

Standard Deviation: A More Easily Interpreted Measure of Variability

What does the population variance mean? In the quiz score example, it's saying that the average of the squares of the errors (the average of the squares of the deviations of the scores from the mean) is 4.4 . What would be the units for this number? Well, if you look at the table, the units for the x_i's and $\bar{x}$ are *quiz points* (out of 10). This means the errors are also in units of points, and so the squares of the errors (and so also their mean) must be in units of *square points*! What does that mean, you ask? Not much!! But if we took the square root, the units would be points again, so would be easier to interpret. This is the idea behind the **(population) standard deviation**, which is denoted,[2] not surprisingly, σ (the square root of σ^2). The formula for the standard deviation is thus:

[2] Recall again that σ is the lower-case Greek letter "s", which you can now think of as standing for "standard deviation."

(population) standard deviation = $\sqrt{\text{(population) variance}}$

$$\sigma = \sqrt{\sigma^2} = \sqrt{\frac{\sum_{i=1}^{n}(x_i - \bar{x})^2}{n}}$$

In our quiz score example, we know the calculation inside the radical is 4.4, so we get that $\sigma = \sqrt{4.4} \approx 2.10$ points. How can we interpret this?

It can be proven mathematically that a vast majority (at least 8/9, or about 88.9%) of probability for a random variable falls within 3 standard deviations ($\pm 3\sigma$) of the mean. For a normal distribution, in fact, about 99.7% of the probability falls within the 3σ range. For most real world probability distributions, 99% is probably a good rough figure to work with intuitively, which will usually be within 1 percentage point of the exact value. In our quiz-score example, $3\sigma = 3(2.10) = 6.30$, so the 3σ interval would be

$$\bar{x} \pm 3\sigma = 7 \pm 3(2.10) = 7 \pm 6.30 \rightarrow [7 - 6.30, 7 + 6.30] = [0.70, 13.30]$$

In this case, in fact, *all* of the data values (scores) fall within the 3σ range, as is often the case with discrete sets of values like this, especially when n is small. Notice that the right endpoint of the interval calculated above (13.30) is outside the domain of the random variable (the domain is presumably the interval [0,10], since there are only 10 questions on the quiz). In terms of probability, then, the interval $[0.70, 13.30]$ is equivalent to (has the same probability as) $[0.70, 10]$.

The **sample standard deviation,** S , would be more appropriate, as before for the sample variance, in the case where the data are being considered a *sample* from a larger population, and is analogously just the square root of the sample variance:

sample standard deviation = $\sqrt{\text{sample variance}}$

$$S = \sqrt{S^2} = \sqrt{\frac{\sum_{i=1}^{n}(x_i - \bar{x})^2}{n-1}}$$

For the quiz scores, since the sample variance was 5.5 , the sample standard deviation is $S = \sqrt{5.5} \approx 2.35$. Thus if the quiz scores were being seen as a sample from a larger group, the best estimate of the population standard deviation for the larger group would be 2.35 and so the most appropriate 3σ interval for the larger group would be

$$\bar{x} \pm 3\sigma = 7 \pm 3(2.35) = 7 \pm 7.05 \rightarrow [7-7.05, 7+7.05] = [-0.05, 14.05] \ ,$$

which is saying that nearly all of the quiz scores for the larger group should lie between -0.05 and 14.05 . If we again assume that all quiz scores are between 0 and 10, then in fact we would actually expect *all* the scores of the larger group to lie between -0.05 and 14.05 .

When working with large values of n, there is very little difference between the population and sample values of the variance and the standard deviation. For very small numbers of data points, the difference can be more significant. The 3σ guideline technically involves the *population* standard deviation. When working with a *sample*, if you are trying to make inferences about the distribution of values in the larger *population*, you should ideally use the *population* mean and standard deviation *of the larger population*, but in their absence, your best estimates of these values are given by the *sample* mean and standard deviation. In simple terms, you use the *best* information you have.

Suppose for a minute that the 5 quiz scores had been $\{8,8,7,6,6\}$. Because of the symmetry, you can see that the mean would again be 7 ($\frac{8+8+7+6+6}{5} = \frac{35}{5} = 7$), and the population standard deviation would be

$$\sigma = \sqrt{\frac{(8-7)^2 + (8-7)^2 + (7-7)^2 + (6-7)^2 + (6-7)^2}{5}} = \sqrt{\frac{4}{5}} \approx 0.89$$

compared to 2.10 for the original data. In this case the 3σ interval would be $7 \pm 3(0.89)$ or $[4.33, 9.67]$, compared to $[0.70, 13.30]$ for the original data. This is a much narrower band of values than for the original data, corresponding to the fact that the data are much more tightly clustered around the mean in this second example. This helps give you a sense of the usefulness of the information contained in the population standard deviation. Just knowing the mean and the population standard deviation of the two sets of quiz scores would give you a pretty good rough sense of where the values should fall in each case.

Variance and Standard Deviation for Continuous Random Variables

Analogous to what we did with the mean, our calculation of the population variance for the origianl quiz-score problem could have been written:

$$\sigma^2 = (4-7)^2\left(\frac{1}{5}\right) + (5-7)^2\left(\frac{1}{5}\right) + (8-7)^2\left(\frac{1}{5}\right) + (9-7)^2\left(\frac{2}{5}\right)$$

This can be generalized as

$$\text{population variance} = \sigma^2 = \sum_{i=1}^{n}(x_i - \bar{x})^2 f(x_i)$$

where $f(x_i)$ is the probability of the value x_i .

As before, the continuous analogy would then be

$$\text{population variance} = \sigma^2 = \int_{-\infty}^{\infty}(x-\mu)^2 f(x)dx$$

where μ is the mean and $f(x)$ is the relative probability function (pdf).

In the case of a continuous random variable X with pdf $f(x)$, then, the (population) variance is given by

$$\text{(population) variance} = \sigma^2 = \int_{-\infty}^{\infty}(x-\mu)^2 f(x)dx \ ,$$

$$\text{where} \quad \mu = E(X) = \int_{-\infty}^{\infty} x\, f(x)\, dx$$

is the mean. Thus, finding the variance in both the finite set of values and continuous cases is a 2-step process: first, you find the mean, and then you do the calculation for the variance. Most calculators have calculations for both the population and sample variances for a finite set of values.

Since the standard deviation is simply the square root of the variance, we define:

$$\text{(population) standard deviation} = \sqrt{\text{(population) variance}}$$

$$\sigma = \sqrt{\sigma^2} = \sqrt{\int_{-\infty}^{\infty}(x-\mu)^2 f(x)dx}$$

Sample Problem 3: Your competitor is about to come out with a new product that will have a large effect on your business, as in Sample Problem 1. You estimate that the probability distribution for the number of months before it is released (X) to be

$$f(x) = \begin{cases} 0.2e^{-0.2x} & \text{for } x \geq 0 \\ 0 & \text{otherwise} \end{cases}$$

To get a measure of the risk and variability of the random variable X (the number of months before the new product is released), find the mean squared deviation from the mean (the variance) and the root-mean-squared deviation from the mean (the standard deviation, which is the square root of the variance). How can these be roughly interpreted in the context of this problem?

Solution: Based on the above discussion, for our competitor's new product release time, the **variance** would be

$$\sigma^2 = \int_{-\infty}^{\infty} (x-\mu)^2 f(x)dx \qquad \text{(definition of variance)}$$
$$= \int_0^{\infty} (x-5)^2 \left[0.2e^{-0.2x}\right]dx \qquad \text{(since domain is } [0,\infty)\text{, \& substituting for } \mu \text{ and } f(x)\text{)}$$

This is even more complicated than the integral for the mean was, but with technology, it doesn't really make a big difference. We can use the same approach as we used for the mean of this distribution in Sample Problem 1, and we get a result of

$$\sigma^2 = 25\text{, so } \sigma = \sqrt{25} = 5$$

The 3σ range would thus be

$$\mu \pm 3\sigma = 5 \pm 3(5) = 5 \pm 15 \to [5-15, 5+15] = [-10, 20]$$

Translation: We can be about 99% sure that the competitor's product will come out within 20 months.

The actual *exact* probability associated with this range would be

$$\int_{-10}^{20} f(x)dx = \int_0^{20} 0.2e^{-0.2x}dx \qquad \text{(since the domain is } [0,\infty)\text{; substituting for } f(x)\text{)}$$
$$= \left[0.2\frac{e^{-0.2x}}{-0.2}\right]_0^{20} \qquad \text{(by the Fundamental Theorem; } \int e^{kx}dx = \frac{e^{kx}}{k} + C\text{)}$$
$$= \left[-e^{-0.2x}\right]_0^{20} \qquad \text{(canceling the 0.2's)}$$
$$= \left[-e^{-0.2(20)}\right] - \left[-e^{-0.2(0)}\right] \qquad \text{(evaluating from 0 to 20)}$$
$$= \left[-e^{-4}\right] - \left[-e^{0}\right] \qquad \text{(multiplying out the exponents)}$$
$$\approx [-0.018] - [-1] \qquad \text{(evaluating the powers of } e\text{)}$$
$$= 1 - 0.018 \qquad \text{(simplifying)}$$
$$= 0.982 \qquad \text{(subtracting)}$$

so about 98.2% of the probability falls within the 3σ range.

Translation: To be more precise, we can be 98.2% sure that the competitor's product will come out within 20 months. □

In the above calculation, the integral $\int_0^{20} 0.2e^{-0.2x}dx$ corresponds to $P(X \le 20)$ for our random variable X. Recall that in Section 4.6, we defined $F(x)$, the **cumulative distribution function (cdf)** to be $P(X \le x)$. In fact, we calculated the cdf for this specific continuous probability distribution, and it came out to be

$$F(x) = \begin{cases} 1 - e^{-0.2x} & \text{for } x \ge 0 \\ 0 & \text{otherwise} \end{cases}$$

If you knew this cdf function, then the calculation above could have also been done as follows:

$$\begin{aligned}
\int_{-10}^{20} f(x)dx &= \int_0^{20} 0.2e^{-0.2x}dx && \text{(since the domain is } [0,\infty)\text{, \& substituting for } f(x)\text{)} \\
&= F(20) && (P(X \le 20) = F(20)) \\
&= 1 - e^{-0.2(20)} && \text{(since } F(x) = 1 - e^{-0.2x}\text{; plugging in 20 for } x\text{)} \\
&= 1 - e^{-4} && \text{(multiplying out the exponent)} \\
&\approx 1 - 0.018 && \text{(evaluating } e^{-4}\text{)} \\
&= 0.982 && \text{(subtracting)}
\end{aligned}$$

Sample Problem 4: Find the mean and the standard deviation of the standard normal distribution, and find the probability of being within the 3σ range.

Solution: As was true for the median, there is a smart and sassy answer to this question for the mean: since the mean is the vertical line on which a sheet shaped like the density function would balance, and since the normal density is symmetric around x=0, the mean must be 0. But let's verify (double-check) this using calculus:

Using the formula for the mean ($\mu = \int_{-\infty}^{\infty} x f(x)dx$), we get:

$$\mu = \int_{-\infty}^{\infty} x f(x)dx = \int_{-\infty}^{\infty} x\left[\frac{1}{\sqrt{2\pi}}e^{-x^2/2}\right]dx \qquad \text{(plugging in for } f(x)\text{)}$$

$$= \int_{-\infty}^{0} x\left[\frac{1}{\sqrt{2\pi}}e^{-x^2/2}\right]dx + \int_{0}^{\infty} x\left[\frac{1}{\sqrt{2\pi}}e^{-x^2/2}\right]dx \qquad \text{(splitting up the integral)}$$

Evaluating each of the two pieces using numerical integration and tables with technology as before, we find that the first integral comes out to -0.399 and the second comes out to 0.399, so the sum is 0, confirming our quick answer.

Unfortunately, for the variance and standard deviation, there is no quick fix, so we go immediately to the formula:

$$\sigma^2 = \int_{-\infty}^{\infty}(x-\mu)^2 f(x)dx = \int_{-\infty}^{\infty}(x-0)^2\left[\frac{1}{\sqrt{2\pi}}e^{-x^2/2}\right]dx = \int_{-\infty}^{0} x^2\left[\frac{1}{\sqrt{2\pi}}e^{-x^2/2}\right]dx + \int_{0}^{\infty} x^2\left[\frac{1}{\sqrt{2\pi}}e^{-x^2/2}\right]dx$$

This time, again using technology, each piece converges to 0.5, so the sum is 1. Thus the variance of the standard normal distribution is 1, and since the standard deviation is just the square root of the variance, the standard deviation is also 1.

Thus the 3σ range in this case is

$$\mu \pm 3\sigma = 0 \pm 3(1) = \pm 3 \rightarrow [-3,3]$$

The probability of being in this range then corresponds to

$$\int_{-3}^{3} f(x)dx = \int_{-3}^{3}\frac{1}{\sqrt{2\pi}}e^{-x^2/2}dx$$

Again numerically, we find the answer using technology to be .9973, which means that about 99.7% of the probability falls within 3 standard deviations of the mean in the standard normal distribution (and in fact, this also holds for *any* normal distribution).

How the Choice of a Descriptive Statistic Can Be Manipulative

How can the different descriptive statistics:

- the mode (or most likely value),
- the median (or 50th percentile),
- the mean (or average or balance point),

- the variance (or the average squared deviation from the mean), and
- the standard deviation (or square root of the variance)

be manipulated? The most common manipulation is the choice between the mode, median, and mean. If a distribution is symmetric about a value, then the median and the mean will be the same, as with the normal distribution. In the case of the normal distribution or a general symmetric unimodal distribution, the mode also coincides with the mean and the median, so there is little latitude for manipulation. But the further a distribution is from being symmetric, the more different all three can be.

When the mean and the median differ, the distribution is said to be **skewed** to one side. **If the mean is larger than the median, we say the distribution is skewed to the right.** This might be the opposite of what you would expect intuitively, since the peak (mode) is on the left side of the graph. Just remember that the direction is determined by the *mean* and not the *mode*. Graphically, this looks like Figure 6.2-5 .

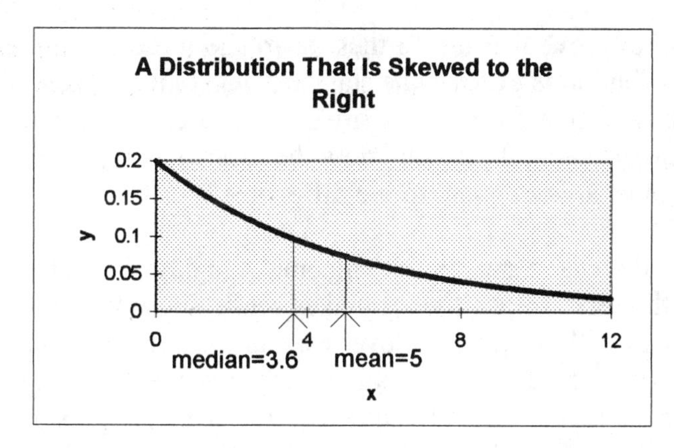

Figure 6.2-5

Skewness usually happens when there is a longer "tail" on one side of the distribution, because the more extreme values in that direction "pull" the mean to that side. As we said earlier, this is something like kids on a see-saw: being further from the middle gives something more influence. From the formula, this happens because the density function is multiplied by x , giving more weight to the probability of more extreme values.

Notice in the distribution in Figure 6.2-5 , which relates to the random variable in Sample Problems 1 and 3: the mode is 0, which also turns out to be the *minimum* value of the entire distribution (not very "central"!)! Thus if you wanted to *scare* your employees about the competitor's new product, you could say that the *most likely* release time was *immediately*. If you wanted to *reassure* them and play down the threat, you could say the *average* (or mean, or expected) release time is in 5 months. And, perhaps as a good *moderate* middle of the road value, you could say the *median* (middle value, 50th percentile) release time is 3.67 months from now.

For our quiz scores {9,9,8,5,4} in Example 1, the mode is 9, the median is 8, and the mean is 7. Thus this distribution is skewed to the *left*, since the mean is *smaller* than the median. Notice that the values 4 and 5, a relative minority of the values, drag *down* the mean. If they had both been 7's, the mean and median would have been the same, 8.

Which measure of central tendency (mode, median, or mean) would you use if you were presenting the quiz score results to the class, and wanted them to feel good? Which would you present if you were talking to crotchety faculty members and wanted to show what a hard grader you are? Which would you present to your partner to most objectively describe the results? These questions give you a further feel for how the choice of different descriptive statistics can be manipulated to serve a specific purpose or agenda.

It is possible to prove that the furthest apart the mean and the median can be for *any* probability distribution is *exactly* one standard deviation.[3] Thus, if you are given the mean and standard deviation for a given situation, you can at least calculate lower and upper bounds on the median. Or, if you know the mean and a median, you know that the standard deviation must at *least* equal to the difference.

In the political arena, a statistic like income is highly skewed in most countries. If you want to make things seem positive, it makes sense to use the mean income, since the income of billionaires will bring up the average significantly. If you want to emphasize the *negative*, it makes sense to use the median income, since this will be much lower than the mean. The median is more reflective of "the average person", rather than "the average income". The mode could be anywhere, but is likely to be even lower than the median, so could be used in various manipulative ways depending on the particular circumstances.

Before trying the exercises, you should:

- Know how to recognize when a finite set of numbers is being thought of a the entire **population** of interest, and when it is being thought of as a **sample** from a *larger* population (to use to make estimates and *inferences* about the larger population), and which formulas for the variance and standard deviation to use in each case.

---

[3] See, for example: Pollack-Johnson, Bruce, and Luis Fernandez, "The Maximum Difference between the Mean and a Median of any Probability Distribution," Working Paper, Dept. of Mathematical Sciences, Villanova University, Villanova, PA 19085.

- Be able to find the mean, (population) variance, and (population) standard deviation of a finite set of numbers, using the following formulas, where the numerical values are $x_1, x_2, \ldots, x_n$:

$$\text{mean} = \bar{x} = \frac{x_1 + x_2 + \ldots + x_n}{n} = \frac{\sum_{i=1}^{n} x_i}{n}$$

$$\text{population variance} = \sigma^2 = \frac{\sum_{i=1}^{n} (x_i - \bar{x})^2}{n}$$

$$\text{population standard deviation} = \sigma = \sqrt{\text{population variance}} = \sqrt{\sigma^2}$$

- Be able to find the mean (expected value), (population) variance, and (population) standard deviation for a continuous random variable given the probability density function $f(x)$ using the following formulas:

$$\text{mean} = \mu = E(X) = \int_{-\infty}^{\infty} x f(x) dx$$

$$\text{(population) variance} = \sigma^2 = \int_{-\infty}^{\infty} (x - \mu)^2 f(x) dx$$

$$\text{(population) standard deviation} = \sigma = \sqrt{\text{(population) variance}} = \sqrt{\sigma^2}$$

- Know how to calculate the Mean Squared Error (MSE)

$$MSE = \frac{SSE}{n} = \frac{\sum_{i=1}^{n} (y_i - \hat{y}_i)^2}{n} = \frac{\sum (\text{actual - predicted})^2}{\text{(number of predictions)}},$$

for a given specific model (or forecasting technique) and a given specific set of data points

- Know how to calculate the sample variance and sample standard deviation for a finite set of numerical values $x_1, x_2, \ldots, x_n$ with mean $\bar{x}$:

$$\text{sample variance} = S^2 = \frac{\sum_{i=1}^{n} (x_i - \bar{x})^2}{n-1}$$

$$\text{sample standard deviation} = S = \sqrt{\text{sample variance}} = \sqrt{S^2} = \sqrt{\frac{\sum_{i=1}^{n} (x_i - \bar{x})^2}{n-1}}$$

- Understand that the mean corresponds to the idea of the average value, and that it is always true that the probability of a random variable falling within three population standard deviations of the mean ($\mu \pm 3\sigma$) is *always at least* 8/9, and commonly around 99%, giving a good way to interpret the standard deviation.
- Understand that graphically the mean corresponds to the balance point of the area under the probability density function: the horizontal location at which a vertical line would just balance a sheet of that shape and uniform thickness and weight.
- Understand the differences between the mean (average), the median (middle value/50th percentile), and the mode (most likely value), and how the choice of which to use can be manipulative.

EXERCISES FOR SECTION 6.2:

Warm Up:

1. For the data values: 3,5,9,6,7 , answer the following questions *by hand* (showing your work), and then verify them using technology:
   a) What is the mean?
   b) What is the population variance?
   c) What is the population standard deviation?
   d) What is the sample variance?
   e) What is the sample standard deviation?
   f) What is the median?

2. For the data values: 0,8,3,6,8 , answer the following questions *by hand* (showing your work), and then verify them using technology:
   a) What is the mean?
   b) What is the population variance?
   c) What is the population standard deviation?
   d) What is the sample variance?
   e) What is the sample standard deviation?
   f) What is the mode?
   g) What is the median?

For Exercises 3-6, find the mean, variance, and standard deviation for the given probability density function:

3. $f(x)=\begin{cases} 20x^3 - 20x^4 \ , & \text{for } 0 \le x \le 1 \\ 0 \ , & \text{otherwise} \end{cases}$

4. $f(x)=\begin{cases} -.0003422x^4+.006845x^3-.04558x^2+.1136x-.0247 \ , & \text{for } 0 \le x \le 10 \\ 0 \ , & \text{otherwise} \end{cases}$

5. $f(x)=\begin{cases} 0.2 \ , & \text{for } 1 \le x \le 3 \\ 0.3 \ , & \text{for } 3 < x \le 5 \\ 0 \ , & \text{otherwise} \end{cases}$

6. $f(x)=\begin{cases} 0.2 \ , & \text{for } 0 \le x \le 5 \\ 0 \ , & \text{otherwise} \end{cases}$

7. For the random variable X with p.d.f. given by $f(x)=\begin{cases}0.01xe^{-.1x} & , \text{ for } x\geq 0\\ 0 & \text{otherwise}\end{cases}$:

   a) Draw a sketch of the p.d.f.
   b) What is the mode?
   c) What is $P(10\leq X\leq 20)$?
   d) What is $P(X\geq 20)$?
   e) What is the median?
   f) What is the mean?
   g) What is the variance?

8. For the random variable X with p.d.f. given by $f(x)=\begin{cases}0.4e^{-.4x} & , \text{ for } x\geq 0\\ 0 & \text{otherwise}\end{cases}$:

   a) Draw a sketch of the p.d.f.
   b) What is the mode?
   c) What is $P(10\leq X\leq 20)$?
   d) What is $P(X\geq 20)$?
   e) What is the median?
   f) What is the mean?
   g) What is the variance?

Game Time:

9. You are interested in buying a specific used car at a written-sealed-bid auction. You have to write down a bid for the car, like everyone else, and the highest written bid gets to buy the car at that price. If you define

X = the highest bid of all the *other* bidders ,

your estimate of the pdf for this random variable is given by:

$$f(x)=\begin{cases}5.041\times 10^{-12}x^4-2.008\times 10^{-8}x^3+.00002965x^2-.01925x+4.636 & , \text{ for } 850\leq x\leq 1150\\ 0 , & \text{otherwise}\end{cases}$$

   a) What is the mode of this distribution? What does this mean?
   b) If you bid \$1000, what is the probability that you will get to buy the car?
   c) What is the median of the distribution? What does this mean? What would be your chances of winning if you bid this value?
   d) If you bid \$1100, what is the probability that you will be outbid?
   e) What is the mean of this distribution? What does this mean? What would be your probability of being able to buy the car if you bid this price?
   f) What is the variance of this distribution? In general terms, what information is this giving you about the highest bid of the other bidders?

10. You have been accepted for admission to one university, but you are still waiting to hear from your first-choice school. To get a scholarship at the first school, you need to respond by the end of this week, and you know that you won't hear from your favorite school until after that time. You have heard that if your GPA is 3.6 or more, you should be able to get into your first choice with a scholarship. You estimate that, if your define

$X=$ your GPA at the end of the current semester ,

the pdf for this random variable is approximately $f(x) = \frac{1}{.1\sqrt{2\pi}} e^{-\frac{1}{2}\left(\frac{x-3.5}{.1}\right)^2}$.

a) What is the mode of this distribution? What does this mean?
b) What is the probability that your GPA will be between 3.0 and 3.5?
c) What is the median of the distribution? What does this mean?
d) What is the probability that you will be accepted at your first choice school with the scholarship?
e) What is the mean of this distribution? What does this mean? What would happen if your GPA ended up at this value?
f) What is the variance of this distribution? In general terms, what information is this giving you about your GPA this semester?

11. You are in the middle of a lawsuit. You have had some experience with similar suits, and estimate that the eventual court settlement will result in a net benefit to you of X hundred thousand dollars, with a probability density function given by:

$$f(x) = \begin{cases} -0.0208x^2 + 0.0417x + 0.208 , & \text{for } -2 \le x \le 4 \\ 0 , & \text{otherwise} \end{cases}$$

a) Write an integral expression for the mean result of the settlement.

b) Evaluate the integral in (a) using the Fundamental Theorem of Calculus to find the mean.

c) Use technology to evaluate the integral in (a). Do you get the same answer?

d) Now find the variance and standard deviation of the settlement.

e) If you just knew the mean and standard deviation of the distribution, what would the 3σ rule of thumb tell you about the possible values of the court settlement?

f) How do the mode, median, and mean compare for this probability distribution? What are the implications for manipulation of the statistics?

12. You are waiting for a bus. From past experience, you have figured that the amount of time you have to wait in minutes, X, has the following distribution:

$$f(x)=\begin{cases}0.1e^{-0.1x}, & \text{for } x\geq 0\\ 0, & \text{otherwise}\end{cases}$$

a) Write an integral expression for the mean waiting time for the bus.
b) Use technology to evaluate the integral in (a).
c) Now find the variance and standard deviation of the waiting time.
d) If you just knew the mean and standard deviation of the distribution, what would the rule of thumb tell you about the possible values of the waiting time?
e) How do the mode, median, and mean compare for this probability distribution? What are the implications for manipulation of the statistics?

13. You work for a company that is developing a drug to help treat people with AIDS. The research is moving along well, but there is a lot of uncertainty in this kind of project. You estimate that the distribution of time (in years) from the present before the drug can go into production, X, is given by the probability density function

$$f(x)=\begin{cases}.25xe^{-.5x}, & \text{for } x\geq 0\\ 0, & \text{otherwise}\end{cases}$$

a) Write an integral expression for the mean time until the drug goes into production.
b) Use technology to evaluate the integral in (a).
c) Now find the variance and standard deviation of the time before the drug goes into production.
d) If you just knew the mean and standard deviation of the distribution, what would the 3σ rule of thumb tell you about the possible values of the time before the drug goes into production?
e) How do the mode, median, and mean compare for this probability distribution? What are the implications for manipulation of the statistics?

14. Suppose you run a construction contracting business, and you are bidding for a big project. You have one major competitor. Based on past experience, you believe that your competitor's bid for the project could be anywhere between $8 million and $10 million, and that all values in between are equally likely.

   a) Define the random variable for this problem.
   b) Define the probability density function for this problem, and sketch its graph.
   c) Write an integral expression for the mean bid from your competitor.
   d) Evaluate the integral in (a) using the Fundamental Theorem of Calculus to find the mean.
   e) Use technology to evaluate the integral in (a). Do you get the same answer?
   f) Now find the variance and standard deviation of the competitor's bid.
   g) If you just knew the mean and standard deviation of the distribution, what would the 3σ rule of thumb tell you about the possible values of the bid?
   h) How do the mode, median, and mean compare for this probability distribution? What are the implications for manipulation of the statistics?

15. To help decide what you want to plant in your garden and when, you want to estimate when the last frost will occur. From historical records, you determine that in your area it could be anywhere between March 28 and April 19, but that any date within that interval is about equally likely to be the last frost.

   a) Define the random variable for this problem.
   b) Define the probability density function for this problem, and sketch its graph.
   c) Write an integral expression for the mean timing of the last frost.
   d) Evaluate the integral in (a) using the Fundamental Theorem of Calculus to find the mean.
   e) Use technology to evaluate the integral in (a). Do you get the same answer?
   f) Now find the variance and standard deviation of the time of the last frost.
   g) If you just knew the mean and standard deviation of the distribution, what would the 3σ rule of thumb tell you about the possible values of the timing of the last frost?
   h) How do the mode, median, and mean compare for this probability distribution? What are the implications for manipulation of the statistics?

16. Suppose you are considering starting a small business selling bagels in your dorm on campus. You buttonhole 40 random students in your dorm and ask what is the most they would pay for a bagel and cream cheese, and the results are as follows:

Max. Price ($)	**[.40,.65)**	**[.65,.90)**	**[.90,1.15)**	**[1.15,1.40)**	**[1.40,1.65)**
Number	9	7	6	4	3

(the other 11 indicated a value below 40 cents).

a) What is the probability of a student picked at random in your dorm being willing to pay between 40 and 65 cents maximum for a bagel? What would be the probability **per unit** in this interval (so the area under a line at that height between .40 and .65 would give the correct probability)?
b) Define a histogram probability density function for this situation. Be sure to define your random variable.
c) Graph your probability density function.
d) Can you find the mean of this distribution? Explain how you got your answer.
e) Can you find the variance and standard deviation of this distribution? Knowing the mean and standard deviation, what would the 3σ rule of thumb indicate about the set of possible values?
f) How do the mode, median, and mean compare for this probability distribution? What are the implications for manipulation of the statistics?

17. International Paper shows information about its relative stock price compared to the Standard and Poor's (S&P) 500 Index for the benefit of its stockholders. The following table gives the stock and index values, assuming that the value invested in the company's common stock and in the S&P 500 was $100 on December 31, 1990, and that all dividends were reinvested.

Value ($) as of 12/31	**1990**	**1991**	**1992**	**1993**	**1994**	**1995**
International Paper	100	136	131	137	156	160
S&P 500	100	131	141	155	157	216

a) Find the percent change for each of the 2 investments during the calendar years 1991-1995.
b) Using the percent change values as data, calculate the mean percent change for each investment.
c) Calculate the variance and standard deviation for each investment.
d) Calculate the 3σ interval for each investment.

18. Public Storage, Inc., shows information about its relative stock price compared to the Standard and Poor's (S&P) 500 Index for the benefit of its stockholders. The following table gives the stock and index values, assuming that the value invested in the company's common stock and in the S&P 500 was $100 on December 31, 1990, and that all dividends were reinvested.

Value ($) as of 12/31	**1990**	**1991**	**1992**	**1993**	**1994**	**1995**
Public Storage, Inc.	100	137	162	277	296	411
S&P 500	100	131	141	155	157	216

a) Find the percent change for each of the 2 investments during the calendar years 1991-1995.
b) Using the percent change values as data, calculate the mean percent change for each investment.
c) Calculate the variance and standard deviation for each investment.
d) Calculate the 3σ interval for each investment.

19. You are responsible for organizing a holiday party at school that has been a tradition for many years. You are trying to forecast how many people will come this year, to know how much food and drink to order. You have data from the last 7 years:

Year	**Attendance**
1	58
2	65
3	63
4	70
5	71
6	74
7	73

a) For the Last Value forecasting method (using the most recent year as the forecast for the next year), show on a table what the predictions from the given data would have been for each year possible.
b) Use your answer to (a) to calculate the error and the squared error corresponding to each prediction you could have made for which you know the actual data value, then calculate the SSE and the MSE.
c) For the Average of the Last 2 Values method of forecasting, show on a table what the predictions from the given data would have been for each year possible.
d) Use your answer to (c) to calculate the error and the squared error corresponding to each prediction you could have made for which you know the actual data value, then calculate the SSE and the MSE.
e) Which method seems better for this situation?
f) Can you fit a model to forecast attendance this year?

g) What would you forecast for the attendance this year, for planning purposes?

20. Suppose your GPA over the last 6 semesters were given by:

Semester	GPA
1	2.8
2	2.7
3	2.9
4	3.1
5	3.0
6	3.3

a) For the Last Value forecasting method (using the most recent semester as the forecast for the next semester), show on a table what the predictions from the given data would have been for each semester possible.
b) Use your answer to (a) to calculate the error and the squared error corresponding to each prediction you could have made for which you know the actual data value, then calculate the SSE and the MSE.
c) For the Average of the Last 2 Values method of forecasting, show on a table what the predictions from the given data would have been for each year possible.
d) Use your answer to (c) to calculate the error and the squared error corresponding to each prediction you could have made for which you know the actual data value, then calculate the SSE and the MSE.
e) Which method seems better for this situation?
f) Can you fit a model to forecast your GPA this semester?
g) What would you forecast for your GPA for this semester?

Section 6.3: R^2, Standard Error, Misuse of Regression, Regression Assumptions

In Section 6.1, we saw how to find a best-fit multivariable function for a given set of data. We talked briefly about how $\mathbf{R^2}$ is a measure of how good the fit of a model is to its data (on a scale of 0 to 1), expressing what fraction of the fluctuations in the output values of the data is explained by the model. We also talked about how some spreadsheet packages give a 95% **confidence interval** or **margin of error** around the given estimate of the parameter's value. If a package does not give a confidence interval, it is likely to give the **standard error** of each parameter (of the constant term and of the coefficients of the variable terms), which can help you estimate a **confidence interval**, as we will explain in this section. As we discussed in Section 6.1, you can use such a confidence interval, after checking that certain statistical assumptions are valid, to determine whether or not the **sign** of that parameter is likely to be correct and to help decide whether or not that term belongs in your model. We also discussed how R^2 is a useful statistic, but should never be the *only* consideration in choosing a model. In general, it is a good idea to use the simplest model possible that has a reasonably high value of R^2. One measure of simplicity is that the **degrees of freedom of the residuals**, which is the number of data points minus the number of parameters in the model. As a modeler, you want this value to be as large as possible. In Section 6.2, we defined the notions of mean, variance (population and sample), and standard deviation (population and sample), both for finite sets of numbers and for continuous random variables. In this section we elaborate further on these ideas, and the **statistical assumptions** underlying some of them, and talk about some pitfalls in using least-squares regression models. One such pitfall is assuming that, because a model suggests a *relationship* between two variables, then one of the variables is necessarily *causing* the other to change. It is possible that the first variable is primarily responsible for changes in the second, or vice-versa, but it is also possible that some third variable is causing both of them to change in related ways, or that there is no simple relationship of causality that can be determined.

Here are some examples of the kind of problems that the material in this section should help you to solve:

- Someone has collected data on sales of handguns, the number of death sentences, and the murder rate, on a state-by-state basis, in the U.S. over a certain period of time. When doing a regression of the murder rate on the other two variables, the coefficients on the two variables had a positive sign. The researcher claims that this means that allowing more handgun sales and increasing death sentences in a particular state will increase the murder rate, and therefore both practices should be curtailed. Is this claim justified?

- You are trying to determine the effect of sleep and exercise on your energy level. You gather data and fit a model. What is the margin of error in using this model to predict your energy level for a specific combination of sleep and exercise, both on average and for a particular day?
- You have collected data on the relationship between violence on TV, violence in movies, and violent crime in teenagers, over a period of decades. When you tried to fit a linear model, the R^2 value came out to be 0.15 . Does this mean there is essentially no relationship between these quantities?

After studying the material in this section, you should:

- Understand in detail exactly what R^2 is measuring, how to calculate it, and how to interpret it.
- Understand more about what is meant by the standard error of a parameter in a regression model, and how it can be used to approximate a confidence interval for the estimate of the true value of the parameter.
- Have a basic understanding of the statistical assumptions underlying the use of confidence intervals.
- Understand that the existence of a model to represent a relationship between quantities does not in itself prove that one phenomenon *causes* another.

R^2: The Coefficient of Determination for a Regression, SSE, TSS, and SSR

Let's start by deepening our understanding of exactly what the value of R^2 means. For simplicity, let's look back at the linear model we used for the demand function in the Coffeehouse concert price problem described in Section 1.1 and discussed elsewhere in the text. Table 1 shows the data on ticket prices and attendance again, to refresh your memory.

Table 1

Ticket Price ($)	Paid Attendance
7	100
8	80
9	70

The best-fit linear function for this data (rounded to 4 significant figures) is given by

$$f(x) = 203.3 - 15x$$

We have already calculated the SSE for this model and data set in Section 5.5 . Table 2 shows these calculations again for your convenience.

Table 2

Actual x (x_{data}) (x)	Actual y (y_{data}) (y)	Predicted y $y_{\text{model}} = f(x_{\text{data}}) = -15x_{\text{data}}+203.3$ ($\hat{y}$)	Errors $y_{\text{data}} - y_{\text{model}}$ ($y-\hat{y}$)	Errors Squared $(y_{\text{data}}-y_{\text{model}})^2$ $(y-\hat{y})^2$
7	100	$-15(7) + 203.3 = 98.3$	$100-98.3 = 1.7$	$1.7^2 = 2.89$
8	80	$-15(8) + 203.3 = 83.3$	$80-83.3 = -3.3$	$(-3.3)^2 = 10.89$
9	70	$-15(9) + 203.3 = 68.3$	$70-68.3 = 1.7$	$1.7^2 = 2.89$
		Sums	0.1	16.67

We see that the SSE is 16.67 .

Now let's think of the y values as a separate, independent set of data values. As we saw in Section 6.2, we can calculate the mean (average) for this data set simply by adding up the values and dividing by the number of values:

$$\bar{y} = \frac{100+80+70}{3} = \frac{250}{3} = 83.33$$

Recall that the population variance of this set of values is given by

$$\sigma^2 = \frac{\sum_{i=1}^{n}(y_i - \bar{y})^2}{n} \approx \frac{(100-83.33)^2 + (80-83.33)^2 + (70-83.33)^2}{3} = \frac{16.67^2 + (-3.33)^2 + (-13.33)^2}{3}$$

$$\approx \frac{277.9 + 11.1 + 177.7}{3} = \frac{466.7}{3} \approx 155.6$$

We are particularly interested here in the *numerator* of the above calculation. This is sometimes called the **Total Sum of Squares (TSS)** for a regression, given by

$$\text{TSS} = \sum_{i=1}^{n}(y_i - \bar{y})^2$$

We can think of the TSS as the **total variation of the y values**. This can also be thought of as the SSE that would correspond to the *constant* model:

$y = \bar{y}$

Graphically, this corresponds to Figure 6.3-1 .

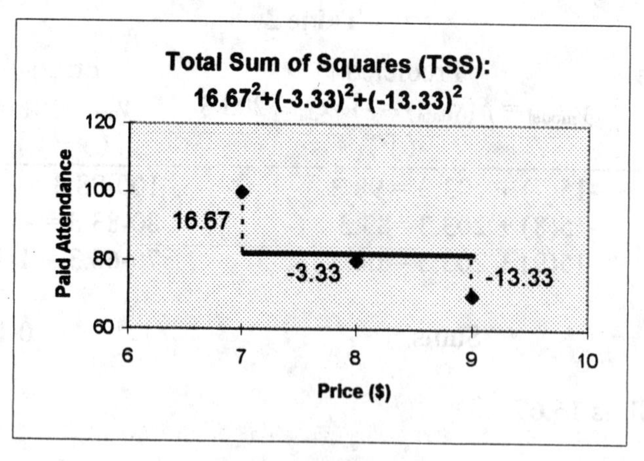

Figure 6.3-1

From our calculation above, we see that our TSS = 466.7 here.

Now, given that our x values in the data set were 7, 8, and 9, we could use the model to find the predicted y values (which we will denote $\hat{y}$) that go with each of these. In fact, we have already done that in Table 2 when we calculated the SSE! From that table, you can see that the predicted values are 98.3, 83.3, and 68.3, respectively. Let's do the same Total Sum of Squares calculations with these values, still using the original mean value of $\bar{y} = 83.3$. We get a value usually called the Sum of Squares (due to the) Regression (or model), or SSR:

$$SSR = \sum_{i=1}^{n} (\hat{y}_i - \bar{y})^2 \approx (98.3 - 83.3)^2 + (83.3 - 83.3)^2 + (68.3 - 83.3)^2 = 15^2 + 0^2 + (-15)^2 = 225 + 0 + 225 = 450$$

This value corresponds to the *total sum of squares* of the y values that we *expect* because of the *model* (the sum of the squared differences between the *predicted* y values and the *mean* of the y values), where the predicted values are based on the x values in the data set. Graphically, it corresponds to Figure 6.3-2 .

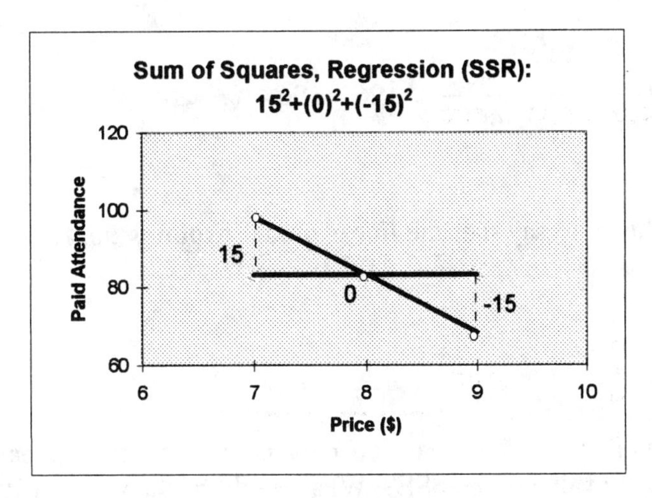

Figure 6.3-2

How does SSE fit into the picture? If you remember, in our example the SSE was 16.7 . Notice that this is exactly the difference between the TSS and the SSR. In fact, this relationship is always true: ***for a least-squares model,* the total variation (TSS) is the sum of the variation explained by the model (SSR) and the variation not explained by the model, which is the SSE. (This need *not* hold for other models that are *not* least squares models!)**

TSS = SSR + SSE

$$\sum_{i=1}^{n}(y_i-\bar{y})^2=\sum_{i=1}^{n}(\hat{y}_i-\bar{y})^2+\sum_{i=1}^{n}(y_i-\hat{y}_i)^2$$

Once again, this relationship is *sure* to hold **only for least-squares models**, even though the relationship in the sigma notation form seems very simple and logical, and would seem to hold for any model. A mathematical proof is beyond the scope of this text; see a statistics book for details.

In fact, for a least-squares model, this relationship can be used to *calculate* the SSR, since the other two calculations are simpler in some ways:

SSR = TSS - SSE

Now, you might have noticed that the SSR (450) was very close to the Total Sum of Squares (TSS = 466.7). This means that most of the variation in the y values of the data is explained by the model. So to get a measure of how well the model fits the data, we can calculate the ratio of SSR to TSS. This is what is called the **coefficient of determination**, or $\mathbf{R^2}$. In other words,

$$R^2 = \frac{SSR}{TSS} = \frac{\sum_{i=1}^{n}(\hat{y}_i - \bar{y})^2}{\sum_{i=1}^{n}(y_i - \bar{y})^2} \approx \frac{450}{466.7} \approx 0.964 = 96.4\%$$

One way to express this is to say that the linear model **explains 96.4% of the variation in the *y* values of the data.**

Degrees of Freedom

Sample Problem 1: If you have two data points and fit a linear least-squares model, what would be the value of the SSE? What would be the value of R^2?

Solution: Since a straight line will fit the two data points perfectly, the SSE will be 0, so the SSR will be the same as the TSS. As a result

R^2 = SSR/TSS = 1 . □

In other words, the model is a **perfect fit** when $\mathbf{R^2 = 1}$. This would also hold when fitting a parabola to three points that do not line up perfectly, or in general when fitting a model with *n* parameters, given *n* data points that do not line up in any special way.

This situation of perfect fit, when the number of data points is the same as the number of parameters to be determined, can be described as having **zero degrees of freedom**. In other words, you have no freedom in picking a model for the data, since a perfect model exists and can be found directly (we will do this in Chapter 9).

In fact, in Section 6.1, we have already defined this notion of degrees of freedom, which we called the **degrees of freedom (d.f.) of the residuals, for a given set of data and model category:**

d.f. of residuals = (# of data points) - (# of model parameters)

In a sense, then, the value of the degrees of freedom of the residuals is telling you how many *additional* data points you have *over and above* the *minimum* number of data points needed to fit a model of the type being considered. As we said earlier, **higher values are better**, because they reflect the fact that more information (data) is being incorporated into the model, or because they reflect simpler models with fewer parameters (the KISS! principle).

From this definition and the calculation formula, we can now see how the degrees of freedom of the residuals for the above case of fitting a linear model (with 2 parameters) to 2 data points would be

$$2 - 2 = 0 \ .$$

This discussion is one way to understand why **R^2 does not tell the *whole* story when deciding what is the best model!** If we only had the first two data points of the concert ticket example, we would have found a perfect model with an $R^2 = 1$, but it would not have been a better model. In fact, it would predict attendance of 60 when the price was $9, while the actual attendance was 70, and our other linear model predicted 68.3 . **In general, it is best to get as much data as possible**, even though this could *lower* the resulting value of R^2. As we discussed in Section 6.1, this boils down to a **trade-off between maximizing the degrees of freedom of the residuals (by using more data or using a simpler model) and maximizing R^2.** This need for judgment in balancing such considerations is largely what makes model formulation as much of an *art* as a *science*.

Some computer software gives a value of an "Adjusted R^2 " value, which adjusts the basic R^2 defined earlier for the number of degrees of freedom. This will lower the R^2 value as more parameters are included in your model. This is one way to try to create *one* measure that reflects both the desire to maximize R^2 and to maximize the degrees of freedom of the residuals.

Standard Error and Confidence Intervals for Parameter Estimates

We have already discussed how confidence intervals for coefficient estimates from a regression analysis can help decide whether the *sign* of a term is clear or not, and whether it makes sense to include that term in our model or not. The default confidence level for most spreadsheet regression output reports is usually 95%, and is a standard middle-of-the-road level for statistical analysis. Some people might also argue for a 99% confidence level (more conservative, since it is more likely to say a term should *not* be included in the model), and others for a 90% level (more inclusive and tentative); we would recommend using either 95% or 99%, giving a preference to simpler models, since they will have higher degrees of freedom of the residuals.

Most computer software packages that do regression give a value of what is called the "**standard error**" for each parameter being estimated in the model, even if they do not automatically give confidence intervals. This can be thought of as an estimate of the **sample standard deviation of the value of that parameter, based on the data**. For the sake of this course, you can think of this value as an approximation of the standard deviation you would get for the estimate of the parameter, if you took *lots* of different samples of the same size as your original data, and calculated the regression model for each. For each parameter, from this entire *set* of regressions you could examine the resulting set of *estimates* of that parameter, and then you could compute the mean and the sample variance and sample standard deviation of that set of estimates. Think of the standard error of the parameter as an estimate of what that sample standard deviation would be, on average.

As a rough guide, based on our rule of thumb for interpreting the standard deviation, you can figure that the approximate 99% confidence interval or margin of error for that parameter estimate is about 3 times the standard error value, consistent with our earlier interpretation of the standard deviation in Section 6.2 . For example, if the estimate of a parameter was 2.7, with a standard error of 0.2, then you would have a roughly 99% level of confidence that the true value was within 3 times 0.2 of 2.7, so between

$$2.7 - 3(.2) = 2.7 - .6 = 2.1 \text{ and } 2.7 + 3(.2) = 2.7 + .6 = 3.3 .$$

On the other hand, if your estimate was 2.7 and the standard error was 1.5, your 99% confidence interval would be roughly between $2.7 - 3(1.5) = 2.7 - 4.5 = -1.8$ and $2.7 + 3(1.5) = 2.7 + 1.5 = 4.2$.

Translation: You could not even have confidence that the **sign** of the parameter was positive!

For a 95% confidence interval, a good rule of thumb is to use plus or minus *two* times the standard error. This is a less conservative value, which is *less* likely to indicate that you can't be sure of the sign of your estimate (*more* likely to *keep* that term in your regression, since the interval is smaller and less likely to include 0).

Suppose the estimate of a parameter was 12, and the standard error of the parameter was 6. Using the 3σ rule of thumb, this would suggest that the rough 99% "confidence interval" would be values within 3(6)=18 of 12; in other words, between 12-18=-6 and 12+18=30. Since the interval [-6,30] includes 0, there is a chance (if we want a rough 99% level of confidence) that the true value of the parameter is actually 0, so we might decide to *remove* that term from our regression. At the rough 95% level, though, we could use the 2σ interval, which would be $12 \pm 2(6)$ or $[0,24]$. In this case, we would be right of the edge of whether 0 was included in the interval or not, so would be on the fence about whether to drop that term or not.

For a smaller standard error (such as 4), the 3σ interval would just barely include 0 (since $12 \pm 3(4)$ would give the interval [0,24]), so we'd be on the fence at the 99% level of confidence about whether or not to drop that term. On the other hand, the 2σ

interval would be $12 \pm 2(4)$ or $[4,20]$, so we would be more than 95% certain that the coefficient was not 0 and would be inclined to *keep* that term in our model.

In the first case, **the ratio of the estimate of the parameter to its standard error** (called the value of the **t-statistic**) was $\frac{12}{6} = 2$ and we were *less* likely to keep the term. In the second case, since we made the standard error smaller, the ratio was $\frac{12}{4} = 3$; *more* than 2, and we were *more* likely to keep the term.

If your spreadsheet does not calculate confidence intervals, one very rough rule of thumb, then, involves looking at the **magnitude of the *ratio* of each coefficient to its standard error**, which could be thought of as the parameter's **coefficient of variation** (in general, this is the mean divided by the standard deviation). If this ratio magnitude is large (such as at least 2 or 3), then at least the sign of the coefficient is pretty sure to be right, and it makes sense to keep that term in your model. **If the ratio magnitude is less than 2, you may want to seriously consider dropping the term from your model. If the ratio magnitude is more than 3, it is probably worth keeping the term in your model. In between 2 and 3, the best course is less clear, so you should probably use other considerations in making your decision, such as R^2, the degrees of freedom of the residuals, and the context of the problem.** In statistics, you will learn more sophisticated tests for your coefficients based on this ratio (the *t*-statistic) and on the number of data points used. These are especially important when you have a small number of data points.

In certain cases, even if this ratio magnitude is *below 2*, we may *still* want to retain a parameter. For example, if we feel the relation of sleep, exercise and energy really *should* be quadratic because there should be a single maximum value, we might choose not to drop the quadratic terms of the model, even if they had a ratio magnitude of less than 2.

In Table 9 (from Sample Problem 5) of Section 6.1, repeated in part here as Table 3,

Table 3

SUMMARY OUTPUT						
Regression Statistics						
Multiple R	0.9896467					
R Square	0.9794005					
Adjusted R Square	0.9720436					
Standard Error	1.910614					
Observations	20					
ANOVA						
	df	*SS*	*MS*	*F*	*Significance F*	
Regression	5	2429.843756	485.9688	133.1259	2.72703E-11	
Residual	14	51.10624446	3.650446			
Total	19	2480.95				
	Coefficients	*Standard Error*	*t Stat*	*P-value*	*Lower 95%*	*Upper 95%*
Intercept	-93.1423	12.64336946	-7.36689	3.53E-06	-120.2596579	-66.0249
c	0.2602388	0.016248958	16.01572	2.13E-10	0.225388226	0.295089
x	0.1459598	0.135347154	1.07841	0.299081	-0.144331272	0.436251
c^2	-9.44E-05	5.48592E-06	-17.206	8.18E-11	-0.000106157	-8.3E-05
cx	4.098E-05	6.00381E-05	0.682533	0.506039	-8.7791E-05	0.00017
x^2	-0.001958	0.000450741	-4.34474	0.000673	-0.002925098	-0.00099

let's look at how the standard error gives a rough estimate of the 95% confidence interval. Looking at the line corresponding to the x term, the estimate of the coefficient is about 0.15, and the standard error is about 0.14. The 2σ interval would then be

$$0.15 \pm 2(0.14) = 0.15 \pm 0.28 \text{ , or } [-0.13,0.43]$$

and the 95% confidence interval given in the Table 3 is [-0.14,0.44]. Pretty close! Notice also that the magnitude of the *t*-statistic (the ratio of the parameter estimate over its standard error) is about 1, which is less than 2, indicating that we are not at all sure of the sign of the coefficient, and so we should seriously consider *dropping* the x term from our model.

In Table 2 (for Sample Problem 1) of Section 6.1, repeated in part here as Table 4,

Table 4

	A	B	C	D	E	F	G
18	SUMMARY OUTPUT						
19							
20	*Regression Statistics*						
21	Multiple R	1					
22	R Square	1					
23	Adjusted R Square	1					
24	Standard Error	2.83585E-13					
25	Observations	12					
26							
27	ANOVA						
28		*df*	*SS*	*MS*	*F*	*Significance F*	
29	Regression	2	779066.6667	389533	4.8E+30	7.1805E-136	
30	Residual	9	7.23783E-25	8E-26			
31	Total	11	779066.6667				
32							
33		*Coefficients*	*Standard Error*	*t Stat*	*P-value*	*Lower 95%*	*Upper 95%*
34	Intercept	700	2.86458E-13	2.4E+15	2E-135	700	700
35	X Variable 1	70	2.80758E-14	2.5E+15	1E-135	70	70
36	X Variable 2	100	4.19395E-14	2.4E+15	2E-135	100	100

the standard errors are all values like 2.86E-13 . This means 2.86 times 10^{-13} = 0.000000000000286 (12 0's), a **very small number**! In fact, these values should be exactly 0, but are not because of roundoff error. For example, if a computer or calculator stores 1/3 as .3333333, then when you calculate 1 - 3(1/3), the machine finds 1 - 3(0.3333333) = 1 - 0.9999999 = 0.0000001, rather than the exact answer of 0. In general, remember that calculators and computers do **not** give exact answers, although they will usually be accurate to 7 or 8 significant figures, if not much more. For this example, the Standard Error for each of the parameters was essentially zero, and the ratio magnitude of each coefficient to its standard error is *huge* (essentially infinite). That means that the model fits the data exactly, as we see from the R-Squared. You can see that the Lower 95% and Upper 95% range for the intercept are *both* 700, the value of the intercept. The same is true for the coefficients.

The notion of a confidence interval is very closely related to the idea of the **margin of error** that we discuss in Section 3.4 and elsewhere. You can think of the margin of error for your optimal decision variable values and for the optimal objective function value as crude 95% or 99% confidence intervals on those values. The only problem is that we don't *know* the *true* values of these quantities, nor even what kind of a *probability distribution* they might have, so we cannot be very rigorous. Still, it is a useful way to think of this concept.

Misuse of Regression Models

As a final comment on the use of regression models, let us consider the example mentioned in the introduction to this section, in which sales of handguns and number of death sentences had positive coefficients in a linear model for the murder rate. One tendency might be to assume that the sales of handguns and the death sentences **caused** the murder rate to go up, and that therefore to lower the murder rate, these variables should be reduced. But it is also very conceivable that causation was in the *opposite* direction: that people carried more guns and that the number of death sentences went up as a *response* to an increase in the murder rate and crime in general. In such a case, it is not at all clear what would be the effect of forcibly curtailing sales of handguns and the number of death sentences. And of course it is also possible that there are causative factors in *both* directions, or in *neither* direction. The only way to decide between these alternatives is to gather more data, and perhaps to try to design one or more *experiments* that would test the implications of the different types of causation. For example, you may want to examine places where policies changed significantly to examine any immediate and/or long-term effects of these changes.

Statistical Assumptions when Using Regression Confidence Intervals*

The use of confidence intervals as we have discussed in this chapter depends on a number of statistical assumptions. These are easiest to explain for a single-variable regression, but are analogous for more complicated regressions as well.

Recall Example 2 from Section 6.2, about shooting 3-point shots in basketball from different angles. The data are given in Table 5:

Table 5

Angle *(°)*	**1st 10** *(%)*	**2nd 10** *(%)*	**3rd 10** *(%)*	**4th 10** *(%)*	**5th 10** *(%)*	**Overall** *(%)*
0	10	20	20	10	30	18
45	70	50	60	60	70	62
90	50	40	60	70	50	54
135	70	80	50	60	60	64
180	0	20	30	40	10	20

In this example, we have 5 data observations at each angle. In Table 5, we have already calculated the overall (mean) shooting percentage for that angle. We could also

---

* This material is quite technical, and can be skipped without loss of continuity, although it is important to understand the material to use regression intelligently and wisely.

calculate the sample standard deviation of the shooting percentage for each angle based on its five observations. These values are shown in Table 6.

Table 6

Angle *(°)*	**Mean** *(%)*	**Sample Standard Deviation (*S*)**
0	18	8.37
45	62	8.37
90	54	11.40
135	64	11.40
180	20	15.80

If we treat the shooting percentage for each of the groups of 10 at each angle as a *separate* data point and do a quartic regression, the regression summary output is shown in Table 7.

Table 7

p	a	a^2	a^3	a^4
10	0	0	0	0
20	0	0	0	0
20	0	0	0	0
10	0	0	0	0
30	0	0	0	0
70	45	2025	91125	4100625
50	45	2025	91125	4100625
60	45	2025	91125	4100625
60	45	2025	91125	4100625
70	45	2025	91125	4100625
50	90	8100	729000	65610000
40	90	8100	729000	65610000
60	90	8100	729000	65610000
70	90	8100	729000	65610000
50	90	8100	729000	65610000
70	135	18225	2460375	332150625
80	135	18225	2460375	332150625
50	135	18225	2460375	332150625
60	135	18225	2460375	332150625
60	135	18225	2460375	332150625
0	180	32400	5832000	1049760000
20	180	32400	5832000	1049760000
30	180	32400	5832000	1049760000
40	180	32400	5832000	1049760000
10	180	32400	5832000	1049760000

SUMMARY OUTPUT

Regression Statistics	
Multiple R	0.8942204
R Square	0.7996301
Adjusted R Square	0.7595561
Standard Error	11.401754
Observations	25

ANOVA

	df	*SS*	*MS*	*F*	*Significance F*
Regression	4	10376	2594	19.9538462	9.38402E-07
Residual	20	2600	130		
Total	24	12976			

	Coefficients	*Standard Error*	*t Stat*	*P-value*	*Lower 95%*	*Upper 95%*
Intercept	18	5.099019514	3.53009043	0.00210329	7.363636618	28.63636
a	2.862963	0.632726588	4.52480268	0.00020649	1.543119041	4.182807
a^2	-0.0622634	0.017648085	-3.5280528	0.00211324	-0.09907662	-0.02545
a^3	0.0005176	0.00015675	3.30210239	0.00355917	0.00019063	0.000845
a^4	-1.443E-06	4.33485E-07	-3.3285313	0.00334954	-2.3471E-06	-5.4E-07

Notice that this is exactly the same function obtained in Example 2 of Section 6.2, when we did a regression on the *mean* values only and obtained a perfect fit. Let's look at the situation graphically, shown in Figure 6.3-3.

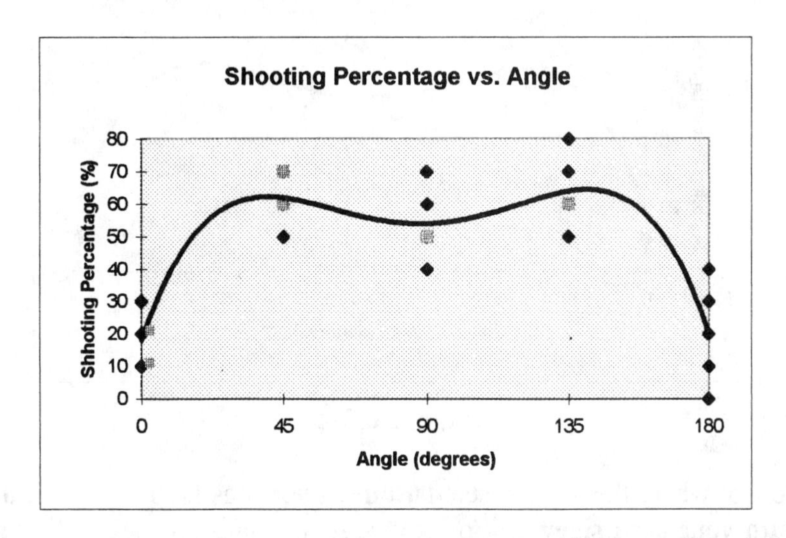

Figure 6.3-3

In Figure 6.3-3, the points marked with squares correspond to places where there were *two* identical data points. Notice that the angles with the largest sample standard deviation in Table 6 (180°) has the data points that are most spread out on the graph, consistent with the basic idea of standard deviation.

As we can see in the graph, we can think of each input value (angle here) as being associated with a *set* or **distribution** of output values. And if we imagined *all* of the possible trials of 10 shots at that angle you could make, we could define a **probability density function** (for instance, a bell-shaped normal distribution pdf) to describe the relative frequencies of the different possible shooting percentages. On Figure 6.3-3, imagine this bell curve for each angle as lining up vertically over the data points there, coming straight out of the page in a third dimension. Figure 6.3-4 is a lame attempt to sketch in these bell curves for our example.

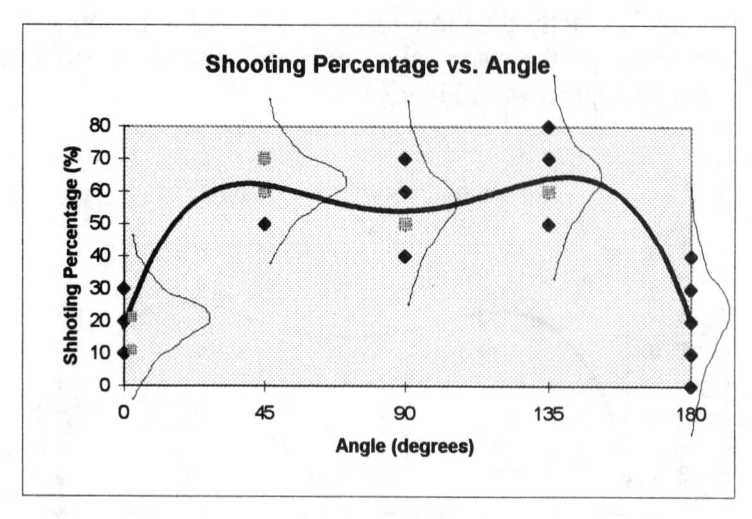

Figure 6.3-4

Notice that where the sample standard deviation was largest (180°), the bell curve (when you turn your head sideways to look at it) is wider and less tall, and where the sample standard deviation was smallest (0° and 45°), the bell curve is narrower and taller.

With this picture in mind, we can now state **the basic assumptions of regression analysis:**

- **At any given input value in the domain, the *distribution* of the output values is a *normal* (Gaussian) probability distribution (bell curve).**
- **At any given input value in the domain, the *mean* of the normal distribution of the output values lies *on* the regression function curve.**
- **At *all* of the input values in the domain, the *standard deviations* (or variances) of the normal distributions are all *the same*** (a property called **homoscedasticity**), **so the bell curves are all the same width and height**.

In a statistics class, you can learn some formal ways to validate (test) these assumptions. For our purposes in this course, since we are only using the confidence intervals in a very rough way, we only need to check that the assumptions are reasonably close to being satisfied. This would not be appropriate for an academic paper, but makes perfect sense when you have a problem you *have* to solve and only have a limited amount of data upon which to base your analysis.

Recall that in Section 1.3 and in Chapters 5 and 6, we have talked about the **errors** between data points and a given model as the vertical distance from the model to the data point, or the actual output minus the predicted output. We also mentioned that these errors are also called **residuals**. At a given input value in the domain, since the residuals are simply the actual values *minus* the model output value there (a constant), the fact that the output values have a normal distribution means that **the residuals also have a normal distribution**. And since the mean of the output values is exactly the value subtracted to get the residuals, **the mean of the residual distribution should be 0**. Furthermore, since subtracting a constant from a random variable doesn't affect the shape of the distribution at all (just shifts it to one side), **the standard deviations of the residual distributions for *all* the input values in the domain should be the same**.

These assumptions of regression can be tested without too much trouble using technology by creating two different kinds of graphs. The first is to **look at a simple histogram graph of the distribution of *all* of the residuals lumped together, to see if the distribution is reasonably close to a bell-shaped curve**. The second is to **look at a plot of the residuals against the predicted output values to see if the means are reasonably close to 0 and the standard deviations reasonably close to each other *everywhere***.

Let's try these validation tests with our example. The histogram of all the residuals is shown in Figure 6.3-5.

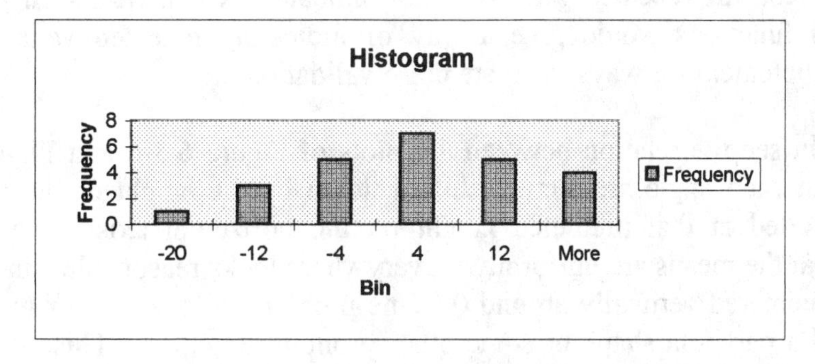

Figure 6.3-5

This may not be a perfect bell-shaped curve, but it certainly seems reasonably close, so is adequate validation of the assumption of a normal distribution for the residuals.

Now let's look at a plot of the residuals against the predicted outputs, shown in Figure 6.3-6.

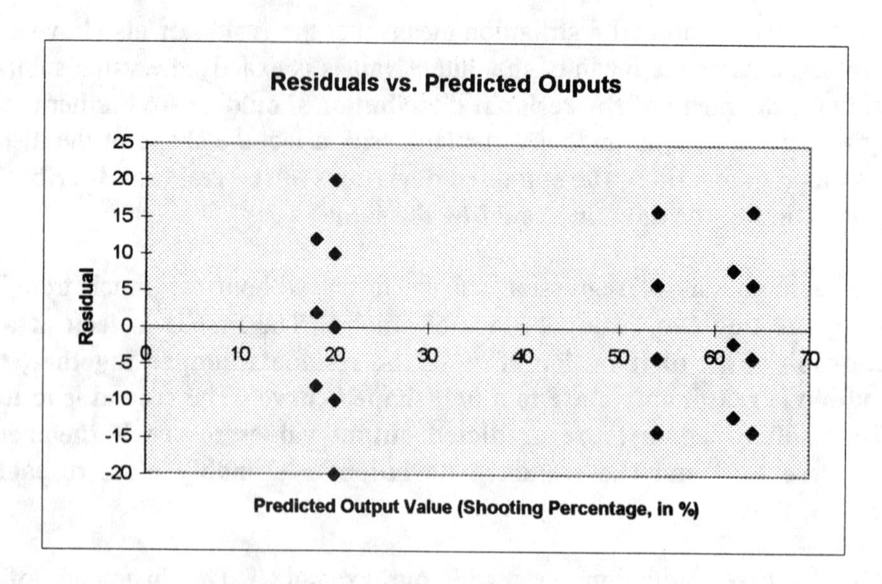

Figure 6.3-6

In this plot, the repeated values are not indicated (which we did in Figure 6.3-3); some plotting functions would give a way of indicating repeated values. See your technology supplement for ways to create these validation plots.

You can see the relation between the plots of Figure 6.3-6 and Figure 6.3-3; the distributions at each angle are simply shifted down by the height of the model at that angle, and plotted at that predicted height on the horizontal axis. We see that the assumption that the means are uniformly 0 everywhere looks reasonable, since everything seems nicely centered vertically around 0. One problem to look out for would be if the points formed a parabola shape or some other pronounced curve. This could happen if you fit a linear model to data that was by nature quadratic, for example, and could suggest you may need more terms in your model. Figure 6.3-7 shows an example.

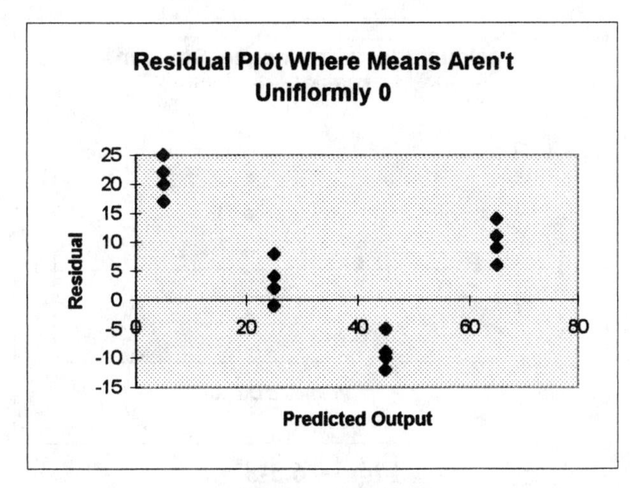

Figure 6.3-7

With regard to the standard deviations being equal, the plot in Figure 6.3-6 does show some differences in the widths of each vertical band, as we in fact saw when we calculated the standard deviations in Table 6. This is an especially useful comparison if there are equal numbers of observations at different predicted output values, as we had in our case. However, in this case, the differences are not large enough to be highly significant. With more advanced statistics, this assumption can be formally tested.

If the number of data points differs widely in different intervals of predicted output values, you can only try to make mental adjustments for the fact that fewer points would tend to cover a narrower band than more points, *even from identical distributions*.

The problem to look out for in this test of equal standard deviations would be something like a horn-shaped pattern, where the standard deviations were clearly quite different in one area compared to another. Figure 6.3-8 shows an example.

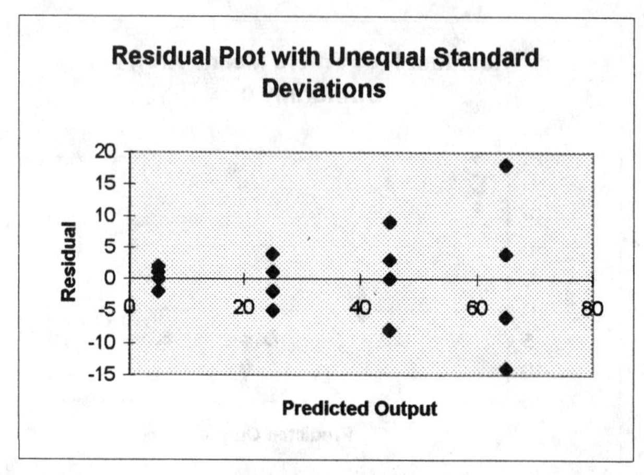

Figure 6.3-8

If there is serious doubt about the standard deviations not being equal, conclusions related to the confidence intervals should be viewed skeptically, and more advanced statistical methods might be called for.

These validation tests are especially important with multivariable models, since it is normally difficult or impossible to *see* the data and model together on a graph to visually judge the fit and appropriateness of a model. An advanced course in statistics can help you with more complex modeling issues, such as performing transformations on your data to transform an exponential relationship into a linear one.

Section Summary

Before trying the exercises, you should:

- Understand that the Total Sum of Squares (TSS) for the y values from a set of data is given by $\text{TSS} = \sum_{i=1}^{n}(y_i - \bar{y})^2$ (the sum of the squares of the deviations of the y values from their mean), and represents the total variation of those y values

- Understand that the Sum of Squares, Regression (SSR) for a set of data and a given model is given by $\text{SSR} = \sum_{i=1}^{n}(\hat{y}_i - \bar{y})^2$ (the sum of the squares of the *differences* between the *predicted* y values and the *mean* of the y values), and represents the variation in the y values that is explained by (due to) the regression model

- Understand that, *if* the model being studied is a *least-squares* model, then TSS=SSR+SSE, but that this is *not* true in general for other models.
- Understand that the coefficient of determination (R^2) for a given data set and a given model is given by $R^2 = \frac{SSR}{TSS} = \frac{\sum_{i=1}^{n}(\hat{y}_i - \bar{y})^2}{\sum_{i=1}^{n}(y_i - \bar{y})^2}$, and represents the fraction (often expressed as a %) of the total variation in the *y* values that is explained by the regression model
- Understand that R^2 is an important way to measure the fit of a model to a set of data, but realize that it can be misleadingly inflated if the number of data points is small, and should not be the *only* criterion in selecting a model (as always, err on the side of simpler models whenever possible)
- Understand that a rough rule of thumb for interpreting the standard deviation is that roughly 99% of all values in the population under consideration normally lie within plus or minus 3 times the standard deviation of the mean, and roughly 95% lie within plus or minus 2 standard deviations.
- Understand that the *standard error of a parameter* in a regression can be thought of as the standard deviation of the estimate of that parameter (as if models were fit to many sets of data from the same population), so the same "plus or minus 3 times the value" rule of thumb can be used to interpret the standard error. This is why, if the magnitude of the ratio of the parameter estimate to the standard error is 3 or more, you can be quite sure that at least the sign of the parameter (direction of the relationship) is correct, and therefore that that term probably belongs in your model. If the ratio has a magnitude of less than 2, you should seriously consider dropping that term. If the ratio magnitude is between 2 and 3, you should use other factors in deciding whether to keep the term.
- Understand that the existence of a model to represent a relationship between quantities does not in itself prove that one phenomenon *causes* another. Further data collection and/or the design of some form of experiment may be necessary to determine this more conclusively.
- Realize that when you use confidence intervals or standard error, you are implicitly assuming that the residuals have a normal (bell-shaped) distribution, that the means of the residuals are uniformly 0 everywhere, and that the standard deviations of the residuals are equal everywhere, and know how to validate these assumptions by a histogram plot of the residuals and a plot of the residuals against the predicted output values.

EXERCISES FOR SECTION 6.3:

Warm Up

1. Consider the following data:

x	y
0	2
1	5
2	6
3	9
4	8

a) Plot the data points and *visually* estimate a linear model with whole number coefficients that would fit it reasonably well.

b) *By hand,* find the mean of the y values ($\bar{y}$) and the TSS. Draw a sketch that relates to TSS.

c) *By hand,* use your model in (a) to calculate the *predicted* y value for each of the x data values given in the table (show your work). Use these to find the SSR. Draw a sketch that relates to SSR.

d) *By hand,* calculate the error and the squared error for each data point with respect to your linear model in (a), and then find the SSE (show your work). What is the relationship between TSS, SSR, and SSE?

e) Use technology to find the least-squares best-fit linear model for the above data.

f) Use technology to find the TSS, SSR, and SSE value for the model in (e). What is the relationship between TSS, SSR, and SSE in this case?

g) What is the R^2 value for the linear least-squares model?

h) Find the R^2 value for the least-squares *quadratic* model for this data. Which model would you use (the least-squares linear model or the least-squares quadratic model)? Why?

2. Consider the following data:

x	y
2	9
4	6
6	4
8	5

a) Plot the data points and *visually* estimate a linear model with whole number coefficients that would fit it reasonably well.

b) *By hand,* find the mean of the y values ($\bar{y}$) and the TSS. Draw a sketch that relates to TSS.

c) *By hand,* use your model in (a) to calculate the *predicted* y value for each of the x data values given in the table (show your work). Use these to find the SSR. Draw a sketch that relates to SSR.

d) *By hand,* calculate the error and the squared error for each data point with respect to your linear model in (a), and then find the SSE (show your work). What is the relationship between TSS, SSR, and SSE?

e) Use technology to find the least-squares best-fit linear model for the above data.

f) Use technology to find the TSS, SSR, and SSE values for the model in (e). What is the relationship between TSS, SSR, and SSE in this case? What conditions make your answer here different from your answer to (d)?

g) Use your answers to part (f) to find the R^2 value for the model in (e).

h) Which R^2 is lower: (e) or (g)? Explain why this should be.

i) Find the R^2 value for the least-squares *quadratic* model for this data. Which of the two least-square models do you think is better? Why?

Game Time

3. You are responsible for organizing a holiday party at school that has been a tradition for many years. You are trying to forecast how many people will come this year, to know how much food and drink to order. You have data from the last 7 years:

Year	Attendance
1	58
2	65
3	63
4	70
5	71
6	74
7	73

a) Use a spreadsheet to fit a quadratic model to this data.

b) What is the R^2 value for this model and data? How can you interpret this value?

c) Interpret the standard error values you obtain. Use the ratio magnitude rule of thumb to determine if the signs of the parameters are meaningful. Could any of them reasonably be 0?

d) Would a linear model make more sense for this situation? Justify your answer.

4. Suppose your GPA over the last 6 semesters were given by:

Semester	GPA
1	2.8
2	2.7
3	2.9
4	3.1
5	3.0
6	3.3

a) Use a spreadsheet to fit a quadratic model to this data.

b) What is the R^2 value for this model and data? How can you interpret this value?

c) Interpret the standard error values you obtain. Use the ratio magnitude rule of thumb to determine if the signs of the parameters are meaningful. Could any of them reasonably be 0?

d) Would a linear model make more sense for this situation? Justify your answer.

5. In Chapter 1, you worked with the following demand data for T shirt sales, based on a market research poll:

Price (in $)	8	10	12	15
Number Sold	302	231	175	123

a) Plot the data points and *visually* estimate a linear model with whole number coefficients that would fit it reasonably well.

b) *By hand,* find the mean of the y values ($\bar{y}$) and the TSS. Draw a sketch that relates to TSS.

c) *By hand,* use your model in (a) to calculate the *predicted* y value for each of the x data values given in the table (show your work). Use these to find the SSR. Draw a sketch that relates to SSR.

d) *By hand,* calculate the error and the squared error for each data point with respect to your linear model in (a), and then find the SSE (show your work). What is the relationship between TSS, SSR, and SSE?

e) Use technology to find the least-squares best-fit linear model for the above data.

f) Use technology to find the TSS, SSR, and SSE values for the model in (e). What is the relationship between TSS, SSR, and SSE in this case? Why is this answer different from your answer to part (d)?

g) Use your answers to part (f) to find the R^2 value for the model in (e).

h) Find the R^2 value for the least-squares *quadratic* model for this data. Which least-squares model seems better? Why?

6. You are trying to determine the best amount of time to spend on studying in one day, with an average workload, to balance your desire to learn and get good grades with the rest of your life (sleeping, eating, social life, extracurricular activities, etc.). You devise a scale to measure the balance for a particular day, from 0 (the worst possible balance between studying and everything else) to 100 (perfect balance). For the first five school nights, you obtain the following data:

Study Hours	2	5	4	6	3
Balance (0-100)	45	75	70	70	55

a) Plot the data points and *visually* estimate a linear model with whole number coefficients that would fit it reasonably well.

b) *By hand,* find the mean of the y values ($\bar{y}$) and the TSS. Draw a sketch that relates to TSS.

c) *By hand,* use your model in (a) to calculate the *predicted* y value for each of the x data values given in the table (show your work). Use these to find the SSR. Draw a sketch that relates to SSR.

d) *By hand,* calculate the error and the squared error for each data point with respect to your linear model in (a), and then find the SSE (show your work). What is the relationship between TSS, SSR, and SSE?

e) Use technology to find the least-squares best-fit linear model for the above data.

f) Use technology to find the TSS, SSR, and SSE values for the model in (e). What is the relationship between TSS,SSR, and SSE?

g) Use your answers to (f) to find the R^2 value for the model in (e).

h) Find the R^2 value for the least-squares quadratic model for this data. Which model would you use? Why?

7. Look for a newspaper or magazine article or column that interprets the results of a regression study, and comment on whether the author seems to be making appropriate conclusions from the results or not.

Overtime[*]

For Exercises 8-15, indicate which regression assumption(s) the given graph relates to, and state whether the graph validates the assumption(s) or not, and why:

8.

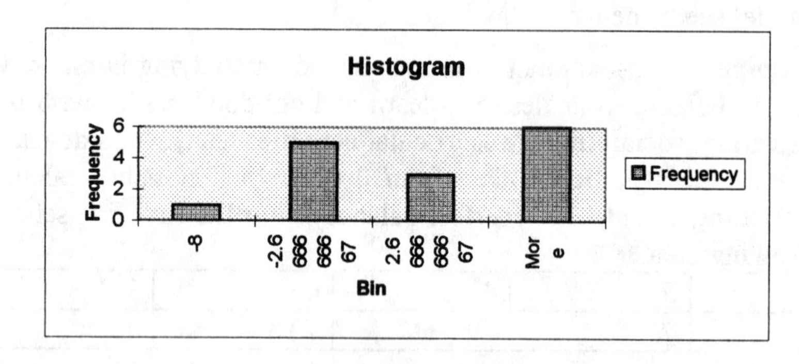

9.

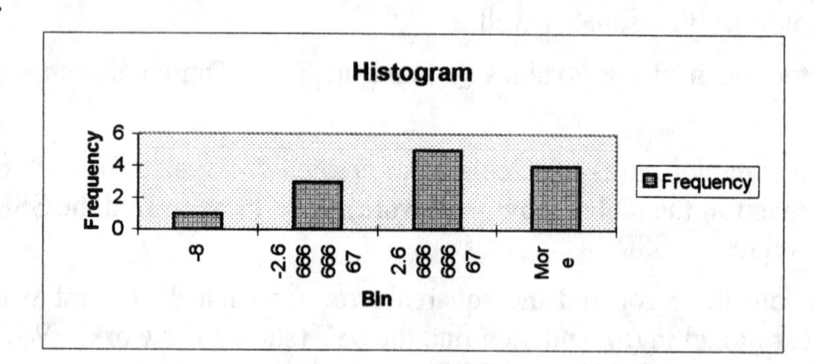

10.

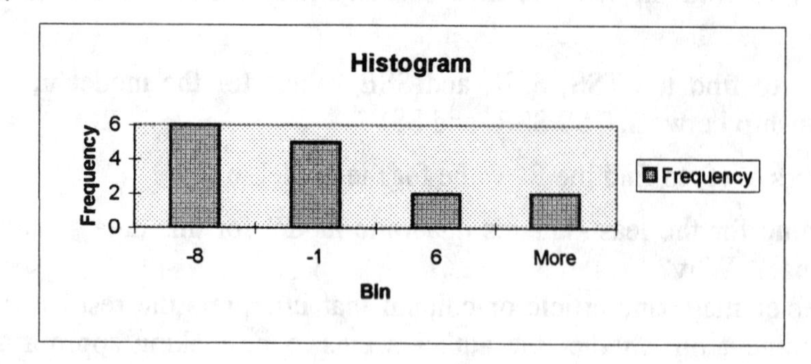

[*] These problems relate to the optional subsection on regression assumptions.

11.

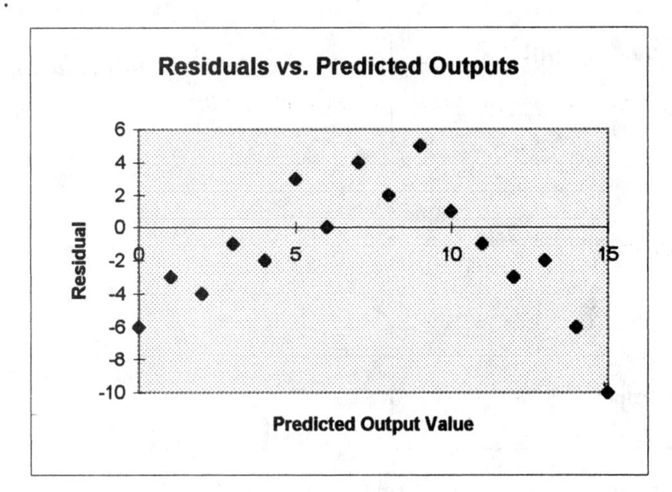
Residuals vs. Predicted Outputs
Residual
6
4
2
0
-2
-4
-6
-8
-10
0
5
10
15
Predicted Output Value

12.

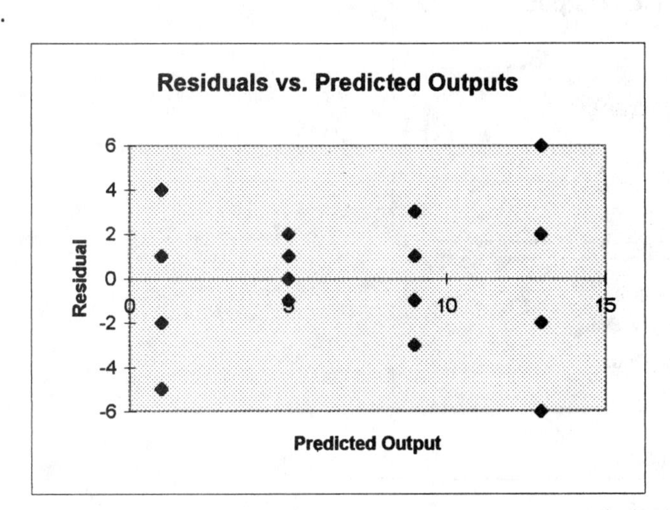
Residuals vs. Predicted Outputs
Residual
6
4
2
0
-2
-4
-6
0
5
10
15
Predicted Output

13.

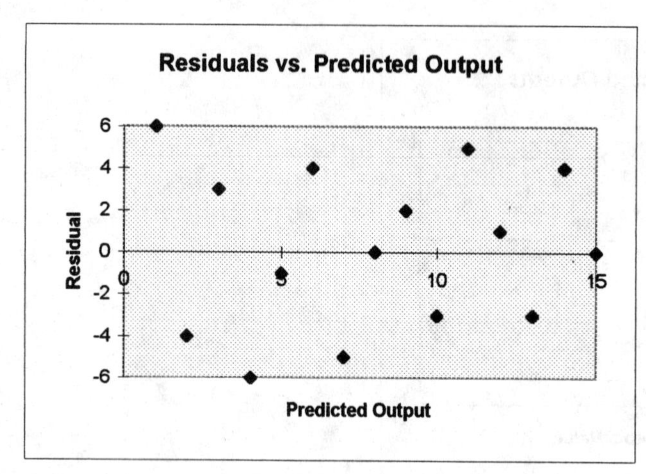
Residuals vs. Predicted Output
Residual
Predicted Output
6
4
2
0
-2
-4
-6
0
5
10
15

14.

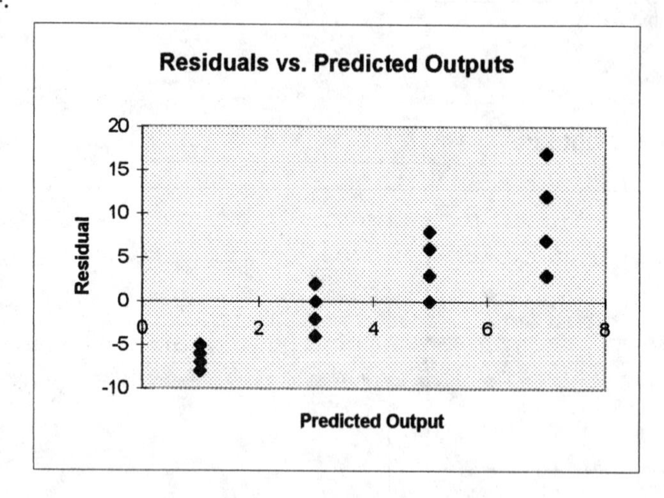
Residuals vs. Predicted Outputs
Residual
Predicted Output
20
15
10
5
0
-5
-10
0
2
4
6
8

15.

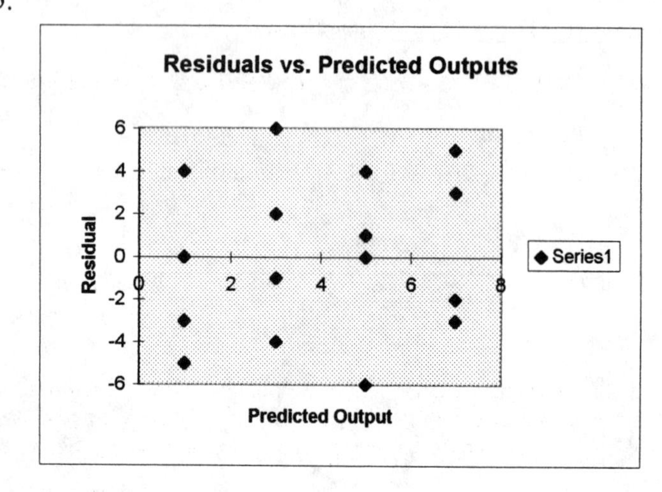
Residuals vs. Predicted Outputs
Residual
Predicted Output
Series1
6
4
2
0
-2
-4
-6
0
2
4
6
8

Section 6.4[*] : Investment Portfolios, Risk/Return Tradeoffs, and Pareto Efficiency

In Sections 6.2 and 6.3, we discussed how the concepts of variance and standard deviation are related to the notions of variability and risk. One of the major areas of the real world that we associate with risk is the stock market, and investments in general. The discipline of Finance has developed extensive theories on how to analyze investment problems and make decisions that incorporate such risk considerations. In this section, we introduce some of these concepts. An **investment portfolio** is a collection of investments. We can calculate rates of return over time for any portfolio, from which we can then calculate the *mean* and sample *variance* (or standard deviation) as measures of **return** and **risk**, respectively. In general, as an investor, we want the highest possible return at the lowest possible risk. Unfortunately, high returns usually come with higher risks, so some way of evaluating **tradeoffs** between risk and return are needed. The discussion here is a first step in that process.

Here are some examples of the kinds of problems the material in this section can help you analyze:

- You are analyzing a number of potential investments. You want to estimate the average (mean) rates of return and the risks (variances) to do a risk/return tradeoff analysis, to help you decide how much to invest in each.
- Your new job offers a retirement plan in which you need to make decisions about how to allocate your investment contributions between the stock market, bonds, foreign markets, and other options. What should you do to meet your goals for retirement?

When you have finished this section, you should:

- Understand the basic idea of a **risk/return tradeoff analysis**.
- Know what it means for one investment to **dominate** (be **Pareto superior** to) another.
- Know what is meant by the **efficient set** of investments out of a given collection of investments.

---

[*] The material in this section is optional and may be skipped without loss of continuity. Even for non-business students, however, if you think you may be saving for retirement at some point in the future, it could be personally useful to understand this material! If nothing else, at least read the few paragraphs under the heading "Saving for Your Retirement" near the end of the section.

Return/Risk Tradeoffs for Investments Using Variance and Standard Deviation

If you know anything about the stock market, you probably know that it is very risky. Technically, if you invest $1000 in the stock market today, you could lose it all tomorrow (for example, if you invested it all in the stock of one company, and an hour later it declared bankruptcy and went belly-up), so *all* of your *principal* is *at risk*. You may have also heard about "blue chip" and utility stocks that are less risky and usually provide a decent return on your money. How can you measure *riskiness* to be able to make comparisons of different stocks or other investments?

Basically, when we talk about *riskiness*, we mean that something is highly *variable* or uncertain, and that there is a significant chance that at least some of the outcomes will not be good for us. We have already seen that variance and standard deviation are measures of variability, and in fact they are frequently used in evaluating the performance of stocks and other investments.

If you look back Sample Problem 9 from Section 1.3 of Volume 1, we calculated the annual returns on Intel stock and the S&P 500 average between 1990 and 1995 with the following results:[1]

Year (1/1-12/31)	1991	1992	1993	1994	1995
Intel (% return)	23	90	38	3	78
S&P 500 (% return)	30	8	11	1	37

Let's try calculating the means, variances,[2] and standard deviations for these two investments.

[1] Notice that the calculations can be done very easily on a spreadsheet. Try them yourself for the practice, and see if you get comparable answers. Most spreadsheets also have special functions for finding the mean and variance of a block of numbers, so this type of analysis we will be doing here is very convenient on a spreadsheet. You may even be able to get tables of data off of the Internet, so that you don't have to type in the values!

[2] In Section 6.2, we discussed the fact that there are in fact two forms of the variance (and so also for the standard deviation), the *population* and the *sample* forms, appropriate for different situations. In this case, the *sample* variance is more appropriate, since it makes more sense to think of these data values as a *sample* from a *larger* population of *possible* return values (for example, including those in the future).

$$\mu_I = \frac{23+90+38+3+78}{5} = \frac{232}{5} = 46.4$$

$$\mu_S = \frac{30+8+11+1+37}{5} = \frac{87}{5} = 17.4$$

$$\sigma_I^2 = \frac{(23-46.4)^2+(90-46.4)^2+(38-46.4)^2+(3-46.4)^2+(78-46.4)^2}{4}$$

$$= \frac{5401.2}{4} = 1350.3$$

$$\sigma_S^2 = \frac{(30-17.4)^2+(8-17.4)^2+(11-17.4)^2+(1-17.4)^2+(37-17.4)^2}{4}$$

$$= \frac{941.2}{4} = 235.3$$

$$\sigma_I = \sqrt{\sigma_I^2} = \sqrt{1350.3} \approx 37$$

$$\sigma_S = \sqrt{\sigma_S^2} = \sqrt{235.3} \approx 15$$

This tells us that the mean return for Intel was about 46% over those 5 years, with a standard deviation of about 37%. Using our rule of thumb, the 3σ interval for the return on Intel would then be

$$46\% \pm 3(37\%) \;=\; 46\% \pm 117\% \;=\; [\text{-}61\%, 163\%]$$

On the other hand, the return on the S&P 500 over those same 5 years averaged about 17%, with a standard deviation of about 15%, so the 3σ interval for the return on the S&P 500 would be

$$17\% \pm 3(15\%) \;=\; 17\% \pm 45\% \;=\; [\text{-}28\%, 62\%]$$

Which investment is riskier? Certainly the Intel return is much more variable, and the potential losses are worse, so it makes sense to say that it is riskier. But it also has a much higher average return! In fact, this is often the case: **the investments with the highest returns are often the ones that are more variable and riskier** (you may have heard of "junk bonds"...). Thus, we say that in general **there is a *tradeoff* between return and risk**. And the judgment about which investment (or what combination of investments) is best depends on who is making the decision and for what purpose. Everyone has a different attitude to risk-taking; this is one aspect of our personalities. There are ways to measure someone's attitude to risk, but they are beyond the scope of this book.

On the other hand, we can certainly say that virtually everyone who values money would prefer investments with the highest possible return and the lowest possible risk, or variability (possibly after ruling out investments they find morally objectionable). For instance, all other things being equal, if investment A had a mean return of 46% with a standard deviation of 1% (for a 3σ interval of [43%,49%]) and Investment B had a mean return of 17% with a standard deviation of 5% (for a 3σ interval of [12%,32%]), it would be pretty clear which one was better! Furthermore, virtually all investors would prefer the performance of A to that of Intel, and would prefer B to the S&P 500, since in each comparison pair the average return is the same, but the risk/variability is smaller.

To see these comparisons visually, we can draw a simple graph (Figure 6.2-3) with return on the vertical axis and risk (expressed using variance, for reasons we will explain later) on the horizontal axis. The variance of Investment A is $1^2 = 1$, and the variance of Investment B is $5^2 = 25$.

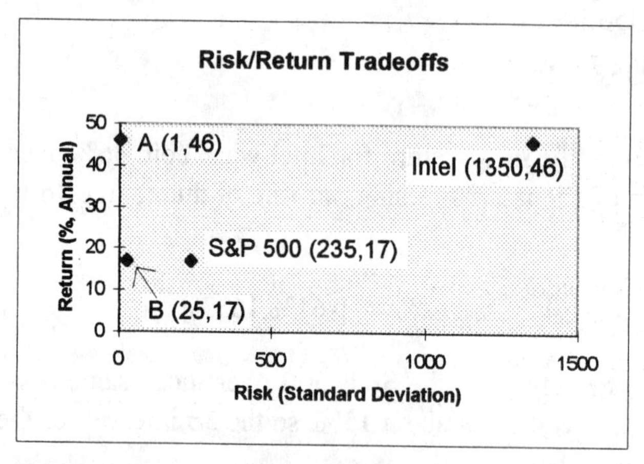

Figure 6.2-3

In the graph, since return is on the vertical axis and *larger* returns are better, better investments are those that are *higher* on the graph. Since risk is on the horizontal axis and *smaller* risk is better, better investments are those further to the *left* on the graph. Put together, we want investments that are as far *up and to the left* as possible. By this criteria, Investment A is clearly the best, since it is furthest up and to the left of the four. The clearest comparison is between Investment A and the S&P 500, since the point for A is *strictly* to the left and *strictly* higher compared to the point for the S&P 500. You could say that Investment A **dominates,** or is **Pareto[3] superior** to, the S&P 500 in this example. In general, **we say one investment dominates, or is Pareto Superior to, a second investment if the return of the first is *at least* as high as the second, the risk is *no more* than the second, and *at least one of these comparisons is a strict inequality*.** Graphically, one point dominates a second if the first is above and to the left (and not identical to) the other.

[3] This term is named after the Italian economist who developed the concept.

In our example, then, Investment A dominates not only the S&P 500, but it also dominates Investment B, since it is also strictly better on both criteria, as we mentioned earlier. Investment A also dominates Intel, since the point for A is above and to the left of the point for Intel. Notice that in this case, however, there is a *tie* on *one* of the criteria (the points line up vertically), but *not on both* (the points are not identical). We could also say that the S&P 500, Intel, and Investment B are each **dominated *by***, or are **Pareto *inferior* to**, Investment A, or that Investment A is the *only* ***non-dominated*** investment in the group.

Notice, however, that if we were *only* considering the two investments from Section 1.3 , the S&P 500 and Intel, neither would dominate the other, since Intel has a better (higher) return, but the S&P 500 has a better (lower) risk. This is the most common situation when making real investment comparisons: usually no single investment dominates all of the others, but there is a **set of non-dominated investments**, sometimes referred to as the **efficient set** out of the comparison group. If the efficient points are *connected* on a graph like that we have set up, we can refer to the **efficient (northwest, or upper left) frontier** of the comparison group. The efficient frontier could also refer to just the *points* corresponding to the efficient set on a graph.

Suppose we had added a fifth investment to our mix, Investment C, with a return of 47% and a standard deviation of 37% (variance of 1350). Recall that Investment A dominated Intel, and the new Investment C is *very similar* to Intel, (1350,47) vs. (1350,46), but now Investment A does *not* dominate Investment C (and C does not dominate A). In this group of 5, the new efficient set would be *both* Investment A *and* Investment C, and the line joining them (or just those two points) could be the efficient frontier. To make this easier to see, the risks and returns are listed in Table 1. This is further illustrated in Figure 6.2-4 .

Table 1

Investment	Risk (Variance)	Mean Return (%)
A	1	46
B	25	17
C	1350	47
Intel	1350	46
S&P 500	235	17

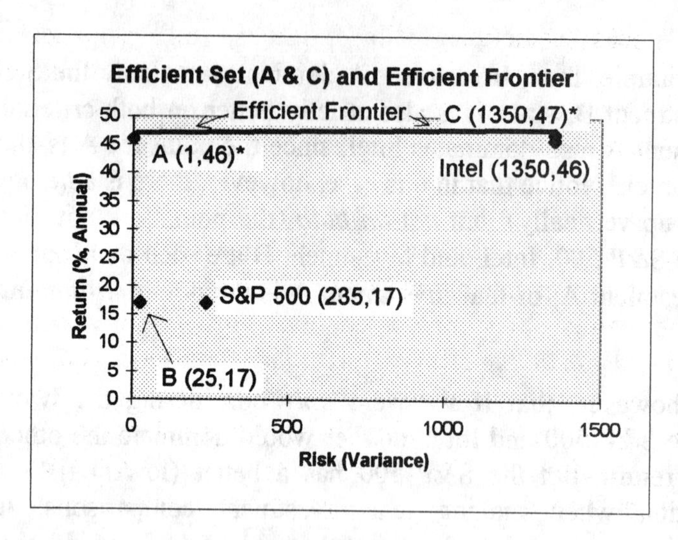

Figure 6.2-4

The points on the efficient frontier line segment could themselves correspond to investment options: in this case, for example, if investments A and C were statistically independent[4], then combinations of investments A and C would have returns and risks that would correspond to points along the efficient frontier line segment. For example, again if A and C were independent, the midpoint of the line segment would correspond to the return and risk of an investment that put 50% into A and 50% into C. This interpretation is only guaranteed to be valid (*given* the independence of A and C) if our risk is measured using *variance*, rather than standard deviation, which is why we have set up our graph this way.

Saving for Your Retirement

As a final note: sometime after your studies in school, you may be faced with making investment decisions if you make contributions to a retirement fund of some kind. The stock market tends to achieve higher returns in the long run (roughly 9-10% over periods of decades) compared to bonds (more like 7-8% over decades), but the stock market is also much riskier. The "conventional wisdom" in investing suggests that, when you are young and saving for a distant retirement, you should have a higher percentage of

[4] In simplistic terms, we say that two random quantities (two random variables) are statistically independent if the value of each is *not connected* or linked to that of the other. Another way of saying this is that their variations are not *correlated*. A *positive* correlation would mean that, on average, when one quantity has a *high* value from its domain, the other is *also* more likely to be *high* (so they tend to vary in the *same* direction). A *negative* correlation means that when the value of one is *high*, the value of the other is more likely to be *low* (so they tend to vary from their means in *opposite* directions). Independence means that there is *neither* type of correlation. For precise definitions, see a statistics book. If investments A and C were *not* independent, combinations of the two would form a *curve*, not a line segment.

your money in stocks (sometimes called **equities**, since stockholders are essentially partial *owners* of the company, like they own their homes), for long-term growth, but as you get closer to retirement, you should move your money into less risky investments, preferably ones where your principal is not at risk if possible, to protect your retirement income.

One crude rule of thumb is to aim to have the percentage of your retirement savings that is in safer investments like bonds or annuities roughly correspond to your age or higher. In other words, at age 20, you could aim for 20% of your investments in bonds and 80% in stocks, while at age 50, you might aim for a more equal mix (50% each). If your personality tends to be cautious (risk-averse), you will probably want a higher percentage of the safer investments, while if you are a risk-taker, you may want a lower percentage of safer investments. A good investment advisor will ask you to fill out a long questionnaire designed to determine your goals and measure your willingness to take risks.

Many people do not receive this kind of advice until long after they have started a retirement investment program, which can mean that their retirement savings turn out much smaller than they could have been otherwise. Of course, there is no guarantee that the future will be similar to the past, so no investment advice is foolproof! Ultimately, you have to take responsibility for your own investment decisions. Understanding probability and statistics is one way to have the best possible chance of making the right decisions.

Section Summary

Before you try the exercises, be sure that you

- Understand that, for investments, higher returns and lower risk (variance) are desirable. If one investment compared to a second is better on both measures (or possibly tied on one measure), we say the first **dominates (is Pareto superior to)** the second, or the second **is dominated by (is Pareto inferior to)** the first. For a given group of investments, those that are non-dominated (by any other in the group) are in the **efficient set**, and can define the **efficient frontier** (the upper left edge) of the graph, if return is on the vertical axis and risk on the horizontal.

EXERCISES FOR SECTION 6.4*

Warm Up

For Exercises 1-4, a set of investments, with the mean return and variance, are given. Graph each as a point on a risk-return graph, and determine which investments dominate which others, which investments are non-dominated, and the efficient set.

1.

Investment	Risk (Variance)	Mean Return (%)
A	538	25
B	321	14
C	374	11
D	23	6
E	476	27

2.

Investment	Risk (Variance)	Mean Return (%)
A	710	29
B	1	3
C	11	8
D	368	19
E	261	12

3.

Investment	Risk (Variance)	Mean Return (%)
A	426	19
B	395	19
C	214	13
D	214	14
E	395	19
F	512	17

4.

Investment	Risk (Variance)	Mean Return (%)
A	328	8
B	156	10
C	163	13
D	142	12
E	170	16
F	139	19

Game Time

5. International Paper shows information about its relative stock price compared to the Standard and Poor's (S&P) 500 Index for the benefit of its stockholders. The following table gives the stock and index values, assuming that the value invested in the company's common stock and in the S&P 500 was $100 on December 31, 1990, and that all dividends were reinvested.

Value ($) as of 12/31	1990	1991	1992	1993	1994	1995
International Paper	100	136	131	137	156	160
S&P 500	100	131	141	155	157	216

a) Find the percent change for each of the 2 investments during the calendar years 1991-1995.
b) Using the percent change values as data, calculate the mean percent change for each investment.
c) Calculate the sample variance and standard deviation for each investment.
d) Calculate the 3σ interval for each investment.
e) Using the mean and variance of the percent return, plot a point on a return/risk tradeoff graph for each investment.
f) Does either investment dominate the other? If so which one? What is another way to express the situation when one investment dominates a second?
g) What is the efficient set out of these two investments?

6. Public Storage, Inc., shows information about its relative stock price compared to the Standard and Poor's (S&P) 500 Index for the benefit of its stockholders. The following table gives the stock and index values, assuming that the value invested in the company's common stock and in the S&P 500 was $100 on December 31, 1990, and that all dividends were reinvested.

Value ($) as of 12/31	1990	1991	1992	1993	1994	1995
Public Storage, Inc.	100	137	162	277	296	411
S&P 500	100	131	141	155	157	216

a) Find the percent change for each of the 2 investments during the calendar years 1991-1995.
b) Using the percent change values as data, calculate the mean percent change for each investment.
c) Calculate the sample variance and standard deviation for each investment.
d) Calculate the 3σ interval for each investment.

e) Using the mean and sample variance of the percent return, plot a point on a return/risk tradeoff graph for each investment.
f) Does either investment dominate the other? If so which one? What is another way to express the situation when one investment dominates a second?
g) On a separate return/risk graph, add a point for Intel from the example in this section, and a point for International Paper (you can use your own calculations or the answer you were given to Exercise 5 above). What is the efficient set out of these four investments?

7. Texas Utilities lists its return on average common stock equity from 1991 through 1995 as follows:

Year (1/1-12/31)	1991	1992	1993	1994	1995
TU (% return)	-6.3	10.9	5.6	8.3	-2.3

a) Calculate the mean return and sample variance for Texas Utilities.
b) Calculate the sample standard deviation, and use it to estimate 95% and 99% confidence intervals (margins of error) for the return on the stock.
c) How would you describe the performance of this stock?
d) Compare the performance of this stock to Intel and the S&P 500 over the same period.
e) Based on this information, how would *you* allocate your money between shares of Texas Utilities, Intel, and an S&P 500 Index mutual fund?

8. Hercules, Inc. lists the total return on its stock from 1990 through 1995, starting with $100 at the end of 1990 and reinvesting all dividends, as given below:

Value ($) as of 12/31	1990	1991	1992	1993	1994	1995
Hercules	100	101	124	144	162	259
S&P 500	100	131	141	155	157	216

a) For each investment, calculate the percent return for 1991-1995, then find the mean and variance of these values.
b) Calculate the sample standard deviation, and use it to estimate 95% and 99% confidence intervals (margins of error) for the return on each investment.
c) Calculate the beta value for Hercules (see Section 1.3). What does it mean?
d) Compare the performance of Hercules stock to Intel and the S&P 500 over the same period.
e) Based on this information, how would *you* allocate your money between shares of Hercules, Intel, and an S&P 500 Index mutual fund?

9. Take all of the data given for the different *real* stocks given in this section, including the exercises, and do an overall risk-return analysis to find the efficient set.

10. When does it make sense to connect the points of two stocks on a risk-return graph with a line segment?

Chapter 6 Summary

Multivariable Regression on Spreadsheets

By organizing your data with a column for the dependent variable first on the left, then a column for each true independent variable, you have the flexibility to create columns for quadratic or higher-order terms using spreadsheet formulas, to allow for nonlinear models as well as linear ones.

From looking at the output report from a regression analysis, you can identify all of your parameters, including the constant term and the coefficients of each term in your model. By looking at the 95% **confidence interval** for each coefficient, you can determine whether the sign of that coefficient is well determined: it is, if 0 is not included in the interval. If 0 is included in the confidence interval, you should seriously consider dropping that term from your model. As an overall measure of the fit of the model to the data, the $\mathbf{R^2}$ value tells what percentage of the total variation in the output data is *explained* by the model; the closer R^2 is to 1, the closer the fit is to being perfect.

Another determinant of the goodness of a model is the **degrees of freedom of the residuals**, which is the number of data points minus the number of parameters in the model (the number of data points *over* the *minimum* needed to fit that category of model). Since more information is always better, and simpler models (with fewer parameters - remember KISS!) are best whenever possible, this value should be as large as possible. Unfortunately, improving R^2 often occurs at the expense of the degrees of freedom of the residuals (such as by fitting a more complicated model to the same data), so the two factors need to be balanced. The **adjusted** $\mathbf{R^2}$ often given does make an adjustment to the value of R^2 to lower it when the number of data points is small.

When formulating and fully defining a model from raw data using regression, remember to be careful about the **domain** you use. The best practice is usually to use the extremes of the values of the independent variables from the data as a starting point, and only expand beyond that after careful consideration of what is reasonable. With multivariable models it is especially helpful to define your variables with letters that remind you of what they stand for.

Descriptive Statistics: Mean, Variance, and Standard Deviation

The **mean** of a finite group of numbers $x_1, x_2, \ldots, x_n$ is simply their average:

$$\text{mean} = \bar{x} = \frac{x_1 + x_2 + \ldots + x_n}{n} = \frac{\sum_{i=1}^{n} x_i}{n}$$

The mean of a continuous random variable (also called the **expected value**) with a pdf of $f(x)$ is a generalization of the idea of the average and is calculated by the formula

$$\text{mean} = \mu = E(X) = \int_{-\infty}^{\infty} x f(x) dx$$

Graphically, the mean can be thought of as the balance point of the density function: it is the location of the vertical line along which the pdf would balance like a see-saw.

The variability of a random quantity can be described using the concept of the **variance**. For finite groups of numbers, there are two possible variance calculations. If the set of numbers is the *entire population* of interest, then you should use the **population variance**, given by the formula

$$\text{population variance} = \sigma^2 = \frac{\sum_{i=1}^{n} (x_i - \bar{x})^2}{n}$$

This is simply the *average* squared deviation of the data values from their mean. However, since the units are the *square* of the original units, an easier interpretation is obtained by taking the *square root* of the variance, called the **standard deviation**. Again, when your numbers represent the entire population of interest you compute the **population standard deviation** using

$$\text{population standard deviation} = \sigma = \sqrt{\text{population variance}} = \sqrt{\sigma^2}$$

The analogous calculations for a continuous random variable are

$$\text{(population) variance} = \sigma^2 = \int_{-\infty}^{\infty} (x - \mu)^2 f(x) dx$$

$$\text{(population) standard deviation} = \sigma = \sqrt{\text{(population) variance}} = \sqrt{\sigma^2}$$

Whether working with finite groups of numbers or continuous random variables, a crude rule of thumb is that roughly 99% of all values (or all probability) should lie within 3 times the standard deviation of the mean, and about 95% should lie within 2 times the standard deviation. This gives a good way to interpret these values, and get a feel for the variability in a situation.

On the other hand, when a finite group of numbers is being considered as a **sample** from a *larger population* about which you want to know more, the appropriate measures of variability to use are the **sample variance and standard deviation**, because they will give you more accurate estimates of the corresponding values for the entire population (higher variability, because of the uncertainty of a sample). The formulas are:

$$\text{sample variance} = S^2 = \frac{\sum_{i=1}^{n}(x_i - \bar{x})^2}{n-1}$$

$$\text{sample standard deviation} = S = \sqrt{\text{sample variance}} = \sqrt{S^2} = \sqrt{\frac{\sum_{i=1}^{n}(x_i - \bar{x})^2}{n-1}}$$

In this situation, you can still use the rule of thumb about 2 or 3 times the standard deviation, since these values are your best estimates of the population values.

Misuse of Descriptive Statistics and Regression

When a distribution is skewed, the mean is to one side of the median (the middle value, or 50th percentile). If the mean is to the right, we say the distribution is skewed to the right. A mode (most likely value) can be almost anywhere. This means that the choice of a measure of "central tendency" can give *very* different impressions, and so can easily be manipulated.

Finding a good regression model relating two variables does *not* mean that one of them *causes* the other. It only means that a relationship exists, but causality could go in *either* or *neither* direction (such as if some *other* variable causes *both*). True causality can only be proved by carefully designed experiments and theoretical reasoning.

Squared Errors: TSS, SSR, SSE, R^2, MSE

The Total Sum of Squares (TSS) for the y values (outputs, dependent variable values) from a set of data is given by TSS $= \sum_{i=1}^{n}(y_i - \bar{y})^2$ (the sum of the squares of the deviations of the y values from their mean), and can be thought of as the *total variation* of those y values. This is like the SSE calculation when the *model* is just the constant horizontal line at the mean.

The Sum of the Squares of the Regression (SSR) for a set of data and a given model is given by $\text{SSR} = \sum_{i=1}^{n} (\hat{y}_i - \bar{y})^2$ (the sum of the squares of the *differences* between the *predicted* y values and the *mean* of the y values), and represents the variation in the y values that is explained by (due to) the regression model. This is like doing the TSS calculation using the *predicted* output values from the model instead of the *actual* outputs, for each of the input values. SSR can be thought of as the variation in output values that is *explained* or *expected* due to the model.

If the model being studied is a *least-squares* model, then TSS=SSR+SSE, but that this is *not* true in general for other models. For least-squares models, then, we can define the coefficient of determination (R^2) for a given data set and a given model as:

$$R^2 = \frac{SSR}{TSS} = \frac{\sum_{i=1}^{n} (\hat{y}_i - \bar{y})^2}{\sum_{i=1}^{n} (y_i - \bar{y})^2}.$$

This represents the *fraction* (often expressed as a %) of the *total variation* in the y values that is *explained* by the regression model. The value ranges from 0 to 1, where 1 represents a perfect fit between the model and the data.

R^2 is an important way to measure the fit of a model to a set of data, but it can be misleadingly inflated if the number of data points is small, and should not be the *only* criterion in selecting a model (as always, err on the side of simpler models whenever possible).

The Mean Squared Error is the average of the squares of the errors in a given situation. This is most useful in comparing forecasting methods' performance historically, especially since they may only be able to be applied over different numbers of data points (so the *average* squared error is fairer than the *sum*). The calculation is given by

$$MSE = \frac{SSE}{n} = \frac{\sum_{i=1}^{n} (y_i - \hat{y}_i)^2}{n} = \frac{\sum (\text{actual - predicted})^2}{(\text{number of predictions})},$$

for a given specific model (or forecasting technique) and a given specific set of data points.

Standard Errors

The *standard error of a parameter* in a regression can be thought of as the standard deviation of the estimate of that parameter (as if models were fit to many sets of data from the same population), so the same "plus or minus 3 times the value" rule of thumb can be used to interpret the standard error. This is why, if the magnitude of the ratio of the parameter estimate to the standard error is 3 or more, you can be quite sure that at least the sign of the parameter (direction of the relationship) is correct, and therefore that that term probably belongs in your model. If the ratio has a magnitude of less than 2, you should seriously consider dropping that term. If the ratio magnitude is between 2 and 3, you should use other factors in deciding whether to keep the term.

Regression Assumptions

When you use confidence intervals or standard error, you are implicitly assuming that the **residuals have a normal (bell-shaped) distribution**, that **the means of the residuals are uniformly 0 everywhere**, and that **the standard deviations of the residuals are equal everywhere**. It is possible to validate these assumptions by a histogram plot of the residuals (seeing if it looks roughly bell-shaped) and a plot of the residuals against the predicted output values (seeing if the means seem uniformly 0 and the standard deviations seem equal everywhere).

Investment Portfolios, Risk-Return Tradeoffs, Pareto Efficiency

For investments, higher returns and lower risk (variance) are desirable. If one investment compared to a second is better on both measures (or possibly tied on one measure), we say the first **dominates (is Pareto superior to)** the second, or the second **is dominated by (is Pareto inferior to)** the first. For a given group of investments, those that are non-dominated (by any other in the group) are in the **efficient set**, and can define the **efficient frontier** (the upper left edge) of the graph, if return is on the vertical axis and risk on the horizontal.

Chapter 7: Matrices and Solving Systems of Equations

Introduction

We all encounter tables of numbers frequently in our lives, whether it is a monthly budget spreadsheet or a table of statistical or other data. In mathematics, such tables are called matrices and the study of them is called matrix theory. **Matrices** [pronounced "**may**-truh-sees"; singular: **matrix**] are rectangular arrays of numbers. Matrix theory helps us to perform calculations involving matrices. In this chapter we show how to perform simple matrix operations such as addition, subtraction, and multiplication of a matrix by a scalar (single number). We show how to multiply matrices, which can greatly reduce the time needed to perform certain kinds of calculations. Finally, we show how to use matrix theory to solve systems of linear equations, which includes finding the inverse of a matrix. In optional Section 7.5, we discuss a category of model called a Markov Chain which can be useful when working with discrete time periods involving a system that can be in a discrete number of possible states, when you are interested in the probabilities of being in different states in the future.

Here is a selection of examples to give you a feel for the kind of problems you should be able to solve after studying this chapter.

- The Nationwide Prefab House Corporation manufactures and ships to its distributors six different models of prefab homes. These homes are different configurations of six modules. You are trying to help the Nationwide Prefab House Corporation automate its orders and shipping. Every week reports come in from the six district distributors specifying the number of orders that they have received for each model of house. The company needs to know how many of each type of module to ship to each distributor. Billing needs to know the total cost of the shipment to each distributor. Shipping needs to know the total weight going to each destination in order to have the proper number of trucks on hand. The company would also like to have a record of how many of each type of module was shipped each week.
- The Nationwide Prefab House Corporation has also asked you to check out a new company that has just started to compete for the prefab home market. From their brochures, you have determined the configuration and prices of the models that they produce. Can you determine what they are charging for each module from this information? If the industry-wide markup on this type of product is 35%, can you determine their approximate cost per module? How do this company's costs seem to compare with your production costs?
- You are the owner of a small auto dealership, called Slim's Car Sales. You employ four salespeople, and sell four different style categories of vehicles. In

each style category there are three price categories (1 vehicle in each price category for each style category). The sales for each of the four salespeople are recorded each month. You want a system to calculate total sales for each salesperson, both for each different style category of cars sold and for each different price category.

- At Slim's Car Sales your sales goal for each salesperson this month was to sell 5 vehicles in each style category within each price category. Calculate a table of discrepancies from these goals where a positive number mean the sales exceeded the goal by that amount, a negative number means sales were below the goal by that amount, 0 means the goal was met exactly.
- At Slim's Car Sales you have planned an advertising blitz for the month of April, and expect to be able to double your last month's sales (in each style category and at each price category). Use a spreadsheet to calculate your sales goals for sedans in April, broken down by salesperson and price category.
- * Suppose you have a summer baby-sitting job taking care of a 4-year-old child every afternoon. The child goes to a program every morning, but if it rains, the program gets canceled, and you have agreed to be with the child on any such mornings. For the last couple of weeks, you have kept track of the weather on consecutive mornings. Today is Monday, and the program was canceled. You are talking with a friend about doing something together on Wednesday morning, or possibly Friday. What are the chances you will be free? In the long run, what are your chances of having a free morning?

After studying this Chapter, in addition to being able to solve problems like those above, you should:

- Understand what a matrix is.
- Know how to add and subtract matrices and multiply a matrix by a scalar by hand and using technology.
- Know how to multiply two matrices by hand and using technology.
- Know how to invert a matrix by hand and using technology.
- Know how to solve systems of linear equations by hand and using technology.
- Understand what a matrix equation is and know how to solve a matrix equation by hand and using technology.
- *Understand what a Markov Chain is, when it is appropriate as a model, and how to calculate probabilities given initial probabilities and a transition matrix.

---

* This relates to optional Section 7.5 .

Section 7.1: Introduction to Matrices and Basic Operations

In this section we introduce the basic terminology and notation commonly used when working with rectangular arrays of numbers, or **matrices**. The matrix operations of addition, subtraction and multiplication by a constant, a single number called a scalar, are explained. We also explain when and how to add and subtract matrices, and how to multiply a matrix by a constant (scalar multiplication) by hand and using technology. Spreadsheets are well suited to many matrix operations since, by nature, they deal with arrays of numbers. They are particularly useful for records that involve repeated matrix operations, such as monthly sales or receipts or weekly dietary intakes.

Here are the kinds of problems that the material in this section will help you solve:

- You are very health conscious. You want to determine the amounts of different nutrients you get from some of your favorite snack foods. You are learning to use a spreadsheet and decide that this is a good way to determine the total amount of each nutrient consumed and keep a record.

- You run a small auto dealership. You want to have a quick and easy way to keep records for each salesperson: monthly sales records for each style category, monthly sales record for each price category, actual sales versus goals for sales, and goals for next month based on actual sales for this month.

When you have finished this section you should:

- Know what is meant by a **matrix**.

- Know what is meant by the **dimension** of a matrix.

- Be able to determine which matrices can be added or subtracted, recognize when you should add or subtract matrices, and know how to carry out these operations by hand and using technology.

- Recognize when you should multiply a matrix by a constant (a scalar) and know how to do this by hand and using one or more technologies.

We encounter rectangular arrays of numbers all the time. One of the most common situations in which this happens is when we are presented with tables of data. The mathematical name for a rectangular array of numbers is a **matrix**. **Matrix algebra**, also called **linear algebra** provides us with a systematic way of working with these arrays of numbers.

For example, the 1992 Statistical Abstract of the United States gives the weekly food costs in dollars at the time, not adjusted for inflation, for families in the U.S., by type of family, from 1983 through 1990.

Table 1

Family Type	1983	1984	1985	1986	1987	1988	1989	1990
2A	**54.70**	**56.50**	**58.30**	**60.40**	**63.50**	**66.70**	**70.40**	**73.80**
2B	**52.00**	**53.90**	**55.70**	**57.80**	**60.70**	**67.50**	**70.40**	**73.80**
3A	**66.90**	**69.20**	**71.30**	**73.90**	**77.50**	**81.60**	**86.40**	**90.50**
3B	**77.80**	**80.50**	**83.00**	**85.80**	**90.10**	**94.80**	**100.60**	**104.90**
4A	**77.80**	**80.50**	**82.90**	**85.80**	**90.00**	**94.90**	**100.60**	**105.20**
4B	**93.50**	**97.10**	**99.80**	**103.30**	**108.30**	**114.10**	**121.00**	**126.20**
4C	**98.00**	**101.60**	**104.60**	**108.20**	**113.50**	**119.40**	**126.60**	**132.30**

2A = Couple 20-50 years old
2B = Couple 51 years and over
3A = Couple with 1 child 1-5 years old
3B = Couple with 1 child 15-19 years old
4A = Couple with 2 children 1-5 years old
4B = Couple with 2 children 6-11 years old
4C = Couple with 2 children 12-19 years old

In mathematics, matrices are normally written within square brackets. For example, the matrix representing the statistical data in Table 1 could be written in the following form:

$$\begin{bmatrix} 54.7 & 56.5 & 58.3 & 60.4 & 63.5 & 66.7 & 70.4 & 73.8 \\ 52 & 53.9 & 55.7 & 57.8 & 60.7 & 67.5 & 70.4 & 73.8 \\ 66.9 & 69.2 & 71.3 & 73.9 & 77.5 & 81.6 & 86.4 & 90.5 \\ 77.8 & 80.5 & 83 & 85.8 & 90.1 & 94.8 & 100.6 & 104.9 \\ 77.8 & 80.5 & 82.9 & 85.8 & 90 & 94.9 & 100.6 & 105.2 \\ 93.5 & 97.1 & 99.8 & 103.3 & 108.3 & 114.1 & 121 & 126.2 \\ 98 & 101.6 & 104.6 & 108.2 & 113.5 & 119.4 & 126.6 & 132.3 \end{bmatrix}$$

Matrix Terminology

When we learn any new thing, such as a game, a musical instrument, a sport, or auto repair, we first have to have a basic understanding of the language or terminology used. Matrix algebra, like most mathematical systems, has its own language and terminology.

The **dimension** (sometimes used in the plural form: dimensions) of a matrix refers to the number of rows and columns in the matrix. Rows go across (horizontal, like a rowboat on a lake viewed from the shore), columns go up and down (vertical, like the columns on a Greek-style building). A matrix that has m rows and n columns is said to have dimension $m \times n$ [read: "em by en"]. The matrix in Table 1 above is a 7×8 matrix because it has 7 rows and 8 columns. Note that, since we are dealing only with the rectangular array of numbers, when figuring out the dimension we do not count the verbal headings or **labels** (such as the family types and years for Table 1) used to identify the rows and columns.

Boldface capital letters, such as **A**, are used to denote matrices. For example, for the data in Table 1, we could define the matrix **A** (when writing by hand, you can put a tilde, or wavy line, underneath the capital letter instead of making it bold: $\underset{\sim}{A}$):

$$\mathbf{A} = \begin{bmatrix} 54.7 & 56.5 & 58.3 & 60.4 & 63.5 & 66.7 & 70.4 & 73.8 \\ 52 & 53.9 & 55.7 & 57.8 & 60.7 & 67.5 & 70.4 & 73.8 \\ 66.9 & 69.2 & 71.3 & 73.9 & 77.5 & 81.6 & 86.4 & 90.5 \\ 77.8 & 80.5 & 83 & 85.8 & 90.1 & 94.8 & 100.6 & 104.9 \\ 77.8 & 80.5 & 82.9 & 85.8 & 90 & 94.9 & 100.6 & 105.2 \\ 93.5 & 97.1 & 99.8 & 103.3 & 108.3 & 114.1 & 121 & 126.2 \\ 98 & 101.6 & 104.6 & 108.2 & 113.5 & 119.4 & 126.6 & 132.3 \end{bmatrix}$$

When a matrix has only one row or one column, it is often called a **vector** and is usually denoted by a lower case boldface letter such as **r** or **p**, although capital letters can also be used. Again, when writing longhand, you can use a tilde, or wavy line, under the letter, to indicate it should be boldface, such as $\underset{\sim}{r}$. For example:

$$\mathbf{r} = \begin{bmatrix} 1 & 3 & 7 & 2 \end{bmatrix} \quad \text{and} \quad \mathbf{s} = \begin{bmatrix} 1 \\ 9 \\ 4 \end{bmatrix}$$

r is a 1×4 **row vector** or **row matrix** and **s** is a 3×1 **column vector** or **column matrix.**

In mathematics, the *entries* of a matrix are indicated by the name of the matrix (using the letter in lower case form, and **not** in bold type), followed by **two subscripts** that indicate the position of the entry. **The first subscript gives the row number and the second gives the column number**. For example, the a_{21} entry is the entry in the **A** matrix that is in the second row and first column, so for the matrix defined above for Table 1, this would be the value 52 (meaning $52.00). In general, a_{ij} is the value in the *i*'th row and the *j*'th column of the matrix **A** .

$$\mathbf{A} = [\, a_{ij} \,] = \begin{bmatrix} a_{11} & a_{12} & \cdots & a_{1j} & \cdots & a_{1n} \\ a_{21} & a_{22} & \cdots & a_{2j} & \cdots & a_{2n} \\ \vdots & \vdots & \ddots & \vdots & & \vdots \\ a_{i1} & a_{i2} & \cdots & a_{ij} & \cdots & a_{in} \\ \vdots & \vdots & & \vdots & \ddots & \vdots \\ a_{m1} & a_{m2} & \cdots & a_{mj} & \cdots & a_{mn} \end{bmatrix}$$

To illustrate specific examples, general forms for a 2×3 matrix (named **B**) and a 3×2 matrix (named **C**) are given by

$$\mathbf{B} = \begin{bmatrix} b_{11} & b_{12} & b_{13} \\ b_{21} & b_{22} & b_{23} \end{bmatrix} \qquad \mathbf{C} = \begin{bmatrix} c_{11} & c_{12} \\ c_{21} & c_{22} \\ c_{31} & c_{32} \end{bmatrix}$$

In many situations, such rectangular arrays (tables) would appear as part of a computer **spreadsheet** in which there are many rows and columns. Spreadsheets are used widely throughout the business world. Knowing how to use a computer spreadsheet is a major asset when job hunting.

Electronic spreadsheets are divided into columns and rows of numbers with labels (usually words or numbers) to identify the contents of each particular row and column for a particular problem. If we ignore the overall title and the labels on the rows and columns we have simply a rectangular array of numbers, or what can be thought of as a matrix (all that is missing are the brackets). The spreadsheet matrix for a particular problem is usually identified by its contents: a title above the table such as Average Family Weekly Food Expense, Monthly Sales, Inflation Rate, etc.

When entered in a spreadsheet, Table 1 might look something like Table 2. In a spreadsheet, each entry has an address, which consists of the column letter and row number. Table 2 shows the column letters and row numbers for the weekly food expenses spreadsheet.

Table 2

	A	B	C	D	E	F	G	H	I
1			Average Family Weekly Food Expense						
2									
3	Year->	1983	1984	1985	1986	1987	1988	1989	1990
4	Family Type								
5	2A	$54.70	$56.50	$58.30	$60.40	$63.50	$66.70	$70.40	$73.80
6	2B	$52.00	$53.90	$55.70	$57.80	$60.70	$67.50	$70.40	$73.80
7	3A	$66.90	$69.20	$71.30	$73.90	$77.50	$81.60	$86.40	$90.50
8	3B	$77.80	$80.50	$83.00	$85.80	$90.10	$94.80	$100.60	$104.90
9	4A	$77.80	$80.50	$82.90	$85.80	$90.00	$94.90	$100.60	$105.20
10	4B	$93.50	$97.10	$99.80	$103.30	$108.30	$114.10	$121.00	$126.20
11	4C	$98.00	$101.60	$104.60	$108.20	$113.50	$119.40	$126.60	$132.30

Columns in spreadsheets are designated by A, B, C, etc., and the rows by 1, 2, 3, etc. Entries are identified by **the column letter first and then the row number**. Notice that this is the *opposite* of the mathematical matrix notation.[1] The entry for the average weekly expenditure on food for Type 4A families during 1983 is in cell B9. The matrix representing weekly food expenses for all 3 categories of Type 4 families (4A, 4B, and 4C) is in the rectangular **block** of cells that goes from B9 in the upper left-hand corner to I11 in the lower right-hand corner. To identify a matrix in a spreadsheet, you can highlight the rectangular block containing the matrix entries or refer to it using two diagonal extreme cells with two periods in between, such as "B9..I11", *not* including the cells for the row and column **labels** (verbal headings of individual rows and columns, such as 1983, 1984, etc.). See your technology supplement for more information about using a spreadsheet program.

Addition and Subtraction of Matrices

Sample Problem 1: Many of us are more health conscious than we used to be, and Federal law now requires labeling on packages to show the nutritional facts of the contents. This usually appears on the package in the form of a table something like Table 3, given here for three different snack foods:

[1] The spreadsheet row and column headings which appear here (the A,B,C,.. and 1,2,3,...along the top and sides) are by default usually *not* included in the printout of a spreadsheet, but it is possible to have them print if you want.

Table 3

	Chocolate Cookie		Hard Pretzel		Dried Apricots	
	Total Amount	% Daily Value	Total Amount	%Daily Value	Total Amount	%Daily Value
Fat	7g	11%	0g	0%	0g	0%
Saturated Fat	5g	25%	0g	0%	0g	0%
Cholesterol	0mg	0%	0mg	0%	0mg	0%
Sodium	70mg	3%	655mg	27%	10mg	0%
Carbohydrates	20g	7%	22g	7%	22g	7%
Fiber	<1g	3%	1g	4%	3g	12%
Sugars	12g		<1g		19g	
Protein	1g		3g		1g	
Total Calories	150		111		90	
Fat Calories	60		0		0	

Chocolate Cookies: Keebler Grasshopper™; Serving Size: 4 cookies
Pretzels: Snyder's Sourdough™; Serving Size: 1 pretzel
Apricots: Ann's House of Nuts™; Serving Size: 5 apricots

We want to know what total amounts of fat, cholesterol, etc., we would consume if we ate one serving each of cookies and pretzels.

Solution: We can consider the column of nutrient amounts per serving for each snack food as a 10×1 column matrix (column vector). Let us define **C** to represent this matrix for the cookies, and **P** for the pretzels. Then

$$\mathbf{C} = \begin{bmatrix} 7 \\ 5 \\ 0 \\ 70 \\ 20 \\ 0.5 \\ 12 \\ 1 \\ 150 \\ 60 \end{bmatrix} \quad \text{and} \quad \mathbf{P} = \begin{bmatrix} 0 \\ 0 \\ 0 \\ 655 \\ 22 \\ 1 \\ 0.5 \\ 3 \\ 111 \\ 0 \end{bmatrix}$$

(where the meaning and units for each row are those specified in the original data table, Table 3).[2]

One serving of cookies has 7g of fat, and one serving of pretzels has 0g of fat, so the total amount of fat from one serving of each is $7\text{g} + 0\text{g} = (7{+}0)\text{g} = 7\text{g}$. We would do a similar calculation for each nutrient, and the results could also be put in a 10 x 1 column matrix. This is the intuitive idea behind **addition of matrices**. In general, it should be clear that addition (**and subtraction**, which is done similarly) **can only be done with matrices of the *same dimension***, in which case the answer is also of the same dimension and its individual values are the sums (or differences) of the corresponding values of the matrices being added (or subtracted). In our example, if we let **S** represent the column vector (matrix) of the sum of the amounts of the nutrients from 1 serving of cookies and 1 serving of pretzels, we would have:

$$\mathbf{S} = \mathbf{C} + \mathbf{P} = \begin{bmatrix} 7 \\ 5 \\ 0 \\ 70 \\ 20 \\ 0.5 \\ 12 \\ 1 \\ 150 \\ 60 \end{bmatrix} + \begin{bmatrix} 0 \\ 0 \\ 0 \\ 655 \\ 22 \\ 1 \\ 0.5 \\ 3 \\ 111 \\ 0 \end{bmatrix} = \begin{bmatrix} 7+0 \\ 5+0 \\ 0+0 \\ 70+655 \\ 20+22 \\ 0.5+1 \\ 12+0.5 \\ 1+3 \\ 150+111 \\ 60+0 \end{bmatrix} = \begin{bmatrix} 7 \\ 5 \\ 0 \\ 725 \\ 42 \\ 1.5 \\ 12.5 \\ 4 \\ 261 \\ 60 \end{bmatrix}$$

Thus, if we eat 1 serving each of cookies and pretzels, we will be getting 7g of fat, 5g of saturated fat, 0g of cholesterol, 725 mg of sodium, etc.

Most technologies have a standard way of entering, adding, and subtracting matrices. See your technology supplement.

In general, then, if both **A** and **B** are $m \times n$ matrices, and $\mathbf{C} = \mathbf{A} + \mathbf{B}$, then **C** also is $m \times n$, and each element $c_{ij} = a_{ij} + b_{ij}$. Similarly, if $\mathbf{D} = \mathbf{A} - \mathbf{B}$, then **D** is also $m \times n$, and each element $d_{ij} = a_{ij} - b_{ij}$.

---

[2] Note that we have interpreted "<1 g" as "0.5g" to make it more precise, since 0 values are also given. This is a rough estimate, because we are not given more detailed information, but should give us an answer that is in the right ballpark.

Sample Problem 2: If $\mathbf{A} = \begin{bmatrix} 5 & -2 & 4 \\ 0 & 4.2 & 1 \end{bmatrix}$, $\mathbf{B} = \begin{bmatrix} 0 & 7 & -4 \\ -3 & 4.8 & 8 \end{bmatrix}$, and if $\mathbf{C} = \mathbf{A} + \mathbf{B}$ and $\mathbf{D} = \mathbf{A} - \mathbf{B}$, find **C** and **D**.

Solution: The calculations are straightforward and intuitive:

$$\mathbf{C} = \mathbf{A} + \mathbf{B} = \begin{bmatrix} 5 & -2 & 4 \\ 0 & 4.2 & 1 \end{bmatrix} + \begin{bmatrix} 0 & 7 & -4 \\ -3 & 4.8 & 8 \end{bmatrix} = \begin{bmatrix} 5+0 & -2+7 & 4+(-4) \\ 0+(-3) & 4.2+4.8 & 1+8 \end{bmatrix} = \begin{bmatrix} 5 & 5 & 0 \\ -3 & 9 & 9 \end{bmatrix}.$$

$$\mathbf{D} = \mathbf{A} - \mathbf{B} = \begin{bmatrix} 5 & -2 & 4 \\ 0 & 4.2 & 1 \end{bmatrix} - \begin{bmatrix} 0 & 7 & -4 \\ -3 & 4.8 & 8 \end{bmatrix} = \begin{bmatrix} 5-0 & -2-7 & 4-(-4) \\ 0-(-3) & 4.2-4.8 & 1-8 \end{bmatrix}$$

$$= \begin{bmatrix} 5 & -9 & 8 \\ 3 & -0.6 & -7 \end{bmatrix}. \quad \square$$

Just as in addition of real numbers, the order of the addition does not matter (commutative property): $\mathbf{A} + \mathbf{B} = \mathbf{B} + \mathbf{A}$. However, in general, $\mathbf{A} - \mathbf{B} \neq \mathbf{B} - \mathbf{A}$ (as with numbers). When subtracting (again, as with numbers), you must be careful in working with parentheses. For example, $\mathbf{A} - (\mathbf{B} - \mathbf{C}) = \mathbf{A} - \mathbf{B} + \mathbf{C}$.

Sample Problem 3: You are the owner of a small auto dealership, called Slim's Car Sales. You employ four salespeople (Brown, Green, Jones, and Smith), and sell four style categories of vehicles (sedans, minivans, sports cars, and sports utility vehicles), each with three price categories (economy, standard, and luxury). Table 4 gives the sales for each of the four salespeople in March (it is now April Fool's Day):

Table 4

	A	B	C	D
1	SLIM'S CAR SALES, INCORPORATED			
2	MARCH SALES RECORD			
3				
4		SEDANS		
5	SALESPERSON	ECONOMY	STANDARD	LUXURY
6	BROWN	1	2	0
7	GREEN	0	1	2
8	JONES	1	3	0
9	SMITH	2	0	1
10				
11		MINIVANS		
12	SALESPERSON	ECONOMY	STANDARD	LUXURY
13	BROWN	2	1	0
14	GREEN	3	0	0
15	JONES	1	2	0
16	SMITH	1	1	2
17				
18		SPORTS CARS		
19	SALESPERSON	ECONOMY	STANDARD	LUXURY
20	BROWN	0	2	1
21	GREEN	2	1	0
22	JONES	1	1	0
23	SMITH	3	0	0
24				
25		SPORTS UTILITY VEHICLES		
26	SALESPERSON	ECONOMY	STANDARD	LUXURY
27	BROWN	2	1	0
28	GREEN	1	1	1
29	JONES	0	2	0
30	SMITH	1	3	0

Find the total number of cars sold by each salesperson for each price category, totaled over all style categories, during March.

Solution: Most of these calculations can easily be carried out by hand, but if the spreadsheet is quite large or the calculations must be done on a regular basis, the job can become very tedious. The power of spreadsheets can help us to do these tasks more easily.

Let's start with the matrices (one for each style category of vehicles) that show the number of Economy, Standard and Luxury vehicles sold by each salesperson. The row labels for each of these matrices are the names of each salesperson and the column labels are the different price categories, so we can refer to each of these matrices as a (*salesperson* $\times$ *price category*) matrix. We call this the *row and column description* of the matrix. We can define the (*salesperson* $\times$ *price category*) matrices for March sales in the four different style categories to be **S** (for sedans), **M** (for minivans), **C** (for sports cars), and **U** (for sports utility vehicles), and we can define **T** to be the matrix of the totals, summed over the style categories. Then we have

$$\mathbf{S} = \begin{bmatrix} 1 & 2 & 0 \\ 0 & 1 & 2 \\ 1 & 3 & 0 \\ 2 & 0 & 1 \end{bmatrix}, \ \mathbf{M} = \begin{bmatrix} 2 & 1 & 0 \\ 3 & 0 & 0 \\ 1 & 2 & 0 \\ 1 & 1 & 2 \end{bmatrix}, \ \mathbf{C} = \begin{bmatrix} 0 & 2 & 1 \\ 2 & 1 & 0 \\ 1 & 1 & 0 \\ 3 & 0 & 0 \end{bmatrix}, \text{ and } \mathbf{U} = \begin{bmatrix} 2 & 1 & 0 \\ 1 & 1 & 1 \\ 0 & 2 & 0 \\ 1 & 3 & 0 \end{bmatrix}, \text{ and so}$$

$$\mathbf{T} = \mathbf{S} + \mathbf{M} + \mathbf{C} + \mathbf{U} = \begin{bmatrix} 1 & 2 & 0 \\ 0 & 1 & 2 \\ 1 & 3 & 0 \\ 2 & 0 & 1 \end{bmatrix} + \begin{bmatrix} 2 & 1 & 0 \\ 3 & 0 & 0 \\ 1 & 2 & 0 \\ 1 & 1 & 2 \end{bmatrix} + \begin{bmatrix} 0 & 2 & 1 \\ 2 & 1 & 0 \\ 1 & 1 & 0 \\ 3 & 0 & 0 \end{bmatrix} + \begin{bmatrix} 2 & 1 & 0 \\ 1 & 1 & 1 \\ 0 & 2 & 0 \\ 1 & 3 & 0 \end{bmatrix}$$

$$= \begin{bmatrix} 1+2+0+2 & 2+1+2+1 & 0+0+1+0 \\ 0+3+2+1 & 1+0+1+1 & 2+0+1+0 \\ 1+1+1+0 & 3+2+1+2 & 0+0+0+0 \\ 2+1+3+1 & 0+1+0+3 & 1+2+0+0 \end{bmatrix} = \begin{bmatrix} 5 & 6 & 1 \\ 6 & 3 & 3 \\ 3 & 8 & 0 \\ 7 & 4 & 3 \end{bmatrix}$$

To do this operation directly on your spreadsheet, the basic idea is to choose a location (block of cells) for your answer matrix first. Then, use a formula to find the answer for one value in the answer matrix (for example, the upper left spot) by adding the appropriate entries from the relevant matrices, and then copy that formula to the entire block (range) of the answer matrix.

For our example, suppose we want to put the values of our Totals matrix (**T**) in the block B35..D38 . Then the upper left cell of that block (B35) should be the sum of the corresponding cells in the individual matrices for each style category, so in B35 we put the formula +B6+B13+B20+B27 (see Table 4 for the original spreadsheet).[3] Then we copy the formula (using the mouse and menus) from B35 to B35..D38 . Note that when you use the copy command, the cell addresses are **relative** unless otherwise specified. This means that the address B6 in the formula stored in B35 is thought of as "that cell in the same column as the current cell, and 29 rows above it." When this is copied to cell C35 , it will thus be correctly interpreted to refer to cell C6 (29 rows above C35 in the same column), and when copied to cell D37 , it will be correctly interpreted to refer to cell D8 . These correct cell addresses will appear in the formulas for the new contents of

---

[3] In order to differentiate a cell address entry from a text or label entry, the cell address is preceded by a +, –, or =.

the cells being copied to. The result (with appropriate row and column labels, and a title) looks like:

	A	B	C	D
33		TOTALS		
34	SALESPERSON	ECONOMY	STANDARD	LUXURY
35	BROWN	5	6	1
36	GREEN	6	3	3
37	JONES	3	8	0
38	SMITH	7	4	3

For more details about these operations on a spreadsheet or on a graphing calculator, please consult your technology supplement. □

Sample Problem 4: Suppose that your sales goals for each salesperson in March were to sell 5 vehicles in each price category (totaled over all four style categories). Calculate a table of discrepancies from these goals, where a positive number mean the sales exceeded the goal by that amount, a negative number means sales were below the goal by that amount, and 0 means the goal was met exactly.

Solution:. You want to construct a (*salesperson* $\times$ *price category*) discrepancy matrix. The goal matrix, which we will call $\mathbf{G}$, is simply the 4×3 matrix of all 5's:

$$\mathbf{G} = \begin{bmatrix} 5 & 5 & 5 \\ 5 & 5 & 5 \\ 5 & 5 & 5 \\ 5 & 5 & 5 \end{bmatrix}.$$

We want to calculate the discrepancy matrix, which we will call $\mathbf{D}$. As defined above, for each salesperson and price category, it will be the total number of vehicles sold minus 5, so we would have

$$\mathbf{D} = \mathbf{T} - \mathbf{G} = \begin{bmatrix} 5 & 6 & 1 \\ 6 & 3 & 3 \\ 3 & 8 & 0 \\ 7 & 4 & 3 \end{bmatrix} - \begin{bmatrix} 5 & 5 & 5 \\ 5 & 5 & 5 \\ 5 & 5 & 5 \\ 5 & 5 & 5 \end{bmatrix} = \begin{bmatrix} 5-5 & 6-5 & 1-5 \\ 6-5 & 3-5 & 3-5 \\ 3-5 & 8-5 & 0-5 \\ 7-5 & 4-5 & 3-5 \end{bmatrix} = \begin{bmatrix} 0 & 1 & -4 \\ 1 & -2 & -2 \\ -2 & 3 & -5 \\ 2 & -1 & -2 \end{bmatrix}$$

Thus, for example, Brown (the salesperson associated with the first row of the matrix) hit the goal exactly for economy vehicles, exceeded the goal by 1 vehicle for medium-priced vehicles, and fell short of the goal by 4 vehicles for luxury vehicles.

To do these calculations on the spreadsheet, if we want to place the discrepancy matrix in the block B43..D46 , we can put the formula +B35-5 into B43 , then copy that formula from B43 to the block B43..D46 . The resulting block, again with labels and a title added, looks as follows:

	A	B	C	D
41		DISCREPANCIES		
42	SALESPERSON	ECONOMY	STANDARD	LUXURY
43	BROWN	0	1	-4
44	GREEN	1	-2	-2
45	JONES	-2	3	-5
46	SMITH	2	-1	-2

□

Multiplication of a Matrix by a Number

Up to this point, the problems have involved only the addition and subtraction of matrices. There are many problems, related to the kinds that we have already done, that involve the multiplication of an array of numbers by a single real number. In matrix algebra, the multiplication of a matrix by a single number is called **scalar multiplication.** **Scalar** is the term in matrix theory for a pure single number (as opposed to a 1×1 matrix), such as 15.3 or -48.

Sample Problem 5: Suppose that you have planned an advertising blitz for the month of April at your car dealership, and expect to be able to double your actual March sales (proportionately in each vehicle style category and in each price category. Use your spreadsheet to calculate your sales goals for all vehicles (total) in April, broken down by salesperson and price level.

Solution: To find the goals for April, we simply need to double all of the entries of the Totals matrix (**T**) that we derived earlier. For our example, if we call the matrix of goals for April **A** , then we denote the scalar multiplication **2T** , so we find that

$$\mathbf{A} = 2\mathbf{T} = 2\begin{bmatrix} 5 & 6 & 1 \\ 6 & 3 & 3 \\ 3 & 8 & 0 \\ 7 & 4 & 3 \end{bmatrix} = \begin{bmatrix} 2(5) & 2(6) & 2(1) \\ 2(6) & 2(3) & 2(3) \\ 2(3) & 2(8) & 2(0) \\ 2(7) & 2(4) & 2(3) \end{bmatrix} = \begin{bmatrix} 10 & 12 & 2 \\ 12 & 6 & 6 \\ 6 & 16 & 0 \\ 14 & 8 & 6 \end{bmatrix}.$$

On the spreadsheet, if we want to put the April goals matrix **A** into block B51..D54 , we simply put the formula +2*B35 into B51 , and copy this formula from B51 to the block B51..D54 . The result, with labels and a title, is:

	A	B	C	D
49		APRIL SALES GOALS		
50	SALESPERSON	ECONOMY	STANDARD	LUXURY
51	BROWN	10	12	2
52	GREEN	12	6	6
53	JONES	6	16	0
54	SMITH	14	8	6

□

Sample Problem 6: Many of us are more health conscious than we used to be, and Federal law now requires labeling on packages to show the nutritional facts about the contents of packaged food. This usually appears on the package in the form of a table like Table 5 (the same as Table 3 from Sample Problem 1, displayed again here for your convenience):

Table 5

	Chocolate Cookie		Hard Pretzel		Dried Apricots	
	Total Amount	% Daily Value	Total Amount	%Daily Value	Total Amount	%Daily Value
Fat	7g	11%	0g	0%	0g	0%
Saturated Fat	5g	25%	0g	0%	0g	0%
Cholesterol	0mg	0%	0mg	0%	0mg	0%
Sodium	70mg	3%	655mg	27%	10mg	0%
Carbohydrates	20g	7%	22g	7%	22g	7%
Fiber	<1g	3%	1g	4%	3g	12%
Sugars	12g		<1g		19g	
Protein	1g		3g		1g	
Total Calories	150		111		90	
Fat Calories	60		0		0	

Chocolate Cookies: Keebler Grasshopper™; Serving Size: 4 cookies
Pretzels: Snyder's Sourdough™; Serving Size: 1 pretzel
Apricots: Ann's House of Nuts™; Serving Size: 5 apricots

Even though we are health conscious, we sometimes don't stop at just one serving of our favorite snack foods. Suppose we want to know what total amounts of fat, cholesterol, etc., we would consume if we ate eight cookies and three pretzels.

Solution: If we wished to find out how much fat, etc., we would consume if we ate 8 cookies and 3 pretzels, we would first have to recognize that 1 serving of cookies is 4 cookies, so 8 cookies is only 2 servings. The table of nutritional information is given per serving (not per cookie), so we need to make this conversion to servings first. (We must always be aware of the *units*!) Then, we simply multiply each matrix by the corresponding number of servings and add the results, $2\mathbf{C} + 3\mathbf{P}$.

If we define **E** to be the matrix that tells how many nutrients we will get from 8 cookies (2 servings of cookies) and 3 pretzels, we find that

$$\mathbf{E} = 2\mathbf{C} + 3\mathbf{P} = 2\begin{bmatrix} 7 \\ 5 \\ 0 \\ 70 \\ 20 \\ 0.5 \\ 12 \\ 1 \\ 150 \\ 60 \end{bmatrix} + 3\begin{bmatrix} 0 \\ 0 \\ 0 \\ 655 \\ 22 \\ 1 \\ .5 \\ 3 \\ 111 \\ 0 \end{bmatrix} = \begin{bmatrix} 2(7) \\ 2(5) \\ 2(0) \\ 2(70) \\ 2(20) \\ 2(0.5) \\ 2(12) \\ 2(1) \\ 2(150) \\ 2(60) \end{bmatrix} + \begin{bmatrix} 3(0) \\ 3(0) \\ 3(0) \\ 3(655) \\ 3(22) \\ 3(1) \\ 3(.5) \\ 3(3) \\ 3(111) \\ 3(0) \end{bmatrix} = \begin{bmatrix} 14+0 \\ 10+0 \\ 0+0 \\ 140+1965 \\ 40+66 \\ 1+3 \\ 24+1.5 \\ 2+9 \\ 300+333 \\ 120+0 \end{bmatrix} = \begin{bmatrix} 14 \\ 10 \\ 0 \\ 2105 \\ 106 \\ 4 \\ 25.5 \\ 11 \\ 633 \\ 120 \end{bmatrix}$$

On many graphing calculators this can be done exactly as you see in the equation above. Once the matrices **C** and **P** have been entered (possibly with different letter names, if name options are limited), **2C + 3P** can be stored in matrix **E** (again, with another name if necessary). In the spreadsheet, if the entry for the fat in one serving of cookies is in B3 and the entry for the fat in one pretzel is in C3 and the total fat entry is to go in F3, enter the formula +2*B3+3*C3 in cell F3 . To do the same for the other nutrients, copy this formula from cell F3 into cells F4..F12 to complete the new column matrix. □

Now that we have defined scalar multiplication, you may find it helpful to think of $\mathbf{A} - \mathbf{B}$ as $\mathbf{A} + (-1)\mathbf{B}$, or $\mathbf{A} + (-\mathbf{B})$. Thus $-\mathbf{B}$ can be thought of as $(-1)\mathbf{B}$.

<u>Section Summary</u>

Before you begin the exercises be sure that you:

- Understand that a matrix is a rectangular array of numbers and that matrices are denoted by boldface capital letters.

- Realize that entries of a matrix are enclosed in square brackets ([ ]) when written out by hand, but do not have these brackets when entered into a spreadsheet.

- Know that the dimension of a matrix refers to the number of rows and columns in the matrix, (*number of rows* $\times$ *number of columns*).

- Know that the entries of a matrix are indicated by the name of the matrix, using the letter in the lower case form and not in bold type, followed by two subscripts that indicate the position of the entry, first the row number and then the column number (so a_{ij} is the entry in the i'th row and j'th column of $\mathbf{A}$).

- Know that matrix addition (or subtraction) can only be carried out on matrices of the same dimension, and that the resulting matrix has the same dimension. Its individual entries are the sums (or differences) of the corresponding matrices being added (or subtracted), and you should know how to add or subtract matrices by hand and using one or more technologies.

- Know that the process of multiplying all of the entries of a matrix by a real number is called scalar multiplication, and know how to carry out a scalar multiplication by hand and using one or more technologies.

EXERCISES FOR SECTION 7.1

Warm Up

Problems 1 - 11 refer to the following matrices:

$$\mathbf{A} = \begin{bmatrix} 1 \\ 2 \\ 3 \end{bmatrix} \quad \mathbf{B} = \begin{bmatrix} 2 & 4 \end{bmatrix} \quad \mathbf{C} = \begin{bmatrix} -1 & 2 \\ 3 & -4 \end{bmatrix}$$

$$\mathbf{D} = \begin{bmatrix} 2 & -1 \\ 4 & 0 \end{bmatrix} \quad \mathbf{E} = \begin{bmatrix} 4 & -3 & 0 \\ 3 & 8 & 5 \end{bmatrix} \quad \mathbf{F} = \begin{bmatrix} 0 \\ 1 \\ -5 \end{bmatrix}$$

$$\mathbf{G} = \begin{bmatrix} 2 & 3 \\ 4 & 7 \\ 1 & 5 \end{bmatrix} \quad \mathbf{H} = \begin{bmatrix} 1 & -1 & 4 \\ 3 & 5 & 6 \\ 1 & 4 & 8 \end{bmatrix} \quad \mathbf{I} = \begin{bmatrix} 1 & 3 \\ 4 & 8 \\ 9 & 7 \end{bmatrix} \quad \mathbf{J} = \begin{bmatrix} 1 & 0 \\ 0 & 1 \end{bmatrix}$$

1. Give the dimension of each matrix.

2. Which pairs of the matrices can be added?

3. **For each pair** of matrices that can be added, give the matrix resulting from the addition.

4. Does order count when adding matrices?

5. For matrix **H,** give the numerical value of h_{23}.

6. For matrix **H,** give the numerical value of h_{31}.

7. Which of the given matrices (**A** through **J**) can be multiplied by a scalar?

8. What is 3**G** ?

9. What is -1**B** ?

10. What is 4**G** - 3**I** ?

11. What is 3**I** - 4**G**?

Game Time

12. In the following spreadsheet, identify the row and column descriptions of each matrix (for example, the Slim's Car Sales data is a *(salesperson × style category)* matrix) and give the dimension of that matrix.

	A	B	C	D	E
1	GIRL SCOUT COOKIE SALE				
2					
3	INDIVIDUAL SALES RECORD			SCOUT: SARAH WOLFE	
4	CUSTOMER	COOKIE TYPE	SAMOAS	THINMINT	TREFOILS
5	JONES		2	1	
6	SMITH		4		
7	BROWN			3	2
8	WHITE		1	2	2
9	BLACK		2	2	2
10	GREEN		4	2	1
11					
12					
13	GIRL SCOUT COOKIE SALE				
14	TROOP SALES RECORD				
15	SCOUT		SAMOAS	THINMINT	TREFOILS
16	SARAH		13	10	7
17	CAROLYN		15	9	10
18	MOLLY		12	13	15
19	AMY		12	9	11
20	SUZANNE		14	18	9
21	DOTTIE		10	18	11

13. In the following spreadsheet, identify the row and column descriptions of the matrix (for example, the Slim's Car Sales data is a *(salesperson × style category)* matrix) and give the dimension of the matrix.

ABC CORPORATION			
INDIVIDUAL OFFICE EXPENSE RECORDS			
MONTHLY EXPENSES FIRST QUARTER 1996			
	JANUARY	FEBRUARY	MARCH
ADVERTISING	652	833	599
CAR EXPENSES	456	305	522
POSTAGE	68	59	73
INSURANCE	379	379	379
CLEANING	80	80	80
OFFICE RENT	750	750	750
UTILITIES	164	145	121
SUPPLIES	173	76	119
TRAVEL	842	598	366
ENTERTAINMENT	109	156	364
TELEPHONE	159	194	209
PRINTING	407	0	85

14. a) Set up a matrix in a spreadsheet that displays the following information: Three different food supplements contain the following amounts of calcium, iron and protein: Food A contains 30 units of calcium, 10 units of iron and 10 units of protein per ounce; food B contains 10 units of calcium, 10 units of iron and 30 units of protein per ounce; food C contains 20 units of each per ounce.

b) What is the dimension of this matrix? What are the row and column descriptions (see Exercise 13)?

c) Create a matrix that shows how many units of calcium, iron and protein you would get if you consumed 3 ounces of each food type (one matrix that is broken down by food type, and another that is for the totals). Be sure to include your row and column labels.

15. You sell merchandise in two different states. Suppose that Pennsylvania charges 6% sales tax on all the items and that New Jersey charges 4.5% sales tax on all the items. You have a 100% markup on all the items that you sell. Your costs for the items are shown in the table below:

	SHEET SETS	COMFORTERS	QUILTS
TWIN	8.95	34.50	43.50
FULL	19.75	43.20	75.00
QUEEN	23.75	53.00	75.00
KING	28.75	63.50	95.50

a) Set up a spreadsheet that shows your costs, base selling price and price with tax in each of the states (four different matrices).

b) What are the dimensions of each of these matrices? What are the row and column descriptions (see Exercise 13) for each of them?

16. You sell merchandise in two different states. You receive weekly orders from your dealers in the two states. You want to keep a record of each dealer's orders, the orders from each state, and the total weekly orders, broken down by size and item. Set up a spreadsheet that will do this in general. Use the sample of the weekly orders from each dealer given below to test your template spreadsheet, and print out the resulting spreadsheet:

CHRIS'S LINEN OUTLET SMITHTOWN, PA.
ORDER SHEET WEEK OF JANUARY 27

	SHEET SETS	COMFORTERS	QUILTS
TWIN	5	4	2
FULL	7	8	3
QUEEN	9	9	5
KING	5	8	3

FRAN'S LINEN OUTLET OURTOWN, PA.
ORDER SHEET WEEK OF JANUARY 27

	SHEET SETS	COMFORTERS	QUILTS
TWIN	6	11	5
FULL	9	3	8
QUEEN	5	9	4
KING	10	10	7

LINEN BOUTIQUE, SPRINGFIELD, N.J.
ORDER SHEET WEEK OF JANUARY 27

	SHEET SETS	COMFORTERS	QUILTS
TWIN	5	3	1
FULL	8	7	8
QUEEN	9	10	5
KING	9	5	7

BED AND BATH OUTLET, NEWTOWN, N.J.
ORDER SHEET WEEK OF JANUARY 27

	SHEET SETS	COMFORTERS	QUILTS
TWIN	5	10	4
FULL	11	5	9
QUEEN	7	8	7
KING	3	8	2

17. You work for a large construction company. Every week the site foremen report to you on the number of each four styles of homes under construction at their sites. These figures will change each week as some homes are completed and new ones are started. You must prepare several reports each week. You must prepare a report each week showing the total number of homes under construction *at each site*, the total

number of each *type* of home under construction, and the *total* number of homes under construction. Set up a spreadsheet to do this in general, enter the given data to test it, and print out the test spreadsheet.

Weekly Foreman's Report

Site____#1____________ Location_Springfield________

Week Starting____6/1__________

Homes Under Construction:

Colonial ___4__ Modern_1__ Cape Cod __3_ Rancher __5__

Weekly Foreman's Report

Site____#2____________ Location_Newtown

Week Starting____6/1__________

Homes Under Construction:

Colonial ___2__ Modern_0__ Cape Cod __5_ Rancher __4__

Weekly Foreman's Report

Site____#3____________ Location_Elmwood

Week Starting____6/1__________

Homes Under Construction:

Colonial __2__ Modern_1__ Cape Cod __3_ Rancher __0__

Weekly Foreman's Report

Site____#4____________ Location_Chestertown

Week Starting____6/1__________

Homes Under Construction:

Colonial ___4___ Modern_2__ Cape Cod __1_ Rancher __3__

Section 7.2: Matrix Multiplication

As we pointed out in Section 7.1, matrix operations can help us to simplify the job of doing repetitive operations such as addition and subtraction of arrays of numbers. There are many occasions when we want to do more complex operations on arrays of numbers, such as multiply elements and then add the results of several of these multiplications. For example, we might want to multiply the number of sodas ordered by the price of a soda and the number of hot dogs ordered by the price of a hot dog and add the results to get the cost of an order. These situations involve the **multiplication of matrices**: for example, matrix **A** (numbers of sodas and hot dogs) times matrix **B** (price of a soda and a hot dog). In this section we define the multiplication of matrices and learn how to carry out such multiplications by hand, using a graphing calculator and using a spreadsheet. We will also learn when a pair of matrices can be multiplied, because, unlike the multiplication of real numbers, not every pair of matrices can be multiplied. We will show that the order of the multiplication *does* matter. For a given pair of matrices, it might be possible to multiply them in one order, but not the other; for example, **A** times **B** may be possible, while **B** times **A** is not possible. We will also learn how to use row and column descriptions to help determine the appropriate order of multiplication, even if both orders are possible.

Below are some examples of the kinds of problems that the concepts in this section will help you solve:

- You run a small company that hand crafts tables and chairs. You have recently expanded production to two locations. Once you have decided how many chairs and table you want to have assembled at each location, you must determine how much material is needed at each location: how many board feet of lumber and how many packages of hardware.
- You and a friend have figured out your averages on tests, quizzes, homeworks and your project. The syllabus for the class gives the weights for each of these components. Calculate your averages.
- The Nationwide Prefab House Corporation manufactures and ships to its distributors six different styles of prefab homes. These homes are different configurations of six basic modules. You are trying to help the Nationwide Prefab House Corporation automate its orders and shipping. Every week reports come in from the six district distributors specifying the number of orders that they have received for each type of house. The company needs reports for the Production, Billing, and Shipping departments, as well as for central management. You want to set up a system so that all that the company has to do each week is to enter the new orders and have a spreadsheet carry out the necessary calculations, a 15 minute job at most. Finally, you have to explain to the person who will be doing

this job how you set up the system and exactly what to do to get the new figures each week.

When you have finished this section you should:

- Know what is meant by matrix multiplication and be able to carry out simple matrix multiplications by hand, and more complex matrix multiplications using a graphing calculator and/or a spreadsheet.
- Know when matrix multiplication is appropriate for solving a problem.
- Know which matrix multiplications can be done.
- Know which matrix multiplications have meaningful results.
- Know what the identity matrix is.

Matrix Multiplication

Sample Problem 1: Let's look at a very simplified example of a logistics or distribution problem. Suppose you run a small business making hand-crafted chairs and tables. You have two locations where these hand-crafted tables and chairs are assembled and finished. These assembly sites are not equipped to store any more materials than are actually needed for each week's production. The chairs each require 3 board feet of wood and 8 packages of hardware, while the tables each require 6 board feet of wood and 4 packages of hardware. Each week you determine how many tables and chairs should be assembled in each location to fill your orders. For next week you want the first location to assemble 9 chairs and 5 tables and the second location to assemble 15 chairs and 2 tables. How can you determine how many board feet of wood and packages of hardware to send to each location?

Solution: The chairs require 3 board feet of wood and the tables require 6 board feet of wood. If you have orders for 9 chairs and 5 tables, how much wood do you need? The answer is found by multiplying 9 chairs times 3 board feet of wood per chair and 5 tables times 6 board feet of wood per table and adding the results: $(9)(3) + (5)(6) = 27 + 30 = 57$. This is really a very simple problem, but it does serve to illustrate a type of multiplication using matrices. Let the quantities of chairs and tables (in that order) at the first location be the 1 x 2 row vector **q** and let the requirements of wood per chair and per table (in that order) be the 2 x 1 column vector **r**:

$$\mathbf{q} = \begin{bmatrix} 9 & 5 \end{bmatrix} \qquad \mathbf{r} = \begin{bmatrix} 3 \\ 6 \end{bmatrix}$$

The **dot product** of **q** and **r,** denoted by **q • r**, or **qr** , is a **real number** (scalar), not a matrix, arrived at by multiplying the first entries in each matrix vector, then the second entries, going from left to right and from top to bottom, and adding the results:

$(9)(3)+(5)(6) = 27+30=57$.

Using matrix notation for the entries, the procedure gives: $q_{11}r_{11}+q_{12}r_{21}$.

Let's begin the rest of our solution by writing up the information in the form of a table (Table 1).

Table 1

	Chairs	Tables		Wood	Hardware
1st Location	9	5	Chairs	3	8
2nd Location	15	2	Tables	6	4

Because this is a very limited problem, it is not too difficult to simply do the calculations by hand. The first location is to assemble 9 chairs and 5 tables, with each chair using 3 board feet of wood and each table using 6 board feet of wood, so it will need:

$(9)(3)=27$ board feet of wood for the chairs and
$(5)(6)=30$ board feet of wood for the tables, for a total of:
$27+30$ or 57 board feet of wood to location one.

In the same way we can determine the packages of hardware:

$(9)(8)=72$ packages of hardware for chairs and
$(5)(4)=20$ packages of hardware for tables, giving:
$72+20=92$ packages of hardware to location one.

The second location will assemble 15 chairs at 3 board feet of wood per chair and 2 tables at 6 board feet per table:

$(15)(3)=45$ board feet of wood for the chairs and
$(2)(6) = 12$ board feet of wood for the tables, a total of:
$45+12 = 57$ board feet of wood to location two.

In the same way we can determine the packages of hardware:

(15)(8) = 120 packages of hardware for chairs and
(2)(4) = 8 packages of hardware for tables, giving:
120 + 8 = 128 packages of hardware to location two.

As mentioned above, this is not too difficult, but it is also not very efficient. If this problem is to be repeated week after week, we need a better way of handling it. Let's look at our table again (Table 2):

Table 2

	Chairs	Tables		Wood	Hardware
1st Location	9	5	Chairs	3	8
2nd Location	15	2	Tables	6	4

We can see that if, for the 1st location, we multiply the chairs needed by the wood required per chair and tables needed by the wood required per table and add the results we get the wood required at location 1, as indicated above. If we multiply the 1st location chairs by hardware required and tables by hardware required, and add the results we get the hardware required at location 1. The same applies for location 2. We can organize our answers into the matrix shown in Table 3:

Table 3

	Wood	Hardware
1st Location	(9)(3) + (5)(6)	(9)(8) + (5)(4)
2nd Location	(15)(3) + (2)(6)	(15)(8) + (2)(4)

If we designate the first matrix **Q** for quantities ordered, the second matrix **R** for requirements and the answer **T** for the totals, we have shown that:

$$\begin{bmatrix} t_{11} & t_{12} \\ t_{21} & t_{22} \end{bmatrix} = \begin{bmatrix} q_{11}r_{11} + q_{12}r_{21} & q_{11}r_{12} + q_{12}r_{22} \\ q_{21}r_{11} + q_{22}r_{21} & q_{21}r_{12} + q_{12}r_{22} \end{bmatrix}$$

This is **matrix multiplication** and is **denoted by QR**. Thus we could write

$$\mathbf{T} = \mathbf{QR} = \begin{bmatrix} 9 & 5 \\ 15 & 2 \end{bmatrix}\begin{bmatrix} 3 & 8 \\ 6 & 4 \end{bmatrix} = \begin{bmatrix} (9)(3)+(5)(6) & (9)(8)+(5)(4) \\ (15)(3)+(2)(6) & (15)(8)+(2)(4) \end{bmatrix} = \begin{bmatrix} 57 & 92 \\ 57 & 128 \end{bmatrix}$$

Using matrix multiplication, we have determined that we must send 57 board feet of wood and 92 packages of hardware to Location 1, and 57 board feet of wood and 128 packages of hardware to Location 2. □

The Order of Matrix Multiplication

Notice that:

• The element in the first row first column position of the **T** matrix (t_{11}) consists of the elements in the first row of the matrix **Q** multiplied by the elements in the first column of the matrix **R** and then added together (finding the dot product).

• The element in the first row second column (t_{12}) consists of the dot product of the first row of the matrix **Q** by the second column of the matrix **R**.

• The element in each position tij is the dot product of its corresponding row, the ith row of Q (the matrix on the left), and its corresponding column in R the jth column of R (the matrix on the right).

Therefore, **two matrices A and B can be multiplied only if the number of columns in matrix A (the length of each row) is equal to the number of rows in matrix B (the length of each column).** In other words, **AB** is only defined if **A** is an $m \times p$ matrix and **B** is a $p \times n$ matrix. **Matrix multiplication will normally make sense only if the labels on the columns of A match the labels on the rows of B.** Using the idea of **row and column descriptions**, we multiplied a (*location* × *furniture type*) matrix by a (*furniture type* × *material*) matrix and the result was a (*location* × *material*) matrix. In our example, it would not make any sense to multiply the matrices in the reverse order. The first entry in the resulting matrix would be:

(wood for chairs) (1st location chairs) = wood for chairs to 1st location
(hardware for chairs)(2nd location chairs) = hardware for chairs to 2nd location

The addition of these results:

(wood for chairs to 1st location) + (hardware for chairs to 2nd location)

does not have a useful meaning.

This shows that, in this problem, multiplying a (*furniture type* × *material*) matrix by a (*location* × *furniture type*) matrix does not give a meaningful result.

Order *does* count in matrix multiplication: in general, AB DOES NOT EQUAL BA.

In our example, the result of multiplying the matrices in the *other* order is:

$$RQ = \begin{bmatrix} 3 & 8 \\ 6 & 4 \end{bmatrix} \begin{bmatrix} 9 & 5 \\ 15 & 2 \end{bmatrix} = \begin{bmatrix} (3)(9)+(8)(15) & (3)(5)+(8)(2) \\ (6)(9)+(4)(15) & (6)(5)+(4)(2) \end{bmatrix} = \begin{bmatrix} 147 & 31 \\ 114 & 38 \end{bmatrix}, \text{ so}$$

$$\mathbf{QR} = \begin{bmatrix} 57 & 92 \\ 57 & 128 \end{bmatrix} \neq \mathbf{RQ} = \begin{bmatrix} 147 & 31 \\ 114 & 38 \end{bmatrix}$$

Spreadsheet programs and most graphing calculators do matrix multiplication, so the tedious job of finding the dot products (multiplying and adding the row and column entries) is done for you. When doing matrix multiplication on a spreadsheet, **do not include the labels in the block that defines either matrix.** If the number of columns of the first matrix does not match the number of rows of the second, you may receive an error message. However, if the matrices are square matrices (equal number of rows and columns) or have properly matching dimensions, the spreadsheet program will not recognize if you have identified the matrices in the incorrect order. Because the calculations are all done for you, it is easy to lose the sense of what you are doing and the spreadsheet program will not check this for you.

It is very important to remember that **the general verbal descriptions as well as the specific individual labels for the columns of the first matrix must match the general verbal descriptions as well as the specific individual labels for the rows of the second matrix. Your resulting matrix will have the verbal description and labels of rows of the first matrix and the verbal description and labels of the columns of the second matrix.**

The Transpose of a Matrix

It is possible in some cases that the entries in your spreadsheet are not set up correctly for matrix multiplication. In the example shown in Table 1 above the entries might have been as shown in Table 4:

Table 4

	Chairs	Tables
1st Location	9	5
2nd Location	15	2
Wood	3	6
Hardware	8	4

These matrices can be multiplied as they are shown, but as we pointed out before, the results would be meaningless ((*location* × *furniture type*) times (*material* × *furniture type*)). It is not necessary to completely change your basic spreadsheet. We want the rows and columns in the second matrix to be interchanged. This is called **transposing** a matrix, and the resulting matrix is called the **transpose** of the original. If the original matrix is designated **A**, its transpose is designated $\mathbf{A}^T$(read "A transpose"). Your spreadsheet package will probably be able to transpose a block of data with a Transpose function of some kind. It copies data to another area of the spreadsheet and transposes the columns and rows. On a spreadsheet, you can and should include the labels of the columns and rows you are transposing so that the new rows and columns will be identified in the transposed matrix. Normally, you can't *replace* an entire block with transposed data; you must copy it to another part of the spreadsheet first.

The transpose of the second matrix in Table 4 is shown in Table 5:

Table 5

	Wood	Hardware
Chairs	3	8
Tables	6	4

You might want to use this command if you have created a fairly large spreadsheet that includes a matrix with numerous entries and later find that you need the transpose for a multiplication. Rather than reenter all the data, you can use the matrix Transpose function. Remember that you must locate the transposed matrix in a different area of your spreadsheet; do not try to overwrite *any part of* the original matrix. Table 6 shows a possible spreadsheet entry for the Furniture Company data with the material requirements block transposed:

Table 6

	A	B	C	D	E	F	G
1	FURNITURE COMPANY						
2							
3	WEEKLY ORDERS						
4		Chairs	Tables				
5	1st Location	9	5				
6	2nd Location	15	2				
7							
8	MATERIALS REQUIRED				TRANSPOSED		
9		Chairs	Tables			Wood	Hardware
10	Wood	3	6		Chairs	3	8
11	Hardware	8	4		Tables	6	4

On many graphing calculators there is an operation for transposing a matrix directly. The transposed matrix can then be stored under a new letter name. See your technology supplement for details.

Sample Problem 2: You and your friends have determined your semester averages for tests, quizzes, homeworks and your project.

	Tests	Quizzes	Homeworks	Projects
Fran	85	82	95	91
Pat	83	87	93	89
Chris	91	85	95	92
Tracy	79	81	87	89

You also have weights the professor gave on the syllabus:

	Tests	Quizzes	Homeworks	Projects	Final Exam
With Final	0.4	0.1	0.1	0.15	0.25

You want to calculate your averages *before* the Final.

Solution: The first thing that you have to do is adjust the weights so that the Final is not included. Since the Final is weighed 25%, the other grades must have a total weight of 75%, and you can determine their relative weights by dividing each weight by 0.75. If you are working in a spreadsheet this is simple: divide the test weight by 0.75 and copy the formula to the other columns.

	Tests	Quizzes	Homeworks	Projects	Final Exam
With Final	0.4	0.1	0.1	0.15	0.25
Without Final	0.533333	0.1333333	0.13333333	0.2	

You want to multiply the test grade by the test weight, the quiz grade by the quiz weight, the homework grade by the homework weight, and the project grade by the project weight and add the results. Let's look at your spreadsheet so far.

	Tests	Quizzes	Homeworks	Projects	Final Exam
Fran	85	82	95	91	
Pat	83	87	93	89	
Chris	91	85	95	92	
Tracy	79	81	87	89	
	Tests	Quizzes	Homeworks	Projects	Final Exam
With Final	0.4	0.1	0.1	0.15	0.25
Without Final	0.533333	0.1333333	0.13333333	0.2	

Let's define the matrix **G** to be the 4×4 (*person* × *grade component*) grade matrix and define **P** to be the 1×4 (*without final* × *grade component*) weight percentage matrix. We want to multiply each grade by its weight and add the results to get the average for the semester going into the final. Since the dimension of **G** is 4×4 and the dimension of **P** is 1×4, we cannot carry out the matrix multiplication **GP**. Mathematically, we *could* carry out the matrix multiplication **PG**, but the row and column verbal descriptions don't match, so the result would not make sense. However, if we get the transpose of **P**, $\mathbf{P}^{\mathrm{T}}$ will have dimension 4×1, so we *can* carry out the matrix multiplication $\mathbf{GP}^{\mathrm{T}}$. Let's look at the transposed matrix:

Transposed Weight Matrix	Without Final
Tests	0.533333
Quizzes	0.133333
Homeworks	0.133333
Projects	0.2

$\mathbf{P}^{\mathrm{T}}$ is a (*grade component* $\times$ *without final*) matrix. This tells us that we want to multiply matrix **G** (*person* $\times$ *grade component*) times $\mathbf{P}^{\mathrm{T}}$ (*grade component* $\times$ *without final*), and that the resulting matrix, which we will call **A**, will be a (*person* $\times$ *without final*) numerical average matrix with dimension 4×1:

	Without Final
Fran	87.1332991
Pat	86.066633
Chris	90.933297
Tracy	82.333304

□

Sample Problem 3: Table 7 shows the nutritional data for three snack foods and Table 8 shows a record of how many of each type of snack you ate during one week. Determine the total fat, saturated fat, etc. you had consumed from these snacks, combined, in the previous week.

Table 7

SNACK FOOD NUTRITION DATA						
	CHOCOLATE COOKIE		HARD PRETZEL		DRIED APRICOTS	
	TOTAL	% DAILY	TOTAL	% DAILY	TOTAL	% DAILY
Fat (g)	7	11	0	0	0	0
Sat Fat (g)	5	25	0	0	0	0
Cholesterol (mg)	0	0	0	0	0	0
Sodium (mg)	70	3	655	27	10	0
Total Carbohy (g)	20	7	22	7	22	7
Fiber (g)	0.5	3	1	4	3	12
Sugars (g)	12		0.5		19	
Protein (g)	1		3		1	
Calories	150		111		90	
Cal/Fat	60		0		0	

Table 8

		CHOC COO	PRETZELS	APRICOTS
Monday		3	5	0
Tuesday		4	0	3
Wednesday		5	6	1
Thursday		0	0	2
Friday		2	4	4
Saturday		3	2	2
Sunday		2	4	6

Solution: To solve this problem using the spreadsheet matrix multiplication command, we must do two things:

We must eliminate the columns showing the % and the spaces between the rows so the quantities form a matrix consisting of the nutritional information and the snacks, a 10×3 (*nutritional component* × *snack*) nutritional content matrix. This is shown in Table 9.

Table 9

	CHOCOLATE COOKIE	HARD PRETZEL	DRIED APRICOTS
	TOTAL	TOTAL	TOTAL
Fat (g)	7	0	0
Sat Fat (g)	5	0	0
Cholesterol (mg)	0	0	0
Sodium (mg)	70	655	10
Total Carbohy (g)	20	22	22
Fiber (g)	0.5	1	3
Sugars (g)	12	0.5	19
Protein (g)	1	3	1
Calories	150	111	90
Cal/Fat	60	0	0

The consumption matrix in Table 8 is a 7×3 (*day* × *snack*) matrix. To be able to multiply the nutritional content matrix (Table 9) by it, we must transpose the consumption matrix so that it is a 3× 7 (*snack* × *day*) matrix. The transpose of the (*day* × *snack*) matrix is shown in Table 10.

Table 10

	Monday	Tuesday	Wednesday	Thursday	Friday	Saturday	Sunday
CHOC COOKIE	3	4	5	0	2	3	2
PRETZELS	5	0	6	0	4	2	4
APRICOTS	0	3	1	2	4	2	6

We can now multiply the (*nutritional component* × *snack*) nutritional content matrix times the (*snack* × *day*) consumption matrix and get a (*nutritional component* × *day*) nutrient consumption matrix, which tells us exactly how much fat, cholesterol, etc., we consumed each day from eating these snacks. The result of the matrix multiplication is shown in Table 11:

Table 11

	Monday	Tuesday	Wednesday	Thursday	Friday	Saturday	Sunday
Fat (g)	21	28	35	0	14	21	14
Sat Fat (g)	15	20	25	0	10	15	10
Cholesterol (mg)	0	0	0	0	0	0	0
Sodium (mg)	3485	310	4290	20	2800	1540	2820
Total Carbohy (g)	170	146	254	44	216	148	260
Fiber (g)	6.5	11	11.5	6	17	9.5	23
Sugars (g)	38.5	105	82	38	102	75	140
Protein (g)	18	7	24	2	18	11	20
Calories	505	870	906	180	704	652	884
Cal/Fat	180	240	300	0	120	180	120

A problem of this magnitude makes it easy to recognize the power of matrix multiplication, particularly when the multiplication can be carried out using technology.
□

Sample Problem 4: The Nationwide Prefab House Corporation manufactures and ships to its distributors six different styles of prefabricated homes. These homes are different configurations of six modules, with shipping information given in Table 12.

Table 12

Description	Cost/Dollars	Weight/Lbs
Module 1: 9x12 all-purpose	3240	972
Module 2: 4x5 bath with fixtures	3175	3175
Module 3: 12x14 all-purpose	5040	1512
Module 4: 10x10 kitchen	7500	2500
Module 5: 3x8 hall	820	216
Module 6: 8x12 garage	2750	875

The different styles and the number of each module in each style are given in the Configuration matrix shown in Table 13 :

Table 13

	Carlson	Devon	Essex	Glenview	Hampton	Norfolk
Module 1	4	3	3	5	5	6
Module 2	2	2	1	3	2	3
Module 3	3	2	2	4	3	3
Module 4	1	1	1	1	1	1
Module 5	4	3	3	6	6	6
Module 6	2	2	1	1	3	3

The orders by region (district) are given in the Orders matrix in Table 14.

Table 14

	Northeast	Southeast	Northcentral	Southcentral	Northwest	Southwest
Carlson	3	4	2	4	5	6
Devon	4	2	3	2	4	5
Essex	2	4	5	6	3	7
Glenview	4	5	0	2	4	4
Hampton	5	4	8	4	2	5
Norfolk	6	1	4	3	2	0

You are trying to help the Nationwide Prefab House Corporation automate its orders and shipping. Every week reports come in from the six district distributors specifying the number of orders that they have received for each type of house. The company has a number of requirements:

1. The company needs to know how many of each type of module to ship to each Distributor.
2. Billing needs to know the total cost of the shipment to each distributor.
3. Shipping needs to know the total weight going to each destination in order to have the proper number of trucks on hand.
4. The company would also like to have a record of how many of each type of module was shipped each week.

The company already has a spreadsheet that shows the configuration of the various models. Set up a spreadsheet so that all that has to be done each week is to enter the new orders and carry out the necessary multiplications, a 15 minute job at most. Explain to the person who will be doing this job how you set up the spreadsheet and exactly what to do to get the new figures each week.

Solution: It makes sense for you to utilize the existing spreadsheet. The company has not yet set up a form for the weekly reports of the orders. You want to set up this form so that when the data is entered into the spreadsheet all of the required reports can be produced quickly and easily. It is very important to plan ahead so that your spreadsheet will be set up correctly.

The following is the kind of initial thinking necessary to set up the spreadsheet in an efficient and easy-to-use form for each of the four desired outputs:

1. You want to be able to determine how many modules should be sent to each distributor. To do this you want to multiply the Configuration matrix by the Orders matrix. Since the Configuration matrix already exists in the spreadsheet, you want to be sure that you can multiply the Configuration matrix times the Orders matrix. Assume that the Configuration matrix is a $(module \times model)$ matrix, as in Table 12. The Orders matrix has distributors and models. In order to carry out the multiplication, it must be a $(model \times distributor)$ matrix (if we put the Configuration matrix on the left, and the Orders matrix on the right). The resulting product matrix will be a $(module \times distributor)$ matrix.

2. To find the cost and weight of the modules shipped to each distributor, you want to be able to multiply the $(module \times distributor)$ matrix times a matrix that shows the cost and weight of each module. The Shipping Information matrix given in Table 12 is a $(module \times shipping\ information\ category)$ matrix. To make the multiplication make

sense, we could transpose the Shipping Information matrix so that it becomes a (*shipping information category* × *module*) matrix, and then we can put it on the left side of the multiplication. The result will be a (*shipping information category* × *distributor*) matrix.

3. To determine how many of each type of module was shipped each week, you simply have to add rows of the (*module* × *distributor*) matrix, which will give the total of each type of module shipped that week.

	A	B	C	D	E	F	G	H
1	Nationwide Prefab House Corporation							
2								
3	Configuration Information							
4		Carlson	Devon	Essex	Glenview	Hampton	Norfolk	
5	Module 1	4	3	3	5	5	6	
6	Module 2	2	2	1	3	2	2	
7	Module 3	3	2	2	4	3	3	
8	Module 4	1	1	1	1	1	1	
9	Module 5	4	3	3	6	6	6	
10	Module 6	2	2	1	1	3	3	
11								
12	Shipping Information							
13		Module 1	Module 2	Module 3	Module 4	Module 5	Module 6	
14	Cost$	3240	3175	5040	7500	820	2750	
15	Weight#	972	3175	1512	2500	216	875	
16								
17	Order Information							
18		Northeast	Southeast	Northcentral	Southcentral	Northwest	Southwest	
19	Carlson	3	4	2	4	5	6	
20	Devon	4	2	3	2	4	5	
21	Essex	2	4	5	6	3	7	
22	Glenview	4	5	0	2	4	4	
23	Hampton	5	4	8	4	2	5	
24	Norfolk	6	1	4	3	2	0	
25								
26	Number of Modules Shipped							
27		Northeast	Southeast	Northcentral	Southcentral	Northwest	Southwest	Totals
28	Module 1	111	85	96	88	83	105	568
29	Module 2	50	41	39	38	41	51	260
30	Module 3	70	59	58	57	57	73	374
31	Module 4	24	20	22	21	20	27	134
32	Module 5	120	94	104	94	89	114	615
33	Module 6	53	36	51	41	37	48	266
34								
35		Northeast	Southeast	Northcentral	Southcentral	Northwest	Southwest	
36	Cost$	1295340	1029015	1117715	1040380	1011105	1298025	
37	Weight#	504777	403807	426922	401049	398634	508485	

Each week, you only need to enter the new order information, perform a couple of matrix multiplications (or have the spreadsheet do them automatically), and you have all of the desired output report information! □

Section Summary

Before you begin working on the exercises be sure that you:

- Know that the element in the *i*'th row, *j*'th column position of the matrix resulting from the multiplication of matrix **A** times matrix **B** consists of the elements in the *i*'th row of the matrix **A** by the elements in the *j*'th column of the matrix **B** and adding the results.
- Know how to carry out simple matrix multiplications by hand, and more complex matrix multiplications using a graphing calculator and/or a spreadsheet.
- Know when matrix multiplication is appropriate for solving a problem: when the problem involves the multiplication of matching elements of two vectors of numbers and the summation of these products.
- Know that matrix multiplications can be done only when the number of columns (size of each row) of the first (left) matrix matches the number of rows (size of each column) of the second (right) matrix.
- Know that matrix multiplications are most likely to have meaningful results when the general verbal description and the individual labels of the columns of the first matrix match the general verbal description and the individual labels of the rows of the second matrix.
- Know that the identity matrix has 1's in all the entries along the main diagonal (from the upper left corner of the matrix to the lower right corner of the matrix) a_{11}, a_{22}, a_{33} ... a_{nn}., and 0's in all the other entries and that the multiplication of any matrix by a matching identity matrix does not change the original matrix.

EXERCISES FOR SECTION 6.2

Warm Up

Problems 1-9 refer to the matrices shown below:

$$A = \begin{bmatrix} 1 \\ 2 \\ 3 \end{bmatrix} \quad B = \begin{bmatrix} 2 & 4 \end{bmatrix} \quad C = \begin{bmatrix} -1 & 2 \\ 3 & -4 \end{bmatrix}$$

$$D = \begin{bmatrix} 2 & -1 \\ 4 & 0 \end{bmatrix} \quad E = \begin{bmatrix} 4 & -3 & 0 \\ 3 & 8 & 5 \end{bmatrix} \quad F = \begin{bmatrix} 0 \\ 1 \\ -5 \end{bmatrix}$$

$$G = \begin{bmatrix} 2 & 3 \\ 4 & 7 \\ 1 & 5 \end{bmatrix} \quad H = \begin{bmatrix} 1 & -1 & 4 \\ 3 & 5 & 6 \\ 1 & 4 & 8 \end{bmatrix} \quad I = \begin{bmatrix} 1 & 3 \\ 4 & 8 \\ 9 & 7 \end{bmatrix} \quad J = \begin{bmatrix} 1 & 0 \\ 0 & 1 \end{bmatrix}$$

1. a) Which of the following matrix multiplications can be carried out: **AA, AB, AC, AD, AE, AF, AG, AH, AI, AJ**?
   b) Give the matrix resulting from each possible multiplication.

2. a) Which of the following matrix multiplications can be carried out: **BA, BB, BC, BD, BE, BF, BG, BH, BI, BJ**?
   b) Give the matrix resulting from each possible multiplication.

3. a) Which of the following matrix multiplications can be carried out: **CA, CB, CC, CD, CE, CF, CG, CH, CI, CJ**?
   b) Give the matrix resulting from each possible multiplication.

4. a) Which of the following matrix multiplications can be carried out: **DA, DB, DC, DD, DE, DF, DG, DH, DI, DJ**?
   b) Give the matrix resulting from each possible multiplication.

5. a) Which of the following matrix multiplications can be carried out: **EA, EB, EC, ED, EE, EF, EG, EH, EI, EJ**?
   b) Give the matrix resulting from each possible multiplication.

6. a) Which of the following matrix multiplications can be carried out: **FA**, **FB**, **FC**, **FD**, **FE**, **FF**, **FG**, **FH**, **FI**, **FJ**?
   b) Give the matrix resulting from each possible multiplication.

7. a) Which of the following matrix multiplications can be carried out: **GA**, **GB**, **GC**, **GD**, **GE**, **GF**, **GG**, **GH**, **GI**, **GJ**?
   b) Give the matrix resulting from each possible multiplication.

8. a) Which of the following matrix multiplications can be carried out: **HA**, **HB**, **HC**, **HD**, **HE**, **HF**, **HG**, **HH**, **HI**, **HJ**?
   b) Give the matrix resulting from each possible multiplication.

9. a) Which of the following matrix multiplications can be carried out: **IA**, **IB**, **IC**, **ID**, **IE**, **IF**, **IG**, **IH**, **II**, **IJ**?
   b) Give the matrix resulting from each possible multiplication.

10. The spreadsheet below shows the sales record of a small company that sold T-shirts at a concert. The concert was held on two successive nights. They sold the shirts outside the concert site before and after the concert. Although the cost to them for the different size shirts varied, they decided not to charge different prices for different sizes. They *did* vary the price by the number of colors in the shirts, however. After the concert on the second night they decided to reduce the price of *all* of the shirts in order to get rid of them.

	A	B	C	D	E	F	G	H	I
1	TEE SHIRT SALES								
2									
3	FIRST DAY								
4									
5	PRE-CONCERT					POST-CONCERT			
6		1 COLOR	2 COLOR	3 COLOR			1 COLOR	2 COLOR	3 COLOR
7	SMALL	10	22	15		SMALL	15	21	35
8	MEDIUM	18	15	24		MEDIUM	22	25	18
9	LARGE	25	42	31		LARGE	22	35	41
10									
11									
12	SECOND DAY								
13									
14	PRE-CONCERT					POST-CONCERT			
15		1 COLOR	2 COLOR	3 COLOR			1 COLOR	2 COLOR	3 COLOR
16	SMALL	15	24	28		SMALL	12	18	24
17	MEDIUM	25	22	18		MEDIUM	25	31	18
18	LARGE	29	35	21		LARGE	25	36	39
19									
20									
21	PRICES:	1 COLOR	2 COLOR	3 COLOR		COST	1 COLOR	2 COLOR	3 COLOR
22		$ 12.50	$ 15.00	$ 18.50		SMALL	4.95	5.59	5.95
23						MEDIUM	5.95	6.59	6.95
24	DISCOUNT PRICE					LARGE	6.95	7.59	7.95
25		$ 10.00	$ 12.50	$ 15.00					
26									

Create a spreadsheet using matrix addition and multiplication that shows the following:

a) The total number of T-shirts sold pre-concert by size and color type.
b) The total revenue from T-shirts sold pre-concert by size.
c) The total number of T-shirts sold post-concert by size and color type.
d) The total number of T-shirts sold by size and color type.

11. From the data shown for the sale of T-shirts in number 10, create a spreadsheet using matrix addition, subtraction, and multiplication that shows the following:

a) The total number of T-shirts sold at the regular price by size and color type.
b) The revenue from the sale of T-shirts at the regular price by size.
c) The total number of T-shirts sold at the discount price by size and color type.
d) The revenue from the sale of T-shirts at the discount price by size.
e) The total revenue from the sale of T-shirts by size.
f) The total cost of the T-shirts by size and color type (use copied formulas).
g) The total cost of the T-shirts by size.
h) The total profit from the sale of T-shirts by size.

12. You work for a large construction company. Every week the site foremen report to you on the number and style of the homes under construction at their sites. These figures will change each week as some homes are completed and new ones are started. The table below shows the number of each style of house under construction at each of four sites in a given week:

Weekly Foreman's Report

	Site 1	Site 2	Site 3	Site 4
Colonial	4	2	2	4
Modern	1	0	1	2
Cape Cod	3	5	3	1
Rancher	5	4	0	3

The table shown below tells how many hours of specialized labor is required for each type of home:

Labor Hours per Specialty per Home Style per week

	Colonial	Modern	Cape Cod	Rancher
Bricklayers	120	29	45	49
Drywallers	63	59	42	45
Electricians	23	35	19	21
Plumbers	19	18	16	17
Carpenters	45	100	78	80
Painters	15	10	9	8

Prepare and print out the following reports:

a) a report telling the total number of hours of each type of worker;
b) a report telling the total worker hours currently used at each site;
c) a report telling the total number of worker hours for that week;
d) a report telling the total number of each type of worker at each site (what assumptions did you make to get your answer?).

13. For the construction company in Exercise 12 you have the hourly pay rate for each type of employee:

Hourly rate:	
Bricklayers	19.50
Drywallers	18.75
Electricians	27.50
Plumbers	25.90
Carpenters	15.00
Painters	10.50

(a) Set up a spreadsheet that will give the total payroll for each site in any given week, and the total payroll for that week. Set up your spreadsheet so that the new

numbers can be entered each week and the totals required will be calculated automatically. Write an explanation of how you have set up your spreadsheet so that you can make up these reports as quickly as possible.
(b) Print a sample of your reports using the given data. Be sure to use labels on your spreadsheet so it can be used as a report.

14. A company is considering a very expensive and perhaps controversial ad campaign. It is controversial in that the plan is to use celebrities in the ads, and some of these celebrities have widespread followings, but also a certain group that really dislike them. Your advertising company does some studies concerning such ads and finds that after each campaign of this type 20% of the people who already use the product will switch to another brand in protest, while 30% of other users will try the advertised product.
    (a) If the company currently has a 14% market share, how big a market share could they expect after running one of this type campaign?
    (b) What could they expect after running two ad campaigns of this type?
    (c) Is there any point at which running further ad campaigns of this type will no longer gain new customers (to 3 significant digits)?

15. The results of the next local election will be of great importance to your company. You are particularly interested in having the candidate for the Blue Party elected. You have had some studies done and found that polls taken have showed the following results:
    For each week that the Blue Party runs extensive ads, 42% of the people who usually vote Blue indicated that they will change to Red because they are tired of being bombarded by advertising, while 44% of the people who usually vote Red indicated that they will be persuaded by the advertising campaign to switch to the Blue Party.
    a) If 35% of the voters usually vote for the Blue Party, what percentage would you expect to indicate that they will vote Blue after *one* week of advertisements?
    b) What percentage would you expect to indicate that they will vote for the Blue Party after *two* weeks of advertising?
    c) How many weeks of advertising would it take to give the Blue Party *over* 51%, to allow a small margin of error?
    d) Do the percentages ever level off at a limit (to 3 significant digits)? After how many weeks of advertising does this occur?

Section 7.3: Systems of Linear Equations and Augmented Matrices

In previous math courses, you should have learned how to solve systems of linear equations (for example, where two lines intersect), probably by substitution and by elimination (adding a multiple of one equation to another equation to cancel out variables). In this section, we will see how this can be done using matrices.

Here are the kind of problems that the material in this section will help you solve:

- You have been trying to determine the price to charge for a product. You want to determine the equilibrium price, the point at which the supply curve and the demand curve meet, so you will be able to sell all the product you order at the best price.
- You are planning to sell T-shirts as a fund raiser for your group. You are looking into the cost of ordering the shirts and the price you should charge for each shirt. How many shirts do you have to sell to cover your costs?
- You have been considering closing down the second site for constructing your hand-crafted chairs and tables. It would cost you money to return the unused wood and hardware to your original site. Can you actually use up all of the materials, and if so, how many chairs and tables should you make to do this?

When you have finished this section you should be able to solve problems like those above and:

- Understand what is meant by a system of linear equations.
- Know how to solve two-variable linear systems graphically.
- Know how to solve multivariable linear systems by elimination and substitution.
- Know how to solve multivariable linear systems using augmented matrices and row reduction (Gaussian elimination).

Finding the Break-Even Point

Sample Problem 1: You are trying to raise money for your organization and you have decided to sell T-shirts. The cost to order the shirts is $6 each, plus a fixed cost of $30 for any order. You have surveyed the potential market and decided that you should charge $12 per shirt. How many shirts do you have to sell before you break even (start to make a profit)?

Solution: The **break-even point** is the point of intersection of the cost and revenue curves. Up until you have reached the break-even point, your costs are higher than your revenue. After you have reached the break-even point your revenue is higher than your costs, and you begin to show a profit. From the definition of the break-even point we know that we want to find the point where the revenue function and the cost function intersect, so this problem can be solved graphically. Let's start by writing out the models for the cost and revenue.

Verbal Definition: $C(s)$ = the cost, in dollars, to order s shirts
Symbol Definition: $C(s) = 6s + 30$ for $s \geq 0$
Assumptions: Certainty and divisibility

Verbal Definition: $R(s)$ = the revenue, in dollars, when s shirts are sold
Symbol Definition: $R(s) = 12s$ for $s \geq 0$
Assumptions: Certainty and divisibility

The graph of these two functions is shown in Figure 1.

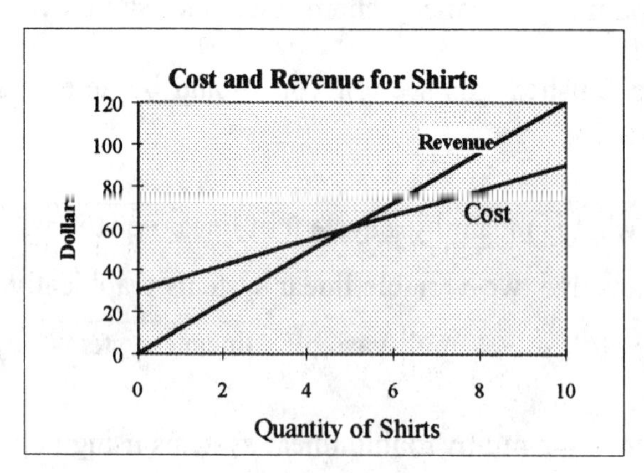

Figure 1

From looking at the graph it appears that the equilibrium point is five shirts. You can use a graphing calculator to graph the two functions and then find the intersection.

We can also solve this problem algebraically. We want to know when the two functions are equal, so we can set the cost function equal to the revenue function and have one equation in one unknown. Before we do that, however, let's rewrite the problem as a system of two linear equations in two variables, where y is the variable corresponding to the vertical axis on the graph:

$y = 6s + 30$

$y = 12s$

The simplest way to solve a system of two equations in two variables is by substitution. In general, that means solving one equation for one of the variables, then substituting the expression obtained *in* for that variable in the *other* equation. In our example above, *both* equations are already solved for y, so our job is particularly easy. If we think of the second equation as the one that is solved for y, and we substitute that expression for y (which we found to be $12s$) in for the y in the first equation, we get:

$12s = 6s + 30$	(Substituting the value of y from the second equation into the first equation)
$12s - 6s = 30$	(Subtracting 6s from both sides.)
$6s = 30$	(Combining like terms.)
$s = 5$	(Dividing both sides by 6.)

So, the break-even point is indeed five shirts, as it appeared from the graph. When five shirts are sold, you have covered your costs ($60) and the next shirt that you sell will earn a profit. □

Finding the Equilibrium Point

Sample Problem 2: A study was undertaken to determine the supply and demand functions for soda for a certain group of students.[1] The demand model is:

Verbal Definition: $D(q)$ = the price (in cents) per soda you would have to charge to sell exactly q sodas
Symbol Definition: $D(q) = -0.401q + 145$ for $0 \le q \le 362$
Assumptions: Certainty and divisibility

The model for the supply function is:

Verbal Definition: $S(q)$ = the price in cents per soda at which the supplier would be willing to supply exactly q sodas
Symbol Definition: $S(q) = 0.418q + 8.33$ for $q \ge 0$
Assumptions: Certainty and divisibility.

An important concept in economics is the **equilibrium point**, the point (combination of price and quantity) at which the demand curve and the supply curve

[1] Data adapted from student project.

intersect. Find the equilibrium point for this situation.

Solution: Since the demand and supply functions in this problem are linear functions, finding where the functions intersect is equivalent to solving a **system of linear equations.** From the definition of the equilibrium point we can see that a graphical solution of this problem should be possible. We graph both functions on one graph and find the intersection of the two lines shown in Figure 2.

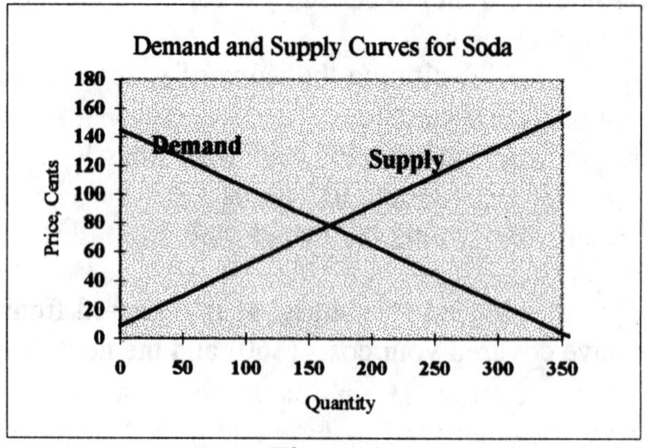

Figure 2

Note that on the graph the intersection of the demand and supply curves appears to be around 170 units, at a price of about 80 cents.

The way that this system of linear equations is set up involves one independent variable, the quantity q. Both the demand price and the supply price can be represented by the variable p, and can be expressed as functions of q. We want to know where the two functions intersect, or where they are equal. First, we will write the two functions as a system of linear equations:

$$p = -0.401q + 145$$
$$p = 0.418q + 8.33$$

To solve this system, we set the two equations equal to each other (since they are both already solved for p) to create one equation in one unknown:

$-0.401q + 145 = 0.418q + 8.33$ (Substituting the value for p from the first equation in the second equation)

$-0.401q - 0.418q = 8.33 - 145$ (subtracting $0.418q$ and 145 from both sides)

$-0.819q = -136.67$ (combining like terms)

$q = 166.87$ (dividing both sides by -.819)

Thus, we see that the equilibrium quantity is approximately 167 sodas. The equilibrium price is found by substituting this value in either the demand or supply function:

$$D(q) = 0.418(167) + 8.33 = 78.136$$

This tells us that the equilibrium price is approximately 78 cents ($0.78) per soda. If the price of the sodas is set at 78 cents per soda, there will be neither a shortage of sodas nor sodas left unsold (everybody is happy). □

Sample Problem 3: You are trying to raise money for your organization and you have decided to sell T-shirts *and* hats. You can order both items from the same company, who will charge you $4 for the hats and $6 for the shirts, plus a fixed cost of $30 for any order. You have surveyed the potential market and decided that you should charge $12 per shirt and $7 per hat. How many shirts and hats do you have to sell before you "break even"?

Solution: This problem is similar to the last one, but we now have two independent variables, the number of T-shirts that we sell and the number of hats that we sell. Let's start by writing out the models for the cost and revenue.

Verbal Definition: $C(s,c)$ = the cost, in dollars, to order s shirts and c caps
Symbol Definition: $C(s,c) = 6s + 4c + 30$ for $s, c \geq 0$
Assumptions: Certainty and divisibility (continuity)

Verbal Definition: $R(s)$ = the revenue, in dollars, when s shirts and c caps are sold
Symbol Definition: $R(s) = 12s + 7c$ for $s, c \geq 0$
Assumptions: Certainty and divisibility (continuity).

This system of equations involves three unknowns,: the quantities of shirts (s) and hats (c) and the dollar amount of the cost and the revenue (which we could call y). A graphical representation of this problem is possible, but considerably more difficult than for problems with only two variables. Let's look at an algebraic solution. We are trying to solve the system of equations

$$y = 6s + 4c + 30$$
$$y = 12s + 7c$$

Once again, we can set the equations equal to each other (this will eliminate the y):

$12s + 7c = 6s + 4c + 30$ (Plugging in for y from the 2nd into the 1st)
$6s + 3c = 30$ (Subtracting $6s+4c$ from both sides and combining like terms.)

You should be able to recognize this as a linear equation in two variables. There are an infinite number of ordered pairs (values for s and c taken together) that will satisfy this equation. Even if we wish to restrict ourselves to integer solutions, there are several that come to mind: for instance, (5,0), (4,2) and (0,10). **A system of equations that has more variables than equations can have an infinite number of solutions.** □

Possible Solutions to Systems of Linear Equations

Some systems of equations that consist of two equations in two unknowns actually consist of different expressions of the *same* equation. In this case, we say that the equations are not independent. Consider the following system:

$$-3x + 6y = -3$$
$$x - 2y = 1$$

If we multiply the second equation by -3, it is exactly the same as the first equation. A system that consists of only one equation in two or more unknowns can have an infinite number of solutions, as we saw in Sample Problem 3. The graph of this system is shown in Figure 3.

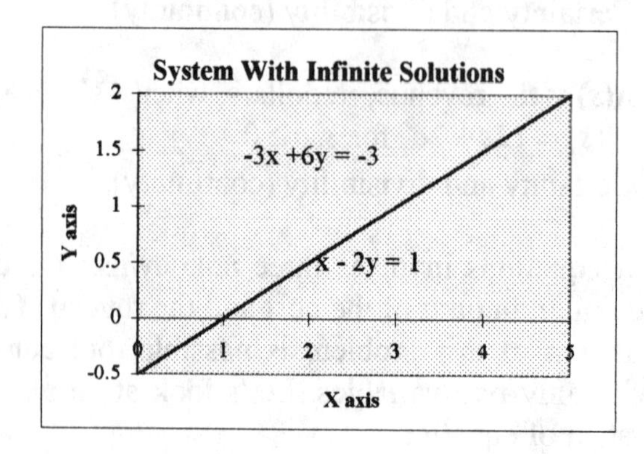

Figure 3

A system of n equations in n+1 unknowns may have infinite solutions.

Some systems that consist of two equations in two unknowns have no solutions. Consider the following system:

$$-10x + 14y = 120$$
$$5x - 7y = 70$$

The graph of this system is shown in Figure 4.

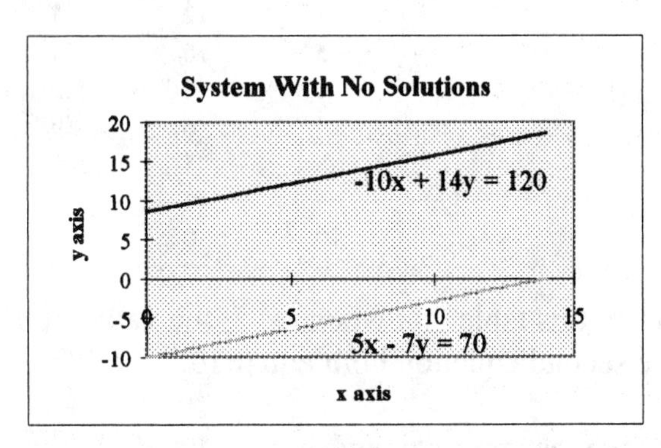

Figure 4

Solving Systems of Equations Numerically

When a *unique* solution to a system of equations exists, it is possible to find that solution numerically using a table. To do this, you first solve each equation for one of the unknowns in terms of the other. Suppose that our system of linear equations is:

$$3c + 6t = 60$$
$$8c + 4t = 100$$

The first equation, solved for c in terms of t, is $c = -2t + 20$ and the second equation is $c = -0.5t + 12.5$. In this form they can be entered into a graphing calculator as functions or entered onto a spreadsheet as formulas. Tables of these function values will show where the input t (or X on the calculator) has the same output (c here, or Y on the calculator) for both functions.

Table 1

t values	-2t+20	-.5t+12.5
0	20	12.5
1	18	12
2	16	11.5
3	14	11
4	12	10.5
5	10	10
6	8	9.5
7	6	9
8	4	8.5
9	2	8
10	0	7.5

We can see from this table that when $t = 5$, the value of c in the first equation and the value of c in the second equation both equal 10.

This method can become somewhat tedious if the solutions are numbers with many decimal places. For the supply and demand functions given in Problem 2, we can read from Table 2 that the equilibrium point is somewhere between 160 and 180

Table 2

0	8.83	145
20	17.19	136.98
40	25.55	128.96
60	33.91	120.94
80	42.27	112.92
100	50.63	104.9
120	58.99	96.88
140	67.35	88.86
160	75.71	80.84
180	84.07	72.82
200	92.43	64.8
220	100.79	56.78

We can use the graph of the functions to make an initial guess to help refine our search using a table.

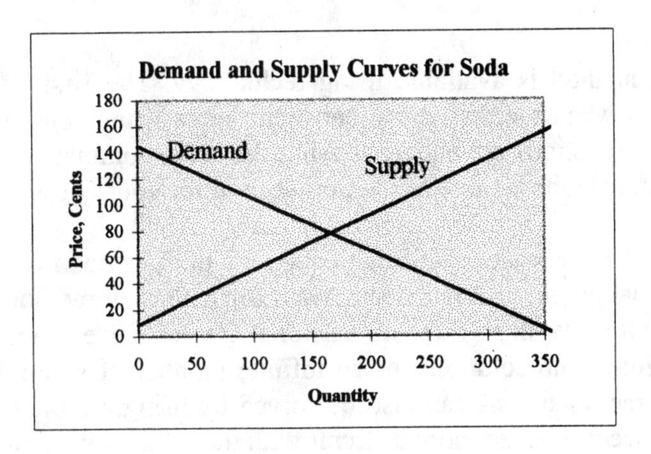

Figure 5

Note on the graph that the intersection of the demand and supply curves appears to be at around 170 units. So we could set the minimum value for our independent variable to be 160 and increase its value by 1s. If this does not show us the intersection, we would then have to refine our entries with even smaller increments by trial and error. As you can see, this is not the most efficient way to find the solution.

Table 2

Quantity	Supply	Demand
160	75.71	80.84
161	76.128	80.439
162	76.546	80.038
163	76.964	79.637
164	77.382	79.236
165	77.8	78.835
166	78.218	78.434
167	78.636	78.033
168	79.054	77.632
169	79.472	77.231
170	79.89	76.83
171	80.308	76.429
172	80.726	76.028
173	81.144	75.627
174	81.562	75.226
175	81.98	74.825
176	82.398	74.424
177	82.816	74.023
178	83.234	73.622
179	83.652	73.221

We can read from the table that the equilibrium quantity lies between 166 and 167 and the price between 78.033 and 78.636.

A third method is available using technology. The first equation will equal the second equation when the difference between the two functions is 0. You can find the point where the functions are equal by using the solve command on the difference. Use the graph to estimate the x value of the intersection for your guess.

We *could* solve systems of three equations in three unknowns graphically, using a three-dimensional graph. These are somewhat difficult to draw, but there are a number of computer programs that draw three dimensional graphs. Once again, the system may have one unique solution, no solutions, or an infinite number of solutions. Systems of three equations in three unknowns can also be solved by using a table of the function values. However, this becomes even more difficult with three functions than with two.

We cannot draw a graph with four variables. Many interesting studies have been made concerning the fourth dimension, fifth dimension and so on, but we cannot practically draw a graph to visualize anything beyond three dimensions. With each added variable and another function to be considered, the use of a table to solve the system becomes less and less efficient, so we need another approach.

Solving Systems of Linear Equation by Row Reduction (Gaussian Elimination)

Sample Problem 4: You have been considering closing down the second site for constructing your hand-crafted chairs and tables. It would cost you money to return the unused wood and hardware to your original sites. Can you actually use up all of the materials, and if so, how many chairs and tables should you make to do this? You know that chairs use 3 board feet of wood and 8 packages of hardware while tables use 6 board feet of wood and 4 packages of hardware. You have 60 board feet of wood on hand and 100 packages of hardware.

Solution 4: We have dealt with this problem in Section 5.2 , so setting up the equations for the quantity of wood and the quantity of hardware should not be a problem. If we let c represent the number of chairs and t represent the number of tables, we can express the problem by the following system of linear equations where we have included the amounts of each material that we have on hand:

$3c + 6t = 60$ wood equation
$8c + 4t = 100$ hardware equation

There are several different analytical methods for solving a system of two linear equations in two unknowns. We will use the reduction (elimination) method because it is readily adapted to matrices and larger systems. We begin by stating what we can do with systems of equations; the **elementary operations** that we can perform **on a system of equations without altering the solution** of the system (the solutions to the new system will be exactly the same as the solutions to the original system).

(1) **We can multiply any equation by a real number.** (If one hamburger and one soda costs \$6, then two hamburgers and two sodas will cost \$12:

If $x + y = 6$, then $2x + 2y = 12$.)

(2) We can add any two equations and replace one of them with the result. (If one hamburger and one soda costs \$6.00 and one hamburger and two sodas cost \$7.00, then two hamburgers and three sodas will cost \$13.00. If we let x = the number of hamburgers and y equal the number of sodas, we can represent this as the following system of equations: If

$x + y = 6$ and
$x + 2y = 7$, then
$2x + 3y = 13$. (This assumes no discount is offered for "bulk" purchases.)

Notice that (1) and (2) together suggest that **we can add a multiple of one equation to another**.

(3) **We can switch the equations around**. In the example above, and in general, the order in which we list the equations doesn't make any difference.

If we were to solve this problem by substitution, we would first solve for one variable; for example:

$3c + 6t = 60$
$3c = 60 - 6t$, (subtracting 6t from both sides)
$c = 20 - 2t$ (dividing through by 3).

(Note that the last equation could be written $c + 2t = 20$.)

We would then plug in (substitute) the expression for c into the other equation:

$8(20 - 2t) + 4t = 100$
$160 - 16t + 4t = 100$ (multiplying through by 8)
$-12t = -60$ (subtracting 160 from both sides & combining like terms)
$t = -60/(-12) = 5$. (dividing both sides by -12)

So we have the solution for t, and now we can plug it back in to the expression for c:

$c = 20 - 2t$
$c = 20 - 2(5)$ (substituting 5 for t)
$c = 20 - 10 = 10$ (multiplying out and subtracting)

Thus we see that $c = 10$. Now let's show how we could obtain the same solution by just using the three equation operations listed above, keeping the equations in the initial format (variables on the left, constants on the right). We will begin by dividing the first equation by 3 (or multiplying by the multiplicative inverse of 3, 1/3). This will leave us with a 1 as the first coefficient. You can think of this as solving the equation for c. To indicate this operation, we will use the notation $\frac{1}{3}R_1 \rightarrow R_1'$. The primed row refers to the *new* row in the next system, and the unprimed row refers to the *old* row in the most recent system.

$3c + 6t = 60$
$8c + 4t = 100$

$(\frac{1}{3}R_1 \rightarrow R_1')$

$1c + 2t = 20$
$8c + 4t = 100$

We can now eliminate the c in the second equation by multiplying the first equation by -8 (the additive inverse of 8) and adding this result to the second equation. This is equivalent to substituting for c in the second equation, as we did above.

$-8c - 16t = -160$ (the 1st equation multiplied by -8 , which we can denote $-8R_1$)
$+\ 8c + 4t = 10$ (the 2nd equation, R_2)
$0c - 12t = -60$ ($-8R_1+R_2 \rightarrow R_2'$

Our system now consists of:

$1c + 2t = 20$
$0c - 12t = -60$

We now divide the second equation by -12 (or multiply by -1/12, the multiplicative inverse), as we did when solving for t earlier, to give us a 1 as the second coefficient.

$(\frac{-1}{12}R_2 \rightarrow R_2')$

$$1c + 2t = 20$$
$$0c + 1t = 5$$

We now want to eliminate the second variable from the first equation, i.e., have a 0 coefficient for t. This is equivalent to plugging back in for t into the first equation, as we did above using the substitution method. From this last equation it is plain that $t = 5$. Then $2t = 10$ and $-2t = -10$. Adding this to the first equation we get:

$$\begin{array}{rll} & 1c + 2t = 20 & (R_1) \\ + & \underline{\quad - 2t = -10} & \underline{(-2R_2)} \\ & 1c + 0t = 10 & (-2R_2 + R_1 \rightarrow R_1') \end{array}$$

Replacing this in our system we have:

$$1c + 0t = 10$$
$$0c + 1t = 5$$

We can now read the solution directly from the system: $c = 10$ and $t = 5$. Substituting in the original equations we see that these are indeed solutions to the system:

$$(3)(10) + (6)(5) = 30 + 30 = 60$$
$$(8)(10) + (4)(5) = 80 + 20 = 100$$

Since this problem involves a system of two equations in two unknowns we can also solve this problem graphically. The graph of this system is shown below:

Figure 6

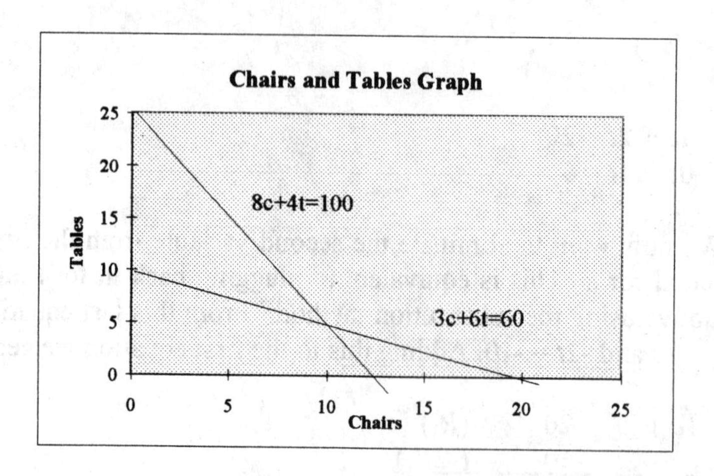

This somewhat laborious explanation of the process used in the reduction method should help us to understand the process better when we use matrices to solve systems of linear equations.

Solving Systems of Equations With Augmented Matrices

Since most systems of equations that represent real world situations involve many more than two or three variables, we cannot rely on either the graphical or numerical approach. We need an analytic method to find the solutions to these systems.

Let's look at the system from for the chairs and tables again.

$3c + 6t = 60$
$8c + 4t = 100$

If we write the coefficients of the two equations in matrix form we call it the **coefficient matrix**. The entries in the first column correspond to the coefficients of the c's and the entries in the second column correspond to the coefficients of the t's.

$$\begin{bmatrix} 3 & 6 \\ 8 & 4 \end{bmatrix}$$

Another matrix associated with this system is called the **augmented matrix**. This is simply the coefficient matrix with a column added representing the constants in the equations (we have shown the corresponding equations to the right, for reference):

$$\begin{bmatrix} 3 & 6 & : & 60 \\ 8 & 4 & : & 100 \end{bmatrix} \qquad \begin{aligned} 3c+6t&=60 \\ 8c+4t&=100 \end{aligned}$$

The dotted line before the last column is just a reminder that the last column consists of the constants (think of the dotted line as an = sign). We can look at this as just a simplified way of writing the system - we have not written down the variables, since the columns help us keep track of where they would go.

Just as there are operations we can perform on *equations* without changing the solution to a system, there are corresponding operations we can perform on the *rows* of the corresponding augmented matrix, called **elementary row operations**:

1. we can multiple a row by a non-zero real number,
2. we can add a non-zero multiple of one row to another row (and place the result in place of the second row), and
3. we can exchange any two rows.

The operations performed on the equations earlier are shown here as row operations on the augmented matrix (again, the corresponding equations are shown at the right, for reference):

$$\begin{bmatrix} 3 & 6 & : & 60 \\ 8 & 4 & : & 100 \end{bmatrix} \qquad \begin{aligned} 3c+6t&=60 \\ 8c+4t&=100 \end{aligned}$$

$R1' = \frac{1}{3}R1$

$$\begin{bmatrix} 1 & 2 & : & 20 \\ 8 & 4 & : & 100 \end{bmatrix} \qquad \begin{aligned} 1c+2t&=20 \\ 8c+4t&=100 \end{aligned}$$

$R2' = -8R1 + R2$

$$\begin{bmatrix} 1 & 2 & : & 20 \\ 0 & -12 & : & -60 \end{bmatrix} \qquad \begin{aligned} 1c+2t&=20 \\ 0c-12t&=-60 \end{aligned}$$

$R2' = -\frac{1}{12}R2$

$$\begin{bmatrix} 1 & 2 & : & 20 \\ 0 & 1 & : & 5 \end{bmatrix} \qquad \begin{aligned} 1c+2t&=20 \\ 0c+1t&=5 \end{aligned}$$

$R1' = -2R2 + R1$

$$\begin{bmatrix} 1 & 0 & : & 10 \\ 0 & 1 & : & 5 \end{bmatrix} \qquad \begin{aligned} 1c+0t&=10 \\ 0c+1t&=5 \end{aligned}$$

We are done now, since

$$\begin{aligned} 1c + 0t &= 10 \\ 0c + 1t &= 5 \end{aligned} \quad \Leftrightarrow \quad \begin{aligned} c &= 10 \\ t &= 5 \end{aligned}$$

Translation: The number of chairs, c, equals 10, and the number of table, t, equals 5. □

The system of equations corresponding to the *final* augmented matrix is equivalent to the system of equations corresponding to the *first* augmented matrix because we have used only elementary row operations. This means that the solutions to the two systems will be exactly the same.

The Identity Matrix

The left-hand side of the final augmented matrix is a special matrix called an **identity** matrix. Notice that re-placing the variables after the coefficients in the final augmented matrix allowed us to read the solution directly; the corresponding system was solved for c and t.

When working with real numbers there is an identity element for multiplication; that is, a number that, when multiplied by any other number, does not change it. That number is, of course, 1. Multiplying any number by 1 leaves the number unchanged: $(3)(1) = 3$, $(0.2357)(1) = 0.2357$, $(\sqrt{79})(1) = \sqrt{79}$, $(\pi)(1) = \pi$. There is also an identity element for multiplication of square matrices of a given size. The multiplicative identity, referred to simply as the **Identity matrix**, of dimension 2 × 2 is:

$$\mathbf{I_2} = \begin{bmatrix} 1 & 0 \\ 0 & 1 \end{bmatrix}$$

as we saw in the augmented matrix above.

Let's see what happens when we multiply another matrix, $\mathbf{R}$, times the identity matrix:

$$\begin{bmatrix} 3 & 9 \\ 6 & 7 \end{bmatrix}\begin{bmatrix} 1 & 0 \\ 0 & 1 \end{bmatrix} = \begin{bmatrix} (3)(1)+(9)(0) & (6)(0)+(9)(1) \\ (6)(1)+(7)(0) & (3)(0)+(7)(1) \end{bmatrix} = \begin{bmatrix} 3+0 & 0+9 \\ 6+0 & 0+7 \end{bmatrix} = \begin{bmatrix} 3 & 9 \\ 6 & 7 \end{bmatrix}$$

In the case of multiplication of a matrix by the identity matrix, $\mathbf{RI_2} = \mathbf{I_2R}$; order does *not* matter. [*Check* on your own that, indeed, in this case, $\mathbf{I_2R} = \mathbf{R}$.] Recall that, *in general*, the multiplication of matrices is not commutative; order *does* matter.

The identity matrix must be a square matrix, and for any given dimension $(n \times n)$, it has 1's in all the entries along the main diagonal (from the upper left corner of the matrix to the lower right corner of the matrix) a_{11}, a_{22}, a_{33} ... a_{nn}., and 0's in all the other entries. For example:

$$\mathbf{I}_3 = \begin{bmatrix} 1 & 0 & 0 \\ 0 & 1 & 0 \\ 0 & 0 & 1 \end{bmatrix} \quad \mathbf{I}_4 = \begin{bmatrix} 1 & 0 & 0 & 0 \\ 0 & 1 & 0 & 0 \\ 0 & 0 & 1 & 0 \\ 0 & 0 & 0 & 1 \end{bmatrix} \quad \text{In general } \mathbf{I}_n = \begin{bmatrix} 1 & 0 & 0 & .. & 0 \\ 0 & 1 & 0 & .. & 0 \\ 0 & 0 & 1 & .. & 0 \\ .. & .. & .. & \ddots & 0 \\ 0 & 0 & 0 & .. & 1 \end{bmatrix}$$

If $\mathbf{A}$ is an $m \times n$ matrix, then $\mathbf{I}_m\mathbf{A} = \mathbf{A}$ and $\mathbf{A}\mathbf{I}_n = \mathbf{A}$. If $\mathbf{A}$ is a square $n \times n$ matrix, then $\mathbf{I}_n\mathbf{A} = \mathbf{A}\mathbf{I}_n = \mathbf{A}$. $\mathbf{I}_m$ and $\mathbf{I}_n$ are the *only* matrices that will have these properties for *all* possible $m \times n$ matrices, $\mathbf{A}$. See a book on linear algebra for a proof of this.

We will see how the identity matrix is used to help solve problems later in this section and in Section 7.4.

Gaussian Elimination/Row Reduction Strategy

A general strategy for solving a system of equations using Gaussian elimination and augmented matrices is to work systematically to get an identity matrix on the left:

1. Proceed one column at a time, from left to right.
2. In each column, first get the 1 in the proper place, using appropriate row operations (usually dividing the row where the 1 belongs by the number that is in that spot, possibly exchanging rows first if needed).
3. Then, get the 0 (or 0's) in the other spots in that column (usually by multiplying the row where you just created a 1 by the negative of the element in another row that you want to cancel out to be 0, and adding to the other row).
4. **If you get stuck, it is likely that you have a problem that does not have a unique solution.** Look back at the above sequence of row operations to see that we have followed the strategy that we just suggested here.

Special Cases: No Unique Solution

There are times when the reduction will not lead to an identity matrix for the coefficient matrix. Two representative possibilities will be discussed in Sample Problems 5 and 6.

Sample Problem 5: Use row operations to reduce the augmented matrix of the system below :

$$5x - 7y = 70$$
$$-10x + 14y = 20$$

Solution 5: We first write the system as an augmented matrix and then follow the procedure to get the reduced form of the augmented matrix:

$$\begin{bmatrix} 5 & -7 & : & 70 \\ -10 & 14 & : & 120 \end{bmatrix}$$ (want a 1 where the 5 is)

$R1' = R1/5$ ↓

$$\begin{bmatrix} 1 & -7/5 & : & 14 \\ -10 & 14 & : & 120 \end{bmatrix}$$ (want a 0 where the -10 is)

$R2' = 10R1 + R2$ ↓

$$\begin{bmatrix} 1 & -7/5 & : & 14 \\ 0 & 0 & : & 12 \end{bmatrix}$$ (can't get 1 in (2,2) spot unless lose 0 in (2,1) spot)

If one row of the reduced matrix has all zeroes in the coefficient matrix part and a nonzero number in the constant matrix part, the equations are inconsistent and there is no solution. (For example, if two lines are parallel, they will never intersect, as shown in the beginning of this section.) In the reduced augmented system above, we know that $0x + 0y$ will always equal 0; it can never equal 12. This example corresponds to Figure 4 from the beginning of this section. □

Sample Problem 6: Use row operations to reduce the augmented matrix of the system below :

$$x - 2y = 1$$
$$-3x + 6y = -3$$

Solution 6: We first write the system as an augmented matrix and then follow the procedure to get the reduced form of the augmented matrix (recall that our strategy says to try to make the first column look like it is the first column of an identity matrix, which would mean $\begin{bmatrix} 1 \\ 0 \end{bmatrix}$ here):

$$\begin{bmatrix} 1 & -2 & : & 1 \\ -3 & 6 & : & -3 \end{bmatrix}$$ (1 in upper left is good; want 0 where -3 is)

↓

$3R1 + R2 \rightarrow R2'$

$$\begin{bmatrix} 1 & -2 & : & 1 \\ 0 & 0 & : & 0 \end{bmatrix}$$ (can't get 1 in the (2,2) spot without losing 0 in the (2,1) spot)

One row of the reduced matrix consists of all zeroes, corresponding to the equation $0x + 0y = 0$. This equation is *always* satisfied, indicating that one of the original equations was actually **dependent** on the others (does not add any *new* restrictions). In this case you will have more variables than equations, so there will be an infinite number of solutions (as long as there are no inconsistencies in the remaining equations). For example, if the two equations are actually just the same line, the intersection is that line, as shown in Figure 3 in the beginning of this section, and any ordered pair of numbers that lie on that line will satisfy the system of equations. □

As you can see, solving even a very simple system of linear equations using Gaussian elimination is tedious at best. You can use your calculator or a spreadsheet to solve relatively small systems using Gaussian elimination. In reality, most systems of equations of any consequence involve a great many equations and many variables. These are solved using computer programs based on the same processes that we have demonstrated above. Even though you will probably not use Gaussian elimination very often, it is important for you to understand how it operates, and being able to use it to solve small systems of equations is the best way to understand it. We will also refer to it in Chapter 9.

Section Summary

Before you begin the exercises be sure that you:

- Understand that a system of linear equations consists of two or more equations in which, when the equations are simplified, each variable appears exactly one time and is raised to the first power.
- Know that you can solve two-variable linear systems graphically, either by hand or using technology, by graphing the equations on a single graph and finding the intersection of the lines, or by using a table.
- Know how to solve multivariable linear systems by substitution, and by elimination - using elementary operations on the equations (multiplying an equation by a constant other than 0, adding a non-zero multiple of one equation to another, and interchanging equations).
- Know that the augmented matrix of a system of linear equations consists of the coefficients of the variables followed by a dotted line and a column consisting

of the constants (right-hand sides).

- Know how to solve multivariable linear systems using augmented matrices and row reduction (Gaussian elimination) by using elementary row operations (multiplying a row by a non-zero constant, adding a non-zero multiple of one row to another, or interchanging two rows), trying to get an identity matrix on the left side of the augmented matrix.
- Know that a good strategy for Gaussian elimination is to proceed column by column from left to right, and in each column, getting the 1 in the right place first, then the 0's.
- Know that, during Gaussian elimination, if you get to a point where an entire row to the left of the partition is 0, then your system is likely to have no unique solution; if the number to the right of the partition is nonzero, you have *no solution*; if it is 0, you probably have an infinite number of solutions.

EXERCISES FOR SECTION 6.3

Warm Up

For Exercises 1-2, solve the systems of linear equations by **both** *substitution and augmented matrices (Gaussian elimination/row reduction):*

1. $\begin{aligned} 3x + y &= 7 \\ 2x + 2y &= -2 \end{aligned}$

2. $\begin{aligned} 5p + 2q &= 36 \\ 8p - 3q &= -54 \end{aligned}$

For Exercises 3-4, solve the systems of linear equations by **both** *elimination (adding multiples equations to each other) and augmented matrices (Gaussian elimination/row reduction):*

3. $\begin{aligned} 4p + 12q &= 6 \\ 2p + 6q &= 3 \end{aligned}$

4. $\begin{aligned} 8m - 6n &= 21 \\ -20m + 15n &= 12 \end{aligned}$

For Exercises 5-11, solve the system of equations by **any** *method in this section:*

5. $\begin{aligned} 0.4x + .2y &= 8 \\ .3x + .6y &= 6 \end{aligned}$

6. $\begin{aligned} .6x + .2y &= 1.4 \\ -.2x - .2y &= .2 \end{aligned}$

7. $\begin{aligned} 4c - 3t - 2 &= 3c - 7t \\ c + 5t - 2 &= 2c + 4 \end{aligned}$

8. $\begin{aligned} 3r + 5t - 7 &= 8r + 10t \\ 7r + 6t - 20 &= 3r + 4t \end{aligned}$

$$x + y = 5$$

9. $2x + y = 9$

$$3x - y = 4$$

$$2x + y + 6z = 3$$

10. $x - y + 4z = 1$

$$3x + 2y - 2z = 2$$

$$5x - 7y + 4z = 2$$

11. $3x + 2y - 2z = 3$

$$3x - 2y + 3z = 4$$

Game Time

12. Find the equilibrium point for the following demand and supply functions :
    Demand function: $p = D(q) = -0.005q + 12$
    Supply function: $p = S(q) = 0.003q + 8$

13. Find the break-even point for the following cost and revenue functions:
    $C(q) = 2.5q + 5000$
    $R(q) = 8.5q$

14. If the amounts of wood and hardware on hand in the example given at the beginning of this section

    $3c + 6t = 60$
    $8c + 4t = 100$

    were 63 instead of 60 and 84 instead of 100, how many chairs and how many tables should be built?

15. If the demand function for a product is given by $p = D(q) = 20 - 0.8q$ (p = price in dollars and q = quantity in thousands) and the supply function is given by $p = S(q) = 4 + 1.2q$, find the equilibrium quantity and price.

16. If the revenue from the sale of a product can be modeled by $R(q) = 8.5q$ and the cost of producing q units can be modeled by $C(q) = 2.5q + 1200$, find the break-even point.

17. If it costs you \$2.60 per cap to manufacture baseball caps with a fixed cost of \$4500, and you can sell all that you make for \$7.50 per cap, what is your break-even point? How many caps must you produce and sell to clear a profit of \$10,000?

18. You have been hired to do a survey for Perfect Pizza, a company that delivers pizza to dorm rooms and to homes. You actually surveyed 250 students. You report that 62.5% more people liked Perfect pizza than liked all other brands combined (the difference between the number preferring Perfect and the number preferring all others was 62.5% of the number preferring all others). However, 16% of *all* the people that you interviewed had no opinion. Exactly how many of the people surveyed liked Perfect Pizza, and how many people (combined) preferred other brands?

19. You have been interviewed for a sales job with a large company. No actual commission rates were mentioned, but it was made clear that sales of at least \$100,000 per month were expected. In fact, the interviewer said that the commission rate was lower on the first \$100,000 sales, and then increased for amounts over this. They showed figures for two of their leading salespeople showing one earned \$8500 on sales of \$175,000 and another earned \$14,800 on sales of \$280,000. Find the percentage commission paid on sales up to \$100,000 and the commission paid on sales over \$100,000.

20. You and a friend have put money into two different investment plans. You put \$1000 in the first plan and \$2000 in the second plan. Your friend put \$1500 in each plan. At the end of one year your total investments were worth \$3124 and your friend's were worth \$3125.75. What was the annual percentage yield on each investment?

Section 7.4: Matrix Equations

We have just finished a section on the solution of systems of linear equations using augmented matrices by hand. Even for small problems involving as little as two or three unknowns with whole number coefficients, this method is slow and tedious. Real world problems are rarely so simple. Is there a faster way? In this section we will learn how to solve systems of equations using the concept of the inverse of a matrix and matrix algebra. We will see how to do small problems by hand (quite similar to the augmented matrix method), and then learn how to solve systems of any size much faster using technology.

You can solve the problems from the last section using the material in this section. Here are some more of the kind of problem that the material in this section will help you solve:

- An experimental farm must test varying diets for its animals, which have very specific dietary needs. There are several different types of feeds to choose from. How should the different feeds be mixed to satisfy the requirements?
- A small scientific observation post must produce its own electricity and water. It uses electricity and water to make electricity and it uses water and electricity to make water. How much water and electricity do they need to make to cover their needs beyond the amounts used in production?
- You have found a system of equations involving the parameters of a category of models whose solution will correspond to the best-fit (least squares) model. How can you find the parameters of this best-fit model?

By the end of this section, in addition to being able to solve the above kinds of problems, you:

- Should know how to set up a matrix equation for a system of linear equations (and *why* it works the way it does).
- Should know how to recognize the notation for the multiplicative inverse of a matrix.
- Should be able to find the multiplicative inverse of a matrix, both by hand for small simple matrices and using technology in general.
- Should know how to solve a matrix equation algebraically, including when given the matrix equations for finding a least-squares model and for Leontief input-output models.
- Have an idea of how computers can encode and decode messages.

Using Matrices to Solve Systems of Linear Equations

A **matrix equation** is simply an equation that has variables representing *matrices* as its elements rather than variables representing *numbers*. Matrices take the place of the coefficients, variables and constants. Systems of linear equations can be written as a matrix equation so that matrix algebra can be used to solve the system. Technology can be used so that solving systems of equations using matrix algebra is quite simple. All of the tedious work is done by the technology. We must look at the process of solving matrix equations in order to understand what the technology is doing and how to enter the commands correctly. You should use whatever technology makes the most sense in a given situation. If available, graphing calculators are particularly convenient for these operations at the present time.

Sample Problem 1: Let's go back to the simple problem of the tables and chairs from the last section to show how we can write a system of linear equations as a matrix equation. You have been considering closing down the second site for constructing your hand-crafted chairs and tables. It would cost you money to return the unused wood and hardware to your original site. Can you actually use up all of the materials, and if so, how many chairs and tables should you make to do this? You know that chairs use 3 board feet of wood and 8 packages of hardware, while tables use 6 board feet of wood and 4 packages of hardware. You have 60 board feet of wood on hand and 100 packages of hardware.

Solution: Recall that our system was written as:

$$3c + 6t = 60$$
$$8c + 4t = 100$$

where c is the number of chairs made and t is the number of tables made.

Just as we did for the Gaussian elimination problem, we write the coefficients of the variables as a **coefficient matrix** and designate it as matrix **A:**

$$\mathbf{A} = \begin{bmatrix} 3 & 6 \\ 8 & 4 \end{bmatrix}$$

Instead of augmenting the matrix with the constants, we make a separate column matrix of the constants, the **constant matrix** ,and designate it matrix **B**:

$$\mathbf{B} = \begin{bmatrix} 60 \\ 100 \end{bmatrix}$$

We also make a separate column matrix of the variables, the **variable matrix,** and designate it matrix **X**:

$$\mathbf{X} = \begin{bmatrix} c \\ t \end{bmatrix}$$

If we multiply the coefficient matrix **A** by the variable matrix **X**, we get

$$\mathbf{AX} = \begin{bmatrix} 3 & 6 \\ 8 & 4 \end{bmatrix} \begin{bmatrix} c \\ t \end{bmatrix} = \begin{bmatrix} (3)(c) + (6)(t) \\ (8)(c) + (4)(t) \end{bmatrix} \text{ or } \mathbf{AX} = \begin{bmatrix} 3c + 6t \\ 8c + 4t \end{bmatrix}$$

From our original equations we know that 3c + 6t = 60 and 8c + 4t = 100. Since **two matrices are equal if, and only if, their entries are equal,** we can write:

$$\begin{bmatrix} 3c + 6t \\ 8c + 4t \end{bmatrix} = \begin{bmatrix} 60 \\ 100 \end{bmatrix} \quad \text{or} \quad \mathbf{AX = B} \ ,$$

In other words, the matrix equation is *equivalent* to the original *system* of linear equations. So, instead of a *system* of equations we have *one* matrix equation, and our problem is reduced to that of solving the matrix equation for **X**, the variable matrix.

To solve a simple algebraic equation such as $3x = 12$, we must simplify the equation so that it reads x = some value; that is, $1x$ = something. We accomplish this by multiplying both sides of the equation by 1/3, the multiplicative inverse of 3 (which could be written 3^{-1}) :

$$\left(\frac{1}{3}\right) 3x = \left(\frac{1}{3}\right) 12$$

This reduces to:

$1x = 4$, or $x = 4$, which is our solution.

We will solve the matrix equation using the same process. You may recall that when we multiply an element by its multiplicative inverse the result is always the multiplicative identity element. The multiplicative identity element for the real numbers is 1. All real numbers except 0 have a multiplicative inverse (the multiplicative inverse of a number is its *reciprocal*; for example, the multiplicative inverse of 3 is 1/3). The identity element for multiplication of $n \times n$ matrices is the **identity matrix, $\mathbf{I}_n$** . Only square matrices have multiplicative inverses and not even all square matrices have multiplicative inverses. If the matrix *does have* an inverse, it is called **nonsingular**. If a matrix *does not* have an inverse, the matrix is called **singular** (sort of backwards from what you might have guessed). We symbolize the multiplicative inverse, referred to as simply the **inverse** of **A**, by $\mathbf{A}^{-1}$ (read: "A inverse").

To solve the matrix equation we must multiply the left and right sides by the multiplicative inverse of **A** (just as we multiplied both sides by 1/3 in solving the simple equation). Recall that **order does count** in matrix multiplication, so we multiply by $\mathbf{A}^{-1}$ on the **left** of both sides:

$$\mathbf{A}^{-1}\mathbf{A}\mathbf{X} = \mathbf{A}^{-1}\mathbf{B} \quad \text{(as in } (1/3)(3x) = (1/3)(9) \text{ for the simple equation)}$$

Notice that the $\mathbf{A}^{-1}$ had to be multiplied on the *left* for the left-hand side (LHS) of the equation, in order to cancel out the **A**, which then meant that it also had to be multiplied on the left for the right-hand side (RHS) since order counts in matrix multiplication.

By definition, when we multiply a number or a matrix by its multiplicative inverse we get the identity: $\mathbf{A}^{-1}\mathbf{A} = \mathbf{I}$ or $\mathbf{I} = \mathbf{A}\mathbf{A}^{-1}$, just as when we multiply a number by its reciprocal, we get 1. Notice that the multiplication can be done in *either* order in this case, although it is not true in general.

As a result, the original equation reduces to:

$$\mathbf{I}\mathbf{X} = \mathbf{A}^{-1}\mathbf{B} \quad \text{(like } 1x = (1/3)(12) = (3^{-1})(12) \text{ for the simple equation)}$$

By definition when we multiply any matrix by the identity, the matrix is not changed: $\mathbf{I}\mathbf{X} = \mathbf{X}$, just as $1x = x$ for the simple equation.

Therefore the original equation reduces to:

$$\mathbf{X} = \mathbf{A}^{-1}\mathbf{B} \quad \text{(analogous to } x = (1/3)(12) = (3^{-1})(12)\text{)}$$

In other words, **the general solution for the matrix equation $\mathbf{A}\mathbf{X} = \mathbf{B}$ is**

$\mathbf{X} = \mathbf{A}^{-1}\mathbf{B}$.

You may be tempted to write $\mathbf{X} = \mathbf{B}/\mathbf{A}$, just as you could write $x = 12/3$ for the simple equation. Unfortunately, there is no separate notation for division of matrices, so this is not correct. You may remember from high school that "division by a fraction is the same as multiplication by its reciprocal." In the same way, the only form of division possible with matrices is multiplication by the inverse.

Our problem is now reduced to finding $\mathbf{A}^{-1}$. Recall that, by definition, $\mathbf{AA}^{-1} = \mathbf{I}$. One way of looking at this is to think of $\mathbf{A}^{-1}$ as the matrix solution (X) to the equation $\mathbf{AX} = \mathbf{I}$. From what we have just said above, the Gaussian elimination we did in the last section can be thought of as solving the equation $\mathbf{AX} = \mathbf{B}$, in which case we reduced [$\mathbf{A} : \mathbf{B}$] to the form [$\mathbf{I}$:] (It turns out that what was on the right is, in fact, the solution for X, which is the matrix product $\mathbf{A}^{-1}\mathbf{B}$). Analogously, now that we are solving the matrix equation $\mathbf{AX} = \mathbf{I}$, we should be able to do it by reducing [$\mathbf{A} : \mathbf{I}$] to the form [$\mathbf{I}$:] . Again, what is on the right will be the solution for X, which this time is the matrix we want, $\mathbf{A}^{-1}$.

This is certainly not as easy as finding the multiplicative inverse of real numbers. In that case we simply divide the identity 1 by that number (find the reciprocal). To find the inverse *matrix*, we must use matrix algebra.

For our 2×2 matrix example, using the augmented matrix concept, the solution of this equation involves reducing the matrix equation

$$\begin{bmatrix} 3 & 6 \\ 8 & 4 \end{bmatrix} \begin{bmatrix} x & y \\ z & w \end{bmatrix} = \begin{bmatrix} 1 & 0 \\ 0 & 1 \end{bmatrix}$$

to:

$$\begin{bmatrix} 1 & 0 \\ 0 & 1 \end{bmatrix} \begin{bmatrix} x & y \\ z & w \end{bmatrix} = \begin{bmatrix} a & b \\ c & d \end{bmatrix}$$

From this system we can see that the solution, $\mathbf{A}^{-1}$, is the matrix

$$\begin{bmatrix} a & b \\ c & d \end{bmatrix}.$$

Thus, to find the inverse of a matrix, we augment the matrix by the identity matrix, then do row operations to get the identity matrix on the left, as before. The augmented matrix of the system of equations for our example is:

$$\left[\begin{array}{cc:cc} 3 & 6 & 1 & 0 \\ 8 & 4 & 0 & 1 \end{array}\right]$$

which is denoted by **[A:I]**.

When we reduce the left matrix to the identity matrix using Gaussian elimination, the inverse of **A** ($\mathbf{A}^{-1}$) will appear on the right-hand side of the augmented matrix.

$$\begin{bmatrix} 3 & 6 & : & 1 & 0 \\ 8 & 4 & : & 0 & 1 \end{bmatrix}$$

$R1' = (1/3)(R1)$

$$\begin{bmatrix} 1 & 2 & : & 1/3 & 0 \\ 8 & 4 & : & 0 & 1 \end{bmatrix}$$

$R2' = -8R1 + R2$

$$\begin{bmatrix} 1 & 2 & : & 1/3 & 0 \\ 0 & -12 & : & -8/3 & 1 \end{bmatrix}$$

$R2' = (-1/12)(R2)$

$$\begin{bmatrix} 1 & 2 & : & 1/3 & 0 \\ 0 & 1 & : & 2/9 & -1/12 \end{bmatrix}$$

$R1' = -2R2 + R1$

$$\begin{bmatrix} 1 & 0 & : & -1/9 & 1/6 \\ 0 & 1 & : & 2/9 & -1/12 \end{bmatrix}$$

Thus, the inverse matrix of **A** is:

$$\mathbf{A}^{-1} = \begin{bmatrix} -1/9 & 1/6 \\ 2/9 & -1/12 \end{bmatrix}$$

One way to understand the concept involved here is that each row operation (or all the row operations together) corresponds to multiplication by a matrix. If we start with **A** and end with **I** on the left, then the matrix used for this multiplication must be $\mathbf{A}^{-1}$, and so we get $\mathbf{A}^{-1}\mathbf{I} = \mathbf{A}^{-1}$ on the right:

$$A^{-1}[A \ : \ I] = [A^{-1}A \ : \ A^{-1}I] = [I \ : \ A^{-1}]$$

We show this is true by carrying out the multiplication $\mathbf{AA}^{-1}$:

$$\mathbf{AA}^{-1} = \begin{bmatrix} 3 & 6 \\ 8 & 4 \end{bmatrix} \begin{bmatrix} -1/9 & 1/6 \\ 2/9 & -1/12 \end{bmatrix} =$$

$$\begin{bmatrix}(3)(-1/9)+(6)(2/9) & (3)(1/6)+(6)(-1/12)\\(8)(-1/9)+(4)(2/9) & (8)(1/6)+(4)(-1/12)\end{bmatrix}=\begin{bmatrix}-3/9+12/9 & 3/6-6/12\\-8/9+8/9 & 8/6-4/12\end{bmatrix}=\begin{bmatrix}1 & 0\\0 & 1\end{bmatrix}$$

We can now use the inverse, $\mathbf{A}^{-1}$, to solve the matrix equation **AX = B**, where:

$$\mathbf{A}=\begin{bmatrix}3 & 6\\8 & 4\end{bmatrix}\quad \mathbf{X}=\begin{bmatrix}c\\t\end{bmatrix}\quad \mathbf{B}=\begin{bmatrix}60\\100\end{bmatrix}\quad \mathbf{A}^{-1}=\begin{bmatrix}-1/9 & 1/6\\2/9 & -1/12\end{bmatrix}$$

As shown above, we can reduce the matrix equation to:

$$\mathbf{X}=\mathbf{A}^{-1}\mathbf{B}$$

$$\begin{bmatrix}c\\t\end{bmatrix}=\begin{bmatrix}-1/9 & 1/6\\2/9 & -1/12\end{bmatrix}\begin{bmatrix}60\\100\end{bmatrix}=\begin{bmatrix}(-1/9)(60)+(1/6)(100)\\(2/9)(60)+(-1/12)(100)\end{bmatrix}=\begin{bmatrix}10\\5\end{bmatrix}\quad \square$$

Doing this by hand is obviously a slow and tedious process. We must first find the inverse of the coefficient matrix (if it exists) and then multiply this inverse times the constant matrix. It certainly does not seem to be much of an improvement over using the augmented matrix. Is there any way to do this faster?

Fortunately, most spreadsheet programs and graphing calculators have a command that will find the inverse of a matrix, if it exists. Most spreadsheet programs will invert up to about a 90x90 matrix. You can then use the matrix multiply command to finish the problem. Many
graphing calculators treat finding an inverse matrix the same as finding the reciprocal of a number, with the "x^{-1}" key. There are other mathematical techniques used to solve very large systems of equations, but they are beyond the scope of this book.

Sample Problem 2: An experimental farm must test varying diets for its animals. The first group of animals requires 38 grams of protein, 6 grams of fat and 16 grams of carbohydrates. The second group of animals requires 19 grams of protein, 2.2 grams of fat and 8 grams of carbohydrates Currently it is working with three basic types of feeds. Type A contains 30% protein, 2% fat and 20% carbohydrates along with other ingredients. Type B contains 50% fat, 6% fat and 10% carbohydrates .and other ingredients. Type C contains 15% protein, 5% fat and 10% carbohydrates plus other ingredients. How much of each type of feed should be used to fill these requirements?

Solution: This problem is fairly well defined. The farm is only going to consider three types of foods. At this time, they are only concerned with the amounts of protein, fat and carbohydrates that the first group of animals get from these foods. The decision or independent variables in this problem will be the quantities of the three feeds. Our next step is to put the information into a table. As suggested, we will list the independent variables across the top of the table, as shown in Table 1:

Table 1

NUTRIENT TABLE				REQUIREMENTS	
	FOOD A	FOOD B	FOOD C		
PROTEIN	0.3	0.5	0.15		38
FAT	0.02	0.06	0.05		6
CARBOHYDRATES	0.2	0.1	0.1		16

Let a = grams of type A, b = grams of type B, and c = grams of type C. We can then write the following system of linear equations for the problem:

$$\begin{aligned} 0.3a + 0.5b + 0.15c &= 38 \\ 0.02a + 0.06b + 0.05c &= 6 \\ 0.2a + 0.1b + 0.1c &= 16 \end{aligned}$$

The coefficient matrix is: $\mathbf{A} = \begin{bmatrix} .3 & .5 & .15 \\ .02 & .06 & .05 \\ .2 & .1 & .1 \end{bmatrix}$

The constant matrix is: $\mathbf{B} = \begin{bmatrix} 38 \\ 6 \\ 16 \end{bmatrix}$

The variable matrix is: $\mathbf{X} = \begin{bmatrix} a \\ b \\ c \end{bmatrix}$

The solution can now be found by multiplying the inverse of the coefficient matrix times the constant matrix: $\mathbf{X} = \mathbf{A}^{-1}\mathbf{B}$. When working in a spreadsheet as illustrated in Table 2 or working with a calculator, we do not need to mention the variable matrix (**X**). We can just *do* the calculation $\mathbf{A}^{-1}\mathbf{B}$.

Table 2

Experimental Farm Food Nutritional Data	Food A	Food B	Food C		Requirements, Grams 1st group	2nd Group
Protein, grams	0.3	0.5	0.15		38	19
Fat, grams	0.02	0.06	0.05		6	2.2
Carbohydrates, grams	0.2	0.1	0.1		16	8
Coefficient Inverse	0.357143	-12.5	5.714286			
	2.857143	0	-4.28571			
	-3.57143	25	2.857143			
					Grams of Each Food	
Coefficient Inverse Times Constant				A	30	25
				B	40	20
				C	60	10

Translation: The experimental farm should mix 30 grams of Food A, 40 grams of Food B, and 60 grams of Food C for the first group of animals. The experimental farm should mix 25 grams of Food A, 20 grams of Food B, and 10 grams of Food C for the second group of animals. □

For the second group of animals only the right hand side constants, desired levels of the nutrients, were different. Once we had the inverse matrix, we found the new solutions almost instantaneously by multiplying the inverse times the new constants. This is one of the benefits of using matrix equations to solve systems of linear equations.

Sample Problem 3: You have been asked to check out a new company that has just started to compete for the prefabricated home market. From their brochures you have determined the information given below:

California Bungalow Model: living room, dining room, 3 bedrooms, 1 bath, kitchen, and one-car garage. Sells for $38,499. The floor plan shows three hallways.
Florida Ranch Model: combined living-dining room, family room, 3 bedrooms,2 baths, kitchen, and one-car garage. Sells for $43,999. The floor plan shows 4 hallways.
Cape Cod Model: living room, dining room, 3 bedrooms, 1 bath, kitchen, and two-car garage. Sells for $47,499. The floor plan calls for 4 hallways.
Mountain Retreat Model: living room, dining room, 4 bedrooms, 1 bath, kitchen, and one-car garage. Sells for $44,999. The floor plan calls for 5 hallways.
Cape Hatteras Model: living room, dining room, family room, 4 bedrooms, 3 baths, kitchen, and two-car garage. Sells for $62,899. The floor plan calls for 6 hallways.
New Mexico Model: living room, dining room, family room, 5 bedrooms, 2 baths, kitchen, and 3-car garage. Sells for $63,999. The floor plan calls for 6 hallways.

Each different kind of room can be built from one of six modules, as given by Table 3:

Table 3

Module	**Size (ft)**	**Types of Rooms**
1	9×12	Bedroom
2	8×8	Bathroom
3	14×14	Living, Dining, Family, Combined Living/Dining
4	10×12	Kitchen
5	3×10	Hallway
6	8×14	Garage (1 Module/Car)

From this information, can you determine what they are charging for each module?

Solution: The problem is fairly well defined. We need to determine the price per module that the other company is charging. We are making the assumption that the company bases the price of the house on the modules that are included; that is, the total cost of each house is entirely determined by the modules that make it up. The variables will then be the price of each module. Let's put the information that we have into a table, again putting the variables across the top of the table:

Table 4

Model	Module 1	Module 2	Module 3	Module 4	Module 5	Module 6	Price $
California	3	1	2	1	3	1	38499
Florida	3	2	2	1	4	1	43999
Cape Cod	3	1	3	1	4	2	47499
Mountain	4	1	2	1	5	1	44999
Hatteras	4	3	3	1	6	2	62899
New Mexico	5	2	3	1	6	3	63999

If we designate the unknown module price (in dollars) of Module 1 as x_1 , Module 2 as x_2, Module 3 as x_3, Module 4 as x_4, Module 5 as x_5, and Module 6 as x_6, we can write the calculated price of the California model as $3x_1 + 1x_2 + 2x_3 + 1x_4 + 3x_5 + 1x_6$, and we know this should be equal to the given price, $38,499. We can write the calculated price of each of the remaining models in the same way, then set each equal to its given price and write the system as:

$$3x_1 + 1x_2 + 2x_3 + 1x_4 + 3x_5 + 1x_6 = 38499$$
$$3x_1 + 2x_2 + 2x_3 + 1x_4 + 4x_5 + 1x_6 = 43999$$
$$3x_1 + 1x_2 + 3x_3 + 1x_4 + 4x_5 + 2x_6 = 47499$$
$$4x_1 + 1x_2 + 2x_3 + 1x_4 + 5x_5 + 1x_6 = 44999$$
$$4x_1 + 3x_2 + 3x_3 + 1x_4 + 6x_5 + 2x_6 = 62899$$

$5x_1 + 2x_2 + 3x_3 + 1x_4 + 6x_5 + 3x_6 = 63999$

From this system we can set up the matrices: the **A** matrix consists of the six rows and columns in Table 4 that correspond to the coefficients of the six equations. The **X** matrix consists of the six variables x_1, x_2, x_3, x_4, x_5, and x_6, and the **B** matrix consists of the constants (the prices in Table 4). If the data in this table is entered in a spreadsheet, we could simply use the spreadsheet command to find the inverse of A, place it on the spreadsheet, and then multiply that times matrix **B**.

Table 5

	A	B	C	D	E	F	G	H	I
1	Model	Module 1	Module 2	Module 3	Module 4	Module 5	Module 6		Price $
2	California	3	1	2	1	3	1		38499
3	Florida	3	2	2	1	4	1		43999
4	Cape Cod	3	1	3	1	4	2		47499
5	Mountain	4	1	2	1	5	1		44999
6	Hatteras	4	3	3	1	6	2		62899
7	New Mexico	5	2	3	1	6	3		63999
8									
9		Inverse of Matrix A - cells B2:G7							
10		2	-2	-1	3.9968E-16	1	0		
11		0.5	7.93016E-17	-0.5	-0.5	0.5	0		
12		2	-3	1.17653E-15	4.10166E-16	2	-1		
13		-3.5	7	1.5	-0.5	-4.5	1		
14		-1.5	1	0.5	0.5	-0.5	0		
15		-1.5	2	0.5	-0.5	-1.5	1		
16									
17		A inverse (cellsB10:G15) times B (cells I2:L7)							
18		4400							
19		4450							
20		6800							
21		2949							
22		1050							
23		1150							

Translation: The price of each module is:

Module 1 price = $4400
Module 2 price = $4450
Module 3 price = $6800
Module 4 price = $2949
Module 5 price = $1050
Module 6 price = $1150 □

Sample Problem 4: If the industry-wide markup on this type of product is 35%, can you determine their approximate *cost* per module? Does this company's costs seem to compare favorably with your production costs of $3240 for a comparable Module 1, $3175 for Module 2, $5040 for Module 3, $7500 for Module 4, $820 for Module 5, and $2750 for Module 6?

Solution: Let's clarify the problem. We determined the prices in Sample Problem 3 and we want to know how they compare to our prices for comparable modules. We are making the assumption that this company uses the industry-wide standard markup of 35%. Recall that markup is (*selling price - cost*) divided by the (*cost*), which we denoted k in Section 5.2 . We want to solve this equation for the cost, since we know the markup and the selling price:

$$k = \frac{p-c}{c}$$

$ck = p - c$ (multiplying both sides by c)

$ck + c = p$ (adding c to both sides)

$c(k+1) = p$ (factoring c from first two terms)

$c = \frac{p}{k+1}$ (dividing both sides by (k+1)

So we want to divide the selling price for each Module by 0.35+1 or 1.35 to determine the estimated cost. Since we have been working in a spreadsheet, we can enter this command once and copy it.

Table 6

Estimated Cost $	Our Cost
3259.259	3240
3296.296	3175
5037.037	5040
2184.444	7500
777.7778	820
851.8519	2750

Translation: Our costs seem to be competitive except for Module 4, the kitchen, and Module 6, a garage. We might check the information further to see if the two modules are really comparable; for example, our kitchen comes complete with appliances, and our garage is finished. □

Using Matrices to Solve Least Squares Regression Problems

Sample Problem 5: In Section 8.4 we will see that solving for the parameters of the least-squares regression model involves solving the following matrix equation:

$(\mathbf{X}^T\mathbf{X})\,\mathbf{M} = \mathbf{X}^T\mathbf{Y}$. The **X** and **Y** matrices are known. Solve for **M**.

Solution: We wish to solve for the **M** matrix, so we need to get it alone. In order to do this, we must multiply both sides of the matrix equation by the multiplicative inverse of $\mathbf{X}^T\mathbf{X}$, denoted by $(\mathbf{X}^T\mathbf{X})^{-1}$:

$(\mathbf{X}^T\mathbf{X})^{-1}(\mathbf{X}^T\mathbf{X})\,\mathbf{M} = (\mathbf{X}^T\mathbf{X})^{-1}(\mathbf{X}^T\mathbf{Y})$ (multiplying both sides by $(\mathbf{X}^T\mathbf{X})^{-1}$ on the left)

$\mathbf{IM} = (\mathbf{X}^T\mathbf{X})^{-1}(\mathbf{X}^T\mathbf{Y})$ (a matrix times its inverse = the identity matrix **I**)

$\mathbf{M} = (\mathbf{X}^T\mathbf{X})^{-1}(\mathbf{X}^T\mathbf{Y})$ (the identity times a matrix leaves it unchanged)

For example, if the **X** matrix is : $\begin{bmatrix} 1 & 2 & 1 \\ 2 & 1 & 1 \\ 3 & 2 & 1 \end{bmatrix}$ and the **Y** matrix is: $\begin{bmatrix} 8 \\ 7 \\ 12 \end{bmatrix}$ our coefficient matrix would be:

$$(\mathbf{X}^T\mathbf{X}) = \begin{bmatrix} 1 & 2 & 3 \\ 2 & 1 & 2 \\ 1 & 1 & 1 \end{bmatrix}\begin{bmatrix} 1 & 2 & 1 \\ 2 & 1 & 1 \\ 3 & 2 & 1 \end{bmatrix} = \begin{bmatrix} 14 & 10 & 6 \\ 10 & 9 & 5 \\ 6 & 5 & 3 \end{bmatrix}$$

Using technology, we can find the inverse of this matrix:

$$(\mathbf{X}^T\mathbf{X})^{-1} = \begin{bmatrix} 0.5 & 1.7E-13 & -1 \\ 1.9E-13 & 1.5 & -2.5 \\ -1 & -2.5 & 6.5 \end{bmatrix}.$$

Next we can calculate $\mathbf{X}^T\mathbf{Y}$:

$$(\mathbf{X}^T\mathbf{Y}) = \begin{bmatrix} 1 & 2 & 1 \\ 2 & 1 & 1 \\ 3 & 2 & 1 \end{bmatrix}\begin{bmatrix} 8 \\ 7 \\ 12 \end{bmatrix} = \begin{bmatrix} 58 \\ 47 \\ 27 \end{bmatrix}$$

and, finally,

$$\mathbf{M} = (\mathbf{X}^T\mathbf{X})^{-1}(\mathbf{X}^T\mathbf{Y}) = \begin{bmatrix} 2 \\ 3 \\ 0 \end{bmatrix}$$

Thus, the **M** matrix would be $\begin{bmatrix} 2 \\ 3 \\ 0 \end{bmatrix}$. □

Using Matrices to Solve Input-Output Problems*

A small scientific observation post must produce its own electricity and water. Records were kept over several months and the average inputs and outputs of these utilities (thought of as *commodities*) are shown in the table below, where the numbers represent the *value* of the commodities **in dollars**:

	Consumption (Inputs)			
	Electric	Water	Final Demand	Total Output
Production (Outputs)				
Electric	1600	1200	2200	5000
Water	500	130	2000	2630
Labor, Other Costs, Profit, etc.	2900	1300		
Total Input	5000	2630		

The first two rows show the use of the production enterprise's output by both production enterprises (including itself), and by the consumer for final use. For example, the dollar value of the total output of electricity was $5000. Out of this $5000 worth of electricity, $1600 worth of it was used to produce the electricity itself (for example, to run a generator), and $1200 worth of it was used to produce all of the water (for example, to run pumps). The remaining $2200 worth of electricity went to consumers for every *other* use of the electricity (called the "final demand").

The first two columns give the dollar value of what each enterprise used as inputs *from* each enterprise (including itself) as well as from all other inputs (such as labor, machinery, other costs, and profit). For example, of the "revenue" of $5000 (the value of the output) from the electricity, $1600 was spent on electricity (for example, to run a generator), $500 was spent on water (for example, to cool generators), and $2900 was "spent" on labor, machinery, other costs, and profit. It is probably more natural to split out the profit from the costs in the $2900 total, but for the problem at hand, this is not necessary, so we have not done it. Recall that $profit = revenue - cost$, which means that $revenue = cost + profit$. This is why the profit must get added in with the costs, to equal the revenue.

---

* This topic is optional, and can be skipped with no loss of logic or flow later in the book.

If we can determine what the final demands will be in the coming months as the project grows, how do we find what the total output of each commodity must be to meet these demands (assuming the interactions of the enterprises involved remains roughly the same proportionately)? Because the demands are interrelated; the problem is not a simple one. That is, if the final demand for one goes up, then more of the second must be produced to support that extra production, and this in turn requires that more of the first be produced, and so on.

The above problem is another example of the use of matrices and matrix algebra to solve problems in economics. This type of problem involves **input-output matrices**, which were developed by Wassily Leontief of Harvard University. Input-output matrices show the supply and demand interrelationships within given sectors of an economy. In a very simple example, early settlers in this country had to retain some of their crops to use as seed for the next season, so some of the output of the farm sector of the economy was also used as an input for the same farm sector. In the same way, industries today often use some of their own output as well as the output of other industries in their production process: an electric company uses electricity and water for cooling, while the water company uses its own water as well as electricity, as in our sample problem. These are internal demands of the economic system. There are also other production factors, such as labor costs, machinery, other costs, and profits, sometimes referred to as **primary inputs**. Obviously, these industries must also produce output to satisfy the consumers for whom they were created. These external demands are called **final demands**. Leontief developed a model of the U.S. economy that included more than 60 industries. The matrix system that Leontief set up shows the values of the outputs of each industry that are used as input by the other industries and the output of that industry available for final use by consumers, hence the name "input-output".

Let's go back to our problem above, which is a very simple example of an input-output model.

Sample Problem 6: Given the above production and consumption information, suppose it has been determined that the total needs for consumers (final demand) in the coming month will be $2400 worth of electricity and $2300 worth of water. How much electricity and water (total dollars' worth) must be produced to fill these final demands?

Solution: If we define E to be the dollar value of the total output of electricity and W to be the dollar value of the total output of water, then the total output for each commodity will be equal to the amount used by *it*, plus the amount used by the *other* commodity, plus the final demand. For example, for electricity, we have:

E = (% used for elec.) (E) + (% used for water) (W) + (final demand) ,

Our given historical data show that, in order to produce \$5000 worth of electricity (output), \$1600 worth of electricity was needed (as input). This represents

$$\frac{\$1600}{\$5000} = 0.32$$

(32%) of its own output. Similarly, to make \$2630 worth of water (output), \$1200 worth of electricity was needed (as input), which represents

$$\frac{\$1200}{\$2630} \approx 0.4563$$

(45.6%) of the dollar value of the output of water. The dollar value of the final demand is \$2400, so our equation becomes:

$$E = 0.32E + 0.4563W + 2400$$

In the same way, making \$5000 worth of electricity required \$500 worth of water, which is $500/5000 = 10\% = 0.10$ of the dollar value of the total output of electricity. Similarly, making \$2630 worth of water used \$130 worth of water, which is $130/2630 \approx 4.943\% = 0.04943$. The final demand for water is for \$2300 worth, so we get an analogous second equation:

$$W = 0.1E + 0.04943W + 2300$$

Writing this as a matrix equation we get:

$$\begin{bmatrix} E \\ W \end{bmatrix} = \begin{bmatrix} 0.32 & 0.4563 \\ 0.1 & 0.04943 \end{bmatrix} \begin{bmatrix} E \\ W \end{bmatrix} + \begin{bmatrix} 2400 \\ 2300 \end{bmatrix}$$

If we designate the variable matrix (the column vector/matrix with E on top and W on bottom) as **X**, the coefficient matrix (the percentages) as **A** and the constant matrix (the new final demands) as **C,** the matrix equation becomes:

$$\mathbf{X} = \mathbf{AX} + \mathbf{C}$$

To solve for **X,** the variable matrix, we must first get all the variables (the **X** matrices) on one side of the equation:

$$\mathbf{X} - \mathbf{AX} = \mathbf{C}$$

Factoring out the **X** from the terms on the left-hand side gives us:

$$(\mathbf{I} - \mathbf{A})\mathbf{X} = \mathbf{C}$$

Note the order of the factorization: **(I - A)X = IX - AX** which simplifies to **X - AX**, which is what we started with. If we had written the **X** on the left, we would have had **X (I -A)**. Multiplying this out, we get **XI - XA = X - XA**, which is not guaranteed to equal what we started with. We have to be very careful working with matrices because the order of multiplication does matter.

To clear the **(I - A)** from the left-hand side we must multiply both sides of the equation by the inverse of **(I - A)** if it exists (**on the *left* on *both* sides**):

$$(\mathbf{I} - \mathbf{A})^{-1} (\mathbf{I} - \mathbf{A})\mathbf{X} = (\mathbf{I} - \mathbf{A})^{-1} \mathbf{C}$$

By definition, any matrix times its inverse is the identity matrix **I**, and the identity matrix **I** times any matrix is that matrix, so this reduces to:

$$\mathbf{IX} = (\mathbf{I}\text{-}\mathbf{A})^{-1} \mathbf{C}$$
$$\mathbf{X} = (\mathbf{I} - \mathbf{A})^{-1} \mathbf{C}$$

This is a general solution for any problem in this form. Small input-output problems such as the one in this problem can be solved by simplifying the original system of equations, combining like terms, and then solving the resulting system by hand using matrices or any of the other methods we have discussed. This is not practical for large input-output problems. The results above are a generalization, or formula, for that process.

In our example the various matrices are given by

$$A = \begin{bmatrix} 0.32 & 0.4563 \\ 0.1 & 0.04943 \end{bmatrix}, \quad X = \begin{bmatrix} E \\ W \end{bmatrix}, \quad \text{and} \quad C = \begin{bmatrix} 2400 \\ 2300 \end{bmatrix}, \text{ so we can calculate}$$

$$I - A = \begin{bmatrix} 1-.32 & 0-.4563 \\ 0-.1 & 1-.04943 \end{bmatrix} = \begin{bmatrix} .68 & -.4563 \\ -.1 & .95057 \end{bmatrix} \quad \text{and}$$

$$(I - A)^{-1} \approx \begin{bmatrix} 1.582 & 0.7595 \\ 0.1665 & 1.132 \end{bmatrix}$$ (These number have been rounded), so

$$X = (I - A)^{-1} C \approx \begin{bmatrix} 5544 \\ 3003 \end{bmatrix}$$

Translation: The scientific observation post must produce about $5544 worth of electricity and $3003 worth of water to satisfy their projected needs. □

Using Matrices for Encryption*

Privacy and security of information has always been important, but with the ability and need to transmit information electronically this has become increasingly important. Is there a fairly easy way to encrypt (code) messages so that they are easy for the recipient to decode but very difficult for anyone else to decipher?

An interesting application of matrices is in the encoding and decoding of messages. The letters of the alphabet can be assigned the corresponding numbers from 1 through 26, and 27 can correspond to a blank space between words as shown in Table 7.

Table 7

A	B	C	D	E	F	G	H	I	J	K	L	M	N	O	P	Q	R	S	T	U	V	W	X	Y	Z	
1	2	3	4	5	6	7	8	9	10	11	12	13	14	15	16	17	18	19	20	21	22	23	24	25	26	27

For example, the message: "THE ANSWER IS YES" is encoded by [1]

20, 8, 5, 27, 1, 14, 19, 23, 5, 18, 27, 9, 19, 27, 25, 5, 19.

A code involving such a direct "translation" of each letter is very simple to break. Letter patterns and frequency of occurrence are good clues. However, if the numbers are multiplied by a secret matrix, this direct method will not work. For our example, we will use the coding matrix:

$$\mathbf{A} = \begin{bmatrix} 3 & -2 \\ 1 & 7 \end{bmatrix}$$

---

* This topic is optional, and can be skipped with no loss of logic or flow later in the book.

[1] Note: we are not encoding any punctuation marks here, and are assuming all letters are upper case. If it were necessary to include other characters, each would be assigned a number, much like the ASCII code system for computers.

We now separate our message into groups of two numbers each, and form 2 x 1 matrices (we will call each one an **X** matrix):

$$\begin{bmatrix} 20 \\ 8 \end{bmatrix} \begin{bmatrix} 5 \\ 27 \end{bmatrix} \begin{bmatrix} 1 \\ 14 \end{bmatrix} \begin{bmatrix} 19 \\ 23 \end{bmatrix} \begin{bmatrix} 5 \\ 18 \end{bmatrix} \begin{bmatrix} 27 \\ 9 \end{bmatrix} \begin{bmatrix} 19 \\ 27 \end{bmatrix} \begin{bmatrix} 25 \\ 5 \end{bmatrix} \begin{bmatrix} 19 \\ 27 \end{bmatrix}$$

(Notice that we have added a space, coded 27, at the end, since the total number of code numbers here must be *even,* to be able to pair them up.)

Now we multiply each of these matrices by the encoding matrix to get a new set of code numbers, **AX** = **B**. For example, for the first matrix,

$$AX = \begin{bmatrix} 3 & -2 \\ 1 & 7 \end{bmatrix} \begin{bmatrix} 20 \\ 8 \end{bmatrix} = \begin{bmatrix} 44 \\ 76 \end{bmatrix} = B$$

The multiplications for *all* of the **X** matrices yield the following matrices (the set of **B** matrices):

$$\begin{bmatrix} 44 \\ 76 \end{bmatrix} \begin{bmatrix} -39 \\ 194 \end{bmatrix} \begin{bmatrix} -25 \\ 99 \end{bmatrix} \begin{bmatrix} 111 \\ 180 \end{bmatrix} \begin{bmatrix} -21 \\ 131 \end{bmatrix} \begin{bmatrix} 63 \\ 90 \end{bmatrix} \begin{bmatrix} 3 \\ 208 \end{bmatrix} \begin{bmatrix} 65 \\ 60 \end{bmatrix} \begin{bmatrix} 3 \\ 208 \end{bmatrix}$$

Thus, the encoded message is:

44, 76, -39, 194, -25, 99, 111, 180, -21, 131, 63, 90, 3, 208, 65, 60, 3, 208.

In order to decode the message, we must reverse the process. We know that we have set up the system so that $\mathbf{AX} = \mathbf{B}$. If we know $\mathbf{B}$ and want to find $\mathbf{X}$, we can solve this equation for $\mathbf{X}$ to get $\mathbf{X} = \mathbf{A}^{-1}\mathbf{B}$. Thus to decode the message, we need to find the inverse of the $\mathbf{A}$ matrix:

$$\mathbf{A}^{-1} \approx \begin{bmatrix} .3043 & .08696 \\ .04348 & .1304 \end{bmatrix}$$

Next, the numbers must be written as a series of 2 × 1 matrices (the **B** matrices) as above. Then we carry out the multiplications, $\mathbf{A}^{-1}\mathbf{B}$, which returns us to the original numerical representation of the alphabet and the message, which in turn can be easily translated back into letters.

Without knowing either the original encoding matrix or its inverse, it is almost

impossible to break this kind of code. The same letter does not have to yield the same code. For example, ABBA in initial numerical form is 1, 2, 2, 1, and when encoded using the same encoding matrix given above, results in: -1, 15, 4, 9. So in this case, the letter A gets encoded as both a -1 and a 9, and B gets encoded as both a 15 and a 4.
□

Section Summary

Before you start the exercises be sure that you:

- Know that to set up a matrix equation for a system of linear equations: the coefficient matrix is matrix **A**, the constant matrix (of right-hand side values) is column matrix **B**, and the variable matrix is column matrix **X**.
- Know that the matrix equation for a system of linear equations is $\mathbf{AX} = \mathbf{B}$ (and why this works).
- Recognize that the multiplicative inverse of a matrix is denoted by a superscripted -1 ($^{-1}$) after the matrix expression.
- Are able to find the multiplicative inverse of a matrix, both by hand for small matrices and using one or more technology.
- Know that the solution to a system of linear equation represented by the matrix equation $\mathbf{AX} = \mathbf{B}$ is found by multiplying both sides by the multiplicative inverse of **A** to solve for **X**: $\mathbf{X} = \mathbf{A}^{-1}\mathbf{B}$.
- Understand why the matrix equation used in least squares regression,

  $(\mathbf{X}^T\mathbf{X})\,\mathbf{M} = \mathbf{X}^T\mathbf{Y}$, has a solution given by $\mathbf{M} = (\mathbf{X}^T\mathbf{X})^{-1}\,(\mathbf{X}^T\mathbf{Y})$.
- *Recognize that, since $\mathbf{X} = \mathbf{AX} + \mathbf{C}$, so $(\mathbf{I} - \mathbf{A})\mathbf{X} = \mathbf{C}$, **the** general solution for an input-output problem is given by $\mathbf{X} = (\mathbf{I} - \mathbf{A})^{-1}\,\mathbf{C}$, where **X** is the total output matrix, **A** is the input-output percentage matrix and **C** is the final demand matrix.
- *Know that a matrix can be used to encrypt a message once it has been translated into numbers (by grouping the numbers into column matrices of equal size, then left-multiplying each of them by the encoding matrix), and that the inverse of that encoding matrix must be used in a similar way in order to recover the original message.

---

* Relates to an optional topic.

EXERCISES FOR SECTION 7.4

Warm Up

*For Exercises 1-2, write out what the A, X, and **B** matrices are, and the general form of the matrix equation, then show the **hand calculations** to find A^{-1} and the solutions:*

1. a) $3x+y=7$, $2x+2y=-2$ b) $3x+y=10$, $2x+2y=5$ c) $3x+y=-3$, $2x+2y=12$

2. a) $5p+2q=36$, $8p-3q=-54$ b) $5p+2q=24$, $8p-3q=11$ c) $5p+2q=0$, $8p-3q=-5$

*For Exercises 3-5, use A^{-1} and **technology** to find the solutions:*

3. a) $4p+12q=6$, $2p+6q=3$ b) $4p+12q=0$, $2p+6q=1$ c) $4p+12q=0$, $2p+6q=0$

4. a) $2x+y+6z=3$, $x-y+4z=1$, $3x+2y-2z=2$ b) $2x+y+6z=1$, $x-y+4z=2$, $3x+2y-2z=3$ c) $2x+y+6z=5$, $x-y+4z=-2$, $3x+2y-2z=8$

5. a) $5x-7y+4z=2$, $3x+2y-2z=3$, $2x-y+3z=4$ b) $5x-7y+4z=4$, $3x+2y-2z=3$, $2x-y+3z=2$ c) $5x-7y+4z=1$, $3x+2y-2z=1$, $2x-y+3z=1$

Game Time

6. You have been considering closing down the second site for constructing your hand crafted chairs and tables. It would cost you money to return the unused wood and hardware to your original sites. You have 84 board feet of wood on hand and 63 packages of hardware. Can you actually use up all of the materials, and if so, how many chairs and tables should you make to do this? You know that chairs use 3 board feet of wood and 8 packages of hardware while tables use 6 board feet of wood and 4 packages of hardware.

7. You have been considering closing down the second site for constructing your hand crafted chairs and tables. It would cost you money to return the unused wood and

hardware to your original sites. You have 120 board feet of wood on hand and 100 packages of hardware. Can you actually use up all of the materials, and if so, how many chairs and tables should you make to do this? You know that chairs use 3 board feet of wood and 8 packages of hardware while tables use 6 board feet of wood and 4 packages of hardware.

8. Discuss the advantages of using matrix equations to solve systems of equations in which the coefficient matrices remain the same, but the constant matrices change.

9. A trucking company owns three types of trucks, quarter-ton, half-ton, and one-ton. The company regularly hauls three different types of machinery: condensers, generators, and transformers. Because of the configuration and weight restraints of the trucks, they have the following capacities: the quarter-ton truck can carry 10 condensers and 20 transformers per load, but cannot carry any generators; the half-ton truck can carry 10 of each per load; and the one-ton can carry 10 condensers, 20 generators, and 10 transformers per load. How many trucks of each type should be sent to haul exactly 150 condensers, 160 generators, and 190 transformers?

10. A trucking company owns three types of trucks, quarter-ton, half-ton, and one-ton. The company regularly hauls three different types of machinery: condensers, generators, and transformers. Because of the configuration and weight restraints of the trucks they have the following capacities: the quarter-ton truck can carry 10 condensers and 20 transformers per load, but cannot carry any generators; the half-ton truck can carry 10 of each per load; and the one-ton can carry 10 condensers, 20 generators, and 10 transformers per load. How many trucks of each type should be sent to haul 120 condensers, 140 generators, and 150 transformers?

11. Discuss the advantage of using matrices to set up the data for problems 9 and 10 if the shipping requirements change on a daily basis.

12. a) Use the matrix $\mathbf{A} = \begin{bmatrix} 2 & -5 \\ 7 & 37 \end{bmatrix}$ to encode "GO AWAY"

    b) Check your answer by decoding the message using the inverse.

13. a) Use the matrix $\mathbf{A} = \begin{bmatrix} 3 & -2 \\ 1 & 7 \end{bmatrix}$ to encode "NO WAY"

    b) Check your answer by decoding the message using the inverse.

14. A company has found that its in-house services, catering and cleaning, consistently use each others' services. The data is shown in the table below:

	Services Performed By	
Services Used by	Catering	Cleaning
Catering S1	0	.164
Cleaning S2	.25	.291
Producing Dept. P1	.50	.218
Producing Dept. P2	.25	.327
	1.00	1.000
Costs to be allocated	\$9313	\$3982

The accounting department needs to know how to allocate the costs of the "service departments" to the producing departments when the service departments render services to one another.

Let matrix **A** represent the final cost allocation *(producing department × service department)* matrix, where the entries represent the fraction of each service department's cost allocable to each production department. Let **S** be the *(service department × service department)* matrix representing the internal use of the service departments (the top 2×2 matrix above) and **P** be the *(producing department × service department)* matrix representing the production departments' fractional use of the service departments (the second 2×2 matrix above). Then **A = P + AS**. This is similar, but not identical, to the input-output problem. Be careful! Your solution, the allocation matrix (**A**), is a 2×2 matrix, and the entries in each column of this matrix must add up to 1.

Section 7.5: Markov[1] Chains[*]

There are situations when we are interested in tracking some phenomenon over time at discrete moments (daily, weekly, etc.) where the phenomenon can only be in one of a discrete set of states at any given time. Furthermore, given that you start in a certain state at a certain time, you don't know exactly where you will end up after one period of time, but you can assess the probability of moving to each of the other possible states, and these **transition probabilities** do not change over time. In such a situation, given information about the state at a particular time, matrices can be used to calculate the probability of being in any of the possible states any number of periods into the future. They can also be used to calculate **steady-state probabilities** for each state (the likelihood of being in each state in the long run), if they exist.

Since the above description is particularly abstract, let's see some examples of situations like this:

- A student's semester Grade Point Average (GPA) can be categorized as being A, B, C, D, or F. From past records at a particular school, transition probabilities over consecutive semesters between each possible pair of grades were calculated. If your GPA this semester is a B, what are the chances it will be an A after three additional semesters? If you become a "permanent student," what is the probability your semester GPA will be an A in the long run?
- Suppose we consider the change in a stock market index on a given day compared to the previous trading day. We call the index "the same" if the change is 0.5% or less from the day before (in either direction), "up" if it increases more than 0.5%, and "down" if it decreases more than 0.5% . You have tracked the performance over the last few months and calculated transition probabilities between the states over consecutive trading days. If the index was "up" yesterday, what are the chances it will be "up" tomorrow? What is the long-term probability it will be "up"?
- You work for a major diet cola manufacturer, and are doing market research. By picking at random some individuals who drink diet cola regularly, you determine on a monthly basis over a period of time whether the individual has been faithful to your brand over the previous month, faithful to your major competitor, or neither (either switching or drinking something else). You then estimate transition probabilities between these three states over consecutive months. If you know the rough proportions of consumers in each state at the present time, what are they

[1] This process is named for Andrei Markov (1856-1922) a Russian mathematician who is credited with introducing the study of this type of situation.

[*] This section is optional, and can be skipped without interrupting the logic or flow of the remainder of the text.

likely to be 6 months from now? If the environment does not change for a while, what proportion of consumers would you expect to be loyal to your brand in the long run?

In addition to being able to solve problems like those above, by the end of this section, you should:

- Understand the concepts of state, period, transition probability, and probability vectors in the context of discrete Markov Chains.
- Know how to find a probability vector after some number of periods, given an initial probability vector and the matrix of transition probabilities, using matrix calculations.
- Know how to find a steady-state probability vector, if one exists, given an initial probability vector and a transition matrix.

Sequential Probability Calculations

Sample Problem 1: Suppose you are part of a special Honors program at your school. Each semester, as part of this program, you take a special course, for which there are only 2 possible grades: Excellent (E) or Satisfactory (S). One of the administrators of the program is a mathematician. She calculated that, of the students who got an E in the course in a given semester, 75% got an E in the next semester as well, and of the students who got an S in the course in a given semester, 50% got an S in the next semester as well. There are 10 people from your class in this program. Last semester, 6 of them got an S in the course, and 4 got an E. If the pattern from the past continues, how many of your classmates would you expect to get an S this semester, and how many an E?

Solution: Let's use our intuition here. If 6 people got an S last semester, the information says we can expect 50% of them, or

$$(0.5)(6) = 3$$

students to again get an S this semester. What about the 4 people who got an E last semester? We expect 75% of them to get an E again, which means that

$$1 - 0.75 = 0.25$$

or 25% of them should get a S. Thus we expect 25% of 4, or

$$(0.25)(4) = 1$$

student to go down from an E to a S. Since students could only get an S or an E last semester (assuming no students drop out or get added), the total number expected to get an S this semester would be

$$(6)(0.5) + (4)(0.25) = 3 + 1 = 4$$

In a sense, we have already done the calculation for the number that we expect to get an E this semester, since there are 10 students, only the two possible grades, and we expect 4 of them to get a S. This means we would expect

$$10 - 4 = 6$$

students to get an E this semester. Notice that we could have also done a calculation similar to that for S as follows:

$$(6)(1\text{-}0.5) + (4)(0.75) = 6(0.5) + 3 = 3 + 3 = 6. \ \square$$

Suppose that, instead of looking at the *number* of students getting each grad, we considered the *fraction* or *proportion* of students in the class. Then instead of saying that 6 students got an S last semester, we would have said that 0.6 (or 60%) of the students got an S last semester, and that 0.4 got an E last semester. The expected fractions for this semester would have then been that 0.4 would get an S and 0.6 would get an E. These fractions could also be interpreted as **probabilities** that a student picked at random from the group would get each grade. The calculations would have been very similar to before:

$$(0.6)(0.5) + (0.4)(0.25) = 0.3 + 0.1 = 0.4$$
$$(0.6)(0.5) + (0.4)(0.75) = 0.3 + 0.4 = 0.6$$

You may have already noticed that these calculations look suspiciously like the results of a matrix multiplication. If so, you're right!

Discovery Question 1: What matrix product would result in the calculations above?

[See if you can find the answer without reading ahead.]

Hint: Try a $1{\times}2$ matrix on the left and a $2{\times}2$ matrix on the right.

[Again, see if you can get the answer on your own, if you haven't yet.]

Answer: One product that works is

$$\begin{bmatrix}0.6 & 0.4\end{bmatrix}\begin{bmatrix}0.5 & 0.5\\0.25 & 0.75\end{bmatrix} = \begin{bmatrix}(0.6)(0.5)+(0.4)(0.25) & (0.6)(0.5)+(0.4)(0.75)\end{bmatrix}$$

$$= \begin{bmatrix}0.3+0.1 & 0.3+0.3\end{bmatrix}$$

$$= \begin{bmatrix}0.4 & 0.6\end{bmatrix}$$

Let's define some notation to help understand what is going on. Let's identify our possible states sequentially in general as 1,2,…,*n*. In our example 1 could correspond to a grade of S and 2 could correspond to a grade of E. Let's define

p_{ij} = the probability of making the transition from state i at any point in time to state j one period later.

In our example, the probability of going from E to S was 0.25, so

$$p_{21} = 0.25$$

(we could also call this p_{ES} to remind us of the states involved). Notice that 0.25 was the entry in the 2nd row, 1st column (the (2,1) entry) of the 2×2 matrix above. In fact that matrix is exactly the matrix of transition probabilities, as defined by p_{ij} above. Notationally, let's call that matrix **P**, sometimes called the **transition probability matrix**. Thus, in our example,

$$\mathbf{P} = \begin{bmatrix}0.5 & 0.5\\0.25 & 0.75\end{bmatrix}$$

Notice that **the rows of the transition matrix add up to one** (100%). This will be true as long as all possible states are included, since when you start from a particular state (i), you've got to go *somewhere*.

What about the 1×2 row matrix [0.6 0.4] ? That was the initial probability (or proportion) row matrix (or row vector). Let's define

$\pi_i(t)$ = the probability of being in state i at time t (t periods after the initial period)

Thus in our example, we would have $\pi_1(0) = 0.6$ and $\pi_2(0) = 0.4$. In fact we could also say that $\pi_1(1) = 0.4$ and $\pi_2(1) = 0.6$. To express this as a row vector, let's call it $\pi(t)$:

$$\pi(t) = [\, \pi_1(t) \;\; \pi_2(t) \;\; \ldots \;\; \pi_n(t) \,]$$

In our example, $\pi(0) = [0.6 \; 0.4]$ and $\pi(1) = [0.4 \; 0.6]$.

How did we calculate the probability of being in state 1 after 1 period? It was

P{ending in 1} =
P{starting in 1}P{moving from 1 to 1} + P{starting in 2}P{moving from 2 to 1}
$= \pi_1(0)p_{11} + \pi_2(0)p_{21}$
$= (0.6)(0.25) + (0.4)(0.25)$

In other words, to find the probability of being in a given state, we multiply each possible state from the period before, multiply by the probability of moving from that state to the given state, and adding the results. This is true not only in going from time 0 to time 1, but from time t to time $t+1$, so we can write in general that

$$\pi_j(t+1) = \pi_1(t)p_{1j} + \pi_2(t)p_{2j} + \cdots + \pi_n(t)p_{nj}$$

Notice that this is just the previous probability vector times column j of the transition matrix, so all of these calculations can be combined by

$$\pi(t+1) = \pi(t)\mathbf{P}$$

In other words, **to get the probability vector at a particular time, you can simply multiply the probability vector from the period before times the transition matrix.**

Probability Calculations Using Matrix Powers

Sample Problem 2: Suppose you have a summer baby-sitting job taking care of a 4-year-old child every afternoon. The child goes to a program every morning, but if it rains, the program gets canceled, and you have agreed to be with the child on any such mornings. For the last couple of weeks, you have kept track of the weather on consecutive mornings, categorizing each morning as "rainy" (meaning the program was canceled, which we will denote R) or "dry" (meaning the program was not canceled and you had a free morning, which we will denote D). The data from these days, in order, was: R,D,D,R,R,D,D,D,R,D . Today is Monday, and the program was canceled. You are talking with a friend about doing something together on Wednesday morning, or possibly Friday. What are the chances you will be free? In the long run, what are your chances of having a free morning?

Solution: Before we start, we should point out that if we jump to apply the theory of Markov Chains here, we are making the assumptions that go with that model. In particular, we are saying that **the probability of being in a certain state depends only on the probabilities of being in each state one period earlier**. This means that there can be no kind of more subtle time lag or momentum effect. In advanced statistics, you could learn how to test this assumption. For now, we will just say that the assumption seems reasonable as a first attempt at a model. Let's call state 1 R (rainy) and state 2 D (dry).

We know the initial probability vector, since today (the initial period) it was rainy. This means $\pi_1(0) = \pi_R(0) = 1$, and so $\pi_2(0) = \pi_D(0) = 0$. Thus,

$$\pi(0) = [\,1 \quad 0\,]$$

How can we estimate the transition probabilities? If you look at the weather data, there were 4 mornings in state R, and for each of them you also know what the state the next morning was. 1 of the R days were followed by another R day, so we can estimate the probability of a transition from R to R (1 to 1) as

$$p_{11} = p_{RR} = \tfrac{1}{4} = 0.25\,.$$

Similarly, 3 of the 4 R days were followed by D days, so

$$p_{12} = p_{RD} = \tfrac{3}{4} = 0.75$$

which could also have been calculated as $1 - 0.25$, since the first (and every) row of transition probabilities must sum to 1.

If we look at transitions from state D, we can make similar estimates, but we cannot use the *last* D observation, since we don't know what came after it. This gives us 5 observations of D days where we know what happened the next day. Analogous to before we can find the other transition probabilities:

$$p_{21} = p_{DR} = 2/5 = 0.40$$
$$p_{22} = p_{DD} = 1 - 0.40 = 0.60$$

(or the latter could have been calculated directly as 3/5). Thus our transition matrix is

$$\mathbf{P} = \begin{bmatrix} 0.25 & 0.75 \\ 0.4 & 0.6 \end{bmatrix}$$

From our earlier discussion, we know that

$$\pi(t+1) = \pi(t)\mathbf{P}$$

Since we know $\pi(0)$, we can use this relation to find $\pi(1)$. Notice that we can then use the *same* relationship to start with $\pi(1)$ and find $\pi(2)$! Now that we know this is just a simple matrix multiplication calculation, we can use a graphing calculator or spreadsheet to do the calculations very quickly, either just displaying the result in passing (possibly writing it down on paper) or storing it to be able to see it again later. We will show the first calculation here by hand, but just give the answer for the second calculation:

$$\begin{aligned}\pi(1) = \pi(0)\mathbf{P} = \begin{bmatrix}1 & 0\end{bmatrix}\begin{bmatrix}0.25 & 0.75\\ 0.4 & 0.6\end{bmatrix} &= \begin{bmatrix}(1)(0.25)+(0)(0.4) & (1)(0.75)+(0)(0.6)\end{bmatrix}\\ &= \begin{bmatrix}0.25+0 & 0.75+0\end{bmatrix}\\ &= \begin{bmatrix}0.25 & 0.75\end{bmatrix}\end{aligned}$$

$$\pi(2) = \pi(1)\mathbf{P} = \begin{bmatrix}0.25 & 0.75\end{bmatrix}\begin{bmatrix}0.25 & 0.75\\ 0.4 & 0.6\end{bmatrix} = \begin{bmatrix}0.3625 & 0.6375\end{bmatrix}$$

So $\pi_1(2) = \pi_R(2) = 0.3625$ and $\pi_2(2) = \pi_D(2) = 0.6375$.
Translation: The chances of your being free on Wednesday morning are about 64%, or roughly a 2/3 chance.

Sample Problem 2 also asked about the probability of your being free in the morning on Friday. How can we answer that? The most obvious way would be to keep calculating the probability vectors for $t = 3$ and 4. Before we do that, let's see if there is another way to make the necessary calculations. Can we express a general future probability vector $\pi(t)$ in terms of the initial information, as represented by $\pi(0)$ and $\mathbf{P}$? We know that

$$\pi(t+1) = \pi(t)\mathbf{P} \text{ and } \pi(1) = \pi(0)\mathbf{P} \text{ and } \pi(2) = \pi(1)\mathbf{P}$$

We have $\pi(2)$ in terms of $\pi(1)$, and $\pi(1)$ in terms of $\pi(0)$, so we should be able to get $\pi(2)$ in terms of $\pi(0)$ by substitution:

$$\pi(2) = \pi(1)\mathbf{P} = [\pi(0)\mathbf{P}]\mathbf{P} = \pi(0)\mathbf{P}^2 .$$

We could then use the same idea for $\pi(3)$:

$$\pi(3) = \pi(2)\mathbf{P} = [\pi(0)\mathbf{P}^2]\mathbf{P} = \pi(0)\mathbf{P}^3 .$$

You have probably already noticed the pattern here - what would be the general expression for $\pi(t)$? Since the exponent of $\mathbf{P}$ is always the same as the number of periods, we get

$$\pi(t) = \pi(0)\mathbf{P}^t$$

This formula would be most helpful if you have a technology that calculates powers of square matrices easily, as do many graphing calculators. Let's try it for finding your probability of being free on Friday, which corresponds to $t = 4$ (4 days after the initial day, Monday). Using a graphing calculator, we find that

$$\pi(4) = \pi(0)\mathbf{P}^4 \approx [\,0.3482 \quad 0.6518\,]$$

(Of course this could have also been calculated by sequentially finding $\pi(3)$ and then $\pi(4)$, as we did before.)
Translation: Your chance of being free on Friday is about 65%, again about a 2/3 chance (slightly higher probability than for Wednesday).

Finding Steady-State Probabilities Sequentially Using Limits

This suggests that we may be getting close to a long-run probability of being free. Let's just calculate the probability vectors for $t = 1,2,3,4, \ldots$, and see if they converge to some limit, as we have seen for limits of secant slopes, limits of sums for subinterval area estimates, and improper integrals. We could use either method of calculation discussed above, and any technology. Rounding off to 4 significant digits, we get:

Table 1

t	$\pi_1(t) = \pi_R(t)$	$\pi_2(t) = \pi_D(t)$
0	1	0
1	0.25	0.75
2	0.3625	0.6375
3	0.3456	0.6544
4	0.3482	0.6518
5	0.3478	0.6522
6	0.3478	0.6522

Notice that the probabilities of R change both up and down as they converge to the limit of 0.3478, and the complementary probabilities for D do the same to converge to 0.6522

Translation: The long-run probability of your being free is about 65%, again roughly a 2/3 chance. Given the margin of error from the data (a very small sample of days), the most realistic answer would probably be something like

"My long-run probability of being free is about 2/3, give or take 10%." □

Notice that in this case, the probability after only 2 periods was already very close to the long-run probability. But what if the convergence had been very slow? Is there any other way to find these long-run probabilities?

Finding Steady-State Probabilities Using Matrix Theory

In the theory of Markov Chains, these long-run probabilities are called **steady-state probabilities**. One way of thinking of these probabilities is that they are the values of a probability vector we could call π satisfying the matrix equation

$$\pi = \pi\mathbf{P}$$

since we essentially have a situation where $\pi(t+1) = \pi(t)$ (the next probability vector is the same as the previous one), and this common value is what we are calling the vector π

Let's write out in more detail what this matrix equation corresponds to for our example:

$$\begin{aligned}
\pi &= \pi\mathbf{P} \\
\begin{bmatrix}\pi_1 & \pi_2\end{bmatrix} &= \begin{bmatrix}\pi_1 & \pi_2\end{bmatrix}\begin{bmatrix}0.25 & 0.75 \\ 0.4 & 0.6\end{bmatrix} \\
\begin{bmatrix}\pi_1 & \pi_2\end{bmatrix} &= \begin{bmatrix}(\pi_1)(0.25)+(\pi_2)(0.4) & (\pi_1)(0.75)+(\pi_2)(0.6)\end{bmatrix} \\
\begin{bmatrix}\pi_1 & \pi_2\end{bmatrix} &= \begin{bmatrix}0.25\pi_1+0.4\pi_2 & 0.75\pi_1+0.6\pi_2\end{bmatrix}
\end{aligned}$$

Now, since matrices being equal means that corresponding elements are equal, we get the following system of equations:

$$\begin{aligned}
\pi_1 &= 0.25\pi_1 + 0.4\pi_2 \\
\pi_2 &= 0.75\pi_1 + 0.6\pi_2
\end{aligned}$$

These could also have been written as

$$\begin{aligned}\pi_R &= 0.25\pi_R + 0.4\pi_D \\ \pi_D &= 0.75\pi_R + 0.6\pi_D\end{aligned}$$

Recall that at the very beginning of this section, we used the logic

$$\begin{aligned}&\text{P\{ending in 1\}} = \\ &\quad \text{P\{starting in 1\}P\{moving from 1 to 1\}} + \text{P\{starting in 2\}P\{moving from 2 to 1\}} \\ &\quad = \pi_1(0)p_{11} + \pi_2(0)p_{21}\end{aligned}$$

Now we are using the same logic and leaving off the time period, since at steady state the ending probability is the same as the starting probability. In general, then we get

$$\pi_1 = p_{11}\pi_1 + p_{21}\pi_2$$

The same logic would hold for state 2, giving us

$$\pi_2 = p_{12}\pi_1 + p_{22}\pi_2$$

Notice that in either case, the coefficients for each equation on the right-hand side come from a *column* of the transition matrix. Thus, in general, for a Markov Chain with n states, the steady state equations would be given by

$$\begin{aligned}\pi_1 &= p_{11}\pi_1 + p_{21}\pi_2 + \ldots + p_{n1}\pi_n \\ \pi_2 &= p_{12}\pi_1 + p_{22}\pi_2 + \ldots + p_{n2}\pi_n \\ &\vdots \\ \pi_n &= p_{1n}\pi_1 + p_{2n}\pi_2 + \ldots + p_{nn}\pi_n\end{aligned}$$

Going back to our example, we are trying to solve the system of equations

$$\begin{aligned}\pi_1 &= 0.25\pi_1 + 0.4\pi_2 \\ \pi_2 &= 0.75\pi_1 + 0.6\pi_2\end{aligned}$$

Notice that this is not yet in standard form, since like terms have not been combined in each equation. When we do this, we get

$$\begin{aligned}0.75\pi_1 - 0.4\pi_2 &= 0 \\ -0.75\pi_1 + 0.4\pi_2 &= 0\end{aligned}$$

Looks odd, doesn't it? One equation is just the negative of the other. Notice that if we added the equations together, we would get $0 = 0$. Or, if we had set up the system as an augmented matrix and done the operation $R_1 + R_2 \rightarrow R_2'$, we would have gotten a bottom row of all 0's. If you recall from Section 7.3, this suggests an infinite number of solutions are possible. In matrix theory, we would say that the two equations are **dependent** (the second does not add any new restrictions on the variables). *But*, we have forgotten one characteristic that we know π_1 and π_2 should have: Since they are probabilities, they should add up to 1! This gives us another equation that π_1 and π_2 should satisfy:

$$\pi_1 + \pi_2 = 1$$

This means we can create a new system of equations, by replacing either of the original 2 equations with this new equation. We will arbitrarily choose the second equation as the one we replace, giving the new system:

$$0.75\pi_1 - 0.4\pi_2 = 0$$
$$\pi_1 + \pi_2 = 1$$

We can now use any of the techniques for solving systems of equations given earlier in this chapter. Let's use the matrix equation method, with

$$\mathbf{A} = \begin{bmatrix} 0.75 & -0.4 \\ 1 & 1 \end{bmatrix}, \ \mathbf{X} = \begin{bmatrix} \pi_1 \\ \pi_2 \end{bmatrix}, \text{ and } \mathbf{B} = \begin{bmatrix} 0 \\ 1 \end{bmatrix}$$

The solution is then

$$\mathbf{X} = \begin{bmatrix} \pi_1 \\ \pi_2 \end{bmatrix} = \mathbf{A}^{-1}\mathbf{B} \approx \begin{bmatrix} 0.347826... \\ 0.652173... \end{bmatrix}$$

Notice that, to 4 significant digits, we get the same answer as before. For most Markov Chains, the two methods will give the same answers[2]. Because this is not guaranteed, it is a good idea to try both. The limit method shows exactly what happens each period, so is the better one to trust if there are any discrepancies, even after verifying your calculations (say, by the sequential matrix multiplication and the matrix power methods).

If we look back at our general system of steady-state equations:

$$\pi_1 = p_{11}\pi_1 + p_{21}\pi_2 + \ldots + p_{n1}\pi_n$$
$$\pi_2 = p_{12}\pi_1 + p_{22}\pi_2 + \ldots + p_{n2}\pi_n$$
$$\vdots$$

[2] To learn more about the conditions under which the methods give the same or different answers, see a book on stochastic processes or Markov Chains, or a general book on Operations Research.

$$\pi_n = p_{1n}\pi_1 + p_{2n}\pi_2 + \ldots + p_{nn}\pi_n$$

you might be able to see that we will always run into the problem of one of the equations being dependent on the others (getting a row of all 0's when trying to solve the system using augmented matrices). If you add the equations all together, notice that on the left-hand side you will get the sum of the probabilities. On the right-hand side, notice that when you combine the like terms for each probability, the coefficient corresponds to the sum of the entries of a *row* of the transition matrix, which we have already observed should always be 1. Thus on the right-hand side, we also get the sum of the probabilities. This is like saying $1 = 1$, and since we could get this by row operations, it implies that one of the equations does not add any new restrictions. Thus in general, we can always remove one of the equations (we will adopt the convention of removing the last one) and replace it with the equation indicating that the sum of the probabilities should be 1. We end up with the following system of equations:

$$\begin{aligned}
\pi_1 &= p_{11}\pi_1 + p_{21}\pi_2 + \ldots + p_{n1}\pi_n \\
\pi_2 &= p_{12}\pi_1 + p_{22}\pi_2 + \ldots + p_{n2}\pi_n \\
&\vdots \\
\pi_{n\text{-}1} &= p_{1n\text{-}1}\pi_1 + p_{2n\text{-}1}\pi_2 + \ldots + p_{nn\text{-}1}\pi_n \\
\pi_1 + \pi_2 &+ \ldots + \pi_n = 1
\end{aligned}$$

This will simplify to the system

$$\begin{aligned}
&(1\text{-}p_{11})\pi_1 - p_{21}\pi_2 - \ldots - p_{n1}\pi_n = 0 \\
&\text{-}p_{12}\pi_1 + (1\text{-}p_{22})\pi_2 - \ldots - p_{n2}\pi_n = 0 \\
&\vdots \\
&\text{-}p_{1n\text{-}1}\pi_1 - p_{2n\text{-}1}\pi_2 - \ldots - (1\text{-}p_{n\text{-}1,\, n\text{-}1})\pi_{n\text{-}1} - p_{nn\text{-}1}\pi_n = 0 \\
&\pi_1 + \pi_2 + \ldots + \pi_n = 1
\end{aligned}$$

Rather than trying to remember the second form, we recommend setting up the first form using the fact that the coefficients on the right-hand side of each equation correspond to a column of the transition matrix and leaving off the last equation, replacing it with the equation indicating that the sum of the probabilities is 1. Then simplify the equations by hand, and use technology to find the final solution.

Calculations Involving Number Vectors instead of Probability Vectors

Sample Problem 3: A study has been made concerning the ways in which a certain group of students' grades change, on average, from the mid-semester grade to the final grade. It has been found that on the average, 72% of students with A's maintain the A, while the other 28% drop down to a B. 14% of the B students raise their grades to an

A, 68% maintain the B, 14% drop to a C and 4% drop all the way to a D. None of the C students got A's but 44% got B's, 43% maintained their C's , 11% dropped to a D and 2% failed. 2 % of the D students raised their grades to a B, 46% raided to a C, 42% stayed at D and 10% dropped to an F. 17% of the students who started with F's raised their grades to a C, 35% went to a D and 48% failed. If 3 students from this target population got A's, 6 received B's, 14 received C's, 5 received D's, and 2 failed at midterm in the fall , what grade distribution could be expected for the final grade for the fall semester? If the same percentages apply to grade transitions from the final grade in the fall to the midterm grade the next spring, and again from midterm spring grades to final spring grades, how many A's, B's and C's should there be at the midterm of the spring semester? At the end of the spring semester? If these students became "permanent students," how many would you expect at each grade level in the long run?[3]

Solution: Here is an example, similar in its context to Sample Problem 1, where data is presented in a prose or written form and is rather difficult to understand. If we present the same information in a tabular form or matrix, it is much easier to read. We will also get an idea of how we can go about answering the questions that are posed.

The starting grades in tabular form are shown in Table 2:

Table 2

A	B	C	D	F
3	6	14	5	2

In matrix form, this would be: $\begin{bmatrix} 3 & 6 & 14 & 5 & 2 \end{bmatrix}$. Notice that the question asks about expected *numbers* of students, rather than probabilities or proportions. **As long as the total number** of students **stays constant,** these are equivalent; **the number vector** (matrix) **is simply the probability or proportion vector times the total number. All of the methods we have discussed still apply, except that the sum of the steady-state values would have to be the total number rather than 1.** Just to keep the distinction clear, let's use the symbol **s** for our student number vector, so

$$\mathbf{s}(0) = \begin{bmatrix} 3 & 6 & 14 & 5 & 2 \end{bmatrix}.$$

The tabular form of the change in grades with percentages written as decimals is shown in Table 3. This gives the transition matrix for this problem.

[3] Data from a faculty study made in 1994.

Table 3

	A	B	C	D	F
A	.72	.28	0	0	0
B	.14	.68	.14	.04	0
C	0	.44	.43	.11	.02
D	0	.02	.46	.42	.10
F	0	0	.17	.35	.48

To determine how many A's there will be at the end of the semester we must multiply the percent of A's that maintain their grades, 72% (or .72) times the original number of A's, 3 and add to that the percent of original B's that raised their grades to A, 14% times the number of B's, 6. No C's, D's or F's got final A's. If we follow the same procedure for the B's, C's, D's and F's we will see again that what is involved is matrix multiplication: the starting grade matrix, a 1×5 (*student* $\times$ *starting grade*) matrix, times the grade change matrix a 5×5 (*starting grade* $\times$ *next grade*) matrix resulting in a 1×5 (*student* $\times$ *next grade*) matrix. This matrix is

$$\mathbf{s}(1) = \mathbf{s}(0)\mathbf{P}$$

$$= \begin{bmatrix} 3 & 6 & 14 & 5 & 2 \end{bmatrix} \begin{bmatrix} 0.72 & 0.28 & 0 & 0 & 0 \\ 0.14 & 0.68 & 0.14 & 0.04 & 0 \\ 0 & 0.44 & 0.43 & 0.11 & 0.02 \\ 0 & 0.02 & 0.46 & 0.42 & 0.10 \\ 0 & 0 & 0.17 & 0.35 & 0.48 \end{bmatrix}$$

$$= [3 \ \ 11.18 \ \ 9.5 \ \ 4.58 \ \ 1.74],$$

meaning that, on average (for example, if you repeated the transitions many times with the same starting grade distribution), we would expect 3 A's, 11.18 B's, 9.5 C's 4.58 D's and 1.74 F's. Obviously, you cannot have fractional numbers of grades in a particular semester, so the numbers only make sense in an average, or expected value, sense.

A major assumption here is that the grade change matrix, the transition matrix, will remain the same, and tells *exactly* how the changes will occur (certainty) on average. As before, and with all Markov Chains, we are also assuming that the values at a particular time period depend *completely* on the values the previous period and the transition matrix, and on nothing else. For example, this means that we are assuming that the transition probabilities are unaffected by grade inflation and by experience in college.

To determine what grades might be expected at the *next* grading period, at spring midterm time (again, assuming that the transition percentages remain constant), this *new* student grade matrix is multiplied by the grade change matrix. The resulting matrix is

$$\mathbf{s}(2) = \mathbf{s}(1)\mathbf{P} = \mathbf{s}(0)\mathbf{P}^2 = [\,3.7252 \ \ 12.714 \ \ 8.0528 \ \ 4.0248 \ \ 1.4832\,],$$

meaning that, on average (if you repeated the transitions many times with the same starting grade distribution), we would expect 3.7252 A's, 12.714 B's, 8.0528 C's 4.0248 D's and 1.4832 F's. Again, the fractional numbers of grades make sense only in an average, or expected value, sense.

At spring semester final grade time, we would expect

$$\mathbf{s}(3) = \mathbf{s}(2)\mathbf{P} = \mathbf{s}(0)\mathbf{P}^3 \approx [\,4.46\ \ 13.31\ \ 7.35\ \ 3.60\ \ 1.28\,]$$

Translation: For final grades this coming spring, we would expect, on average, about 4.46 A's, 13.31 B's, 7.35 C's, 3.60 D's, and 1.28 F's. Notice that the high end grades are growing and the low end grades are decreasing.

What about in the long run? If we keep up the above process, we get

Table 3

t	$s_A(t)$	$s_B(t)$	$s_C(t)$	$s_D(t)$	$s_F(t)$
0	3	6	14	5	2
1	3.73	12.71	8.05	4.02	1.48
2	4.46	13.31	7.35	3.60	1.28
3	5.08	13.61	6.90	3.30	1.12
4	5.56	13.77	6.58	3.08	1.01
5	5.93	13.88	6.35	2.92	0.92
6	6.21	13.95	6.17	2.80	0.86

These seem to be converging, but *very* slowly! Let's try the steady-state equation method. The initial system of equations, using what looks like the transpose of the transition matrix is

$$\begin{aligned}
s_1 &= 0.72s_1 + 0.14s_2 + 0s_3 + 0s_4 + 0s_5 \\
s_2 &= 0.28s_1 + 0.68s_2 + 0.44s_3 + 0.02s_4 + 0s_5 \\
s_3 &= 0s_1 + 0.14s_2 + 0.43s_3 + 0.46s_4 + 0.17s_5 \\
s_4 &= 0s_1 + 0.04s_2 + 0.11s_3 + 0.42s_4 + 0.35s_5 \\
s_5 &= 0s_1 + 0s_2 + 0.02s_3 + 0.10s_4 + 0.48s_5
\end{aligned}$$

If we remove the last equation, simplify, and add the equation that the sum should be 30, we get

$$
\begin{aligned}
0.28s_1 - 0.14s_2 - 0s_3 - 0s_4 - 0s_5 &= 0 \\
-0.28s_1 + 0.32s_2 - 0.44s_3 - 0.02s_4 - 0s_5 &= 0 \\
-0s_1 - 0.14s_2 + 0.57s_3 - 0.46s_4 - 0.17s_5 &= 0 \\
-0s_1 - 0.04s_2 - 0.11s_3 + 0.58s_4 - 0.35s_5 &= 0 \\
s_1 + s_2 + s_3 + s_4 + s_5 &= 30
\end{aligned}
$$

Remember that we are working with numbers rather than probabilities here, so the sum is 30 (the total number of students, which we assume stays constant) rather than 1. We could have left off the 0 terms completely, but with them in, you can compare this to the simplified form of the system we wrote out in general above. Notice that there should be nonnegative coefficients down the diagonal from the upper left to lower right, with nonpositive coefficients everywhere else except the bottom equation, where the coefficients are all 1.

From matrix theory, we know that we can use $\mathbf{A}^{-1}\mathbf{B}$ to get the solution to this system, and the solution comes out to be

$$\mathbf{s} = [\,7.06 \;\; 14.12 \;\; 5.66 \;\; 2.47 \;\; 0.69\,]$$

Translation: In the long run, we would expect 7.06 A's, 14.12 B's, 5.66 C's, 2.47 D's, and 0.69 F's. This seems consistent with our sequential calculations, but does suggest it might take a very long time to get to the solution. □

Section Summary

Before you try the exercises, be sure that you

- Understand the basic kind of situation for which a Markov Chain is an appropriate model: discrete time periods, a system that can be in exactly one state from a discrete set of states at any time, and transitions between every pair of states from one time period to the next that do not change over time.
- Realize that using a simple Markov Chain model as presented here makes the implicit assumption that the probabilities of being in any of the states at a given point in time depends only on the probabilities one period before and the fixed transition probabilities.
- Know that $\mathbf{P}$, sometimes called the transition probability matrix, with entries p_{ij}, which are the probabilities of making the transition from state i to state j in one period at *any* time.
- Know that the rows of $\mathbf{P}$ sum to 1 (100%), and why.

- Understand the basic logic that the probability of being in a given state at a given time is the sum of the probabilities of being in each of the states one period before times the probability of a transition from each state to the given state in one period: $\pi_j(t+1) = p_{1j}\pi_1(t) + p_{2j}\pi_2(t) + \ldots + p_{nj}\pi_n(t)$, where the transition probabilities on the right-hand side correspond to the j'th column of the transition matrix.
- Understand why, because of the above relationship, $\pi(t+1) = \pi(t)\mathbf{P}$, and how this can be used sequentially to calculate probabilities of being in the different states at any point in time, both by hand and using technology.
- Understand how, using substitution, the above relationship also means that $\pi(t) = \pi(0)\mathbf{P}^t$, and how this can also be used to calculate probabilities at any given point in time, especially using technology.
- Understand that steady-state, or long-run, probabilities can be calculated by sequentially calculating probabilities until they converge to a limit, or by solving the system of equations determined by the matrix equation $\pi = \pi\mathbf{P}$, with one equation removed (and why one is redundant), simplified and supplemented by an equation saying that the sum of the steady-state probabilities should be 1 (or, if you are using total numbers instead of probabilities, the sum should be the total number, which is assumed to be fixed). The right-hand side coefficients of the initial matrix equation correspond to the transpose of the transition matrix.

EXERCISES FOR SECTION 7.5

Warm Up

1. If $\mathbf{P} = \begin{bmatrix} .8 & .2 \\ .4 & .6 \end{bmatrix}$ and $\pi(0) = [\,0 \;\; 1\,]$,
   a) Find $\pi(1)$ and $\pi(2)$.
   b) Find $\pi(4)$ to 4 significant digits.
   c) Use the sequential procedure to find the long-run probabilities to 2 significant digits.
   d) Formulate the system of equations you would solve to find the steady-state probabilities.
   e) Use matrix theory to solve the system you formulated in part (d). When rounded to 2 significant digits, does it agree with your answer to part (c)? Explain.

2. If $\mathbf{P} = \begin{bmatrix} 0.3 & 0.7 \\ 0.15 & 0.85 \end{bmatrix}$ and $\pi(0) = [\,0.8 \;\; 0.2\,]$,
   a) Find $\pi(1)$ and $\pi(2)$.
   b) Find $\pi(4)$ to 4 significant digits.
   c) Use the sequential procedure to find the long-run probabilities to 2 significant digits.
   d) Formulate the system of equations you would solve to find the steady-state probabilities.
   e) Use matrix theory to solve the system you formulated in part (d). When rounded to 2 significant digits, does it agree with your answer to part (c)? Explain.

3. If $\mathbf{P} = \begin{bmatrix} 0.8 & 0.2 & 0 \\ 0.2 & 0.5 & 0.3 \\ 0.1 & 0.2 & 0.7 \end{bmatrix}$ and $\pi(0) = [\,0 \;\; 1 \;\; 0\,]$,
   a) Find $\pi(1)$ and $\pi(2)$.
   b) Find $\pi(4)$ to 4 significant digits.
   c) Use the sequential procedure to find the long-run probabilities to 2 significant digits.
   d) Formulate the system of equations you would solve to find the steady-state probabilities.
   e) Use matrix theory to solve the system you formulated in part (d). When rounded to 2 significant digits, does it agree with your answer to part (c)? Explain.

4. If $\mathbf{P} = \begin{bmatrix} 0.1 & 0.6 & 0.3 & 0 \\ 0 & 0.2 & 0.5 & 0.3 \\ 0 & 0 & 0.4 & 0.6 \\ 0.2 & 0.6 & 0.2 & 0 \end{bmatrix}$ and $\pi(0) = [\,0.3\ \ 0.3\ \ 0.2\ \ 0.2\,]$,
   a) Find $\pi(1)$ and $\pi(2)$.
   b) Find $\pi(4)$ to 4 significant digits.
   c) Use the sequential procedure to find the long-run probabilities to 2 significant digits.
   d) Formulate the system of equations you would solve to find the steady-state probabilities.
   e) Use matrix theory to solve the system you formulated in part (d). When rounded to 2 significant digits, does it agree with your answer to part (c)? Explain.

Game Time

5. Suppose your school has a simple grading system that just uses E (Excellent), S (Satisfactory), and U (Unsatisfactory) for one required core course each semester. Your freshman orientation group bonded, and has kept in touch. Your first semester, 5 of you got E on the course, 7 got S, and 3 got U. You have seen statistics on how people tend to do on this course from one semester to the next. Starting from E, 70% get E again the next semester, 20% get S, and 10% get U. Starting from S, 30% move to E, 50% stay at S, and 20% drop to U in the next semester. Starting from U, 10% jump to E, 60% rise to S, and 30% stay at U for the next semester.
   a) In your orientation group, what distribution of grades would you expect in the second semester?
   b) What distribution of grades would you expect for your group in the fourth semester?
   c) What would you expect the long-run grade distribution to be for your group?
   d) Verify your calculation in part (c) by using a different method.
   e) What assumptions are you making in the analysis for this question? Do you think they are reasonable? Explain.

6. A company is considering a very expensive and perhaps controversial ad campaign. It is controversial in that the plan is to use celebrities in the ads, and some of these celebrities have widespread followings, but also a certain group that really dislike them. Your advertising company does some studies concerning such ads and finds that after each campaign of this type, 20% of the people who already use the product will switch to another brand in protest, while 30% of other users will try the advertised product.

(a) If the company currently has a 14% market share, how big a market share could they expect after running one of this type campaign?
(b) What could they expect after running two ad campaigns of this type?
(c) Is there any point at which running further ad campaigns of this type will no longer gain new customers (to 3 significant digits)?

7. The results of the next local election will be of great importance to your company. You are particularly interested in having the candidate for the Blue Party elected. You have had some studies done and found that polls taken have showed the following results:

For each week that the Blue Party runs extensive ads, 42% of the people who usually vote Blue indicated that they will change to Red because they are tired of being bombarded by advertising, while 44% of the people who usually vote Red indicated that they will be persuaded by the advertising campaign to switch to the Blue Party.

a) If 35% of the voters usually vote for the Blue Party, what percentage would you expect to indicate that they will vote Blue after *one* week of advertisements?
b) What percentage would you expect to indicate that they will vote for the Blue Party after *two* weeks of advertising?
c) How many weeks of advertising would it take to give the Blue Party *over* 51%, to allow a small margin of error?
d) Do the percentages ever level off at a limit (to 3 significant digits)? If so, after how many weeks of advertising does this occur?

Chapter 7 Summary

Basic Matrix Definitions and Operations

A **matrix** is a rectangular array of numbers, normally enclosed in square brackets and referred to using boldface *upper*-case letters. If matrix $\mathbf{A} = [a_{ij}]$ has m rows and n columns, we say the **dimension** of $\mathbf{A}$ is $m \times n$. Element a_{ij} refers to the value in the i'th row and j'th column of the matrix $\mathbf{A}$. Remember that spreadsheets identify cells by listing the *column* first (usually with a letter) and then the row (usually with a number). A matrix with only one row can be called a **row matrix** or **row vector** and a matrix with only one column can be called a **column matrix** or **column vector**; either is often denoted by a boldface *lower*-case letter. Instead of boldface, when writing matrices and vectors by hand, you can put a wavy tilde under the letter.

Two **matrices being equal** means that they must be the same dimension, and the corresponding elements must all be equal. In order to **add or subtract** two matrices, they must have the same dimension, and each element of the answer matrix is the sum or difference of the corresponding elements in the two matrices. **Scalar multiplication** refers to multiplying each element of a matrix by a single pure real number, which can be called a scalar.

Matrix Multiplication

The **dot product** of a row vector by a column vector (which each must have the same number of elements) is calculated by multiplying the corresponding elements and adding the results. Two matrices can be multiplied only if the size of the rows (number of columns) in the left matrix equals the size of the columns (number of rows) in the right matrix. If matrix $\mathbf{A}$ is $m \times k$ and matrix $\mathbf{B}$ is $k \times n$, and if matrix $\mathbf{C}$ is the result of the **matrix multiplication,** $\mathbf{C} = \mathbf{AB}$, then $\mathbf{C}$ will have dimension $m \times n$, and the element in the i'th row and j'th column of $\mathbf{C}$ (element c_{ij}) will be the dot product of the i'th row of $\mathbf{A}$ by the j'th column of $\mathbf{B}$. **Matrix multiplication is not commutative in general; there is no guarantee that AB = BA.** In fact, one direction could be possible and the other *impossible*, just because of the dimensions.

When working with real-world problems involving matrix multiplication, use general **verbal descriptions** for the rows and columns of each matrix to help determine whether or not a matrix product will make sense (even if it is mathematically possible). The individual row and column **labels** should normally also match.

The **transpose** of a matrix is the matrix that results from interchanging the rows and columns, and is denoted $\mathbf{A}^T$. When matrices have been entered in a spreadsheet, it will sometimes be necessary to transpose one of them to be able to multiply them.

Solving Systems of Linear Equations Using Augmented Matrices

If a system of linear equations has the same number of variables as equations, it normally has a unique solution. To try to find a solution, you can create an **augmented matrix** with the matrix of coefficients from the left-hand side of the system beside the column matrix of right-hand side constants, separated by a dotted line: $[\ \mathbf{A} : \mathbf{B}\ ]$. The **identity matrix** of dimension $n \times n$, usually written $\mathbf{I_n}$ or simply $\mathbf{I}$, is a matrix with 1's on the main diagonal (from the upper left to the lower right) and 0's everywhere else. If the dimensions match, the identity matrix times any other matrix yields the other matrix (analogous to multiplying a real number by 1). If the left side of an augmented matrix is $\mathbf{I}$, then the system has a unique solution, which can be read off from the values to the right of the partition (dotted line).

Three kinds of **elementary row operations** can be performed on an augmented matrix to result in an equivalent system (with identical solutions):

1) Interchanging two rows $(R_i \leftrightarrow R_j)$
2) Multiplying a row by a nonzero constant $(kR_i \rightarrow R_i')$
3) Adding a nonzero multiple of one row to another $(kR_i + R_j \rightarrow R_j')$

These row operations correspond to operations that can be done on systems of equations directly to yield equivalent systems. The process of applying such row operations to an augmented matrix to try to get an augmented matrix with $\mathbf{I}$ on the left is called **Gaussian elimination** or **row reduction**. A recommended strategy is to

1) Work column by column, from left to right.
2) In a given column, try to get the 1 in the proper place first (such as by interchanging rows or by multiplying a value that is not 1 or 0 and its row by its reciprocal), then the 0's (using the newly-created 1 in that column, and multiplying it and its row by the negative of the value you want to cancel out to get 0 and adding the result to the other row), which can usually be done all at once, if there are more than one.

If, in this process, **you obtain a row of all 0's to the left of the partition line**: If the number to the right of the partition line for that row is *not* 0, then the system has no feasible solutions (no solution), such as with two parallel lines that are not the same. If the number to the right *is* 0, then the system is likely to have an infinite number of solutions (such as two equations representing the same line).

Solving Linear Systems Using Matrix Equations

A system of linear equations can be expressed as the matrix equation $\mathbf{AX} = \mathbf{B}$, where $\mathbf{A}$ is the coefficient matrix, $\mathbf{X}$ is the column matrix of the variables, and $\mathbf{B}$ is the column matrix of the right-hand side constants. The **multiplicative inverse** of a square matrix (often called its inverse for short), if it exists, is denoted $\mathbf{A}^{-1}$, and is the unique square matrix of the same dimension which, when multiplied by the original matrix on *either* side, yields the identity matrix: $\mathbf{AA}^{-1} = \mathbf{A}^{-1}\mathbf{A} = \mathbf{I}$. Thus to solve the matrix equation we can do the following:

$$\mathbf{AX} = \mathbf{B}$$
$$\mathbf{A}^{-1}(\mathbf{AX}) = \mathbf{A}^{-1}(\mathbf{B})$$
$$\mathbf{IX} = \mathbf{A}^{-1}(\mathbf{B})$$
$$\mathbf{X} = \mathbf{A}^{-1}(\mathbf{B})$$

To find $\mathbf{A}^{-1}$ by hand, set up the augmented matrix $[\,\mathbf{A} : \mathbf{I}\,]$, and do row operations until you get $\mathbf{I}$ on the left (if possible). The matrix on the right will be $\mathbf{A}^{-1}$. The row operations, all together, correspond to multiplying each side by $\mathbf{A}^{-1}$. This is also true for row reduction as discussed above, which corresponds to reducing $[\,\mathbf{A} : \mathbf{B}\,]$ to $[\,\mathbf{I} : \mathbf{A}^{-1}\mathbf{B}\,]$ by multiplying by $\mathbf{A}^{-1}$.

Markov Chains

Markov Chains can be used to model situations involving discrete time periods and a discrete set of possible states. They assume that the probability of a transition from each state to each other possible state from one period to the next is constant over time. The matrix of these transition probabilities is denoted $\mathbf{P}$, where p_{ij} is the probability of the transition from state i to state j. $\pi(t)$ can be used to denote the row vector of probabilities of being in each state after t time periods. In general, $\pi(t+1) = \pi(t)\mathbf{P}$. Given $\pi(0)$, $\pi(t)$ can also be found from $\pi(t) = \pi(0)\mathbf{P}^t$. To find steady-state probabilities for the states, you can sequentially calculate $\pi(t)$ for $t = 1, 2, 3, \ldots$ until the values converge, or you can solve the system of equations given by $\pi = \pi\mathbf{P}$, with one equation removed, simplified, and augmented by an equation forcing the probabilities to sum to 1. The use of a Markov Chain assumes that the probabilities of the states at a given time depend *only* on the probabilities the period before and the transition probabilities.

Chapter 8: Unconstrained Optimization of Multivariable Functions

Introduction

In Chapters 5 and 6 we learned how to develop mathematical models for problems with several variables. In Chapter 7 we learned how to use matrices to help us solve systems of linear equations involving two or more variables. In this chapter we see how we can use the matrix concepts from Chapter 7 to help us optimize the models that we developed in Chapters 5 and 6. In Section 8.1 we generalize the idea of a derivative to the multivariable function case, called a **partial derivative**, where we examine the rate of change of the output with respect to *one* input variable, holding the *other* input variables constant. In Section 8.2, we then build on the methods we learned in Chapter 3 concerning the optimization of single-variable functions by setting the derivative equal to 0 to develop a method for **optimizing multivariable functions**. In Section 8.3 we see how **non-linear optimizations** can be done using **spreadsheets**, and in Section 8.4 we learn how to **test critical points** found using calculus (or other test points) to see if they are local or global extrema. Up to this point in the text, we have relied on technology to help us find the best-fit model for sets of data. In Section 8.5, we use multivariable optimization and matrix equations to see *how* the best-fit model is found.

Here is a selection of examples to give you a feel for the kind of problems you should be able to solve after studying this chapter:

- You are the owner of a small company that produces just two products, chairs and tables. You have developed a model for the profits from the sale of the chairs and tables as a function of the number of each made per month. How will your profit increase if you increase the number of chairs by one? How much will your profit increase per additional table made?
- You have developed a model for your energy level as a function of both the amount of exercise that you get in a day and the amount of calories that you consume during the day. How much exercise should you get each day and how many calories should you consume to optimize your energy level?
- You play golf and have been having trouble with your approach shots. You want to figure out the amount of backswing to use on your approach to get the ball as close to the pin as possible. You've gotten some data and now want to find the best-fit model. How can you determine the model that minimizes the sum of the squares of the errors?

After studying this chapter, in addition to being able to solve problems like those above you should:

- Know what a partial derivative means graphically and in a real-world context.
- Know how to find the partial derivatives of multivariable functions.
- Know how to optimize unconstrained multivariable functions by hand.
- Know how to optimize nonlinear functions using a spreadsheet.
- Know how to find critical points, and test them to see if they are local or global extrema.
- Know how to minimize SSE to find the best-fit least-squares regression model given a set of data and a model type.

Section 8.1: Rates of Change of Multivariable Functions

In Chapters 5 and 6 we saw how to formulate multivariable functions, both from verbal descriptions and from using least-squares regression to fit models. Often, we formulate such functions because we want to optimize them - that is, find the maximum or minimum values over some domain. As we saw earlier in this course with single-variable functions, we do this by finding the derivative of the function (the instantaneous rate of change of the dependent variable with respect to the independent variable), setting it equal to 0, and solving for the independent variable. This corresponds to finding where the tangent line to the curve is horizontal, since this will occur at the local extrema of a smooth curve. In this section, we see how to generalize the idea of a derivative to define **a partial derivative** (instantaneous rate of change) of the dependent variable with respect to *one* of the independent variables, holding all of the others constant.

Here are the kinds of problems that the material in this section will help you solve:

- You are the owner of a small company that produces just two products, chairs and tables. You have developed a model for the profits from the sale of the chairs and tables as a function of the number of each made per month. How will your profit increase if you increase the number of chairs by one? How will your profit increase per additional table?
- You have developed a model for your energy level as a function of both the amount of exercise that you get each day and the amount of calories that you consume during the day. Compared to a certain level of exercise and calorie consumption in a day, what would be the marginal effect of another minute of exercise? What would be the effect on your energy level of each additional calorie consumed?

After studying this chapter, in addition to being able to solve problems like those above, you should:

- Know what a partial derivative means graphically.
- Be able to interpret the meaning of a partial derivative in a real-world situation.
- Know how to find the partial derivatives of multivariable functions.
- Know the symbols used to denote the partial derivatives of multivariable functions.

Sample Problem 1: Suppose again that you own the small furniture manufacturing company discussed in Chapters 5, 6, and 7. You have established that, if you make an average of c chairs and t tables each month, the model for your cost (C, in dollars) is:

Verbal Definition: $C(c,t)$ = the average cost per month[1], in dollars, to make c chairs and t tables per month

Symbol Definition: $C(c,t) = 700 + 70c + 100t$, for $c,t \geq 0$

Assumptions: Certainty and divisibility. Certainty implies that the relationship is exact. Divisibility implies that any fractional values of dollars, chairs, and tables are possible.

You are currently making an average of 7 chairs and 4 tables each month. How would increasing the number of chairs made each month affect your cost? Specifically, how much would your cost increase if you made one more chair per month? 2 more? What is the rate of change of the cost with respect to the number of chairs (keeping the number of tables fixed at 4)?

Solution: We can see that the current average cost per month is

$$\begin{aligned} C(7,4) &= 700 + 70(7) + 100(4) \\ &= 700 + 490 + 400 \\ &= 1590. \end{aligned}$$

In other words, the average cost per month is \$1590 when we make 7 chairs and 4 tables. If we made one more chair (for a total of 8) per month, the cost would become

$$\begin{aligned} C(8,4) &= 700 + 70(8) + 100(4) \\ &= 700 + 560 + 400 \\ &= 1660 \end{aligned}$$

(an increase of $\$1660 - \$1590 = \$70$). Adding one more again ($c = 9$), the cost would be

$$\begin{aligned} C(9,4) &= 700 + 70(9) + 100(4) \\ &= 700 + 630 + 400 \\ &= 1730\ . \end{aligned}$$

(again, an increase of \$70 for the additional chair). It makes sense that, because of the linear cost function, each additional chair costs \$70, so the **instantaneous rate of change**

---

[1] In Section 3.5 , we talked about average cost *per unit produced,* while here (as in Section 1.2 when we formulated the problem of how much cheese to buy at the store each trip) we are finding the average cost *per unit of time*.

of the cost with respect to the number of chairs is 70 (the cost increases at a rate of $70 per chair), as long as we hold the number of tables constant. We call this the **partial derivative of C with respect to c**, and use the notation $\frac{\partial C}{\partial c}$, or $C_c(c,t)$.

Notice that if we hold t fixed at 4, we get

$$C(c,4) = 700 + 70c + 100(4) = 700 + 70c + 400 \text{ .}$$ [2]

This is really just a function of one variable, c, since 4 is a constant, so we could take the simple derivative, as we did in Chapter 3:

$$\begin{aligned} D_c\, C(c,4) = D_c(700 + 70\text{c} + 400) &= 0 + 70 + 0 \\ &= 70 \text{ .} \end{aligned}$$

If we hold t fixed at any other constant value, say k, the cost would be

$$C(c,k) = 700 + 70c + 100(k) = 700 + 70c + 100k \text{ .}$$

Since k is just a constant, as before, this is really just a function of one variable, c, and so we can take the simple derivative:

$$\begin{aligned} D_c\, C(c,k) = D_c(700 + 70c + 100k) &= 0 + 70 + 0 \\ &= 70 \end{aligned}$$

$D_c\, C(c,k)$ can also be written $C_c(c,k)$, so this could also be written:

$$C_c(c,k) = 70$$

Keep in mind here that c is a variable, but k is a constant. The derivative of a constant (like 700 or $100k$) is 0, and the derivative of a constant times a variable is the constant (a linear function has a constant rate of change, which is the slope).

When taking a partial derivative with respect to one variable, we simply **treat all of the other variables as constants** and take the simple derivative with respect to the one variable. Thus,

$$C_c(c,t) = \frac{\partial}{\partial c}(700 + 70c + 100t) = 0 + 70 + 0 = 70 \text{ .}$$

[2] We could have simplified this to $1100 + 70c$, but left it in this form to better see the pattern when generalizing.

As before, we think of the t as a constant, just like when there was a k in the same spot, so the partial derivative of $100t$ is 0, since it is the simple derivative of a constant. Notice that this particular partial derivative is the same for **any** value of t, because the function is linear (each variable only occurs *by itself* in any term, and only to the *first* power). It is also the same for *any* value of c because of the linear form.

Translation: The average cost per month is increasing at the rate of \$70 per chair produced. □

Sample Problem 2: Given the same model from Sample Problem 1, what is the rate of change of the cost with respect to the number of *tables* made if the number of *chairs* is kept constant?

Solution: In a similar way, the partial derivative of C with respect to t would correspond to thinking of c as a constant. Taking the simple derivative as if t were the only variable, we get

$$C_t(c,t) = \frac{\partial}{\partial t}(700 + 70c + 100t) = 0 + 0 + 100 = 100 .$$

This time, the $70c$ is treated as a constant, so its derivative is 0, and the $100t$ is like a linear function. As before, the partial derivative is the same (100) for any value of c and t, because of the linear form.

Translation: The average cost per month is increasing at the rate of \$100 per table produced. □

Sample Problem 3: Once more, as owner of the furniture company, suppose you have established that the model for your profit is:

Verbal Definition: $P(c,t)$ = the profit, in dollars, on average, when c chairs and t tables are made in a month

Symbol Definition: $P(c,t) = 140c + 200t - 4c^2 + 2ct - 12t^2 - 700$, for $c,t \geq 0$

Assumptions: Certainty and divisibility. Certainty implies that the relationship is exact. Divisibility implies that any fractional values of dollars, chairs, and tables are possible.

Again assuming your current production rate is 7 chairs and 4 tables each month, find the partial derivative of your profit with respect to the number of chairs, both in general and at the current rate of production, and interpret the latter value.

Solution: Proceeding as we did in the first example, to find the partial derivative with respect to c, we simply treat t as a constant, and take the simple derivative, thinking of c as the variable. Thus we get

$$\begin{aligned} P_c(c,t) &= \frac{\partial}{\partial c}(140c + 200t - 4c^2 + 2ct - 12t^2 - 700) \\ &= 140 + 0 - 8c + 2t - 0 - 0 \\ &= 140 - 8c + 2t \ . \end{aligned}$$

Notice that since we treat t as a constant, the derivatives of $200t$ and $-12t^2$ are both 0 since both expressions are also constants. Also we can think of the term $2ct$ as $(2t)c$, where $(2t)$ is a constant, so the derivative with respect to c is $2t$. Notice that here, unlike in the last example, the partial derivative *does* depend on the values of c and t. This is because the profit function is *not linear,* it is a **nonlinear function**, which you can recognize because of the quadratic terms (such as $-4c^2$ and $-12t^2$) and the cross-product term ($2ct$). In fact, this function is **quadratic**, because the sums of the exponents of the variables in any of the terms, and the individual exponents themselves, are 0, 1, or 2.

Now, to find the partial derivative at the current rate of production, we simply substitute 7 in for c and 4 in for t:

$$\begin{aligned} P_c(7,4) &= 140 - 8(7) + 2(4) \\ &= 140 - 56 + 8 \\ &= 92 \ . \end{aligned}$$

This means that the instantaneous rate of change of the profit with respect to the number of chairs produced is \$92/chair at the current production rate.

Translation: Compared to making 7 chairs and 4 tables per month, if you made 1 more chair per month (and kept the number of tables fixed at 4), your average profit per month would go up by *approximately* \$92 . As in the single-variable derivative case, this is a *linear approximation* of a *curve* using the tangent line, so is only approximate. In general, the further from the starting point you move, the more the error increases.

Since you are currently manufacturing and selling 7 chairs and 4 tables each month, your profit is given by

$$\begin{aligned} P(7,4) &= 140(7) + 200(4) + 2(7)(4) - 4(7)^2 - 12(4)^2 - 700 \\ &= 980 + 800 + 56 - 196 - 192 - 700 \\ &= 748 \end{aligned}$$

Translation: This means that, on average, your profit should be about \$748 if you make 7 chairs and 4 tables in a month.

If you made 1 more chair each month ($c = 8$), the partial derivative predicts that this profit would increase by \$92, for a profit of \$748 + \$92 = \$840 . The actual profit would be

$$\begin{aligned} P(8,4) &= 140(8) + 200(4) + 2(8)(4) - 4(8)^2 - 12(4)^2 - 700 \\ &= 1120 + 800 + 64 - 256 - 192 - 700 = 836 . \end{aligned}$$

This is an error of \$840 - \$836 = \$4 over the actual value, so you see that the approximation from the partial derivative is pretty good, but not perfect. For 2 more chairs ($c = 9$), the partial derivative predicts a profit of $748 + 2(92) = 748 + 184 = 932$, while the actual profit would be $P(9,4) = 916$, showing how the error of the approximation gets worse (increases from 4 to 16) as you move farther from the original point. □

To see what partial derivatives correspond to graphically (at least for a function of *two* variables, as in this example), consider the graph in Figure 8.1-1 .

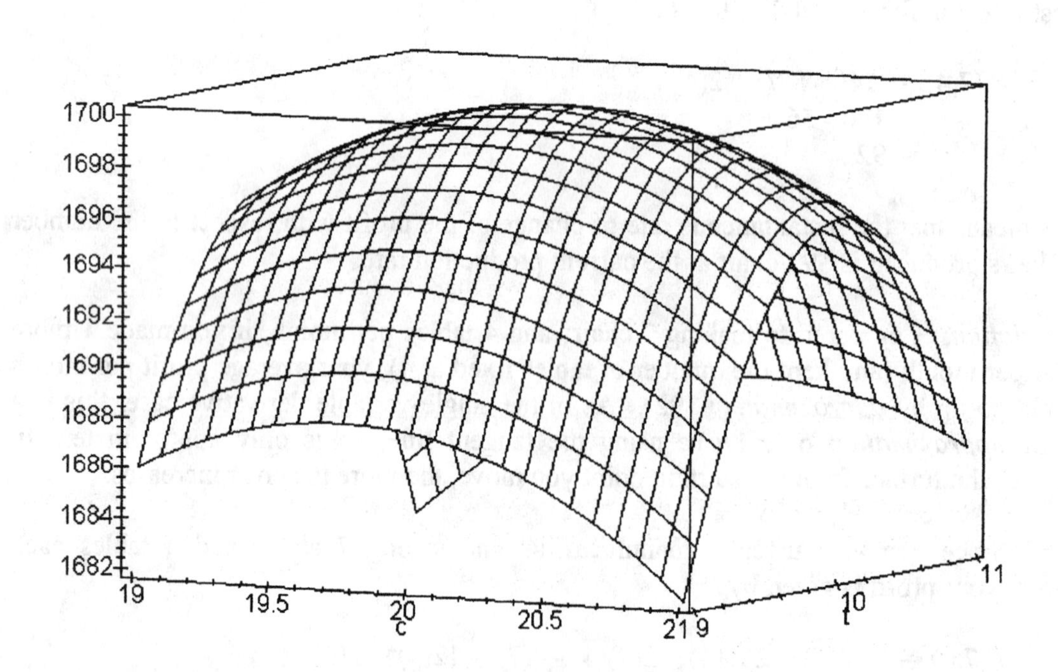

Figure 8.1-1

In this graph, the curve you see at the bottom of the surface, just over the c value line (sort of the left front of the graph) corresponds to the curve you get when you set $t = 9$ (notice that 9 is the value on the near end of the t value line). Plugging 9 in for t, we see that this curve has the equation

$$\begin{aligned} P(c,9) &= 140c + 200(9) - 4c^2 + 2c(9) - 12(9)^2 - 700 \\ &= 140c + 1800 - 4c^2 + 18c - 972 - 700 \\ &= 128 + 158c - 4c^2 . \end{aligned}$$

This makes sense, since it looks like a parabola opening downward. In fact, this is the curve you get if you "slice" the curved surface, which is the graph of the whole profit function, using the plane that is perpendicular to the t axis, passing through it at $t = 9$. **If you think of the surface as the skin of an apple and the plane as a knife, the curve is the curve you would see formed by the edge of the skin along the cut slice.**

The derivative of this sliced curve is the slope of the tangent to the curve, and this is the graphical meaning of the partial derivative. Specifically,

$$\begin{aligned} P_c(c,9) &= D_c\,(128 + 158c - 4c^2) \\ &= 0 + 158 - 8c \\ &= 158 - 8c . \end{aligned}$$

Equivalently, using the general form we got for $P_c(c,t) = 140 - 8c + 2t$, we see that

$$\begin{aligned} P_c(c,11) &= 140 - 8c + 2(9) \\ &= 140 - 8c + 18 \\ &= 158 - 8c . \end{aligned}$$

When $c = 20$ (near the peak of the parabola we have been discussing in Figure 8.1-1), we find the derivative with respect to c is:

$$\begin{aligned} P_c(20,9) &= 158 - 8(20) \\ &= 158 - 160 = -2 . \end{aligned}$$

Look again at the curve at the front edge of the graph in Figure 8.1-1, the curve defined for values of c when t is 9. The tangent drawn to this curve at the point $c = 20$ is very nearly horizontal. When $c = 19$, we get that:

$$\begin{aligned} P_c(19,9) &= 158 - 8(19) \\ &= 158 - 152 = 6 \end{aligned}$$

This is a fairly steep positive slope, which again is consistent with the graph. As before, this tells us that the marginal effect on your profit from an additional chair beyond 19 (holding the number of tables fixed at 9) is approximately \$6 , whereas the marginal profit at 20 chairs is -\$2. In fact, the marginal profit will go even more negative at $c = 21$:

$$P_c(21,9) = 158 - 8(21)$$
$$= 158 - 168 = -10$$

Translation: This indicates that if you are committed to making 9 tables per month, your optimal number of chairs is approximately 20 (slightly less). □

From this example, we hope you can see why **finding the rates of change of functions that involve more than one input variable can be very helpful in making management decisions.**

Sample Problem 4: Recall the small coffeehouse problem in Section 5.5 (Sample Problem 2). We wanted to find a linear model, in the form $y = mx+b$, for the attendance at a concert as a function of the price charged (a demand function). When trying to fit this linear model to the demand data we found that the expression for the SSE was given by

$$\text{SSE}(m,b) = (100\text{-}7m\text{-}b)^2 + (80\text{-}8m\text{-}b)^2 + (70\text{-}9m\text{-}b)^2 \ .$$

Find the rate of change of the SSE with respect to m and b.

Solution: Recall that the Chain Rule can help take the derivative of complicated functions like this. Think of the quantity in parentheses as the "inner function" or u , and multiply the outer derivative times the inner derivative. For example:

$$D_x\left[(100-7x)^2\right] = 2(100-7x)^{2-1}(-7) = 2(100-7x)^1(-7)$$ [3]

If we take the partial derivative with respect to m , we treat b as if it were a constant, so we use the same idea of the Chain Rule to get:

$$\text{SSE}_m(m,b) = 2\,(100\text{-}7m\text{-}b)^1\,(\text{-}7) + 2\,(80\text{-}8m\text{-}b)^1\,(\text{-}8) + 2\,(70\text{-}9m\text{-}b)^1\,(\text{-}9)$$
$$= -1400 + 98m + 14b - 1280 + 128m + 16b - 1260 + 162m + 18b$$
$$= -3940 + 388m + 48b \ .$$

[3] As in Sample Problem 1, we could have simplified this further to $-14(100\text{-}7x) = -1400+98x$, but left it as we did to help make it easier to see the numerical patterns to facilitate finding a general formula.

Thus, when b is held constant and m is allowed to vary, the SSE is changing at the instantaneous rate of $-3940 + 388m + 48b$ per unit increase in the value of m.

Similarly, for the partial derivative with respect to b, we treat m as a constant and b as the variable, to get

$$\begin{aligned} SSE_b(m,b) &= 2\,(100\text{-}7m\text{-}b)^1\,(-1) + 2\,(80\text{-}8m\text{-}b)^1\,(-1) + 2\,(70\text{-}9m\text{-}b)^1\,(-1) \\ &= -200 + 14m + 2b - 160 + 16m + 2b - 140 + 18m + 2b \\ &= -500 + 48m + 6b \end{aligned}$$

As above, this means that when m is held constant and b is allowed to vary, the SSE changes at the instantaneous rate of $-500 + 48m + 6b$ per unit increase in the value of b.

For example, suppose we were considering the model

$$y = -1.5x + 200$$

(so $m = -1.5$ and $b = 200$). Then

$$SSE_b(-1.5,200) = -500 + 48(-1.5) + 6(200) = -500 - 72 + 1200 = 628\,.$$

Translation: If we hold the slope of our model constant at $m = -1.5$, and increase the value of our y-intercept b by 1 unit, from 200 to 201, our SSE should go up by approximately 628. Put differently, increasing the value of b from 200 will make our SSE worse (*increase* the error) if we hold $m = -1.5$. We could give a similar interpretation of the partial derivative with respect to m. □

Cobb-Douglas Production Functions and Marginal Productivity

Sample Problem 5: Suppose you own a landscaping business. You measure your productivity by the number of lawns you maintain for customers. You are often faced with the choice of whether to spend money for new workers, or for new machinery. To answer the question, you estimate data from other similar businesses you know of in your area. The variables involved are the number of full-time workers (l , for labor), the dollar value of the machinery and equipment in \$1000's (c , for capital), and the number of lawns maintained (P , for productivity). Using your data, you estimate that a good model for P in terms of l and c is given by:

Verbal Definition: $P(l,c)$ = the number of lawns that can be maintained, on average, for 1 year when l is the number of full-time workers and c is the dollar value of the machinery and equipment (in \$1000s)

Symbol Definition: $P(l,c) = 16l^{0.068}c^{0.26} \quad l,c \geq 0$

Assumptions: Certainty and divisibility. Certainty implies that the relationship is exact. Divisibility implies that any fractional values of numbers of lawns, full-time workers and dollars are possible.

The form of this function is called a **Cobb-Douglas production function**, named after the economists who developed it. The **marginal productivity of labor** simply means the rate of change of the productivity function (P) with respect to the number of workers (l), holding c constant. Thus it corresponds to the partial derivative of P with respect to l. In a similar way, the **marginal productivity of capital** is the partial derivative of P with respect to c. Find the marginal productivity of labor and capital, both in general and at your current configuration of 3 workers and \$50,000 worth of equipment and machinery. If workers cost \$15,000 per season and you have an additional \$30,000 to spend on either labor or capital for the coming season, which resource would give you more productivity for your money?

Solution: To find the partial derivatives, the calculations are:

$$\frac{\partial P}{\partial l} = \left(16c^{0.26}\right)\left(0.068l^{-0.932}\right)$$

$$\frac{\partial P}{\partial c} = \left(16l^{0.068}\right)\left(0.26c^{-0.74}\right)$$

If your business currently has 3 full-time workers ($l = 3$) and \$50,000 worth of machinery and equipment ($c = 50$), then the rate of change of production with respect to labor, capital being kept constant, is given by:

$$\begin{aligned}\frac{\partial P}{\partial l} &= (16)(50)^{0.26}(0.068)(3)^{-0.932} \\ &\approx (16)(2.8)(0.068)(0.36) \\ &\approx 1.1\end{aligned}$$

Translation: Each increase in the number of workers above 3 (keeping capital value at \$50,000) will increase productivity by approximately 1.1 lawns. This is called the **marginal productivity of labor**.

The rate of change of production with respect to capital, labor being kept constant, called the **marginal productivity of capital,** is

$$\frac{\partial P}{\partial c} = (16)(3)^{0.068}(0.26)(50)^{-0.74}$$

$$\approx (16)(1.1)(0.26)(0.055)$$

$$\approx 0.25$$

Translation: Each additional \$1000 *over* the current \$50,000 spent on machinery and equipment (keeping the number of full-time workers at 3) will increase productivity by approximately one quarter of a lawn.

The problem states that you can hire a new worker full-time for one full season for \$15,000 . Furthermore, it says that you have \$30,000 in your budget for this year that you could spend on machinery and equipment, or for 2 new workers. Since the marginal productivity of labor is 1.1 lawns per worker given your current configuration, the added productivity from 2 new workers that the \$30,000 could pay for would be only about $2(1.1) = 2.2$ lawns. But since the marginal productivity of capital is about 0.25 lawns per \$1000 capital invested in machinery and equipment, for \$30,000 you could increase productivity by about $30(0.25) = 7.5$ lawns. Clearly, the capital investment is *much* more productive ($7.5/2.2 \approx 3.5$ times as much!) and should be the better way to go. □

To make the units for labor and capital more compatible, it is common to express the labor variable in money units as well (such as \$1000s spent on labor). This would automatically give the marginal productivity per \$1000 invested for both resources, making the direct comparison easier. Similarly, productivity could be measured in money units (revenue generated, such as in \$100s), in which case the marginal productivity values would correspond to rates of return on each investment.

Section Summary

Before you begin the exercises, be sure that you:

- Know that a partial derivative of a multivariable function with respect to one of the independent variables is the instantaneous rate of change of that function with respect to that one independent variable, holding all of the other independent variable values constant.
- Know that to find the partial derivative of a multivariable function with respect to one of the independent variables, you treat all other independent variables as constants, and take the derivative using the same methods as used for single-variable functions.

- Know that the partial derivative of a function, $F(x, y)$, with respect to one of its independent variables, say x, is denoted by $F_x(x, y)$ or $\frac{\partial F}{\partial x}$.

- Know that, for a function of two variables such as $F(x, y)$, its graph is a three-dimensional surface, and holding one of the variables (such as y) constant at a particular value (say, $y = k$), is like slicing the surface with the plane $y = k$ (perpendicular to the y axis, through the value k on the y axis), resulting in a two-dimensional curve. The partial derivative $\frac{\partial F}{\partial x}$ is then the slope of that two-dimensional curve at a given x value.

EXERCISES FOR SECTION 8.1:

1. $f(x,y)=3x-5y+21$
   a) Find $f_x(x,y)$
   b) Find $f_y(x,y)$
   c) Evaluate $f_x(3,2)$
   d) Evaluate $f_y(1,-3)$

2. $f(s,t)=-0.2s+3.9t+12$
   a) Find $f_s(s,t)$
   b) Find $f_t(s,t)$
   c) Evaluate $f_s(0.5,4.2)$
   d) Evaluate $f_t(-1.2,0.28)$

3. $R(x,y)=-3x^2-2y^2+3xy+4y-20$
   a) Find $\frac{\partial R}{\partial x}$
   b) Find $\frac{\partial R}{\partial y}$
   c) Evaluate $\frac{\partial R}{\partial x}$ for $x=3, y=1$
   d) Evaluate $\frac{\partial R}{\partial y}$ for $x=2, y=7$

4. $P(x,y)=-2x^2-y^2+2xy-4y+3x+24$
   a) Find $\frac{\partial P}{\partial x}$
   b) Find $\frac{\partial P}{\partial y}$
   c) Evaluate $\frac{\partial P}{\partial x}$ for $x=5, y=0$
   d) Evaluate $\frac{\partial P}{\partial y}$ for $x=0.5, y=2.1$

5. $C(f,d)=-50+3.5f+7.2d+3.5fd-4.7f^2-3.9d^2$
   a) Find $C_f(f,d)$
   b) Find $C_d(f,d)$

c) Evaluate $C_f(180,170)$
d) Evaluate $C_d(170,180)$

6. $E(c,r) = -2.8c^2 - 3.7r^2 + 4.3cr + 3.0c - 2.1r + 10$
a) Find $E_c(c,r)$
b) Find $E_r(c,r)$
c) Evaluate $E_c(30,20)$
d) Evaluate $E_r(45,15)$

7. $c(l,w) = 2lw + \frac{1}{w} + \frac{1}{l}$
a) Find $c_l(l,w)$
b) Find $c_w(l,w)$
c) Evaluate $c_l(2,1)$
d) Evaluate $c_w(1,2)$

8. $A(r,t) = 1000e^{rt}$
a) Find $A_r(r,t)$
b) Find $A_t(r,t)$
c) Evaluate $A_r(0.065,1)$
d) Evaluate $A_t(0.065,1)$

9. $c(l,w,h,\lambda) = 1.46lw + 1.24lh + 1.24wh - \lambda lwh + \lambda 7$
a) Find $c_l(l,w,h,\lambda)$
b) Find $c_w(l,w,h,\lambda)$
c) Find $c_h(l,w,h,\lambda)$
d) Find $c_\lambda(l,w,h,\lambda)$

10. $V(l,w,h,\lambda) = lwh - 2\lambda w - 2\lambda h - \lambda l + 108\lambda$
a) Find $V_l(l,w,h,\lambda)$
b) Find $V_w(l,w,h,\lambda)$
c) Find $V_h(l,w,h,\lambda)$
d) Find $V_\lambda(l,w,h,\lambda)$

11 You have developed the following model for the cost of building a box:

Verbal Definition: $c(l,w)$ = the cost in dollars to build a box of length l feet and width w feet.

Symbol Definition: $c(l,w) = 1.46lw + \left(\frac{8.68}{w}\right) + \left(\frac{8.68}{l}\right)$ for $l, w, > 0$

Assumptions: Certainty and divisibility. Certainty implies that the relationship is exact. Divisibility implies that any fractional values of dollars and feet are possible.

a) Find the rate at which the cost is changing with respect to the width in general.
b) Find the rate at which the cost is changing with respect to the length in general.
c) Evaluate $c_l(2,5)$. Write what this *means* in a complete sentence.
d) Evaluate $c_w(4,3)$ Write what this *means* in a complete sentence.

12. The following model has been developed for the volume of a shipping box that is limited in its dimensions by requiring that the length plus the girth (distance around the box in the direction of the label) must be no more than 108 inches:

Verbal Definition: $V(w,h)$ = the volume in cubic inches for a box of width w inches and height h inches.

Symbol Definition: $V(w,h) = 108wh - 2w^2h - 2wh^2$ for $w, h > 0$
$w + h < 54$

Assumptions: Certainty and divisibility. Certainty implies that the relationship is exact. Divisibility implies that any fractional values of inches and cubic inches are possible.

a) Find the rate at which the volume of the box is changing with respect to the width in general.
b) Find the rate at which the volume of the box is changing with respect to the height in general.
c) Evaluate $V_w(3.5,2.5)$. Write what this *means* in a complete sentence.
d) Evaluate $V_h(4.5,2.5)$. Write what this *means* in a complete sentence.

13. A profit model for hamburgers and hot dogs is given by

Verbal Definition: $\pi(b,d)$ = the profit, in dollars, from the sale of hamburgers and hot dogs at b dollars per hamburger and d dollars per hot dog.

Symbol Definition: $\pi(b,d) = -300b^2 + 80bd + 1220b + 1890d - 2500 - 500d^2$ for $b,d \geq 0$

Assumptions: Certainty and divisibility. Certainty implies that the relationship is exact. Divisibility implies that any fractional values of dollars are possible.

a) Find the rate at which the profit is changing with respect to the price of the hamburgers in general.
b) Find the rate at which the profit is changing with respect to the price of the hot dogs in general.
c) Evaluate $\pi_b(2.50, 2.00)$. Write what this *means* in a complete sentence.
d) Evaluate $\pi_d(2.50, 2.00)$. Write what this *means* in a complete sentence.

14. Total profit from the sale of sodas and pretzels is given by:

Verbal Definition: $\pi(s,p)$ = the profit, in dollars, from the sale of sodas at s dollars per soda and pretzels at p dollars per pretzel

Symbol Definition: $\pi(s,p) = 2780s - 1000s^2 - 300sp + 3482p - 2500p^2 - 1340$, $s,p \geq 0$

Assumptions: Certainty and divisibility. Certainty implies that the relationship is exact. Divisibility implies that any fractional values of dollars are possible.

a) Find the rate at which the profit is changing with respect to the number of sodas sold in general.
b) Find the rate at which the profit is changing with respect to the number of pretzels sold in general.
c) Evaluate $\pi_s(1.25, 1.00)$. Write what this *means* in a complete sentence.
d) Evaluate $\pi_p(1.25, 0.65)$. Write what this *means* in a complete sentence.

15. You are selling official and replica sports jerseys. You have already formulated a demand function and from it derived a profit model, given below:

Verbal Definition: $\pi(o,r)$ = the profit, in dollars when o official jerseys and r replica jerseys are sold.

Symbol Definition: $\pi(o,r) = -4.98o^2 + 3.97or + 55.3o - 2.99r^2 + 58.0r$ for $o,r \geq 0$

Assumptions: Certainty and divisibility. Certainty implies that the relationship is exact. Divisibility implies that any fractional values of dollars and jerseys are possible.

a) Find the rate at which the profit is changing with respect to the number of official jerseys sold in general.
b) Find the rate at which the profit is changing with respect to the number of replica jerseys sold in general.
c) Evaluate $\pi_0(7,10)$. Write what this *means* in a complete sentence.
d) Evaluate $\pi_r(8,8)$. Write what this *means* in a complete sentence.

16. Suppose you have developed the following model for the distance of your shot put throw as a function of the number of warm-up throws and laps you do in practice before a meet:

Verbal Definition: $d(t,l)$ = the expected distance in feet of your shot put throws after t warm-up throws and l laps

Symbol Definition: $d(t,l) = 4.35t + 5.21l - 0.330t^2 - 0.323tl - 0.507l^2 + 11.5$ for $t,l \geq 0$

Assumptions: Certainty and divisibility. Certainty implies that the relationship is exact. Divisibility implies that any fractional values of feet and warm up throws and laps are possible.

a) Find the rate at which the expected distance is changing with respect to the number of warm-up throws in general.
b) Find the rate at which the expected distance is changing with respect to the number of warm-up laps in general.
c) Evaluate $d_t(5,3)$. Write what this *means* in a complete sentence.
d) Evaluate $d_l(6,3)$. Write what this *means* in a complete sentence.

17. The following model has been developed for energy level:

Verbal Definition: $E(c,l)$ = the overall energy level on an average day, on a scale of 0-100, when c minutes are spent on cardiovascular exercises and l minutes are spent on weight lifting.

Symbol Definition: $E(c,l) = 1.22c + 4.41l + 0.0343c^2 - 0.167cl + 0.0589l^2 - 6.74$ for $c, l \geq 0$

Assumptions: Certainty and divisibility. Certainty implies that the relationship is exact. Divisibility implies that any fractional values of energy level and minutes are possible.

a) Find the rate at which the energy level is changing with respect to the minutes spent on cardiovascular exercises in general.
b) Find the rate at which the energy level is changing with respect to the minutes spent on weight lifting in general.
c) Evaluate $E_c(30,30)$. Write what this *means* in a complete sentence.
d) Evaluate $E_l(30,30)$. Write what this *means* in a complete sentence.

18. Suppose you have a friend who runs a small landscaping business. He currently works by himself (has 1 worker) and has $30,000 worth of equipment and machinery. If the productivity function is given by

Verbal Definition: $P(l,c)$ = the number of lawns that can be maintained, on average, when l is the number of full-time workers and c is the dollar value of the machinery and equipment (in $1000s)

Symbol Definition: $P(l,c) = 16l^{0.068}c^{0.26}$ $l, c \geq 0$

Assumptions: Certainty and divisibility. Certainty implies that the relationship is exact. Divisibility implies that any fractional values of lawns, full-time workers and dollars are possible.

a) Calculate the marginal productivity of labor and capital for your friend's business.
b) Interpret your answers to part (a).
c) If your friend has an additional $15,000 to spend on either labor (this would pay for 1 more full-time worker in addition to himself) or capital for next year, what would your answer to part (a) suggest would be the better use of his money (if he has to use it all for labor or all for capital)?
d) Now use the productivity function itself to answer the question in part (c).

19. The productivity model of a major chemical manufacturing company is assessed using a Cobb-Douglas production function:

Verbal Definition: $P(l,c)$ = the dollar value (in millions of dollars) of production when l is the expenditure (in millions of dollars) on labor for the coming year and c is the expenditure (in millions of dollars) on capital investment for the coming year

Symbol Definition: $P(l,c) = 75l^{0.35}c^{0.65} \quad l,c \geq 0$

Assumptions: Certainty and divisibility (continuity).

Assumptions a) Compare the marginal productivity of labor and capital when $l = 2000$ and $c = 1000$.

b) Interpret your answers.

20. The volume of a cylinder is given by: $V = \pi r^2 h$, where r is the radius (say, in inches), h is the height or altitude (also in inches), and V is the volume (in cubic inches).

a) At what rate is the volume changing with respect to the height? Can you explain why this makes sense?
b) At what rate is the volume changing with respect to the radius? Can you explain why this makes sense?

21. A company makes two types of portable CD players, the deluxe random-select 6-disk model and the regular 6-disk model. They have determined that the weekly revenue and cost functions (both in dollars) are given by:

$R(d,r) = 100d + 120r - .02dr - .07d^2 - .03r^2$ $\qquad d$ = number of deluxe

$C(d,r) = 10d + 6r + 32000$ $\qquad r$ = number of regular

($d,r \geq 0$ for both functions)

a) Find the profit function $P(d,r)$ (also in dollars).
b) Find $P_d(150,200)$ and interpret the results.
c) Find $P_r(150,200)$ and interpret the results.

Section 8.2: Finding Local Extrema of Multivariable Functions

In the last section we continued our discussion of multivariable functions. We learned how to find the rate of change of the output when one of the inputs was changing and the other inputs were being held constant. We called this rate of change the partial derivative of the function. We studied the case of the furniture manufacturer and we calculated how the profit would change when the quantities of chairs and tables changed. This was quite interesting and certainly useful, but it would be much more useful to know exactly what quantities of chairs and tables should be sold to maximize the profit. We will now use these partial derivatives to help find the input values that will optimize the output, by finding the potential **local extrema**. The basic idea is that, analogous to setting the derivative equal to 0 and solving for the input variable in single-variable optimization, here we set *all* of the *partial* derivatives equal to zero and solve the resulting *system* of equations for *all* of the input variables.

Here are the kinds of problems that the material in this section will help you solve:

- You are the owner of a small company that produces just two products, chairs and tables. You have developed models for the prices you would have to charge to sell different numbers of each type of furniture. Assuming you will use these demand functions to determine the prices you will charge, you have also formulated a model for the total profit from the sale of the chairs and tables as a function of the number of each made each month. How many chairs and tables should you make to maximize your profit?
- You have developed a model for your energy level in a day as a function of the amount of exercise you get that day and the amount of calories that you consume during the day. On average, how much exercise should you get each day and how many calories should you consume each day to maximize your energy level?
- You run a small coffeehouse that has featured the same act for three years running. You have kept records of the price charged per ticket and the number of tickets sold. You want to determine the best-fit linear model for the number of tickets sold as a function of the price charged per ticket. What are the best slope and y-intercept values to minimize the SSE?

After studying this section, in addition to being able to solve problems like those above, you should:

- Know how to find all the partial derivatives of a multivariable function.

- Understand what setting the partial derivatives equal to 0 corresponds to graphically for functions of two variables, and why this helps find potential optimal solutions.
- Know ways to solve the system of equations resulting from setting the partial derivatives equal to 0.
- Know that the solution(s) to the resulting system of equations are called **critical points**, and are *candidates* for optimal solutions (possible local and global extrema, meaning maximum or minimum points), but are not necessarily optimal.
- Know how to *interpret* the results of the solution to the system of equations resulting from setting the partial derivatives equal to 0.

Sample Problem 1: You are the owner of a small company that produces just two products, chairs and tables. You have developed models for the prices you should charge for each type of furniture to be able to sell a given number of each, and from these demand functions you have derived a model for the total profit from the sale of the chairs and tables as a function of the number of each you make each month.

Verbal Definition: $P(c,t)$ = the profit, in dollars, on average, when c chairs and t tables are made each month

Symbol Definition: $P(c,t) = 140c + 200t + 2ct - 4c^2 - 12t^2 - 700$, for $c,t \geq 0$

Assumptions: Certainty and divisibility. Implied assumptions are that you will charge the prices needed to sell c chairs and t tables as determined by the demand functions, and that consumers will buy exactly the numbers predicted by the demand functions. Divisibility implies that any fractional number of dollars, chairs, and tables are possible.

Find the number of chairs and tables to make on average each month to maximize the expected profit.

Solution: Recall that in our study of smooth functions with one input variable we were able to determine the input value that gave the maximum (or minimum) output. Looking at the graph of the curve, this global extremum, if it was also a local extremum (rather than an endpoint), was a point at which the tangent line to the curve was horizontal.

If a tangent line is horizontal, it has a slope of zero. (This is different from a vertical line, which has *no slope*, or an *undefined slope*, since the slope formula causes division by 0.) The slope of the tangent line at a point on a curve that represents a function is another name for the instantaneous rate of change of the output with respect to the input, or the derivative of the function. Therefore, if we find the derivative of the function and find an input value for which the derivative is equal to zero, this point could give a maximum (or minimum) of the function. Thus for the function $y = f(x)$, we can find the derivative $f'(x)$, set the derivative equal to 0, $f'(x) = 0$, and solve for x . At any value of x where $f'(x) = 0$, $f(x)$ may have a local maximum or minimum, but it could also be a point of inflection. We have said that a point where the derivative equals 0 is called a **critical point**, and it is a *candidate* for a local or global extremum (maximum or minimum).

Look at the graph of the function for the profit from the sale of the chairs and tables, shown in Figure 8.2-1 . Instead of a **curve** defined by a function of one variable, we now have a **surface**, a three-dimensional figure. To find the maximum point of this surface, we must find the **plane** that is tangent to it and horizontal (parallel to the c-t plane here, or the x-y plane in standard form). In other words, we want to "balance" a level *plane* on top of the surface. This is *analogous* to finding a horizontal tangent *line* for a function of one variable.

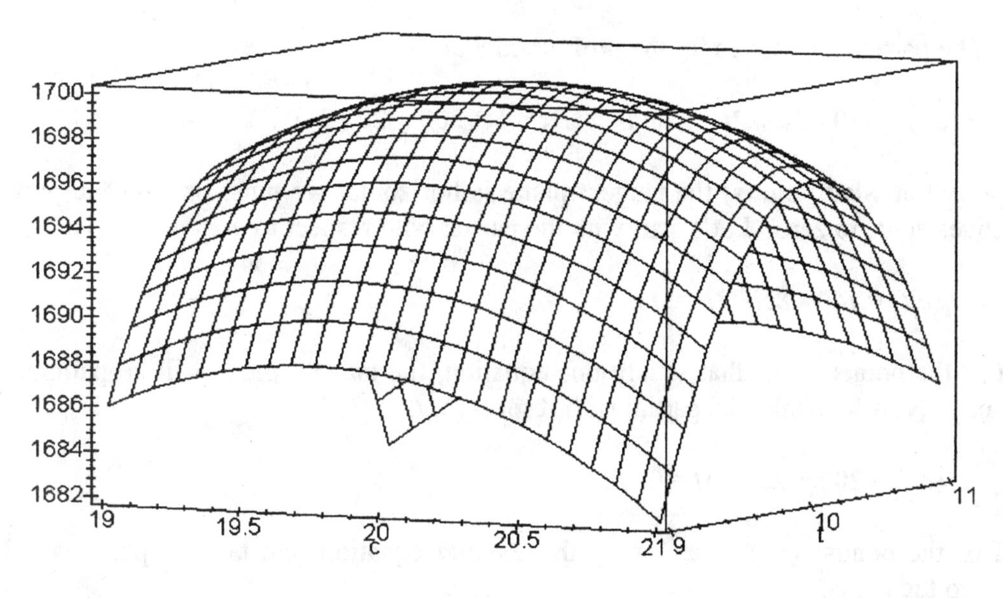

Figure 8.2-1

We want to find a point where the tangent plane is horizontal (for example, the top of the box frame around the graph in Figure 8.2-1 appears to be very nearly a horizontal tangent plane at the maximum point). In the last section, we saw that, for functions of two variables, by selecting a fixed value of a variable being held constant, we *sliced* the surface with a vertical plane corresponding to the fixed value, producing a curve. The partial derivative is then the *slope* of that sliced curve. If you think about it, you will see that, **if the tangent plane is horizontal, then the tangent lines to the sliced-off curves perpendicular to the independent variable axes at the optimal point will also be horizontal, and so will have slopes equal to 0**.

In Figure 8.2-1 , we can see that the maximum point on the surface occurs somewhere around the values $c = 20$ and $t = 10$. If we hold c fixed at 20, we slice the surface to get a parabola looking a lot like the one on the near right face of the box, but going through the middle of the box. It looks reasonable that at the point on that sliced-off parabola where t is 10, the tangent to the sliced-off parabola is horizontal, and so has a slope of 0. This means that the partial derivative with respect to t, evaluated at approximately $c = 20$ and $t = 10$, should be equal to 0. Similarly, if we hold t fixed at 10, we slice the surface the other way, and again at the maximum point the tangent will be horizontal, and so the partial derivative with respect to c will be 0 there.

For single-variable functions, we set *the* derivative equal to 0 and solved for the single input variable. For multivariable functions, we **set *all* of the partial derivatives equal to 0 and solve for *all* of the input variables**. This means that we **solve a *system* of simultaneous equations.**

The partial derivatives for the profit model are:

$$P_c(c,t) = 140 - 8c + 2t \quad \text{and} \quad P_t(c,t) = 200 + 2c - 24t \ .$$

To find out at what point(s) the tangent plane is horizontal we must first set the partial derivatives equal to zero. Let's start with the partial with respect to c:

$$P_c(c,t) = 140 - 8c + 2t = 0$$

At all of the points (c,t) that satisfy this equation, the tangent plane will be parallel to the c-axis. Now let's take the partial with respect to t:

$$P_t(c,t) = 200 + 2c - 24t = 0 \ .$$

At all of the points (c,t) that satisfy this *second* equation, the tangent plane will be parallel to the t-axis.

At all of the points (c,t) that satisfy *both* of these equations, the tangent *plane* will be parallel to *both* the c-axis and the t-axis, and so the entire tangent *plane* will be horizontal, since the function is smooth.

As in the single-variable case, we call such a point a **critical point**; in the case of smooth multivariable functions, this means **a point where *all* of the partial derivatives are equal to 0**. As before, critical points are **potential local and global extrema**.

Since both of these equations must be satisfied, we must solve the system of equations. These equations are linear, so we can use any of the methods that we discussed in Sections 7.3 and 7.4 . The equations are:

$$140 - 8c + 2t = 0$$
$$200 + 2c - 24t = 0$$

The equivalent system of linear equations is:

$$-8c + 2t = -140$$
$$2c - 24t = -200$$

If we use matrix equations to solve the system we have:

$$\mathbf{A} = \begin{bmatrix} -8 & 2 \\ 2 & -24 \end{bmatrix}, \ \mathbf{X} = \begin{bmatrix} c \\ t \end{bmatrix}, \text{ and } \mathbf{B} = \begin{bmatrix} -140 \\ -200 \end{bmatrix}$$

By hand or using technology, we find that $\mathbf{X} = \mathbf{A}^{-1}\,\mathbf{B} = \begin{bmatrix} 20 \\ 10 \end{bmatrix}$.

Therefore, the only critical point for the profit function is to make 20 chairs and 10 tables each month. The corresponding average monthly profit would be:

$$\begin{aligned} P(20,10) &= 140(20) + 200(10) - 4(20)^2 + 2(20)(10) - 12(10)^2 - 700 \\ &= 2800 + 2000 - 1600 + 400 - 1200 - 700 = 1700 \end{aligned}$$

Translation: If you make 20 chairs and 10 tables per month, the corresponding profit would be \$1700 . This *looks* like the optimal solution to maximize the profit, but we won't really *know* until we learn how to *test* it in Section 8.4 . □

Sample Problem 2: You have determined that your energy level on a typical day as a function of your food intake and your exercise can be modeled by:

Verbal Definition: $E(c,x)$ = Your energy level, on average, on a scale of 0-100, for a day when you consumed c calories and got x minutes of exercise

Symbol Definition: $E(c,x) = -93.1 + 0.260c + 0.146x - 0.0000944c^2 + 0.0000410cx - 0.00196x^2$ for $1000 \le c \le 2400,\ 0 \le x \le 120$.

Assumptions: Certainty and divisibility. Certainty implies that the relationship is exact. Divisibility implies that any fractional values of energy level, calories and minutes are possible.

How many calories should you consume and how much exercise should you get each day to maximize your energy for that day?

Solution: In order to solve this problem, we need to find the partial derivatives:

$$\frac{\partial E}{\partial c} = 0.260 - 0.0001888c + 0.0000410x$$

$$\frac{\partial E}{\partial x} = 0.146 + 0.0000410c - 0.00392x$$

Setting the two partial derivatives equal to zero, we have:

$$0.260 - 0.0001888c + 0.0000410x = 0$$
$$0.146 + 0.0000410c - 0.00392x = 0$$

The resulting system is:

$$-0.0001888c + 0.0000410x = -0.260$$
$$0.0000410c - 0.00392x = -0.146$$

For this problem we can definitely see the advantage of setting up a matrix equation and solving the matrix equation using technology rather than trying to do elimination, substitution, Gaussian elimination, or matrix inversion by hand.

$$\mathbf{A} = \begin{bmatrix} -0.0001888 & 0.0000410 \\ 0.0000410 & -0.00392 \end{bmatrix} \quad \text{and} \quad \mathbf{B} = \begin{bmatrix} -0.260 \\ -0.146 \end{bmatrix}, \quad \text{with} \quad \mathbf{X} = \begin{bmatrix} c \\ x \end{bmatrix},$$

$$\mathbf{X} = \mathbf{A}^{-1}\,\mathbf{B} \approx \begin{bmatrix} 1390 \\ 51.8 \end{bmatrix}$$

Thus the only critical point corresponds to consuming approximately 1400 calories per day, on average, and exercising about 52 minutes per day, on average. Your corresponding energy level would then be:

$$E(1390,51.8) =$$

$$-93.1+.260(1390)+.146(51.8)-.0000944(1390)^2+.0000410(1390)(51.8)-.00196(51.8)^2$$

$$\approx 91.2 \ .$$

Translation: If you eat about 1390 calories and exercise about 51.8 minutes,[1] you can expect your energy level to be about 91.2 on a scale of 0 to 100. This appears to be the global maximum, but, again, we need to test it to see if it is indeed a local or global maximum. We will do this in Section 8.4 . □

Minimizing SSE for Least Squares Regression

Sample Problem 3: For the coffeehouse problem discussed in Section 5.5 and in Chapters 1-3, the data on concert price and attendance were:

Price	Attendance
7	100
8	80
9	70

The SSE for the linear model $y = mx + b$ and the given data was found to be

$$\text{SSE}(m,b) = (100-7m-b)^2 + (80-8m-b)^2 + (70-9m-b)^2$$

Find the critical points, that is, the candidates for combinations of values of m and b that minimize the SSE (to eventually find the least-squares *linear* regression model).

Solution: The partial derivatives of this function are:

$$\begin{aligned} \text{SSE}_m(m,b) &= 2(100-7m-b)(-7) + 2(80-8m-b)(-8) + 2(70-9m-b)(-9) \\ &= -3940 + 388m + 48b \\ \text{SSE}_b(m,b) &= 2(100-7m-b)(-1) + 2(80-8m-b)(-1) + 2(70-9m-b)(-1) \\ &= -500 + 48m + 6b \ . \end{aligned}$$

Once again, we set the partials equal to 0:

$$\begin{aligned} \text{SSE}_m(m,b) &= -3940 + 388m + 48b = 0 \quad \text{and} \\ \text{SSE}_b(m,b) &= -500 + 48m + 6b = 0 \ , \end{aligned}$$

---

[1] For our final conclusion in this problem, we would probably round off more, but at this stage we are looking for the mathematical optimal solution, so we retain more precision than we expect to need in the final conclusion.

The equivalent system is:

$$\begin{aligned} 388m + 48b &= 3940 \\ 48m + 6b &= 500 \ . \end{aligned}$$

Setting this up as a matrix equation and solving we have:

$$\mathbf{X} = \begin{bmatrix} m \\ b \end{bmatrix} = \begin{bmatrix} 388 & 48 \\ 48 & 6 \end{bmatrix}^{-1} \begin{bmatrix} 3940 \\ 500 \end{bmatrix} \approx \begin{bmatrix} -15 \\ 203.33 \end{bmatrix} .$$

Translation: The only critical point is given by $m = -15$ and $b \approx 203.33$, so the candidate for the optimal least-squares linear demand model is

$$y = f(x) = mx+b = (-15)x + (203.33) = 203.33 - 15x \ .$$

This would mean that, if the coffeehouse charged \$8 , the model would predict paid attendance of $f(8) = 203.33 - 15(8) = 203.33 - 120 = 83.33$. Recall that the actual attendance when the price was \$8 was 80, so this model is reasonably close to the actual value, and does seem likely to indeed be the optimal linear model. As above, we will test this critical point for optimality in Section 8.4 .

A note on accuracy: for a model like this, it probably does not make much sense to carry any decimal places in the constant, so $f(x) = 203 - 15x$ would probably be a good working model, although it would have a slight bias on the low side of 0.33 . □

Saddle Points

For a point to be an optimal point, we find all of the partial derivatives of the function, set the partial derivatives equal to zero, and solve the resulting system of equations. Once again, an optimal solution *must* satisfy these conditions if the surface is smooth (that is, they are **necessary conditions** for optimality), but just because they are satisfied does not mean a point is optimal. The following example shows this:

Sample Problem 4: Consider the function $z = f(x,y) = x^2 - y^2$. Find its critical points.

Solution: The partial derivatives of this function are:

$$\begin{aligned} f_x(x,y) &= 2x \\ f_y(x,y) &= -2y \end{aligned}$$

Setting the partial derivatives equal to zero, we get the following system:

$$2x = 0$$
$$-2y = 0$$

The only solution for this system is the point (0,0), so this is the unique critical point. The graph of the function is shown in Figure 8.2-2 .

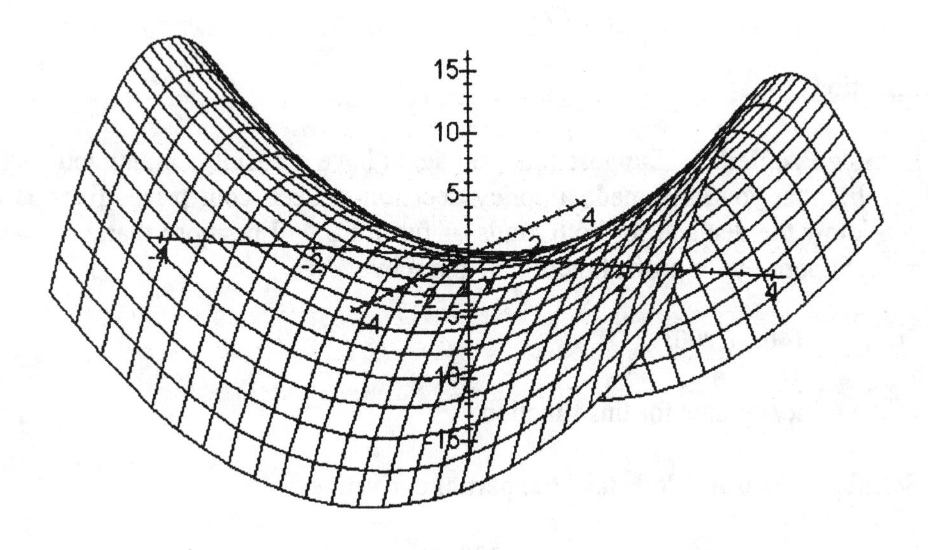

Figure 8.2-2

Notice that this surface is shaped like a *saddle*. The critical point we found, (0,0), corresponds to the point where a mouse would sit on the saddle. At this **saddle point**, the surface is flat (the tangent plane is horizontal there), but it will be neither a maximum nor a minimum (neither the highest nor the lowest point for the region around it). Look at the graph carefully to see why: If you slice the saddle through the origin with a vertical plane *along* the *length* of the horse (parallel to the x-axis; perpendicular to the y-axis), the saddle point is a local minimum. The sliced curve is approximately the parabola opening up that is formed by the top of the surface in Figure 8.2-2 , which you can see has a minimum at the origin. But if you slice the saddle through the origin with a vertical plane *across* the *width* of the horse (parallel to the y-axis; perpendicular to the x-axis), it is a local maximum. If you look at the grid lines going through the origin, you can see the near side of the one cutting across the width of the horse, which forms half of a parabola opening down, whose maximum is at the origin. The back half is hidden behind the saddle.

For smooth functions of two variables, any point that is a local minimum when sliced in one direction and a local maximum when sliced in another direction (has this same general saddle shape) can be called a saddle point.

We can thus see that the point (0,0) is neither a (local or global) maximum point nor a minimum point in 3-D, even though it is a solution to the system of equations resulting from setting both the partial derivatives equal to zero. This is why we need to *test* every critical point, as we will discuss in Section 8.4.

Linear Functions

Sample Problem 5: Suppose that you again have a business manufacturing chairs and tables, but that you have made a policy decision to keep your prices fixed at a level where you know the demand for both kinds of furniture will be more than you can make in a month. Suppose your profit function is given by:

$$P(c,t) = 140c + 200t \ .$$

What are the critical points for this function?

Solution: As usual, let's take the partial derivatives:

$$P_c\,(c,t) = 140 \quad \text{and} \quad P_t\,(c,t) = 200 \ .$$

If we set these equal to 0, we get

$$140 = 0 \quad \text{and} \quad 200 = 0 \ .$$

But these are both impossible! This means that this function has *no* critical points. In fact, this will be true of *all* linear functions except a constant function (such as $z = k$), for which *every* point is a critical point (see if you can explain why!). What does that mean? In this problem, it means that our optimal decision about how many chairs and tables to make will depend heavily on the *domain* of the function; on the *constraints* or restrictions the situation imposes on the decision variables (such as what raw materials you have on hand, how many people are working, etc.). We will focus on this kind of linear optimization problem in Chapter 9. □

General Functions of Two Variables

Sample Problem 6: Suppose you are again making chairs and tables, but now determine that your profit (in dollars) in a particular week is given by the function

$$P(c,t) = -7c^3 + 82c^2 - 5ct - 5t^2 - 182c + 84t - 130$$

Find the critical points for this function.

Solution: As always, let's take the partial derivatives:

$$\frac{\partial P}{\partial c} = -21c^2 + 164c - 5t - 182$$
$$\frac{\partial P}{\partial t} = -5c - 10t + 84$$

Setting these equal to 0, we get:

$$-21c^2 + 164c - 5t - 182 = 0$$
$$-5c - 10t + 84 = 0 \ .$$

How can we solve this system of equations? This time, we do *not* have a *linear* system of equations, so we can't use matrices to solve the system. That brings us back to substitution and elimination. Substitution is the most generally applicable, so let's try it. Since the second equation looks a lot simpler, let's solve for one variable there first. Which one should we solve for? That will be the one we eliminate by substitution into the first equation. Since the first equation has only one linear term involving t, but two terms, one being quadratic, involving c, it will be a little easier to substitute in for t than for c, so let's do that. That means we want to solve the second equation for t first:

$$-5c - 10t + 84 = 0$$
$$-10t = 5c - 84 \qquad \text{(adding } 5c - 84 \text{ to both sides)}$$
$$t = \frac{5c-84}{-10} = -0.5c + 8.4 \qquad \text{(dividing both sides by 10 and simplifying)}$$

Now we can substitute $t = -0.5c+8.4$ in for t in the first equation:

$$-21c^2 + 164c - 5t - 182 = 0$$
$$-21c^2 + 164c - 5(-0.5c+8.4) - 182 = 0 \qquad \text{(substituting)}$$
$$-21c^2 + 164c + 2.5c - 42 - 182 = 0 \qquad \text{(multiplying out)}$$

$-21c^2 + 166.5c - 224 = 0$ (combining like terms)

We are left with a quadratic equation to solve. We could use the solve process on your calculator or spreadsheet, or we could use the quadratic formula. Just for the sake of review, let's try the quadratic formula, which says that if $ax^2 + bx + c = 0$, then

$$x = \frac{-b \pm \sqrt{b^2 - 4ac}}{2a}$$

In our case, $a = -21$, $b = 166.5$, and $c = -224$, and our variable is c instead of x . In this situation, we are coming close to using the same symbol (c) to stand for two different things. This is not a good idea in general! But it's a little late to change our symbols in mid-problem, so let's think of the c from the quadratic formula discussion as if it were an upper case C . In any case, we find that our solution for our variable representing the number of chairs is given by

$$c = \frac{-166.5 \pm \sqrt{(166.5)^2 - 4(-21)(-224)}}{2(-21)} = \frac{-166.5 \pm \sqrt{27722.25 - 18816}}{-42}$$

$$= \frac{-166.5 \pm \sqrt{8906.25}}{-42} \approx \frac{-166.5 \pm 94.373}{-42} \approx 3.964 \pm 2.247 = 1.717 \text{ or } 6.211$$

This suggests that we are going to have *two* critical points! Recall that this is possible for cubic functions of one variable, so it makes sense that it can happen in a cubic function of two variables as well. Now we can find the values of t that correspond to these two values of c by substituting the c values into the expression we got for t in terms of c :

$t = -0.5c + 8.4$.

Thus, if $c = 1.717$, we find that

$t = -0.5(1.717) + 8.4 = -0.8585 + 8.4 = 7.5415$,

and if $c = 6.211$, we find that

$t = -0.5(6.211) + 8.4 = -3.1055 + 8.4 = 5.2945$.

This means that our two critical points are approximately $(1.72, 7.54)$ and $(6.21, 5.29)$. If we plug these into our function and evaluate using technology, we find that

$P(1.72, 7.54) \approx 48.2$ and $P(6.21, 5.29) \approx 366$

This *suggests* that the first critical point might be a local minimum and the second a local maximum, but for all we know one or both could also be a saddle point! Again, we need to be able to test these critical points to find out, and we will do that in Section 8.4 . □

Example 1: The volume of a box of width w inches and height h inches, and a maximum allowable total measurement (length plus girth) of 108 inches, which we previously formulated in Section 5.2 (see Sample Problems 7, 8, and 10), is given by:

$$V(w,h) = 108wh - 2w^2h - 2wh^2.$$

The partial derivatives of this function are:

$$V_w(w,h) = 108h - 4wh - 2h^2 \text{ and}$$
$$V_h(w,h) = 108w - 2w^2 - 4wh$$

Setting these partials equal to zero will again *not* result in a system of *linear* equations, so cannot be solved quickly using matrix theory.

Such systems can often be solved by substitution or elimination, as we did in Sample Problem 6 above. This particular example would be quite difficult, since to solve one of the equations for one variable, you would need to use the quadratic formula (involving a square root of the other variable), which expression would then be plugged into the other equation. In this course, we do not expect you to be able to solve systems of equations as complicated as this, although you could use a spreadsheet to help you, as we will see in Section 9.4. In Section 8.3 we shall see how to solve optimization problems like this using optimization procedures available on most spreadsheets (without taking partial derivatives first).

<u>Functions of Three or More Variables</u>

Sample Problem 7: Find the critical points for the function

$$S(r,h,\lambda) = 2\pi r^2 + 2\pi rh - (\lambda)(\pi r^2h - 32) .$$

Solution 7: For your edification: λ is the Greek letter "lambda." We now have three variables, and this function is not only *not* quadratic, it's not even cubic. If multiplied out, the third term would be

$$(\lambda)(\pi r^2h) = \pi r^2h\lambda .$$

Now, π is the constant you remember fondly from high school geometry ($\pi \approx 3.14$), so is not a variable. But the other three letters ($r, h,$ and λ) are all variables. Both h and λ are raised to the *first* power, and r is raised to the *second* power. The sum of the exponents of all the variables in that term is $2+1+1 = 4$. This means we would say the **degree** of the term is 4. In general, **the degree of a term is the sum of the exponents of its variables. When a function is made up of a sum of terms, each of which is just a product of variables to whole-number powers times a coefficient, we call the function a *polynomial*. The degree of a polynomial is the highest degree of all the terms with nonzero coefficients.** We can see this even more clearly by multiplying out our function:

$$S(r,h,\lambda) = 2\pi r^2 + 2\pi rh - (\lambda)(\pi r^2h - 32) = 2\pi r^2 + 2\pi rh - \pi r^2h\lambda + 32\lambda \ .$$

Each term is a product of variables to integer powers with a coefficient. The degrees of the terms are 2, 2, 4, and 1, in order from left to right, and all have nonzero coefficients, so the degree of the polynomial is 4.

To find the critical points, as usual we take the partial derivatives and set them equal to 0:

$$S_r(r,h,\lambda) = 4\pi r + 2\pi h - 2\pi rh\lambda = 0$$
$$S_h(r,h,\lambda) = 2\pi r - \pi r^2\lambda = 0$$
$$S_\lambda(r,h,\lambda) = -\pi r^2h + 32 = 0$$

How can we solve this system? Let's try substitution again. There are many possibilities for which variable to solve for first. We could solve for h or λ fairly easily in the first equation, or for r or λ in the second equation, or for h in the third equation. The simplest of these choices seems like it might be to solve the second equation for λ, since there are only two terms and we can divide through by π. Let's see what happens. The equation is

$$2\pi r - \pi r^2\lambda = 0$$

As we said above, we can divide both sides by π (since it is not 0) to get

$$2r - r^2\lambda = 0$$

Now, you might be tempted to say "Let's do the same with the r!" If we did that, we would get:

$$2 - r\lambda = 0$$
$$-r\lambda = -2 \qquad \text{(subtracting 2 from both sides)}$$

Next, to solve for λ, you would want to divide both sides by $-r$. This would leave you with

$$\lambda = \frac{-2}{-r} = \frac{2}{r}$$

This expression for λ only makes sense if $r \neq 0$. What happens if r is 0? Well, if we look back at the original equation in the form

$$2r - r^2\lambda = 0 ,$$

we see that $r = 0$ *does* satisfy the equation, and that in fact λ could take on *any* value in conjunction with $r = 0$.

How did we lose all those solutions? *By dividing both sides of the equation by a variable, which could be 0!* A better approach is to *factor* the equation:

$$2r - r^2\lambda = 0$$
$$r(2 - r\lambda) = 0$$

In this form, we can remember that **$uv = 0$ is equivalent to $u = 0$ or $v = 0$**. This means that

$$r(2 - r\lambda) = 0 \quad \text{is equivalent to} \quad r = 0 \text{ or } (2 - r\lambda) = 0 .$$

In other words, the solution to the equation is

$$r = 0 \text{ (and } \lambda \text{ } \textit{any} \text{ real number)} \quad \text{or} \quad r \neq 0 \text{ and } \lambda = \frac{2}{r}$$

The lesson from this is **if you divide both sides of an equation by an expression involving a variable, you can *lose* some solutions. To avoid this, *factor* and solve rather than dividing. Or, if necessary, at least explore what happens when the expression being divided by is equal to 0. If this results in one or more solutions to the original equation, those are the solutions that would otherwise be lost.**

Notice that, technically, we should have used this idea when solving the equation

$$r\lambda = 2 ,$$

since we divided both sides by r to solve for λ. In this case, $r = 0$ is *not* a solution, so we rule *out* any solution with $r = 0$. This is the same as saying that our solution $\lambda = \frac{2}{r}$ assumes $r \neq 0$. We knew this anyway, however, because we could *see* the r in the denominator of the fraction, and know that division by 0 is undefined.

If we go back to our original system of equations, we can see that the $r = 0$ solution will *not* satisfy the third constraint, $-\pi r^2 h + 32 = 0$, so the only solution that *could* satisfy the entire system is $\lambda = \frac{2}{r}$ (with $r \neq 0$). Notice that we can also solve the third constraint easily for h in terms of r:

$$-\pi r^2 h + 32 = 0$$

$$-\pi r^2 h = -32 \qquad \text{(subtracting 32 from both sides)}$$

$$h = \frac{-32}{-\pi r^2} = \frac{32}{\pi r^2} \qquad \text{(dividing both sides by } \pi r^2\text{)}$$

STOP!! Notice that we divided both sides of the equation by an expression involving the variable r, so we need to follow the advice above. This time, however, the value that makes the expression that is being divided *by* equal to 0 is $r = 0$, but this value cannot be part of a solution, as we have already mentioned. Once again, this simply means that our solution of $h = \frac{32}{\pi r^2}$ assumes that $r \neq 0$.

Where are we? Well, we have solved for both λ and h in terms of r, so we could now plug both of these expressions back into the first equation, and see if we can solve it for r:

$$4\pi r + 2\pi h - 2\pi r h \lambda = 0$$

$$4\pi r + 2\pi\left(\frac{32}{\pi r^2}\right) - 2\pi r\left(\frac{32}{\pi r^2}\right)\left(\frac{2}{r}\right) = 0 \qquad \text{(plugging in for } h \text{ and } \lambda\text{)}$$

$$4\pi r + \frac{64}{r^2} - \frac{128}{r^2} = 0 \qquad \text{(multiplying out)}$$

$$4\pi r - \frac{64}{r^2} = 0 \qquad \text{(combining like terms)}$$

$$4\pi r = \frac{64}{r^2} \qquad \text{(adding } \frac{64}{r^2} \text{ to both sides)}$$

At this point, you probably want to multiply both sides by r^2. In a sense, this is the *opposite* of the dividing by an expression involving a variable. In that case, you can *lose* solutions. In this case, since the denominator is r^2, we know that r can't be 0, but if we multiply through by r^2, this will no longer be apparent. **When multiplying both sides of an equation by an expression involving a variable, "solutions" can be *added* that do not satisfy the original equation (sometimes called "*spurious roots*"). To make sure your solutions are correct, be sure to check them in the original equation.**

If we do multiply through by r^2, we get:

$$4\pi r^3 = 64$$

$$r^3 = \frac{64}{4\pi} = \frac{16}{\pi} \qquad \text{(dividing both sides by } 4\pi \text{ and simplifying)}$$

$$r = \sqrt[3]{\frac{16}{\pi}} \qquad \text{(taking the cube root of both sides)}$$

As the advice above suggested, we need to check this back in the original equation to make sure it is not a spurious root. Essentially, this means we need to make sure we will not get a 0 in a denominator. Since one form of the original equation was

$$4\pi r = \frac{64}{r^2},$$

we see that our solution, $r = \sqrt[3]{\frac{16}{\pi}} \approx 1.72$, does *not* make the denominator 0, so it is in fact the only solution to the equation.

We finally have part of our solution! All that remains now is to substitute this value of r back into the expressions for λ and h:

$$\lambda = \frac{2}{r} = \frac{2}{\sqrt[3]{\frac{16}{\pi}}} \approx 1.16 \quad \text{and} \quad h = \frac{32}{\pi r^2} = \frac{32}{\pi\left(\sqrt[3]{\frac{16}{\pi}}\right)^2} \approx 3.44$$

Thus our only critical point is (1.72,3.44,1.16). As usual, this would need to be tested to see if it is a local or global extremum. □

Section Summary

Before you begin the exercises be sure that you:

- Know that to optimize a smooth multivariable function, you can first find all of the partial derivatives of the multivariable function (one for *each* independent variable).
- Know that to optimize a smooth multivariable function, you can set the partial derivatives equal to zero and solve the resulting system of equations for the independent variables.
- Know ways to solve the system of equations resulting from setting the partial derivatives equal to 0, including substitution and elimination, as well as matrix theory for linear systems.
- Know that, for a smooth function of two variables, if a point satisfies these conditions (all partial derivatives equal to 0), then the curves sliced by vertical planes perpendicular to each of the two independent variables' axes will have horizontal tangent lines, and that these two horizontal tangent *lines* will determine a **horizontal tangent *plane*** to the surface at the point.
- Know that any solutions of the system of equations resulting from setting the partial derivatives equal to 0 represent points (called **critical points**), each of which is a candidate to be a locally optimal solution (a local extremum), and therefore could also be a global optimum. If it is indeed the global optimum (we will discuss tests in Section 8.4), then plugging in these values of the independent variables into the function will give the optimal value of the dependent variable.
- Recognize that not all of the solutions of the resulting system will *necessarily* be either a maximum or minimum point, but could, for example, be saddle points, and that they could correspond to a *minimum* when you wanted to find a *maximum*, or vice versa.
- Know that, when solving an equation, if you *divide* both sides of the equation by an expression involving a variable, you need to check to see if that variable being 0 could be part of a solution, since otherwise you could lose solutions (you might not find *all* possible solutions).
- Know that, when solving equations, if you *multiply* both sides of the equation by an expression involving a variable, you need to *check* all of your solutions to make sure they satisfy the *original* equation, since otherwise you could end up with spurious roots.

EXERCISES FOR SECTION 8.2

Warm Up

1. $R(x,y) = -3x^2 - 2y^2 + 3xy + 4y - 20$
   a) Find all critical points for the function.
   b) Find the value of the function at each critical point.

2. $P(x,y) = -2x^2 - y^2 + 2xy - 4y + 3x + 24$
   a) Find all critical points for the function.
   b) Find the value of the function at each critical point.

3. $C(f,d) = -50 + 3.5f + 7.2d + 3.5fd - 4.7f^2 - 3.9d^2$
   a) Find all critical points for the function.
   b) Find the value of the function at each critical point.

4. $E(c,r) = -2.8c^2 - 3.7r^2 + 4.3cr + 3.0c - 2.1r + 10$
   a) Find all critical points for the function.
   b) Find the value of the function at each critical point.

5. $S(m,b) = (5-m-b)^2 + (6-2m-b)^2 + (9-3m-b)^2$
   a) Find all critical points for the function.
   b) Find the value of the function at each critical point.

6. $S(m,b) = (5-8m-b)^2 + (4-6m-b)^2 + (6-4m-b)^2 + (9-2m-b)^2$
   a) Find all critical points for the function.
   b) Find the value of the function at each critical point.

7. $f(x,y,z) = 4x^2 + y^2 - 8xz - 2yz + 20z$
   a) Find all critical points for the function.
   b) Find the value of the function at each critical point.

8. $f(x,y,z,\lambda) = x + y + z - xyz\lambda + 8\lambda$
   a) Find all critical points for the function.
   b) Find the value of the function at each critical point.

9. $f(x,y) = 3x - 5y + 21$
   a) Find all critical points for the function.
   b) Find the value of the function at each critical point.

10. $f(s,t) = -.2s + 3.9t + 12$
    a) Find all critical points for the function.
    b) Find the value of the function at each critical point.

Game Time

11. The profit function for a firm selling specialized flooring materials can be modeled by the expression:

$$\pi(f,d) = -22f^2 + 40.2df + 201.8f - 26d^2 + 170.9d - 237.4$$

where f = the price in dollars for one square yard of the Florentine tiles and d = the price in dollars for one square yard of the Delmonico tiles. Find the critical points for this profit function, and the profit each corresponds to.

12. The cost of manufacturing a box with restrictions on its measurements of 108 inches for the total length plus the girth, at $2 per square foot for the sides and top and $3 per square foot for the bottom, is modeled by:

Verbal Definition: $C(w,h)$ = the cost, in dollars, to manufacture a box of width w inches and height h inches, given that the length plus the girth equals 108 inches.

Symbol Definition: $C(w,h) = 5w(108 - 2w - 2h) + 4hw + 4h(108 - 2w - 2h)$ for $w + h < 54$, $w,h \geq 0$

Assumptions: Certainty and divisibility. Certainty implies that the relationship is exact. Divisibility implies that any fractional values of dollars and inches are possible.

    a) Find the critical points of the cost function.
    b) Find the cost of the box at each critical point.

13. If a box costs $2 per square foot for the sides and top and $3 per square foot for the bottom, and if it must have a volume of 6 cubic feet, then the cost of the box is modeled by:

Verbal Definition: $C(w,l)$ = the cost, in dollars, to make a box of width w feet and length l feet, given that the volume is exactly 6 cubic feet

Symbol Definition: $C(w,l) = 5wl + \frac{24}{l} + \frac{24}{w}$, for $w,l > 0$.

Assumptions: Certainty and divisibility. Certainty implies that the relationship is exact. Divisibility implies that any fractional values of dollars, feet, and cubic feet are possible.

a) Find the critical points of the cost function.
b) Find the cost of the box at each critical point.

14. If the cost to produce Fudge Cakes and Cream Cakes can be modeled by :

Verbal Definition: $C(f,c)=$ cost, in dollars, to make Fudge Cakes sold at f dollars per cake and Cream Cakes sold at c dollars per cake

Symbol Definition: $C(f,c)=21.71f+3.54c$ for $f,c\geq 0,$

Assumptions: Certainty and Divisibility. Certainty implies that the relationship is exact. Divisibility implies that any fractional values of dollars and cakes are possible.

and the revenue from the sales of both of the cakes can be modeled by:

Verbal Definition: $R(f,c)$ = the revenue, in dollars, when Fudge Cakes are sold at f dollars each and Cream Cakes at c dollars each, assuming the quantity made of each is calculated from the demand function.

Symbol Definition: $R(f,c)=-0.43f^2+19.56f+1.45fc-1.93c^2+13.97c$ for $f,c\geq 0,$

Assumptions: Certainty and Divisibility. Certainty implies that the relationship is exact. Divisibility implies that any fractional values of dollars and cakes are possible

a) Find the critical points of the **profit** function.
b) Find the profit at each critical point.

15. The demand of two hockey jerseys, an officially sanctioned jersey and a replica can be modeled by:

Verbal Definition: $O(o,r)$ = the price to charge, in dollars, per official jersey, to sell o official jerseys and r replica jerseys.

Symbol Definition: $O(o,r)=-4.98o+3.03r+74.5$ for $o,r\geq 0$

Assumptions: Certainty and divisibility. Certainty implies that the relationship is exact. Divisibility implies that any fractional values of dollars and jerseys are possible.

Verbal Definition: $R(o,r)$ = the price to charge, in dollars, per replica jersey, to sell o official jerseys and r replica jerseys.

Symbol Definition: $R(o,r)=0.936o-2.99r+70.4$ for $o,r\geq 0$

Assumptions: Certainty and divisibility. Certainty implies that the relationship is exact. Divisibility implies that any fractional values of dollars and jerseys are possible.

The officially sanctioned jerseys cost the retailer $19.23 each and the replicas cost $12.38 each. There is a one-time order charge of $50.00.

a) Formulate a model for the *profit* and *fully define* it.
b) Find the critical points of the profit function.
c) Find the profit at each of the critical points.

16. The profit from the sale of CD players can be modeled by:

Verbal Definition: $P(d,r)=$ the profit, in dollars, from the sale of d deluxe and r regular CD players, assuming the price is fixed based on the demand function.

Symbol Definition: $P(d,r)=90d+114r-.02dr-.07d^2-.03r^2-32000$ for $d,r\geq 0$,

Assumptions: Certainty and divisibility. Certainty implies that the relationship is exact. Divisibility implies that any fractional values of dollars and CD players are possible.

a) Find the critical points of the profit function.
b) Find the profit at each critical point.

17. If the profit function for the furniture manufacturer in Sample Problem 1 changes to

$$P(c,t)=150c+200t+2ct-4c^2-12t^2-700\ ,$$

a) Find the new critical points.
b) Find the new profit at each new critical point.

18. If the profit function for the furniture manufacturer in Sample Problem 1 changes to

$$P(c,t)=150c+220t+2ct-4c^2-12t^2-700\ ,$$

a) Find the new critical points.
b) Find the new profit at each new critical point.

19. If the model for the profit from the sale of divinity and fudge is:

Verbal Definition: $P(f,d)$ = the profit, in dollars, from the sale of chocolate fudge at f dollars per pound and divinity fudge sold at d

dollars per pound, where the quantity made is determined from the price according to the demand function

Symbol Definition: $P(f,d) = -18.2f^2 + 199.14f + 36.0fd - 26.0d^2 + 168.8d - 237.4$ for $f,d \ge 0$

Assumptions: Certainty and divisibility. Certainty implies that the relationship is exact. Divisibility implies that any fractional values of dollars and pounds of fudge are possible.

a) Find the critical points of the profit function.
b) Find the profit at each critical point.
c) Comment on your solution.

Section 8.3: Optimization Using a Spreadsheet

In the preceding sections we discussed the optimization of multivariable functions using calculus. To find critical points, or potential optimal solutions, of a particular function, we found the partial derivatives of the function, set the partial derivatives equal to zero and solved the resulting system of equations. If the system was linear, we could use matrix theory and matrix equations to solve it. However, we found that solving the systems of equations when they were not linear was sometimes very difficult. We also found, even for linear systems, that trying to solve by calculus alone was not enough to obtain an optimal solution. In this section we see how we can use technology to help us find optimal solutions in these cases.

Below are the kinds of problems that the material in this section will help you solve:

- You use the United States Postal Service to send a great many packages. You want to order some shipping boxes that will have the maximum capacity and still stay within the limits prescribed by the Post Office. What should the dimensions of the box be?
- You have determined that your batting average over the six months of the baseball season can be modeled by a cubic function. Can you use a spreadsheet to project your highest batting average for this season?
- You have determined the profit function for two of your products based on the prices that you charge for each of the products. When you tried to optimize your profit using calculus, the results did not make sense. Can a spreadsheet help you to find reasonable answers?
- Most airlines restrict the size of luggage that you can carry on, usually by designating a limited overall measurement. What dimensions should the luggage be to maximize the volume and stay within airline restrictions?

When you have finished this section you should be able to answer these questions and:

- Know how to enter data into a spreadsheet so that you can use its optimizing capabilities.
- Know how to use the optimizing function on your computer spreadsheet.
- Know when constraints must be identified, and how to enter simple constraints within a spreadsheet optimizer.
- Know that the optimizing function may need to be rerun with different starting values to get the global maximum or minimum over the entire domain.

Most spreadsheet programs have a feature that allows you to solve or optimize a problem. To use this very powerful feature, **you must first *set up* the spreadsheet**. This involves, at a minimum, selecting a block of adjacent cells to serve as your ***variable* cells**, and typing in the formula for your objective function in a different cell (the ***target* or *solution* cell**), using the variable cells instead of your original variable names.

Once you have set up the spreadsheet, you are ready to *run* the optimization. To start this procedure, select the optimization process on your spreadsheet (typically called "Solver" or "Optimizer" or some such name). The dialogue box for the optimizing feature will typically ask you to do several things:

1. Identify the cell where the function or formula that you wish to optimize, variously called the **target cell** or **solution cell**, is located.
2. Identify the cell or cells in which the numbers that can be changed, the **variable cell(s)**, are located.
3. Indicate whether you want to **maximize or minimize** the objective function.

It is important that you keep these things in mind as you set up your spreadsheet. Remember: the target or solution cell must contain the objective function formula, which must be a function of the variable cells. For more details, see your technology supplement.

Sample Problem 1: For the coffeehouse problem discussed in Section 5.5 and in Chapters 1-3, the data on concert price and attendance were as shown in Table 1.

Table 1

Price	Attendance
7	100
8	80
9	70

The SSE for the linear model $y = mx + b$ and the given data was found to be

$$\text{SSE}(m,b) = (100-7m-b)^2 + (80-8m-b)^2 + (70-9m-b)^2$$

Find the values of m and b that minimize the SSE (to find the least-squares *linear* regression model).

Solution: To find the best-fit linear model, we want to minimize the total squared error. In the last section we saw that we can use calculus to solve this problem: we can take the partial derivatives, set them equal to zero, and solve the resulting system of equations. The function that we want to minimize is not linear, but the system resulting from setting the partial derivatives equal to zero was a linear system, so we were able to us matrix equations to solve it. We can also use the built-in feature of the spreadsheet program to solve this problem.

Table 2

	A	B	C	D	E	F	G	H	I
1	Coffeehouse: Linear Demand			Min D3=+(100-7*A3-B3)^2+(80-8*A3-B3)^2+(70-9*A3-B3)^2					
2	m	b		SSE					
3	-15	200		50		(Initial)			

Table 2 is the original setup of the spreadsheet, *before* the optimization. Note that the formula for SSE is typed out to show what is actually entered in cell D3, and we have also added verbal labels to help understand the setup of the spreadsheet. These are not necessary for using the optimizer, but are very helpful when you go back after a period of time to see what you have done or want to communicate the results to someone else. As a time-saver, if you type in the formula as in cell D1, you can then simply highlight the formula itself (everything to the right of the = sign, including the first + sign), copy it, move the cursor to cell D3, and paste the formula in there. That way, you only need to type the actual formula once.

To invoke the optimizer/solver feature on the spreadsheet, you first select the optimizer, then identify the contents of D3 as the function to be optimized (the target cell), the contents of A3..B3 as the variable cells and indicate that we wanted to *minimize* the function. Note that the cell to be optimized, D3, contains references to the variable cells, A3 and B3. **To *do* the optimization**, you click on the appropriate box within the dialog box, usually labeled "Solve" or "Go". Table 3 shows the spreadsheet as it appears *after* the optimization.

Table 3

	A	B	C	D	E	F	G	H	I
1	Coffeehouse: Linear Demand			Min D3=+(100-7*A3-B3)^2+(80-8*A3-B3)^2+(70-9*A3-B3)^2					
2	m	b		SSE					
3	-15	203.3333		16.66667		(Final)			

Notice that the solution values have *replaced* the initial values we put in for the variables. If you do not put in initial values for the variables, it will assume that they are 0, which could result in an error statement, for example if there is division involved in your formula, and an initial value of 0 results in a division by 0. After completing the optimization, most spreadsheets also give you the option of displaying the solution values *or* the initial values of the variables. The default option is normally to display the solution.

Table 3 is telling us that our optimal value of m is -15 , our optimal value of b is 203.33 , and the minimum SSE is 16.67 . Writing out the equation for the least squares linear regression line, we have:

$$y = mx + b = -15x + 203.33 ,$$

as we saw in Section 8.2 . When using the spreadsheet optimizer, you can normally be fairly confident that the answer you obtain is at least a local optimum (if any exist), but you can't be certain it is globally optimal, so some testing, which we discuss in Section 8.4 , is still a good idea. □

Sample Problem 2: You have to build some containers to ship special pellets. To conform with customs regulations and take advantage of duty pricing the containers must hold exactly 35 pounds of pellets. In the past, you shipped the pellets in 1 foot by 1 foott by 1 foot containers , which held 5 pounds of pellets, but you are now getting larger orders and the small containers are not economical. Because of the weight of the contents, the bases of the containers must be built of stronger (and more expensive) material than the sides and top. The base of the container costs $0.84 per square foot and the sides and top cost $0.62 per square foot. You have developed the following model (see Sample Problem 8 of Section 5.2) for the cost of building a box:

Verbal Definition: $c(l,w)$ = the cost, in dollars, to build a box of length l feet and width w feet.

Symbol Definition: $c(l,w) = 1.46lw + \left(\frac{8.68}{w}\right) + \left(\frac{8.68}{l}\right)$ for $l, w, > 0$

Assumptions: Certainty and divisibility. Certainty implies that the relationship is exact. Divisibility implies that any fractional values of dollars, feet, and cubic feet are possible.

What dimensions should you make the box to minimize the cost?

Solution : The function that defines the relationship between the cost and the length and width is not linear. We can take the partial derivatives and set them equal to zero, but the resulting system of equations is not easy to solve because it is not linear either. Let's try using a spreadsheet to find the solution.

Notice that this problem is a little different than Sample Problem 1 because there are **domain restrictions** on the decision variables. We will discuss such constrained optimization at great length in Chapter 9. For the most part in this chapter, our approach will be to try to **find the unconstrained solution to the problem first. If that unconstrained solution is within the domain, then it is also the solution to the**

constrained problem, so we don't need to concern ourselves with the constraints further.

For this problem, then, we will simply enter the length and width as the variables and the cost as the objective function and use initial or starting values of 1 foot for both the length and width. The initial setup of the spreadsheet before running the optimizer is shown in Table 4.

Table 4

	A	B	C	D	E	F	G
1	Pellet Box			Min D3=+1.46*A3*B3+8.68/A3+8.68/B3			
2	Length (ft)	Width (ft)		Cost ($)			
3	1	1		18.82		(Initial)	

We are ready to invoke the optimizer/solver feature on the spreadsheet. We identify the objective function or target cell as D3, indicate that it is to be minimized, and that the length and width, cells A3..B3, are the variables. The cell to be optimized, D3, contains references to the variable cells. Be sure that you always do this. The spreadsheet cannot optimize a function unless it contains references to the cells that are identified as the variables. The optimized spreadsheet is shown in Table 5.

Table 5

	A	B	C	D	E	F	G
1	Pellet Box			Min D3=+1.46*A3*B3+8.68/A3+8.68/B3			
2	Length (ft)	Width (ft)		Cost ($)			
3	1.811598	1.811598		14.37426		(Final)	

The solution given is that the length and width of the box should both be approximately 1.81 feet and the that the minimum cost is approximately $14.37. Notice that both variable values in the solution are positive, so **the solution is in the domain.** Notice also that the solution does not tell us anything about the *height* of the box. It would be necessary to go back to the original problem to determine the height. The original problem stated that the box must have a volume of 7 cubic feet, and that the cost of the base was $0.84 per square foot and the cost of the sides and top was $0.62 per square foot. It was determined that the height times the width times the length must equal 7,

$$lwh = 7,$$

so the height must be equal to 7 divided by the width times the length:

$$h = \frac{7}{lw}.$$

Using this information we can solve for the height:

$$h \approx \frac{7}{(1.81)(1.81)} \approx 2.14$$

So the box should have a square bottom, approximately 1.81 feet on each side, and a height of approximately 2.14 feet. Once again, this is most likely a local minimum, but we need to test it to see if it is also the global minimum, as we discuss in Section 8.4 . □

Sample Problem 3: The post office has strict regulations about the size package that you can mail first class. You want to be able to ship the maximum amount of goods while staying within the size limits. You called the Post Office and they gave you the following information: The total measurement of each package (the girth, measured around the package, plus the length) must be no more than 108 inches. Find the dimensions of the box that will maximize the volume.

Solution: First we must formulate a model for this problem. As we saw in Sample Problem 9 of Section 5.2 , we can eliminate one of the variables and get the following model:

Verbal Definition: $V(w,h)$ = the volume in cubic inches for a box of width w inches and height h inches.

Symbol Definition: $V(w,h) = 108wh - 2w^2h - 2wh^2$ for $w, h > 0$ and $w + h < 54$

Assumptions: Certainty and divisibility. Certainty implies that the relationship is exact. Divisibility implies that any fractional values of cubic inches and inches are possible

Notice that there are again domain restrictions. As before, our strategy will be to **optimize the unconstrained problem, then check to make sure the domain constraints are satisfied**. Having made a mental note of this, we are now ready to enter our data in a spreadsheet. We designate our variables (the width and the height) using cells A3 and B3, respectively, and enter the objective function (the volume) formula in cell D3, as shown in Table 6.

Table 6

	A	B	C	D	E	F	G
1	Post Office Box			Max D3=+108*A3*B3-2*A3^2*B3-2*A3*B3^2			
2	Width (in)	Height (in)		Volume (in^3)			
3	1	1		104		(Initial)	

We now invoke the optimizer/solver and identify the objective function or target cell as that containing the volume, D3. We want to maximize the volume; the variable cells are those containing the width and height, A3 and B3. The spreadsheet with the optimal solution is shown in Table 7.

Table 7

	A	B	C	D	E	F	G
1	Post Office Box			Max D3=+108*A3*B3-2*A3^2*B3-2*A3*B3^2			
2	Width (in)	Height (in)		Volume (in^3)			
3	18	18		11664		(Final)	

This solution tell us that the width and height should both be 18 inches, which gives a maximum volume of 11,664 inches. Notice that both solution values are positive, and their sum is 36, which is less than 54, as required in the domain, so the unconstrained solution is also the constrained solution. To determine the *length* of the shipping box, we must go back to our derivation of the model in Section 5.2 . We determined there that

$$l = 108 - 2w - 2h,$$

so our length would be

$$l = 108\text{-}2(18)\text{-}2(18) = 36$$

inches. Checking the solution we see that the volume is given by

$$\text{Volume} = lwh = (36)(18)(18) = 11{,}664$$

cubic inches. As before, we should still check to make sure this is a global maximum. □

The two problems above showed how we can use a spreadsheet to find solutions for multivariable functions. Will this work for us all of the time?

Sample Problem 4: If the profit from the sale of homemade fudge and caramels can be modeled by:

Verbal Definition: $P(f,c)$ = the profit, in dollars, from the sale of f pounds of fudge and c pounds of caramels

Symbol Definition: $P(f,c) = +20.1f + 36.7c - 0.430f^2 - 1.42cf - 0.468c^2 - 119$ for $f, c \geq 0$

Assumptions: Certainty and divisibility. Certainty implies that the relationship is exact. Divisibility implies that any fractional values of dollars and pounds are possible.

Find the quantities of fudge and caramels that should be sold to maximize the profit.

Solution : The information can be entered in a spreadsheet such as that shown in Table 8.

Table 8

	A	B	C	D	E	F	G	H	I
1	Fudge and Caramels			Max D3=+20.1*A3+36.7*B3-0.43*A3^2-1.42*A3*B3-0.468*B3^2-119					
2	Fudge (lbs)	Caramels (lbs)		Profit ($)					
3	10	10		217.2		(Initial)			

We have used starting values of 10 pounds each for the fudge and caramels, and see that this will give us a profit of about $217.20. Can we do better than that? Let's try to maximize the profit using the spreadsheet. We invoke the optimizer/solver feature and identify the profit in D3 as the objective function we wish to maximize, while the quantities of fudge and caramels, A3 and B3, respectively, are the variables.

Table 9

	A	B	C	D	E	F	G	H	I
1	Fudge and Caramels			Max D3=+20.1*A3+36.7*B3-0.43*A3^2-1.42*A3*B3-0.468*B3^2-119					
2	Fudge (lbs)	Caramels (lbs)		Profit ($)					
3	-5.37E+08	1036114040		1.64E+17		(Final)			

Our spreadsheet gives the results shown in Table 9.[1] What's wrong here? The results show that a very large negative number, -537 million pounds of fudge, should be made and sold, and then the profit would be $\$1.64 \times 10^{17}$ (164 quadrillion dollars, more than the combined GDP of all countries in the world)! These answers are not reasonable at all, although the maker of the fudge and caramels sure wishes the profit figure was! Obviously, you cannot have a negative number of pounds of fudge. Nor is it reasonable to think that you are going to make over a billion pounds of caramels! Not if they're really *home-made* candies!

---

[1] Other spreadsheets are likely to give different solutions in a case like this. The details and numbers may be different, but the basic result is comparable.

As we said earlier, since our model has constraints defining a restricted domain, we must check our unconstrained solution to see if it lies within the domain. In our model, the only restrictions given were that the variables have to be nonnegative. Since our spreadsheet solution gave a negative value to the fudge variable, it is clearly *not* in the domain; we say **the solution is infeasible**.

Let's go back to our spreadsheet and add a constraint to prevent the infeasibility we obtained. Now, in addition to identifying our objective function or target cell and the variable cells, we will add the constraint that the number of pounds of fudge, A3, must be greater than or equal to zero. **To enter a constraint in a spreadsheet optimizer**, select the optimizer procedure so that the dialog box appears. Within the dialog box there should be an area that relates to constraints and you want to indicate that you wish to *add* a constraint. The general format for entering a constraint is to enter a cell or a number for the left-hand side (LHS) and the right-hand side (RHS) of the constraint. The cell can contain a number, or a formula involving the variable cells. In addition, you need to indicate the *kind* of constraint you are adding (≤, ≥, or =). *After* you have indicated the target cell, variable cells, *and any constraints you want to add*, you can then *do* the optimization as before.

For our problem, the results of the optimization with a nonnegativity constraint on the fudge variable is shown in Table 10.

Table 10

	A	B	C	D	E	F	G	H	I
1	**Fudge and Caramels**			Max D3=+20.1*A3+36.7*B3-0.43*A3^2-1.42*A3*B3-0.468*B3^2-119					
2	Fudge (lbs)	Caramels (lbs)		Profit ($)		s.t. A3 >= 0			
3	0	39.20940165		600.4925		(Final)			

Translation: You can see that we actually got a solution when we added the non-negativity constraint: we should drop the fudge completely and make approximately 39 pounds of caramels to make a profit of approximately $600 . Lets use the spreadsheet to try some different combinations of fudge and caramel quantities and see how the profit changes. We'll start with equal combinations and keep increasing them until the profit starts to decrease, then we'll start with another combination (Table 11). This is accomplished on the spreadsheet by copying the formula in the target cell (D3) to the block of cells D4..D15 .

Table 11

	A	B	C	D
2	Fudge (lbs)	Caramels (lbs)		Profit ($)
3	0	39.20940165		600.4925
4	5	5		107.05
5	10	10		217.2
6	20	20		89.8
7	30	30		-501.2
8	30	0		97
9	20	0		111
10	10	0		39
11	0	10		201.2
12	0	20		427.8
13	0	30		560.8
14	0	40		600.2
15	0	50		546

We certainly didn't try all of the possible combinations, or even a really good sampling of combinations, but we can see that the solution that the spreadsheet gave looks optimal. It may not have been the solution we wanted to find, but at least it is possible. The decision to do away completely with the fudge could depend on many other things. However, we should still test it for global optimality, as with the others.

If this model was obtained by obtaining data about different prices and quantities sold, then fitting a demand model, then deriving functions for the revenue, cost, and profit, it should probably have more restrictions, at least based on the data input values, if not on the size of the kitchen. As we have said before, when defining models based on regression, it is normally best to restrict the domain to the intervals defined by the input data, unless you can justify enlarging it. In this case, if the initial solution of 10 pounds of each type of candy was the actual amount being produced currently by the business, then a total volume of about 39 pounds (compared to the initial 20 pounds) sounds doable, although possibly requiring more workers or longer hours. As long as any related costs have been incorporated into the profit function, this solution seems reasonable. □

Sample Problem 5: Your company plans to make conical cups, such as the type that sometimes come with bottled water dispensers. They are interested in forming cups from round papers, 10 inches in diameter. What should the dimensions of these cups be to get the maximum volume?

Solution: The first step in solving this problem is to be certain that you understand exactly what is involved. The cup will look something like Figure 8.3-1 .

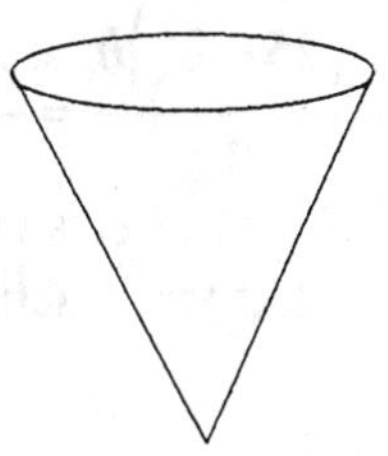

Figure 8.3-1

Since this cup is to be made by folding up a round piece of paper that is 10 inches in diameter, each side of the cup will be one half of that diameter, or 5 inches. Next you have to use the formula for the volume of a cone:

$$V = \frac{\pi r^2 h}{3}.$$

The height of the cone is measured from the base to the top, and the radius is one half the distance across the top. Figure 8.3-2 shows a sketch that marks in the known information and the variables.

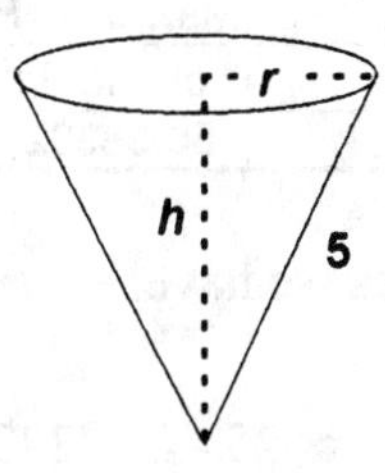

Figure 8.3-2

Since the radius, height, and slant height (5) form a right triangle, we know that we can write an equation involving the three values. Using the Pythagorean theorem, we can find the radius of the cone (actually we only need r^2 to substitute) in terms of the height:

$$h^2 + r^2 = 25 \text{ , or } r^2 = 25 - h^2.$$

If we now substitute for r^2 into the volume formula, we have a model for the volume of the cone, or cup, in terms of the height:

Verbal Definition: $V(h)$ = the volume, in cubic inches, of a cup made from a round piece of paper 10 inches in diameter, with height h inches

Symbol Definition: $V(h) = \frac{\pi r^2 h}{3} = \frac{\pi(25-h^2)h}{3} = \frac{\pi(25h-h^3)}{3}$ for $0 < h < 5$

Assumptions: Certainty and divisibility. Certainty implies that the relationship is exact. Divisibility implies that any fractional values of cubic inches and inches are possible.

We are now ready to enter the information into our spreadsheet: the volume to be maximized is in C3, and the variable, the height, is in cell A3 (Table 12).:

Table 12

	A	B	C	D	E
1	Paper Cup		Max C3=+PI()*(25*A3-A3^3)/3		
2	Height (in)		Volume (in^3)		
3	1		25.13274		(Initial)

Following the optimization, the result is shown in Table 13.

Table 13

	A	B	C	D	E
1	Paper Cup		Max C3=+PI()*(25*A3-A3^3)/3		
2	Height (in)		Volume (in^3)		
3	2.886751		50.38332		(Final)

Using our formula for the radius we have:

$$r^2 = 25 - h^2 = 25 - 2.886751^2 \approx 25 - 8.333333 = 16.666667$$
$$r \approx 4.08.$$

Translation: This says that if we make a conic cup of height approximately 2.89 inches and radius of approximately 4.08 inches, the cup will hold approximately 50.38 cubic inches. This is a very short, wide cup. Let's use the spreadsheet to check our solution by copying the volume formula from cell C3 to cells C5 to C8 and trying several values for the height, with the results shown in Table 14.

Table 14

	A	B	C	D	E
1	Paper Cup		Max C3=+PI()*(25*A3-A3^3)/3		
2	Height (in)		Volume (in^3)		
3	2.886751		50.38332		(Final)
4					
5	1		25.13274		
6	2		43.9823		
7	3		50.26548		
8	4		37.69911		

The maximum volume seems to lie somewhere between 2 and 3. In order to be more exact, we would have to enter values that lie between 2 and 3. This does seem to validate our solution, however. For a more complete optimality test, we can use the techniques that we will introduce in Section 8.4 . □

Sample Problem 6: An airline allows one carry-on bag, the sum of whose three dimensions must be no more than 50 inches. Furthermore, the length can be no more than 26 inches, the width no more than 24 inches, and the height no more than 10 inches, so that it can fit under the seat. What dimensions would maximize the volume you can carry on?

Solution: One model for this problem is:

Verbal Definition: $V(l,w,h)$ = the volume, in cubic inches, for a bag of length l, height h and width w

Symbol Definition: $V(l,w,h) = lwh$ subject to

$$l + w + h \le 50$$
$$0 \le l \le 26$$
$$0 \le w \le 24$$
$$0 \le h \le 10$$

Assumptions: Certainty and divisibility. Certainty implies that the relationship is exact. Divisibility implies that any fractional values of cubic inches and inches are possible.

This problem has many constraints, and we will look at it again in Section 9.4 . However, with a little thought, we can at least see if we can set it up to be solvable as an unconstrained problem. For example, common sense tells us that if the sum of the dimensions has to be no more than 50 inches, and if 50 inches is less than the sum of the maximum values for each of the three dimensions, then we would expect that the optimal solution would use the full 50 inches, so we could assume that the equation

$$l + w + h = 50$$

will hold. From this, we could solve for one of the variables (say, l) and substitute back into the volume expression. The result would be the following model:

Verbal Definition: $V(w,h)$ = the volume, in cubic inches, for a bag of width w and height h, assuming that the sum of the three dimensions (length, width, and height) is 50 inches

Symbol Definition: $V(w,h) = (50\text{-}w\text{-}h)wh$ subject to $0 \leq 50\text{-}w\text{-}h \leq 26$

$0 \leq w \leq 24$

$0 \leq h \leq 10$

Assumptions: Certainty and divisibility. Certainty implies that the relationship is exact. Divisibility implies that any fractional values of cubic inches and inches are possible.

Let's try maximizing this with no constraints, then check to see if the resulting solution is feasible. The initial spreadsheet, using the upper limits for the width and the height as the starting values, is shown in Table 17, and the solution is shown in Table 18

Table 17

	A	B	C	D	E	F
1	Luggage			Max D3=+(50-A3-B3)*A3*B3		
2	Width (in)	Height (in)		Volume (in^3)		
3	24	10		3840		(Initial)

Table 18

	A	B	C	D	E	F
1	Luggage			Max D3=+(50-A3-B3)*A3*B3		
2	Width (in)	Height (in)		Volume (in^3)		
3	16.66667	16.66666		4629.63		(Final)

The solution in Table 18 does look like a maximum, but the height value in the solution of 16.67 is *over* its upper limit of 10. The value for the width (also 16.67) is nicely between its limits (0 to 24). And we still need to check the length, which comes out to be

$$l = 50 - w - h = 50 - 16.67 - 16.67 = 16.66$$

which is also nicely between its limits of 0 to 26.

Since the height is over its upper limit, our solution is *infeasible*. But since that is the *only* problem with the solution, let's try imposing the upper limit as a constraint. The result is shown in Table 19.

Table 19

	A	B	C	D	E	F	G	H
1	**Luggage**			Max D3=+(50-A3-B3)*A3*B3				
2	Width (in)	Height (in)		Volume (in^3)		s.t. B3<=10		
3	20	10		4000		(Final)		

The spreadsheet in Table 19 indicates that the optimal solution has $w = 20$, $h = 10$, and $V = 4000$, and we can calculate the length to be

$$l = 50 - w - h = 50 - 20 - 10 = 20 .$$

Now all of our values satisfy all of the domain constraints, so this solution would seem to be optimal, although again we should do some testing to be sure it is the global maximum. But common sense suggests that this is indeed the solution.

Translation: Our carry-on bag should be 20"x20"x10", for a maximum volume of 4000 cubic inches. □

We have just seen that the spreadsheet solver/optimizer is a very powerful tool. Will it always find the optimal solution? Or are there any problems of which we should be aware?

Sample Problem 7: You have tracked your average scores on 40 point quizzes and the number of hours spent on preparation for the quiz, including homework and review, and come up with this model:

Verbal Definition: $A(x)$ = expected number correct on 40 point quiz after x hours of preparation (on average).

Symbol Definition: $A(x) = x^3 - 9x^2 + 24x$, for $0 \le x \le 6$

Assumptions: Certainty and divisibility. Certainty implies that the relationship is exact. Divisibility implies that any fractional values of number correct and hours are possible.

What is your optimal amount of study time?

Solution You entered the data in a spreadsheet using the initial value of 3 for the study time, as shown in Table 20, and got the results shown in Table 21.

Table 20

	A	B	C	D	E
1	**Quiz Study Time**		Max C3=+A3^3-9*A3^2+24*A3		
2	Time (hrs)		Score (out of 40)		
3	3		18		(Initial)

Table 21

	A	B	C	D	E
1	Quiz Study Time		Max C3=+A3^3-9*A3^2+24*A3		
2	Time (hrs)		Score (out of 40)		
3	2		20		(Final)

The solution looks fine: the optimal study time (2) is between the required limits (0 to 6), although the score on the quiz is not too encouraging. Let's see what happens if we try a higher initial value for the study time, like 5. The initial spreadsheet is shown in Table 22 and the resulting solution is shown in Table 23.

Table 22

	A	B	C	D	E
1	Quiz Study Time		Max C3=+A3^3-9*A3^2+24*A3		
2	Time (hrs)		Score (out of 40)		
3	5		20		(Initial)

Table 23

	A	B	C	D	E
1	Quiz Study Time		Max C3=+A3^3-9*A3^2+24*A3		
2	Time (hrs)		Score (out of 40)		
3	2.68E+08		1.93E+25		(Final)

What happened? The solution suggests that you should study for 268 million hours, and your quiz score will be 19,300,000,000,000,000,000,000,000 out of 40! That might be enough points to get you a Ph.D.! Of course, the problem is that the study time solution value is over its upper limit of 6, so we need to re-solve it with that upper limit put in as a constraint. The initial solution is shown in Table 24, and the resulting solution is shown in Table 25.

Table 24

	A	B	C	D	E	F
1	Quiz Study Time		Max C3=+A3^3-9*A3^2+24*A3			
2	Time (hrs)		Score (out of 40)		s.t. A3<=6	
3	5		20		(Initial)	

Table 25

	A	B	C	D	E	F
1	Quiz Study Time		Max C3=+A3^3-9*A3^2+24*A3			
2	Time (hrs)		Score (out of 40)		s.t. A3<=6	
3	6		36		(Final)	

This time the final value of x is 6, with an objective function value of 36, which is clearly better than the first solution we got. To see what happened, let's graph the function over its domain:

Figure 8.3-3

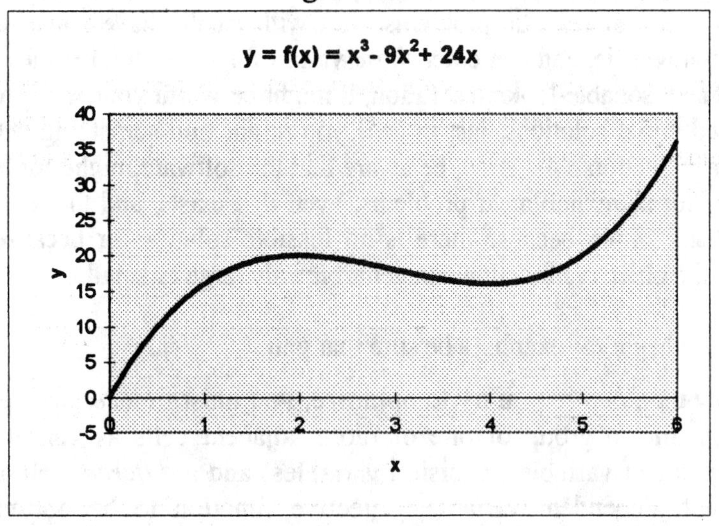

Now we can see that there is a local (relative) maximum at $x = 2$, which is what the first "optimization" obtained, but the global maximum over the domain is actually $x = 6$, as the second search found. You can also see this by entering some values for x in your spreadsheet, as shown in Table 26.

Table 26

	A	B	C	D	E	F
1	Quiz Study Time		Max C3=+A3^3-9*A3^2+24*A3			
2	Time (hrs)		Score (out of 40)		s.t. A3<=6	
3	6		36		(Final)	
4						
5	0		0			
6	1		16			
7	2		20			
8	3		18			
9	4		16			
10	5		20			
11	6		36			

Translation: If you were not careful, you might have decided that you could have gotten the best score by preparing for just two hours, and that the best score that you could get, on average, would be 20 points out of 40 points. The actual fact, assuming that your model is valid, is that you can get, on average, a maximum score of 36 points after preparing for 6 hours. □

The point of this exercise is to show that nonlinear optimization can be very tricky, and in general, uses many advanced techniques and concepts to try to find global optimal solutions. However, for realistic problems, you will usually have some idea of where an optimal solution might lie, and can enter *those* values for your initial values of the variables. Even if you get a reasonable-looking solution, it might be worth your while to try a few other randomly located initial solutions, just in case you might find something better. We pursue this idea further in Section 8.4 . Also, be aware that the software might not be able to *find* an optimal solution for some nonlinear problems, even if it exists, and that there is *no solution* for some problems, either because there is no feasible solution, or because the solution is unbounded and the objective function can get arbitrarily large or small.

Before you begin the exercises be sure that you:

- Know that, *before* you try to optimize, you must first *set up* the spreadsheet by designating a group of one or more adjacent cells as variable cells (inputs, independent variables, decision variables) and a *different* cell as the target cell (output, dependent variable, objective function to be optimized), with the formula (your objective function) entered into the target cell in terms of the variable cell(s).
- Know that it is a good idea to label your variable and target cells and the problem you are solving, as well as writing out the *formula* for the target cell and any limit constraints you have imposed, so someone else (or you, later) can understand what is going on from a printout.
- Test your formulas by entering some very simple starting values for the independent variable(s).
- Enter some reasonable starting values for the independent variable(s), as this will shorten the search time of the optimization.
- Know how to invoke the optimizer or solver feature on the spreadsheet and how to identify the solution or target cell and the variable cells.
- Always identify exactly what it is you want the optimizer or solver to do; that is, select maximize or minimize from the menu.
- Recognize when lower or upper limit constraints on the input variable(s) must be entered, and how to enter them.
- Are aware that you can sometimes get incorrect answers, such as a local maximum that is not the global maximum, so that sometimes the optimizing

function must be rerun with different starting values to get the global maximum or minimum.

- **Realize that the spreadsheet procedure that optimizes functions is a powerful tool, but must be used cautiously.**

EXERCISES FOR SECTION 8.3

Warm Up

1. $R(x,y) = -3x^2 - 2y^2 + 3xy + 4y - 20$
   a) Use a spreadsheet to find values of x and y that will maximize the function, if they exist.
   b) Use a spreadsheet to find values of x and y that will minimize the function, if they exist.

2. $P(x,y) = -2x^2 - y^2 + 2xy - 4y + 3x + 24$
   a) Use a spreadsheet to find values of x and y that will maximize the function, if they exist.
   b) Use a spreadsheet to find values of x and y that will minimize the function, if they exist.

3. $C(f,d) = -50 + 3.5f + 7.2d + 3.5fd - 4.7f^2 - 3.9d^2$
   a) Use a spreadsheet to find values of f and d that will maximize the function, if they exist.
   b) Use a spreadsheet to find values of f and d that will minimize the function, if they exist.

4. $E(c,r) = -2.8c^2 - 3.7r^2 + 4.3cr + 3.0c - 2.1r + 10$
   a) Use a spreadsheet to find values of c and r that will maximize the function, if they exist.
   b) Use a spreadsheet to find values of c and r that will minimize the function, if they exist.

5. $S(m,b) = (5 - m - b)^2 + (6 - 2m - b)^2 + (9 - 3m - b)^2$
   a) Use a spreadsheet to find the values of m and b, if they exist, that will minimize the function.
   b) Find the minimum value of the function.

6. $S(m,b) = (5 - 8m - b)^2 + (4 - 6m - b)^2 + (6 - 4m - b)^2 + (9 - 2m - b)^2$
   a) Use a spreadsheet to find the values of m and b, if they exist, that will minimize the function.
   b) Find the minimum value of the function.

7. For $f(x,y) = 3x - 5y + 21$,

   a) Use a spreadsheet to find values of x and y that will maximize the function, if they exist.

   b) Use a spreadsheet to find values of x and y that will minimize the function, if they exist.

8. For $f(s,t) = -0.2s + 3.9t + 12$

   a) Use a spreadsheet to find values of s and t that will maximize the function, if they exist.

   b) Use a spreadsheet to find values of s and t that will minimize the function, if they exist.

9. $f(x,y) = x^2 - y^2$

   a) Use a spreadsheet to find values of x and y that will maximize the function, if they exist.

   b) Use a spreadsheet to find values of x and y that will minimize the function, if they exist

10. $g(r,s) = -3s^2 + 5p^2 + 3sp - 4s$

    a) Use a spreadsheet to find values of s and p that will maximize the function, if they exist.

    b) Use a spreadsheet to find values of s and p that will minimize the function, if they exist

11. $f(x,y) = 2x^2 - x^3 + 8y^2 - 4y^3 + 15$

    a) Use a spreadsheet to find values of x and y that will maximize the function, if they exist.

    b) Use a spreadsheet to find values of x and y that will minimize the function, if they exist

12. $c(d,f) = 4d^3 - 2d^2 + 3f^3 - 7f^2 + 3df - 20$

    a) Use a spreadsheet to find values of d and f that will maximize the function, if they exist.

    b) Use a spreadsheet to find values of d and f that will minimize the function, if they exist.

13. $f(x,y,z) = 4x^2 + y^2 - 8xz - 2yz + 20z$

    a) Use a spreadsheet to find the values of x, y and z, if they exist, that will minimize the function.

    b) Find the minimum value of the function.

14. $f(x,y,z,\lambda) = x + y + z - xyz\lambda + 8\lambda$
    a) Use a spreadsheet to find the values of x, y, z, and λ, if they exist, that will minimize the function.
    b) Find the minimum value of the function.

Game Time

15. The volume of a box, with restrictions on its measurements of 100 inches for total length plus girth, is modeled by:

    Verbal Definition: $V(w,h)$ = the volume in cubic inches of a box of width w inches and height h inches
    Symbol Definition: $V(w,h) = 100wh - 2w^2h - 2wh^2$ for $w,h \geq 0$
    Assumptions: Certainty and divisibility. Certainty implies that the relationship is exact. Divisibility implies that any fractional values of cubic inches and inches are possible.

    a) Use a spreadsheet to find the dimensions of the box that maximize the volume.
    b) Find the maximum volume of the box.

16. If a box costs \$2 per square foot for sides and top and \$3 per square foot for the bottom and it must have a volume of 6 cubic feet, the cost of the box is modeled by:
    Verbal Definition: $C(w,l)$ = the cost, in dollars to make a box of width w feet and length l feet, given that the volume must be 6 cubic feet
    Symbol Definition: $C(w,l) = 5wl + \frac{24}{l} + \frac{24}{w}$ for $w,l > 0$
    Assumptions: Certainty and divisibility. Certainty implies that the relationship is exact. Divisibility implies that any fractional values of dollars and feet are possible.

    a) Use a spreadsheet to find the dimensions of the box that minimize the total cost.
    b) Find the minimum cost of the box.

17. One of the exercises for Section 1.5 found a model for unemployment in the United States. One possible model is:
    Verbal Definition: $U(x)$ = the unemployment (thousands) in the United States civilian population x years after Dec. 31, 1983.
    Symbol Definition: $U(x) = -13.72x^3 + 305.7x^2 - 1901x + 10704$, $x \geq 0$
    Assumptions: Certainty and divisibility. Certainty implies that the relationship is exact. Divisibility implies that any fractional values of unemployed and years are possible.

Use a spreadsheet to determine, according to the model, when unemployment was at its highest and lowest levels between 1983 and 1994.

18. Another possible model for the unemployment is:

Verbal Definition: $U(x)$ = the unemployment (thousands) in the US civilian population x years after Dec. 31, 1983.

Symbol Definition: $U(x) = -3.910x^4 + 72.30x^3 - 286.9x^2 - 586.6x + 10262,\ x \geq 0$

Assumptions: Certainty and divisibility. Certainty implies that the relationship is exact. Divisibility implies that any fractional values of unemployed and years are possible.

Use a spreadsheet to determine, according to this model, when unemployment was at its highest and lowest levels between 1983 and 1994.

19. One of the problems in Section 1.5 referred to the danger of destroying a crop if the temperature fell below freezing for more than 15 minutes. The model for the temperature was:

Verbal Definition: $T(x)$ = temperature in degrees Fahrenheit x hours after 21:00 (9:00 P.M.)

Symbol Definition: $T(x) = 0.0581x^3 - .7587x^2 + 1.0317x + 42.848,\ 0 \leq x \leq 12$

Assumptions: Certainty and divisibility. Certainty implies that any fractional values of degrees and hours are possible.

Use a spreadsheet to determine, according to the model, when the minimum temperature occurs.

20. If the cost of producing each chair in the furniture problem can be reduced to \$60 while the cost of producing the tables remains at \$100 (that is, if the coefficient of c in the profit function increases by 10, from 140 to 150), so that the profit function is now:

$$P(c,t) = 150c + 200t + 2ct - 4c^2 - 12t^2 - 700,$$

use a spreadsheet to find out how many chairs and tables should be made to maximize the profit. What is the maximum profit?

21. If the cost of producing each table can be reduced to \$80 (so the coefficient of t in the profit function increases by 20, from 200 to 220) and the cost of producing the chairs remains at \$60, so that the profit function is now

$$P(c,t) = 150c + 220t + 2ct - 4c^2 - 12t^2 - 700,$$

use a spreadsheet to find out how many chairs and table should be made to maximize the profit. What is the maximum profit?

22. Previously we developed a model for the profit from the sale of tickets to a local sporting event as a function of the price charged per ticket:

Verbal Definition: $\pi(p)$ = profit, in dollars, from the sale of tickets at p dollars per ticket.

Symbol Definition: $\pi(p) = 3165.6(p-1)(.94919)^p - 1000,\ p \geq 0$.

Assumptions Certainty and divisibility. Certainty implies that the relationship is exact. Divisibility implies that any fractional values of dollars are possible.

Use a spreadsheet to find the price per ticket that will maximize the profit.

23. If the model for the profit from the sale of divinity fudge and chocolate fudge is:

Verbal Definition: $P(f,d)$ = the profit, in dollars, from the sale of chocolate fudge at f dollars a pound and divinity fudge sold at d dollars a pound

Symbol Definition: $P(f,d) = -18.2f^2 + 199.14f + 36.0fd - 26.0d^2 + 168.8d - 237.4$ for $f,d \geq 0$

Assumptions: Certainty and divisibility. Certainty implies that the relationship is exact. Divisibility implies that any fractional values of dollars are possible.

a) Use a spreadsheet to find the prices to charge for the chocolate fudge and divinity fudge to maximize the profit.
b) Comment on the answer that you got.

24. You have been practicing your three point shot in basketball, particularly from your favorite spot at an angle of 45 degrees. For each one hour practice session, you have recorded the average number of shots you made from 10 tries. These are the results:

One Hour Practice Session	Average Shots Made
1	4
2	6
3	7
4	9

a) Find the SSE for a linear model for this data.
b) Use a spreadsheet to minimize the SSE.
c) Write the linear model for the data.

25. You have been very discouraged with your marks on certain quizzes. You know that you can do better, so you have really been studying hard to try and reduce the number of answers that you get wrong. You recorded the number of hours you studied for each quiz and the number of questions that you got wrong:

Hours Studied	Incorrect Answers
0	15
1	10
2	7
4	2

a) Find the SSE for a linear model for this data.
b) Use a spreadsheet to minimize the SSE.
c) Write the linear model for the data.

Section 8.4: Testing for Local and Global Extrema

In Section 8.2, we learned how to find **critical points** by setting the partial derivatives equal to 0 and solving for the independent variables. But these points are only *candidates* for the optimum we are seeking: they could be maxima, minima, inflection points, or **saddle points**, and it is very important to be able to determine which they are. In this section we will discuss some ways of determining whether any of the critical points we find are in fact local or global optima, at least with a very high degree of confidence.

Here are some examples of the kinds of problems that the material in this section will help you solve:

- You are the owner of a small company that produces just two products, chairs and tables. You have developed a model for the profits from the sale of the chairs and tables as a function of the number of each made each month (assuming you price them according to a demand function to be able to sell them all). How many chairs and tables should you make to maximize your profit?
- You have developed a model for your energy level each day as a function of the amount of exercise you get that day and the amount of calories that you consumed during the day. On average, how much exercise should you get each day and how many calories should you consume each day to maximize your energy level?
- You run a small coffeehouse that has featured the same act for three years running. You have kept records of the price charged per ticket and the number of tickets sold. You want to determine the best-fit linear model for the number of tickets sold as a function of the price charged per ticket. What are the best slope and y-intercept values of the linear model to minimize the SSE?

By the end of this section, in addition to being able to solve problems like those above, you should:

- Be able to use a spreadsheet to test values around a critical point to determine to your satisfaction whether it is a local and/or global maximum or minimum.
- Be able to look at the coefficients of a multivariable quadratic function to see whether the critical point obtained is clearly *not* the kind of extremum you are looking for.

Testing to See If a Critical Point of a Function of Two Variables is a Local Extremum

Sample Problem 1: You are the owner of a small company that produces just two products, chairs and tables. You have developed a model for the profits from the sale of the chairs and tables as a function of the number of each you make each month:

Verbal Definition: $P(c,t)$ = the profit, in dollars, on average, when c chairs and t tables are made each month and the price is set according to the demand function

Symbol Definition: $P(c,t) = 140c + 200t - 4c^2 + 2ct - 12t^2 - 700$, for $c,t \geq 0$

Assumptions: Certainty and divisibility. Certainty implies that the relationship is exact. Divisibility implies that any fractional values of dollars, chairs and tables are possible.

In Section 8.2, we found the partial derivatives of the profit function, set them equal to 0, solved the resulting system of linear equations for c and t, and obtained a solution of:

$$c = 20 \text{ chairs} \quad \text{and} \quad t = 10 \text{ tables}, \text{ with a profit of } P(20,10) = \$1700.$$

Test whether this solution is a local maximum or not.

Solution : In single-variable calculus (Chapter 3 of Volume 1), we obtained critical points by setting the derivative equal to 0 and solving for the independent variable (usually x). It was easy to know whether a critical point was a local or global extremum: you could graph it using technology and *see* for yourself. Alternatively, you could use the second derivative test to check the concavity to help determine if you had a local maximum or a local minimum, though this does not *always* give a definitive answer. With functions of two variables, it is at least *possible* to graph the function, but most basic graphing calculators and spreadsheets will *not* do this for you. And for functions of three or more variables, it is not even *possible* to draw a simple graph. So you are likely to be in a situation where you need to test a critical point in some other way.

Suppose we do not have a way to graph the profit function for this problem. What can we do to test the critical point? Fortunately, there is a great deal of computing power in a spreadsheet, even if they do not yet do 3D plots of general functions of two variables. The simplest idea of how to test a critical point is to simply **evaluate other points all around that critical point, especially pairs of points that are on opposite sides of it.** This way, we can see whether the critical point is a maximum or minimum compared to the points on either side of it. Recall that **a local maximum is a point at which the value of the function being studied is greater than or equal to the value of the function for *any* other point in *some* neighborhood or region all around it.** In other words, it is like the top of a hill, even though there may be higher hills some distance from it. The definition of a local minimum is analogous; just substitute "less" in place of "greater" in the definition of a local maximum. A **local extremum** is simply a point that is *either* a local maximum *or* a local minimum. Sometimes you will see the term ***relative*** **extremum** (maximum or minimum) instead of *local* extremum, but they have exactly the same meaning.

What would be reasonable and easy values to check for our problem, that are around the point (20,10)? If we think of graphing (20,10) on a standard 2D graph of the domain of our profit function, we could construct a square around it, as in Figure 8.4-1, checking all the points that are 1 unit less or more in either or both variable values:

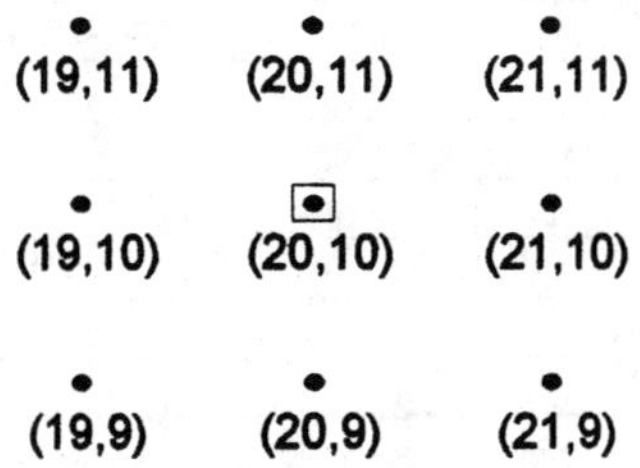

(19,11)	(20,11)	(21,11)
(19,10)	(20,10)	(21,10)
(19,9)	(20,9)	(21,9)

Figure 8.4-1

Notice that, just like on a standard graph, the points that line up vertically have the same value for the first variable (what we usually think of as x), and those that line up horizontally have the same value for the second variable (what we usually think of as y). In other words, the columns have the same first variable value, and the rows have the same second variable value.

Now, if we find the value of $P(c,t)$ for each pair of points on opposite sides of (20,10), such as (19,10) and (21,10), then we can see if (20,10) has the largest profit. This is basically just plugging values into a function or formula, which is very easy to do on a spreadsheet, so let's try it. Table 1 shows one way to do this:

Table 1

	A	B	C	D	E	F	G	H
10		1st var is:	c	& 2nd var is:	t		fn is:	P(c,t)
11	Critical Point: 1st =		20	& 2nd =	10			
12								
13								
14	B18=+140*B$21+200*$E18-4*B21^2+2*B21*$E18-12*$E18^2-700							
15								
16	fn(1st,2nd)				2nd var			
17	1682	1688	1686		11			
18	1696	1700	1696		10			
19	1686	1688	1682		9			
20								
21	19	20	21		<= 1st var			
22								
23								
24	(c,t)							
25	(19,11)	(20,11)	(21,11)					
26	(19,10)	(20,10)	(21,10)					
27	(19,9)	(20,9)	(21,9)					

All we have done here is type in by hand the c and t values, put the formula for the profit function:

+140*B$21+200*$E18 -4*B21^2+2*B21*$E18-12*$E18^2-700

into cell B18, and copied the formula from B18 to A17..C19 . In the formula we have replaced the c values with "B" references and the t values with "E" references. Notice that we have used a *combination* of **relative** and **absolute** addressing. **The $ means to treat the letter or number after it *exactly as is*, absolute, while *no* $ means to use the *position* of the letter or number after it *relative to the cell where the formula is located*.** For example, B$21 located in cell B18 means "take the value in the *21st* row of *this* column." If we had put B21 into B18, it would mean "take the value in *this* column, *3 rows down from here*." Putting $E18 in B18 means "take the value in column *E* of *this* row." Using this system, the formula given above cleverly picks off the values of c and t that we want for each entry in the matrix. The headings, ordered pairs, and the printed-out formula for B18 are only typed in to make the spreadsheet easier to understand and interpret, but are not necessary. For later tests, we will leave out the matrix of ordered pairs, hoping that this example will familiarize you with how to interpret our format.

Reading along the second row of the "fn(1st,2nd)" profit matrix (in block A17..C19), notice that, indeed, (20,10) has a higher profit than either (19,10) or (21,10), so when we slice the surface by fixing the value of t at 10, the local maximum does appear to occur at (20,10). This analysis corresponds to looking at the *second* row of the grid in Figure 8.4-1 and of the matrices in Table 1.

Let's think about what we just did: we held t fixed at 10. This would be like plugging in 10 for t into the profit function:

$$P(c,10) = 140c + 200(10) + 2c(10) - 4c^2 - 12(10)^2 - 700 \;=\; -4c^2 + 160c + 100\ .$$

As we observed before, this is simply a parabola opening downward, which we know because of the negative coefficient of the squared term. Notice that these properties would also hold for *any* value of t that we might want to plug in. Let's look at the graph of this function again in Figure 8.4-2.

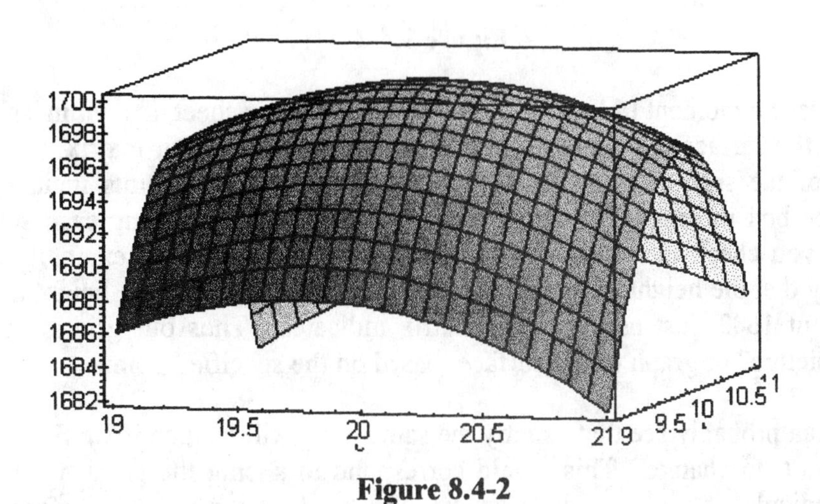

Figure 8.4-2

Fixing the value of t at 10 corresponds to slicing the surface with a vertical plane perpendicular to the t axis (sort of *along* the *length* of the surface), resulting in a sliced curve that is essentially the arc at the *top* of Figure 8.4-2 . We can see that indeed this is a parabola opening down, and that the tangent to it would be horizontal at the top, which is where $c = 20$ (see Figure 8.4-3).

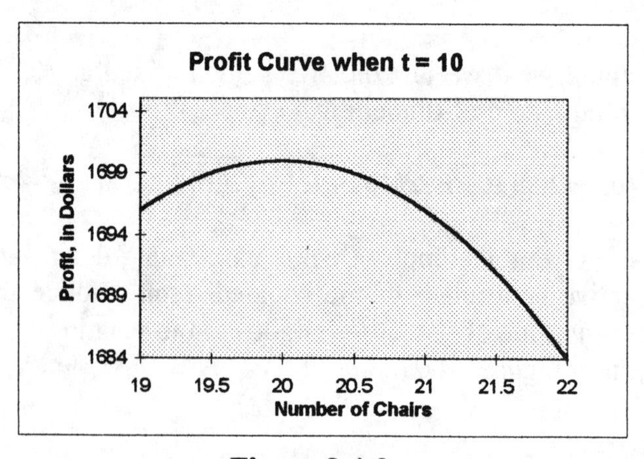

Figure 8.4-3

Let's take a moment to look back at our testing spreadsheet, shown in Table 1, and the graph of the surface in Figure 8.4-2 . The values in our profit matrix correspond to the *heights* of the surface at the corresponding points. These points include the four corners of the box in Figure 8.4-2 and the horizontal midpoints along each side face of the box. If you check the corners, you can see this directly. For example, the graph shows clearly that the height at the point (19,9) is about 1686, and the height at the point (21,9) is about 1682, just as the profit matrix indicates. Thus our grid is giving us a numerical "picture" or graph of the surface, based on the specified points.

You can probably see that exactly the same thing will happen if we fix the value of c and allow t to change. This would correspond to slicing the graph with a vertical plane perpendicular to the c axis (*perpendicular* to the *length* of the surface in Figure 8.4-2). Since the coefficient of t^2 is negative, we will again get parabolas opening down (Figure 8.4-4).

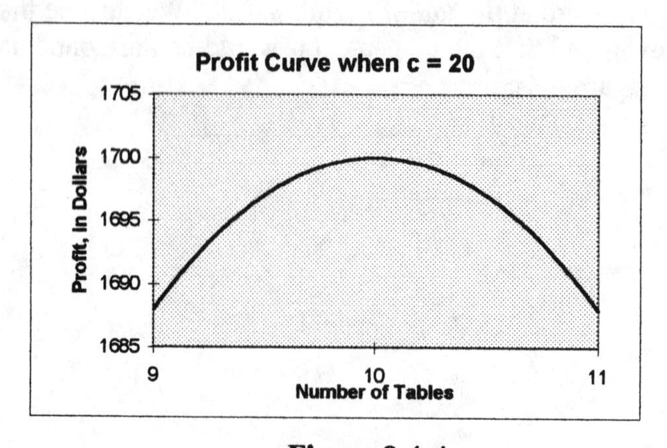

Figure 8.4-4

From this discussion, we know that keeping *either* variable fixed should make (20,10) locally best along the corresponding sliced curve. From the profit matrix on the spreadsheet, this corresponds to looking at just the second row and just the second column of the profit matrix. On the graph, it corresponds to slicing the surface perpendicular to each axis and through the point (20,10). However, from looking at the profit matrix, we can see that (20,10) is *also* locally best when we slice *diagonally*, which, of course, the graph also confirms. Thus, from considering *only* the profit matrix, the value of the profit at (20,10), 1700, is larger than at *any* of the other grid points. This means we have strong evidence that it is a local maximum. For a quadratic model of a real-world situation, this is normally enough.[1]

These observations suggest some general principles for **testing critical points of quadratic functions**:

- If the sign of a squared term coefficient is negative, then your critical point will be a local and global maximum *among all the solutions in which you hold all other variables but that one constant, and allow that one to vary.*
- If the sign of a squared term coefficient is positive, then your critical point will be a local and global minimum *among all the solutions in which you hold all other variables but that one constant, and allow that one to vary.*

Since we have not claimed that the above strategy of checking grid points *guarantees* that we have a local maximum, let's try one more strategy to convince ourselves: we'll pick some **random points** from the box around (20,10) graphed in Figure 8.4-1. Most spreadsheets have a function to generate random numbers between 0 and 1. In our case, we would really like numbers between -1 and 1 as the amounts to be added to (20,10), since this will allow points on *all* sides of the critical point. There's a slick trick you can do to accomplish this. Suppose the random number generator function is called RAND(), and it gives values between 0 and 1. If you multiply RAND() by 2, you will get values between 0 and 2, at which point you can subtract 1 to get values between -1 and 1. In general, you can **multiply RAND() by the *width* of the interval you want to generate, then add the lower value you want** (-1 in our case).. Here are the results:

[1] For a method that can prove this absolutely using calculus, see the end of this section.

Table 2

	A	B	C	D	E	F	G	H
33	C38=+140*G38+200*H38-4*G38^2+2*G38*H38-12*H38^2-700							
34								
35		fn name:	P(c,t)		variable names ==>		c	t
36								
37			fn Value		Variables ==>		1st	2nd
38			1700		<== Test Point ==>		20	10
39			1698.826				20.54492	10.07623
40			1698.333				20.64797	10.09756
41			1694.772				19.58906	10.58265
42			1696.609		Other		20.5968	9.641907
43			1688.928		Points		20.82624	10.90542
44			1688.028				20.05912	9.00666
45			1697.899				20.64269	10.25432
46			1693.306				20.85106	10.63784

To achieve this spreadsheet we first typed in the headings and the test point coordinates: 20 for c and 10 for t. Then in A33 we entered the formula

C38=+140*L4+200*M4 -4*L4^2+2*L4*M4-12*M4^2-700 ,

copied this to cell C38, and erased the "**C38=**", similar to what we did in Table 1. Next, we copied the formula from C38 to the block C39..C46 . Finally, in G39 we put the formula

+G$38+2*RAND()-1

and copied it to the block G38..H46 . Notice that by using relative addressing for the column (G with no $ in front) and direct addressing for the row ($38), the formula works for *both* columns, since it says "take the value in the *38th* row of the *this* column and ...". When this is copied to a cell in column H, it will be changed to +H$38+2*RAND()-1 .

Once again, we see that (20,10) is the best of all of these random points, giving us even more confidence that it is the local maximum. If you want to see *other* random points, simply perform a manual recalculation of the spreadsheet (for example, hit F9 in Excel). Repeat as many times as you like to convince yourself you have a local maximum. To be able to *prove* we have a local and global maximum, see the end of this section. □

Sample Problem 2: If $f(x,y) = x^2 - y^2$, find the critical point and test to see if it is a local or global extremum.

Solution: As usual, let's take the partial derivatives to find the critical point:

$f_x(x,y) = 2x = 0$, so $x = 0$, and
$f_y(x,y) = 2y = 0$, so $y = 0$.

Our critical point is (0,0) . Let's test it:

Table 3

	A	B	C	D	E	F	G	H
10		1st var is:	x	& 2nd var is:	y		fn is:	f(x,y)
11	Critical Point: 1st =		0	& 2nd =	0			
12								
13								
14	B18=+B21^2-E18^2							
15								
16	fn(1st,2nd)				2nd var			
17	0	-1	0		1			
18	1	0	1		0			
19	0	-1	0		-1			
20								
21	-1	0	1		<= 1st var			

We see that the critical point (0,0) is *not* a local extremum. This time, if we hold y fixed (at 0) and vary x (move along the y axis), it is a local *minimum*. On the other hand, if we hold x fixed and vary y , it is a local *maximum*. And, perhaps even more curiously, when we move along either diagonal, the function is *level* (stays at 0)! If we look at a graph of this function just over this domain ($-1 \le x \le 1$ and $-1 \le y \le 1$) , it looks a bit like the shape of a horse's saddle, so the critical point is called a **saddle point** (Figure 8.4-5).

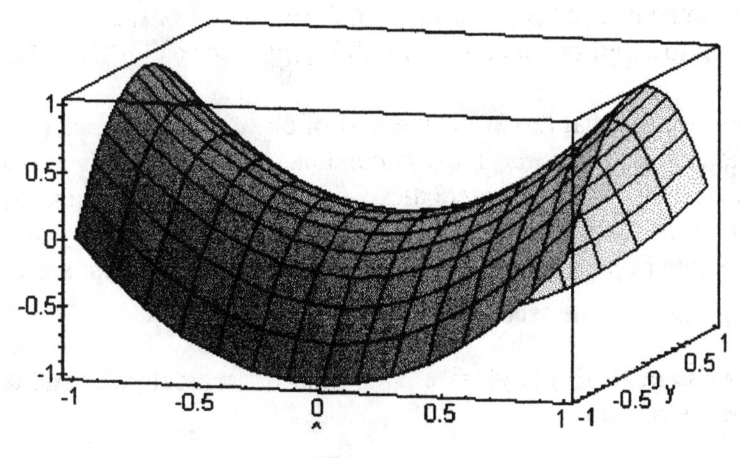

Figure 8.4-5

If you slice the saddle *perpendicular* to the horse, you get a parabola opening *down*, so the saddle point (where a mouse would sit on the saddle) is a local *maximum*. But if you slice the saddle *parallel* to the length of the horse, you get a parabola opening *up*, so the saddle point is a local *minimum*. This is exactly what we observed for this function in our test matrix. Notice also that the middle of the surface (where a mouse would sit) is essentially flat, so the tangent plane is *horizontal* at (0,0).

This emphasizes the importance of the signs of the squared-term coefficients: **in a multivariable quadratic function, if any two squared-term coefficients have *opposite* signs (and are not both 0's), then a critical point, if one exists, will *not* be a local extremum**.

We will note here that the partial derivatives of a multivariable quadratic function are all linear, and there is one partial derivative for each variable. This means you are solving n equations in n unknowns to find critical points, which means that you could get a unique solution, no solution, or an infinite number of solutions, as we saw in Chapter 7. The case of an infinite number of solutions will only happen if one or more of the squared terms have coefficients of 0. For quadratic functions in 2 variables, this means that the surface is a horizontal plane, or that there is an entire line at the highest or lowest value of the function.

Let's take a look at a problem where the signs of the coefficients of the squared terms *don't* tell the whole story.

Sample Problem 3: Calculate the critical point for $f(x,y) = x^2 + y^2 - 10xy$, and test to see if it is a local or global minimum.

Solution : First, let's use calculus to find the critical point:

$$f_x(x,y) = 2x - 10y = 0$$
$$f_y(x,y) = 2y - 10x = 0$$

Using substitution, from the first equation,

$$x = 5y .$$

Plugging this in for x into the second equation, we get

$$2y - 10(5y) = 0 \quad \text{, so}$$
$$2y - 50y = 0 \quad \text{, so}$$
$$-48y = 0 \quad \text{, so}$$
$$y = 0$$

Plugging back into the expression we got for x, we get

$$x = 5y = 5(0) = 0$$

Thus, our (unique) critical point is once again $(0,0)$. Let's use the spreadsheet as discussed in Sample Problem 1 to test this critical point to try to determine if it is a local or global extremum. The result is shown in Table 4.

Table 4

	A	B	C	D	E	F	G	H
10		1st var is:	x	& 2nd var is:	y		fn is:	f(x,y)
11	**Critical Point: 1st =**		0	**& 2nd =**	0			
12								
13								
14	B18=+B21^2+E18^2-10*B$21*$E18							
15								
16	**fn(1st,2nd)**				**2nd var**			
17	12	1	-8		1			
18	1	0	1		0			
19	-8	1	12		-1			
20								
21	-1	0	1		**<= 1st var**			

We see that (0,0) is not a local extremum, and therefore cannot be a global extremum either. Notice that slicing on one diagonal, $y = x$ (from lower left to upper right in the matrix), shows (0,0) to be a local *maximum* on the sliced curve, but for the other diagonal it is a local *minimum*. This time, however, the signs of the squared-term coefficients were both positive! What happened?

The answer lies in the cross-product term, the $-10xy$. This is a product of two variable terms, similar to the squared terms, which can be thought of as xx and yy. But since it has a coefficient of -10, if x and y are both positive (or both negative), the result of $-10xy$ will be "very negative," *overpowering* the value of $x^2 + y^2$. In other words, moving along the 45° line, (0,0) looks like a local maximum. To see this another way, the equation of the 45° line is $y = x$, so we could substitute x in for y in the original equation to get

$$x^2 + y^2 - 10xy = x^2 + (x)^2 - 10x(x) = x^2 + x^2 - 10x^2 = -8x^2 ,$$

which is a parabola opening down.

If you move along either axis, however (keep one of the variables 0), then the $-10xy$ term drops out (is 0), so the critical point looks like a local minimum. This would also happen if you move along the *other* diagonal, $y = -x$, since x and y will have opposite signs, so $-10xy$ will be positive and will just *reinforce* $x^2 + y^2$. In this case, we could also see it by substituting $-x$ in for y in the original equation to get

$$x^2 + y^2 - 10xy = x^2 + (-x)^2 - 10x(-x) = x^2 + x^2 + 10x^2 = 12x^2 ,$$

which is a parabola opening up.

Let's look at the graph in Figure 8.4-6.:

Figure 6

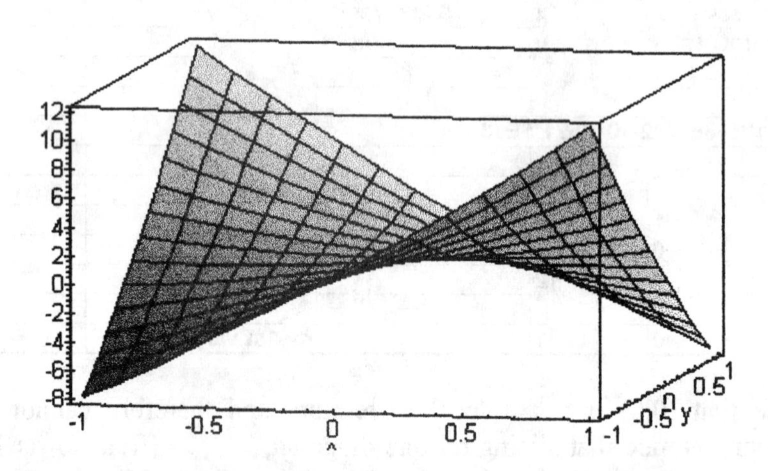

Notice again how the values in the test matrix correspond to the graph. Also, notice that the surface is again a saddle, but twisted at a diagonal angle.

The short answer to our problem, then, is that *the unique critical point (0,0) is not a local or global extremum.* □

The main point here is that **in a multivariable quadratic function, having all squared terms' coefficients be the same sign *does not guarantee* a critical point will be an *overall* local extremum (it will be in certain directions, but not necessarily in *all* directions). On the other hand, to *be* a local extremum, the signs of the squared-term coefficients *must* all be the same (or 0).** Mathematicians would say that the same-sign condition is *necessary*, but not *sufficient*, for being a local extremum.

Once we have seen that the critical point is not a local extremum, there is no need to test further using random points around it.

With round numbers such as 1700 and 0 for the value of the function at a critical point, it is easy to look at the table and see if the function values of the surrounding points are better or worse. What if the function value for our critical point is not such an even number? Let's look at an example:

Sample Problem 4: In Section 8.2, Sample Problem 2, we found that the critical point for

$$E(c,x) = 0.260c + 0.146x - 0.0000944c^2 + 0.0000410cx - 0.00196x^2 - 93.1$$

was $c = 1390$ and $x = 51.8$. Test this solution to see if it is a local and global maximum.

Solution: If we apply our method from before, the spreadsheet will appear as in Table 5.

Table 5

	A	B	C	D	E	F	G	H
10		1st var is:	c	& 2nd var is:	x		fn is:	E(c,x)
11	Critical Point: 1st =		1390	& 2nd =	51.8			
12								
13								
14	B18=+0.260*B$21+0.146*$E18-0.0000944*B$21^2+0.0000410*B$21*$E18-0.00196*$E18^2-93.1							
15								
16	fn(1st,2nd)				2nd var			
17	91.16364	91.16347	91.1631		52.8			
18	91.16571	91.16549	91.16509		51.8			
19	91.16385	91.1636	91.16315		50.8			
20								
21	1389	1390	1391		<= 1st var			

This time, notice that it is rather hard to determine from a quick look whether the function values are larger or smaller than the function value for the critical point. Let's make this a lot easier to see. We can do this by making another matrix in which we simply calculate the value of the function at the *other* point minus the value at the *critical* point. That way, values that are bigger than at the critical point will have "+" signs and values that are lower will have "-" signs. Table 6 shows what happens.

Table 6

	A	B	C	D	E	F	G	H	I
10		1st var is:	c	& 2nd var is:	x		fn is:	E(c,x)	
11	**Critical Point: 1st =**		1390	**& 2nd =**	51.8				
12									
13									
14	B18=+0.260*B$21+0.146*$E18-0.0000944*B$21^2+0.0000410*B$21*$E18-0.00196*$E18^2-93.1								
15									
16	**fn(1st,2nd) = fn(Other) :**				**2nd var**		**fn(Other) - fn(Crit.Pt.):**		
17	91.16364	91.16347	91.1631		52.8		-0.00185	-0.00203	-0.00239
18	91.16571	91.16549	91.16509		51.8		0.000214	0	-0.0004
19	91.16385	91.1636	91.16315		50.8		-0.00164	-0.00189	-0.00234
20									
21	1389	1390	1391		**<= 1st var**	**=>**	1390	1390	1390

To create this spreadsheet, we first simply typed in the headings and copied the values of c below the matrix on the right. Then, in cell H18 we entered the formula

+B18-$B$18 .

Once again, we are using a combination of **relative** and **absolute** addresses. This formula being located in cell H18 means "take the value in the cell that is in *this* row, *6 columns to the left*, and subtract from it the value that is in (absolute) cell *B18*." When copied to cell G17, this will be saved as +A17-$B$18 , since A17 is in the same row as, and 6 columns to the left of, G17. In other words, A17 is in the same position *relative to* G17 as B18 is to H18.

Studying the new matrix of *differences*, notice that all but one of the new points have negative signs, so are lower. Now, if we set up the cells containing the values of the two variables with general formulas, we can easily *change* the point we are testing to see if the point (1389,51.8) would be better. For example, in cell A21, we could put the formula +$C$11-1 . This says "take the contents of cell C11 (absolute address), and subtract 1 from it, and put the answer in A21." Similarly, we can enter the formula $C$11 into cell B21 and $C$11+1 into cell C21, then copy A21..C21 to G21..I21 . In cell E17, we could put the formula +$E$11+1 , and, similarly, +$E$11 into E18 and +$E$11-1 into E19. If we put these formulas in, we can then simply change the values in cells C11 and E11 (the values of c and x) to test other possible local extrema. Table 7 shows the result if we now test (1389,51.8).

Table 7

	A	B	C	D	E	F	G	H	I
10		1st var is:	c	& 2nd var is:	x		fn is:	E(c,x)	
11	**Critical Point: 1st =**		1389	**& 2nd =**	51.8				
12									
13									
14	B18=+0.260*B$21+0.146*$E18-0.0000944*B$21^2+0.0000410*B$21*$E18-0.00196*$E18^2-93.1								
15									
16	**fn(1st,2nd) = fn(Other) :**				**2nd var**		**fn(Other) - fn(Crit.Pt.):**		
17	91.16362	91.16364	91.16347		52.8		-0.00208	-0.00207	-0.00224
18	91.16573	91.16571	91.16549		51.8		2.5E-05	0	-0.00021
19	91.16392	91.16385	91.1636		50.8		-0.00179	-0.00185	-0.00211
20									
21	1388	1389	1390		**<= 1st var**	**=>**	1388	1389	1390

Now, we *still* have one point, (1388,51.8), that is better than our test point, so let's try testing *it*. Table 8 shows the result.

Table 8

	A	B	C	D	E	F	G	H	I
10		1st var is:	c	& 2nd var is:	x		fn is:	E(c,x)	
11	**Critical Point: 1st =**		1388	**& 2nd =**	51.8				
12									
13									
14	B18=+0.260*B$21+0.146*$E18-0.0000944*B$21^2+0.0000410*B$21*$E18-0.00196*$E18^2-93.1								
15									
16	**fn(1st,2nd) = fn(Other) :**				**2nd var**		**fn(Other) - fn(Crit.Pt.):**		
17	91.16342	91.16362	91.16364		52.8		-0.00231	-0.00211	-0.00209
18	91.16557	91.16573	91.16571		51.8		-0.00016	0	-2.5E-05
19	91.1638	91.16392	91.16385		50.8		-0.00193	-0.00181	-0.00188
20									
21	1387	1388	1389		**<= 1st var**	**=>**	1387	1388	1389

Now *that's* more like it! *Finally* we have a point that seems to be a local maximum. What happened here? Well, you may have noticed that we rounded our solution off to 3 significant digits. So even though the value given was $c = 1390$, it could have been anything *between* 1385 and 1395. When we were adding and subtracting 1 to the 1390, we were actually finding the more accurate answer!

But suppose we don't care about the more accurate answer, and 3 significant digits is more than enough? The problem is that our increment was 1 unit, which was too precise for our solution value. An increment of 10 or more would have fit better with the precision of our answer. Similarly, an increment of 0.1 would make sense for the exercise variable, x, since that reflects the decimal place to which we rounded our answer. So let's set up our spreadsheet in a way that allows *any* increment for each variable as seen in Table 9.

Table 9

	A	B	C	D	E	F	G	H	I
10		1st var is:	c	& 2nd var is:	x		fn is:	E(c,x)	
11	Critical Point: 1st =		1390	& 2nd =	51.8				
12	Increments ==>		10		0.1				
13									
14	B18=+0.260*B$21+0.146*$E18-0.0000944*B$21^2+0.0000410*B$21*$E18-0.00196*$E18^2-93.1								
15									
16	fn(1st,2nd) = fn(Other) :				2nd var		fn(Other) - fn(Crit.Pt.):		
17	91.15907	91.16547	91.15298		51.9		-0.00643	-2.6E-05	-0.01251
18	91.15913	91.16549	91.15297		51.8		-0.00636	0	-0.01252
19	91.15916	91.16548	91.15292		51.7		-0.00633	-1.3E-05	-0.01258
20									
21	1380	1390	1400		<= 1st var	=>	1380	1390	1400

To accomplish this, we have put the increments in the cells below the values of the variables, with appropriate labels. Then, in each formula, instead of adding or subtracting 1, we add or subtract the value in the cell containing the increment. For example, in cell A21, we have put the formula "+$C$11-$C$12" , and in cell C21 we put +$C$11+$C$12 (cell B21 stays as it was). As before, we can then copy A21..C21 to G21..I21 . Similarly, we put $E$11+$E$12 in E17 and $E$11-$E$12 in E19 (E18 stays as it was).

As we expected, now the solution rounded to 1390 *does* test be a local maximum. At last! This is the test table that we really want. Except ... what if we wanted to be just a little more sure, and do some randomized testing? Recall that before, we multiplied RAND() by 2 and subtracted one. Now, if our increment is Δ, we would like to multiply RAND() by 2Δ and then subtract Δ, so that we would be adding values between $-\Delta$ and Δ. Let's try it! The results are shown in Table 10.

Table 10

	A	B	C	D	E	F	G	H
33	C38=+0.260*G38+0.146*H38-0.0000944*G38^2+0.0000410*G38*H38-0.00196*H38^2-93.1							
34								
35		fn name:	**f(x,y)**		variable names ==>		x	y
36					Increments ==>		10	0.1
37		i	fn Value	fn(Other$_i$)-fn(Test)	Variables ==>		1st	2nd
38			91.16549	0	<== Test Point ==>		1390	51.8
39		1	91.16533	-0.000163408			1390.465	51.79201
40		2	91.16498	-0.000513735			1391.21	51.83406
41		3	91.15924	-0.006254844			1380.087	51.85671
42		4	91.15738	-0.008112826	Other		1397.783	51.82047
43		5	91.16564	0.000149158	Points		1387.309	51.77259
44		6	91.15966	-0.005831912			1380.324	51.7175
45		7	91.16072	-0.00477401			1395.644	51.7159
46		8	91.15441	-0.011083027			1399.331	51.87297

This time, in cell G39, we put the formula "+G$38+2*G$36*RAND()-G$36" , and copied this to cells G39..H46 . This means "take the value in row *38* of *this* column, add 2 times the value in row *36* of *this* column times a random number between 0 and 1, and then subtract the value in row *36* of this column."

As before, in cell A33 we typed in "**C38=**" followed by the formula for the function $E(c,x)$, using G38 and H38 in place of c and x , respectively. We then copied the contents of cell A33 to cell C38, and deleted the "**C38=**" at the beginning, so the mathematical formula remained. Next, we copied the formula in C38 to the block C39..C46, as before.

To compute the differences, similar to what we did for the grid test, in cell D38 we put the formula +C38-$C$38 and copied it to cells D39..D46 . This means "take the value of the cell to your left and subtract the value in the (absolute) cell C38."

Again, we see that there exists a point, (1387.309,51.77259), that is better than our test point, but for the same reason as before (the exact maximum is closer to 1388 than to 1390). To pursue this further, we could make *it* the *new* test point (just type those values into the top row of the table of values, in cells G38 and H38).[2] This in fact gives a method for zeroing in on a local extremum, analogous to using a table or zooming in on a graph to find the maximum of a function of one variable. The result of one iteration of this process is shown in Table 11.

[2] You could also do this by copying the cells with the new solution and pasting their *values* into the cells for the test point. If you do a simple copy and paste, you will not get what you want, since you will be copying the *formulas* in the cells. In Excel, for example, copying the *values* is done using the "Paste Special" command. See your technology supplement for details.

Table 11

	A	B	C	D	E	F	G	H
33	C38=+0.260*G38+0.146*H38-0.0000944*G38^2+0.0000410*G38*H38-0.00196*H38^2-93.1							
34								
35		fn name:	f(x,y)		variable names ==>		x	y
36					Increments ==>		10	0.1
37		i	fn Value	fn(Other$_i$)-fn(Test)	Variables ==>		1st	2nd
38			91.16564	0	<== Test Point ==>		1387.309	51.77259
39		1	91.16172	-0.003916591			1394.875	51.72374
40		2	91.15992	-0.005719393			1380.502	51.74979
41		3	91.16021	-0.00543528			1395.998	51.69352
42		4	91.16573	8.81832E-05	Other		1388.778	51.7676
43		5	91.15921	-0.006429923	Points		1396.678	51.758
44		6	91.16106	-0.004576559			1395.408	51.79742
45		7	91.16334	-0.002296772			1393.371	51.6826
46		8	91.15513	-0.010506429			1377.761	51.77842

From Table 11, you can see that the next test point to try would be (1388.778,51.7676), since it is the only point with a higher function value. This procedure can then continue, always choosing for the new test point the point of the Other Points which has the best objective function value (in this example, the one whose difference is most positive; be aware, though, that you need to be careful when values are written in scientific notation!). Sometimes, your answer may *appear* optimal, but if you recalculate the random points (usually by hitting the F9 key), you may be able to find a better point. As your answers converge, you can also make your increments smaller, to refine your search. One such series of iterations resulted in the spreadsheet shown in Table 12.

Table 12

	A	B	C	D	E	F	G	H
33	C38=+0.260*G38+0.146*H38-0.0000944*G38^2+0.0000410*G38*H38-0.00196*H38^2-93.1							
34								
35		fn name:	f(x,y)		variable names ==>		x	y
36					Increments ==>		0.0001	0.0001
37		i	fn Value	fn(Other$_i$)-fn(Test)	Variables ==>		1st	2nd
38			91.16575	0	<== Test Point ==>		1388.36	51.76601
39		1	91.16575	-8.49809E-12			1388.36	51.76607
40		2	91.16575	-1.71667E-11			1388.36	51.76592
41		3	91.16575	-5.6275E-12			1388.36	51.76596
42		4	91.16575	-8.98126E-12	Other		1388.36	51.76594
43		5	91.16575	-1.56035E-11	Points		1388.36	51.7661
44		6	91.16575	-1.93552E-11			1388.36	51.76591
45		7	91.16575	-1.59162E-12			1388.36	51.76604
46		8	91.16575	-1.16813E-11			1388.36	51.76609

To avoid this kind of complication, we could test the *precise* answer to our problem, rather than the rounded form. Carrying out the matrix calculation in Sample Problem 2 of Section 8.2 to more places, the precise solution is

$$c = 1388.360204 \quad \text{and} \quad x = 51.76601234 \ .$$

Let's try this in our table. Table 13 shows the result.

Table 13

	A	B	C	D	E	F	G	H
33	C38=+0.260*G38+0.146*H38-0.0000944*G38^2+0.0000410*G38*H38-0.00196*H38^2-93.1							
34								
35		fn name:	f(x,y)		variable names ==>		x	y
36					Increments ==>		10	0.1
37		i	fn Value	fn(Other$_i$)-fn(Test)	Variables ==>		1st	2nd
38			91.16575	0	<== Test Point ==>		1388.36	51.76601
39		1	91.15732	-0.008429194			1397.798	51.72211
40		2	91.16524	-0.000508784			1390.64	51.69091
41		3	91.16534	-0.000406238			1390.426	51.7408
42		4	91.16569	-5.1631E-05	Other		1387.675	51.69722
43		5	91.15844	-0.007308374	Points		1397.157	51.75524
44		6	91.16485	-0.000898382			1391.427	51.71728
45		7	91.16517	-0.000580306			1390.842	51.77915
46		8	91.15737	-0.00838027			1397.793	51.85244

In this case, recalculating dozens of times still yields no better solutions, even if we make our increments smaller. Of course, if our spreadsheet precision is greater than the "optimal" solution we enter, we could find an even more precise solution, as happened in Table 10.

Our conclusion so far is that the solution $c = 1390$ and $x = 51.8$ represents a local maximum for the function $E(c,x)$. Is it also global? As mentioned earlier, this can be proved using calculus, and we will discuss how to do this at the end of this section. However, the calculus check for functions of 3 or more variables involves complicated matrix concepts we have not covered in this text, so it is helpful to know some other strategies for checking global optimality.

Without such a firm guarantee, we could make a slight modification of our randomized testing above that would help us gain more confidence that the solution is the global maximum. All we need to do is to **increase the increment, Δ, so that virtually all reasonable solutions within the domain could be chosen**. What does that mean in this problem? From the original data for this problem in section 5.6, the calories ranged from 1000 to 2000, and the exercise time varied from 0 to 120 minutes. Let's set our increment to 700 for calories (c) and to 80 for exercise (x). The latter will allow negative values for exercise, which makes no sense realistically, but is perfectly legitimate mathematically, so it will not hurt us to look at them. The resulting spreadsheet is as follows:

Table 14

	A	B	C	D	E	F	G	H
33	C38=+0.260*G38+0.146*H38-0.0000944*G38^2+0.0000410*G38*H38-0.00196*H38^2-93.1							
34								
35		fn name:	f(x,y)		variable names ==>		x	y
36						Increments ==>	700	80
37		i	fn Value	fn(Other$_i$)-fn(Test)		Variables ==>	1st	2nd
38			91.16549	0	<== Test Point ==>		1390	51.8
39		1	58.01911	-33.14638285			1966.312	87.13979
40		2	81.88604	-9.279452168			1075.675	54.5136
41		3	83.13653	-8.028962666			1309.003	112.5308
42		4	83.75008	-7.415409857	Other		1552.675	103.3451
43		5	53.83828	-37.32721618	Points		1908.904	-20.4019
44		6	69.39395	-21.77153964			1855.764	31.95878
45		7	57.22325	-33.94224579			791.6628	59.9517
46		8	73.91575	-17.24974062			971.4454	26.23013

Table 14 shows us that, out of these 8 randomly chosen solutions ranging widely over the domain, none is better than our critical point. Of course, if we wanted to, we could use the more refined critical point we used in Table 13 to be surer that nothing close to the test point would show up as slightly better. We can also keep recalculating the spreadsheet to view dozens or even hundreds of solutions. After ten such recalculations, if no better solutions are uncovered, we can have a reasonable level of confidence that our solution is also the global maximum.

So our final conclusion for Sample Problem 4 is that (1390,51.8) is the unique local and global maximum within the reasonable domain for the problem. In other words, eating about 1400 calories and exercising somewhere around 52 minutes should maximize the person's energy level, on average. From looking at the original data in Table 5 of Section 5.2, this seems quite reasonable. □

Testing Points with Three or More Variables

What can we do if our multivariable function has *three or more variables*? Let's look at an example.

Sample Problem 5: In Section 5.5, Problem 1, we looked at a data set for hours of practice at basketball and the number of shots made out of 15 at the free-throw line:

Table 15

Hours (x)	**Shots Made (y)**
1	3
2	8
3	11
4	10

In that section, after Sample Problem 3, we discussed how to derive a model for the SSE as a function of the parameters of a quadratic model $f(x) = ax^2 + bx + c$:

Verbal Definition: $SSE(m,b)$ = the SSE for the model $f(x) = ax^2 + bx + c$ and the given data

Symbol Definition: $SSE(m,b) = (3\text{-}a\text{-}b\text{-}c)^2+(8\text{-}4a\text{-}2b\text{-}c)^2+(11\text{-}9a\text{-}3b\text{-}c)^2+(10\text{-}16a\text{-}4b\text{-}c)^2$

Assumptions: Certainty and Divisibility. Certainty implies that the relationship is exact. Divisibility implies that any fractional values of SSE and the parameters are possible.

Taking partial derivatives, setting them equal to 0, and solving for the parameters gives a solution of $a = -1.5$, $b = 9.9$, and $c = -5.5$. Is this critical point the point that minimizes the SSE, over all possible quadratic models?

Solution: Clearly, we can use the same ideas given above in this section, but they can get more complicated. If you look back at Figure 8.4-1 , you can see that the pattern of subtracting or adding (or neither) the increment to each variable can be thought of as associating a -1, +1, or 0, respectively. When there are two variables, there are these three possibilities for the first variable, and then, *for each of those possibilities*, there are three possibilities for the second variable. This makes a total of $(3)(3) = 3^2 = 9$ permutations (including the original test point). Another way to look at this calculation is that it is analogous to the *area* of the square in Figure 8.4-1 (thinking of it as 3 units by 3 units).

By extension, if we have three variables, there will be $3^3 = 27$ possibilities, $3^4 = 81$ for four variables, and in general, 3^n possibilities for n variables. This *could* be done, either by hand or in an organized and fast way with software commands, but we will recommend for this course that we simply use the random point procedure when we have three or more variables. The extension to more variables is very simple in this case. Let's try it with our given critical point; the results are shown in Table 16.

Table 16

	A	B	C	D	E	F	G	H	I
33	C38=+(3-G38-H38-I38)^2+(8-4*G38-2*H38-I38)^2+(11-9*G38-3*H38-I38)^2+(10-16*G38-4*H38-I38)^2								
34									
35		fn name:	**SSE(a,b,c)**		variable names ==>		a	b	c
36					**Increments ==>**		0.1	0.1	0.1
37		**i**	**fn Value**	**fn(Other$_i$)-fn(Test)**	**Variables ==>**		**1st**	**2nd**	**3rd**
38			0.2	0	**<== Test Point ==>**		-1.5	9.9	-5.5
39		1	0.302372	0.102371525			-1.53126	9.94936	-5.49056
40		2	1.658464	1.458464071			-1.41281	9.824217	-5.52302
41		3	0.858535	0.658534998			-1.52681	9.823167	-5.43595
42		4	2.261478	2.06147837	**Other**		-1.59795	9.9614	-5.44462
43		5	0.215269	0.015268556	**Points**		-1.4736	9.812677	-5.45468
44		6	2.746561	2.546560659			-1.55927	9.827594	-5.55499
45		7	1.63479	1.434789836			-1.53518	9.830747	-5.59436
46		8	0.276013	0.076013271			-1.50859	9.955862	-5.43838

Notice that we have used increments based on the number of written digits (varying the last one), and our solution does seem to be a local minimum, even after 10 recalculations. To make sure it is also a global minimum, let's try much larger increments, like 100, as shown in Table 17.

Table 17

	A	B	C	D	E	F	G	H	I
33	C38=+(3-G38-H38-I38)^2+(8-4*G38-2*H38-I38)^2+(11-9*G38-3*H38-I38)^2+(10-16*G38-4*H38-I38)^2								
34									
35		fn name:	**SSE(a,b,c)**		variable names ==>		a	b	c
36					**Increments ==>**		100	100	100
37		**i**	**fn Value**	**fn(Other$_i$)-fn(Test)**	**Variables ==>**		**1st**	**2nd**	**3rd**
38			0.2	0	**<== Test Point ==>**		-1.5	9.9	-5.5
39		1	10677.24	10677.04271			-5.42527	-12.9333	61.0148
40		2	1299716	1299715.419			-41.9506	-45.4592	-53.9651
41		3	4934802	4934801.6			92.81854	69.87883	70.75591
42		4	2356936	2356935.727	**Other**		47.0101	108.2613	49.3044
43		5	45445.88	45445.68201	**Points**		27.24006	-73.6978	-89.6449
44		6	2654878	2654877.74			-87.2067	28.37796	-76.8702
45		7	3616693	3616693.007			80.89245	86.14887	-39.8258
46		8	253486.5	253486.2898			1.252671	94.06256	-9.88617

Again, after many recalculations, it appears that the critical point is a global minimum as well as a local minimum. □

The Second Derivative Test for Functions of Two Variables*

In Section 3.3, we saw that we could test a critical point to see if it was a local extremum by plugging its input value into the second derivative, which is the derivative of the derivative. If the result was strictly positive, the critical point was a local minimum, and if the result was strictly negative, the critical point was a local maximum.

What would a second derivative correspond to for a multivariable function? Since derivatives of multivariable functions are *partial* derivatives, with respect to one of the variables, a *second* derivative would mean a partial derivative *of* a partial derivative. Analogous to our notation of $f_x(x, y)$ for the partial derivative with respect to x, we will use the notation

$$f_{yx}(x, y)$$

to mean that **we first take the partial derivative with respect to y, then we take the partial derivative of *that result* with respect to x**, so it is the partial with respect to x *of* the partial with respect to y. The easiest way to remember this is that the order of the subscripts tells the order in which you take the partial derivatives. Another notation you may sometimes see for this same second partial derivative is

$$\frac{\partial}{\partial x}\left\{\frac{\partial}{\partial y}[f(x,y)]\right\} = \frac{\partial^2}{\partial x \partial y}[f(x,y)] = \frac{\partial^2 z}{\partial x \partial y} = f_{yx}(x,y) = f_{yx} ,$$

where $z = f(x, y)$. When the partial is taken with respect to the same variable both times, this can also be written

$$\frac{\partial}{\partial x}\left\{\frac{\partial}{\partial x}[f(x,y)]\right\} = \frac{\partial^2}{\partial x^2}[f(x,y)] = \frac{\partial^2 z}{\partial x^2} = f_{xx}(x,y) = f_{xx} .$$

In either case, the notation acts as if the $\partial, \partial x$, and ∂y symbols are variables being multiplied together.

The **Second Derivative Test for functions of two variables** says that:

If (x, y) is a critical point (so both partial derivatives equal 0 there):
If $f_{xx} > 0$ and $f_{xx}f_{yy} - f_{xy}^2 > 0$, then (x, y) is a local minimum;
If $f_{xx} < 0$ and $f_{xx}f_{yy} - f_{xy}^2 > 0$, then (x, y) is a local maximum;
Otherwise, the test is inconclusive.

---

* This topic is optional and can be skipped without loss of continuity.

In the above statement, it is assumed that all of the second partial derivatives are evaluated at the critical point (x, y).

Notice that the double derivative with respect to x, f_{xx}, is like the second derivative of a single-variable function, and so *that* part of the test corresponds to the single-variable Second Derivative Test (when the double derivative is positive, the critical point is a local minimum, and when the double derivative is negative, the critical point is a local maximum). The only *addition* is the complicated condition, which is the *same* for both cases, that

$$[f_{xx}(x, y)][f_{yy}(x, y)] - [f_{xy}(x, y)]^2 > 0$$

Let's see how this condition applies to some simple functions.

Sample Problem 6: Use the Second Derivative Test to check critical points for the functions

(a) $f(x, y) = x^2 + y^2 - 10xy$
(b) $f(x, y) = x^2 + y^2$
(c) $f(x, y) = -x^2 - y^2$
(d) $f(x, y) = -x^2 - y^2$
(e) $P(c, t) = -7c^3 + 82c^2 - 5ct - 5t^2 - 182c + 84t - 130$

Solution: (a) $f(x, y) = x^2 + y^2 - 10xy$, so the partial derivative with respect to x is

$$f_x(x, y) = \frac{\partial}{\partial x}\left(x^2 + y^2 - 10xy\right) = 2x + 0 - 10y = 2x - 10y \quad ,$$

as we saw in Sample Problem 3. If we now take the partial derivative of *this* with respect to x, we get the double derivative with respect to x:

$$f_{xx}(x, y) = \frac{\partial}{\partial x}(2x - 10y) = 2 - 0 = 2$$

This is positive, which suggests that, if anything, we may have a local minimum. In Sample Problem 3, we found that the only critical point of this function was the origin, (0,0). Let's now find the other first and second partial derivatives and check the rest of the condition for a local minimum:

$$f_{xy}(x, y) = \frac{\partial}{\partial y}(2x - 10y) = 0 - 10 = -10$$

$$f_y(x, y) = \frac{\partial}{\partial y}\left(x^2 + y^2 - 10xy\right) = 0 + 2y - 10x = 2y - 10x$$

$$f_{yx}(x, y) = \frac{\partial}{\partial x}(2y - 10x) = 0 - 10 = -10$$

$$f_{yy}(x, y) = \frac{\partial}{\partial y}(2y - 10x) = 2 - 0 = 2$$

$$f_{xx}f_{yy} - f_{xy}^2 = [\, f_{xx}(x, y)][\, f_{yy}(x, y)] - [f_{xy}(x, y)]^2 = [2][2] - [-10]^2 = 4 - 100 = -96$$

Thus, since -96 is not positive, we cannot conclude that the critical point (0,0) is a local minimum. In fact, in Sample Problem 3, we saw that it was a saddle point. This example helps to see the need for the complicated condition.

Notice that **the two mixed second partial derivatives, f_{xy} and f_{yx}, were the same. This will be true in general for continuous, smooth functions.** In fact, the complicated condition actually involves the *product* of these, but can be simplified to the *square* of either one, as we have done.

(b) In this case, $f(x, y) = x^2 + y^2$. $f_x = 2x$ and $f_y = 2y$, so the only critical point is again the origin, (0,0). In this case, both double derivatives are 2, and both mixed second derivatives are 0, so $f_{xx} = 2 > 0$ and

$$f_{xx}f_{yy} - f_{xy}^2 = (2)(2) - (0)^2 = 4 - 0 = 4 > 0 ,$$

so (0,0) is a local minimum. This surface turns out to be a simple bowl shape, so the result makes sense.

(c) In this case, $f(x, y) = -x^2 - y^2$. $f_x = -2x$ and $f_y = -2y$, so the only critical point is again the origin, (0,0). In this case, both double derivatives are -2, and both mixed second derivatives are 0, so $f_{xx} = -2 < 0$ and

$$f_{xx}f_{yy} - f_{xy}^2 = (-2)(-2) - (0)^2 = 4 - 0 = 4 > 0 ,$$

so (0,0) is a local maximum. This surface turns out to be a simple rounded hilltop (the mirror image of a bowl), so the result makes sense.

(d) In this case, $f(x, y) = x^2 - y^2$. $f_x = 2x$ and $f_y = -2y$, so the only critical point is again the origin, (0,0). Again, both mixed derivatives are 0. This time, $f_{xx} = 2 > 0$ and $f_{yy} = -2$, so

$$f_{xx}f_{yy} - f_{xy}^2 = (2)(-2) - (0)^2 = -4 - 0 = -4 < 0 ,$$

so neither of the conditions in the Second Derivative Test applies, and we can not conclude that the critical point (0,0) is a local extremum. In fact, as we saw in Sample Problem 2, it is a saddle point.

(e) In this case, $P(c,t) = -7c^3 + 82c^2 - 5ct - 5t^2 - 182c + 84t - 130$. In Sample Problem 6 of Section 8.2 , we found that this function had two critical points: (1.72,7.54) and (6.21,5.29). Let's calculate the first and second partial derivatives:

$$\begin{aligned} P_c(c,t) &= -21c^2 + 164c - 5t - 0 - 182 + 0 - 0 = -21c^2 + 164c - 5t - 182 \\ P_t(c,t) &= 0 + 0 - 5c - 10t - 0 + 84 - 0 = -5c - 10t + 84 \\ P_{cc}(c,t) &= -42c + 164 - 0 - 0 = -42c + 164 \\ P_{ct}(c,t) &= 0 + 0 - 5 - 0 = -5 \\ P_{tc}(c,t) &= -5 - 0 + 0 = -5 \\ P_{tt}(c,t) &= 0 - 10 + 0 = -10 \end{aligned}$$

Let's check the complicated condition first, since it applies to both cases. Notice that the variables are different here. Just remember that the complicated condition is that **the product of the double derivatives minus the product of the mixed derivatives (or either mixed derivative squared) must be positive.** In this case, that means

$$P_{cc}P_{tt} - P_{ct}^{\,2} = (-42c + 164)(-10) - (-5)^2 = 420c - 164 - 25 = 420c - 189 > 0 \,,$$

which is equivalent to

$$420c > 189 \,, \text{ or } c > \frac{189}{420} = 0.45$$

Both of our critical points satisfy this condition, so let's check the simple double derivative condition at each point. At (1.72,7.54),

$$\begin{aligned} P_{cc}(1.72,7.54) = P_{cc}(c,t) &= -42c + 164 = -42(1.72) + 164 = -72.24 + 164 \\ &= 91.76 > 0 \,, \end{aligned}$$

so (1.72,7.54) is a local minimum.

At (6.21,5.29),

$$\begin{aligned} P_{cc}(6.21,5.29) = P_{cc}(c,t) &= -42c + 164 = -42(6.21) + 164 = -260.82 + 164 \\ &= -96.82 < 0 \,, \end{aligned}$$

so (6.21,5.29) is a local maximum. □

This Second Derivative Test can be generalized to general multivariable functions, but involves complex matrix concepts that are beyond the scope of this text. To learn more, see an advanced multivariable calculus text.

Section Summary

Before moving on to the exercises, be sure that you:

- Know that if you are trying to optimize a quadratic function and if some of the signs of your squared term coefficients are strictly positive and some are strictly negative, your critical point (if one exists) *will not* be a local or global extremum.
- Know that if you are trying to optimize a quadratic function, even if the squared term coefficient signs are all positive and you are minimizing, or all negative and you are maximizing, you still need to test further to see if a critical point is a local or global extremum.
- Know that for functions of 2-variables, to see if a critical point is probably a local extremum or not, you can test points adjacent to the critical point by finding points on a square grid around it, corresponding to all possible permutations of subtracting or adding a small increment to each variable, or leaving it unchanged (8 points in addition to the original critical point). If the function value at the critical point is better than the function values at the points on the grid around it, then the critical point is probably a local extremum. If not, it probably isn't. You can make the increment size the value of the smallest significant digit's place (for example, if your answer was rounded to 5.36 , use .01).
- Know how to do the above grid testing of a critical point on a spreadsheet, including a matrix of differences resulting from subtracting the function values at the grid points minus the function value at the critical point.
- Know how to use both relative and absolute addresses on a spreadsheet.
- Know that for functions of any number of variables, you can specify an increment for each variable and add or subtract to the value of that variable in the critical point a random amount up to that increment. To test for local optimality, use relatively small increments (although for rounded test points, it may be best to use the value of the smallest significant digit times 10 or 100). To test for global optimality, choose a value for the increment so that the values of that variable that are generated will range over all of its reasonable values in the domain.

- **Know how to do the randomized testing discussed above using a spreadsheet, including a column to calculate the function value of each randomly generated other point minus the function value of the test point.**
- **Understand how the testing procedures can be used to search for local and global optimal solutions by trial and error.**
- **[*] Understand how to apply the Second Derivative Test for functions of two variables: for a critical point of $f(x, y)$ satisfying $f_{xx}f_{yy} - f_{xy}^2 > 0$: if $f_{xx} > 0$, then the critical point is a local minimum, and if $f_{xx} < 0$, then the critical point is a local maximum. Otherwise, the test is inconclusive.**

---

[*] This comes from the optional subsection on the Second Derivative Test for functions of two variables.

EXERCISES FOR SECTION 8.4:

For any of the exercises involving testing, hand in a printout to show at least one spreadsheet block of cells for each test you conduct.

Warm Up

1. The only critical point of $R(x,y) = -3x^2 - 2y^2 + 3xy + 4y - 20$ is $x = 0.8,\ y = 1.6$.
   a) Based on the squared-term coefficients, can you easily determine that the critical point is *not* an extremum? Briefly explain.
   b) Based on the squared-term coefficients, if the critical point is an extremum, what kind of extremum (maximum or minimum) would it have to be?
   c) Use increments of 1 for both variables to test if the critical point is probably a local extremum, and if it is, which kind (maximum or minimum).
   d) Now use an increment based on the last significant digit of each variable value of the critical point to do a similar test. Do you get the same result?

2. The only critical point of $P(x,y) = -2x^2 - y^2 + 2xy - 4y + 3x + 24$ is $x = -0.5$, $y = -2.5$.
   a) Based on the squared-term coefficients, can you easily determine that the critical point is *not* an extremum? Briefly explain.
   b) Based on the squared-term coefficients, if the critical point is an extremum, what kind of extremum (maximum or minimum) would it have to be?
   c) Use increments of 1 for both variables to test if the critical point is probably a local extremum, and if it is, which kind (maximum or minimum).
   d) Now use an increment based on the last significant digit of each variable value of the critical point to do a similar test. Do you get the same result?

3. The unique critical point of $S(m,b) = (5-m-b)^2 + (6-2m-b)^2 + (9-3m-b)^2$ is at $m \approx 2.00,\ b \approx 2.67$.
   a) What would be the signs of the squared-term coefficients if you multiplied out the expression and simplified? What does this tell you with respect to what kind of extremum the critical point could be?
   b) Use increments based on the last significant digit of the value of each variable at the critical point to test whether the critical point seems to be a local extremum.
   c) Now use the same increments with *random* comparison points to give stronger evidence if you found the critical point to be an extremum. If you find any better points, try testing *them* and repeat until you find a solution that is not improved upon after ten recalculations. How does this solution compare to a precise calculation of the critical point using calculus and matrices?

4. For the function $S(m,b)=(5-8m-b)^2+(4-6m-b)^2+(6-4m-b)^2+(9-2m-b)^2$, the unique critical point is $m = -0.7, b = 9.5$.
   a) What would be the signs of the squared-term coefficients if you multiplied out the expression and simplified? What does this tell you with respect to what kind of extremum the critical point could be?
   b) Use increments based on the last significant digit of the value of each variable at the critical point to test whether the critical point seems to be a local extremum.
   c) Now use the same increments with *random* comparison points to give stronger evidence if you found the critical point to be an extremum. If you find any better points, try testing *them* and repeat until you find a solution that is not improved upon after ten recalculations. How does this solution compare to a precise calculation of the critical point using calculus and matrices?

5. The unique critical point of $C(f,d)=50-3.5f-7.2d-3.5fd+4.7f^2+3.9d^2$ is at $f \approx 0.860$ and $d \approx 1.31$
   a) Based on the squared-term coefficients, can you easily determine that the critical point is *not* an extremum? Briefly explain.
   b) Based on the squared-term coefficients, if the critical point is an extremum, what kind of extremum (maximum or minimum) would it have to be?
   c) Use increments based on the last significant digit of the value of each variable at the critical point to test whether the critical point seems to be a local extremum.
   d) Given that values of f could reasonably range between 0 and 5, and values of d could reasonably range between 0 and 8, find an appropriate increment for each variable to use to generate *randomized* comparison points to test whether the critical point seems to be a global extremum.

6. The unique critical point of $E(c,r)=-2.8c^2-3.7r^2+4.3cr+3.0c-2.1r+10$ is at $c \approx 0.574$ and $r \approx 0.0497$.
   a) Based on the squared-term coefficients, can you easily determine that the critical point is *not* an extremum? Briefly explain.
   b) Based on the squared-term coefficients, if the critical point is an extremum, what kind of extremum (maximum or minimum) would it have to be?
   c) Use increments based on the last significant digit of the value of each variable at the critical point to test whether the critical point seems to be a local extremum.
   d) Given that values of c could reasonably range between 0 and 4, and values of r could reasonably range between 0 and 0.5, find an appropriate increment for

each variable to use to generate *randomized* comparison points to test whether the critical point seems to be a global extremum.

7. For the function $g(s,t) = 4s^2 - 3st - 2t^2 + 5s - 6t + 4$
   a) Based on the squared-term coefficients, can you easily determine that a critical point (if one exists) is *not* an extremum? Briefly explain.
   b) Find any critical points and test them for local optimality.
8. For the function $E(s,n) = -2s^2 + sn + 5n^2 + 3s - 4n + 7$
   a) Based on the squared-term coefficients, can you easily determine that a critical point (if one exists) is *not* an extremum? Briefly explain.
   b) Find any critical points and test them for local optimality.
9. For the function $H(s,a) = -2s^2 + 12sa - a^2 + 5s - 3a + 6$:
   a) Based on the squared-term coefficients, can you easily determine that a critical point (if one exists) is *not* an extremum? Briefly explain.
   b) Find any critical points and test them for local optimality.
10. For the function $P(l,t) = l^2 - 6lt + 2t^2 + 3l + 4t + 7$:
    a) Based on the squared-term coefficients, can you easily determine that a critical point (if one exists) is *not* an extremum? Briefly explain.
    b) Find any critical points and test them for local optimality.
11. $f(x,y,z) = 4x^2 + y^2 - 8xz - 2yz - 20z$
    a) Is this a quadratic function?
    b) What do the signs of the squared-term coefficients tell you about what kind of extremum a critical point could be?
    c) Find any critical points and test them for local optimality.
12. $f(x,y,z,\lambda) = x + y + z - xyz\lambda + 8\lambda$
    a) Is this a quadratic function?
    b) Find any critical points and test them for local optimality.

Game Time

13. The profit function (in thousands of dollars) for a firm selling specialized flooring materials can be modeled by:

Verbal Definition: $\pi(f,d)$ = the profit, in dollars when f is price in dollars for one square yard of the Florentine tiles and, d is price in dollars for one square yard of the Delmonico tiles..

Symbol Definition: $\pi(f,d) = -22.0f^2 + 40.2df + 202f - 26.0d^2 + 171d - 237$, $0 \le f,d \le 50$

Assumptions: Certainty and divisibility. Certainty implies that the relationship is exact. Divisibility implies that any fractional values of dollars are possible.

a) Find any critical points and test them for local and global optimality.
b) What is your conclusion and advice to the firm?

14. The cost of manufacturing a box with restrictions on its measurements of 108 inches for total length plus girth, at \$0.02 per square inch for sides and top and \$0.03 per square inch for the bottom is modeled by:
Verbal Definition: $C(w,h)$ = the cost, in dollars to manufacture a box of width w inches and height h inches
Symbol Definition: $C(w,h) = 5w(108-2w-2h)+4hw+4h(108-2w-2h)$ for $w,h \geq 0$ and $w+h \leq 54$
Assumptions: Certainty and divisibility. Certainty implies that the relationship is exact. Divisibility implies that any fractional values of dollars and inches are possible.
a) Find any critical points and test them for local and global optimality.
b) What is your conclusion and advice?

15. If a box costs \$2 per square foot for sides and top and \$3 per square foot for the bottom and it must have a volume of 6 cubic feet, the cost of the box is modeled by:
Verbal Definition: $C(w,l)$ = the cost, in dollars, to make a box of width w feet and length l feet
Symbol Definition: $C(w,l) = 5wl + \frac{24}{l} + \frac{24}{w}$ for $w,l > 0$
Assumptions: Certainty and divisibility. Certainty implies that the relationship is exact. Divisibility implies that any fractional values of dollars and feet are possible
a) Find any critical points and test them for local and global optimality.
b) What is your conclusion and advice?

16. If the cost to produce Fudge Cakes and Cream Cakes can be modeled by :
Verbal Definitions: $C(f,c)$ = cost, in dollars, to make Fudge Cake sold at f dollars per cake and Cream Cakes sold at c dollars per cake
Symbol Definition: $C(f,c) = 21.71f + 3.54c$ for $f,c \geq 0$,
Assumptions: Certainty and Divisibility. Certainty implies that the relationship is exact. Divisibility implies that any fractional values of dollars are possible

and the revenue from the sales of both of the cakes can be modeled by:
Verbal Definitions: $R(f,c)$ = the revenue, in dollars, when Fudge Cakes are sold at f dollars each and Cream Cakes at c dollars each.
Symbol Definition: $R(f,c) = -0.43f^2 + 19.56f + 1.45fc - 1.93c^2 + 13.97c$ for $f,c \geq 0$,
Assumptions: Certainty and Divisibility. Certainty implies that the relationship is exact. Divisibility implies that any fractional values of dollars are possible.

a) Formulate and fully define a profit model.
b) Find any critical points for the profit function and test them for local and global optimality.
c) What is your conclusion and advice?

17. The demand for two hockey jerseys, an officially sanctioned jersey and a replica can be modeled by:

Verbal Definition: $PO(o,r)$ = the price, in dollars, per official jersey when o official jerseys and r replica jerseys are sold.
Symbol Definition: $PO(o,r) = -4.98o + 3.03r + 745$ for $o, r \geq 0$
Assumptions: Certainty and divisibility. Certainty implies that the relationship is exact. Divisibility implies that any fractional values of dollars and jerseys are possible.

Verbal Definition: $PR(o,r)$ = the price, in dollars, per replica jersey when o official jerseys and r replica jerseys are sold.
Symbol Definition: $PR(o,r) = 0.936o - 2.99r + 70.4$ for $o, r \geq 0$
Assumptions: Certainty and divisibility. Certainty implies that the relationship is exact. Divisibility implies that any fractional values of dollars and jerseys are possible.

The officially sanctioned jerseys cost the retailer $19.23 each and the replicas cost $12.38 each. There is a onetime order charge of $50.00.
a) Formulate and fully define a profit model.
b) Find any critical points for the profit function and test them for local and global optimality.
c) What is your conclusion and advice?

18. The profit from the sale of CD players can be modeled by:
Verbal Definition: $P(d,r)$ = the profit, in dollars, from the sale of d deluxe and r regular CD players
Symbol Definition: $P(d,r) = 90d + 114r - .02dr - .07d^2 - .03r^2 - 32000$ for $d, r \geq 0$,
Assumptions: Certainty and divisibility. Certainty implies that the relationship is exact. Divisibility implies that any fractional values of dollars, and CD players are possible.
a) Find any critical points for the profit function and test them for local and global optimality.
b) What is your conclusion and advice?

19. If the profit function for the furniture manufacturer changes to

$$P(c,t) = 150c + 200t + 2ct - 4c^2 - 12t^2 - 700 \ ,$$

a) Find the new critical points and test them for local and global optimality.
b) What is your conclusion and advice?

20. If the profit function for the furniture manufacturer changes to

$$P(c,t) = 150c + 220t + 2ct - 4c^2 - 12t^2 - 700 \ ,$$

a) Find the new critical points.
b) What is your conclusion and advice?

21. If the model for the profit from the sale of divinity fudge and chocolate fudge is:

Verbal Definition: $P(f,d)$ = the profit, in dollars, from the sale of chocolate fudge at f dollars a piece and divinity fudge sold at d dollars a piece

Symbol Definition: $P(f,d) = -18.2f^2 + 199.14f + 36.0fd - 26.0d^2 + 168.8d - 237.4$ for $f,d \geq 0$

Assumptions: Certainty and divisibility. Certainty implies that the relationship is exact. Divisibility implies that any fractional values of dollars are possible. The profit function assumes that the quantity made is calculated from the demand function.

a) Find the critical points for the profit function and test them for local and global optimality.
b) Comment on your results.

Overtime*

22-31.* Apply the Second Derivative Test for functions of two variables to the functions and critical points given in Exercises 1-10.

* These problems use concepts from the optional subsection on the Second Derivative Test for functions of two variables.

Section 8.5: The Method of Least Squares

In the first semester we began this course by finding functions or models that described the relationship between two variables. First we created scatter plots of data. Then, by examining the shape of the scatter plot, we selected the type of model that best fit the scatter plot: linear, quadratic, cubic, exponential or logistic. Once we had decided on the type of model that would best describe the data, we used technology to find this best-fit model. We discussed the fact that the best-fit model is that for which a measure of the *difference* between the actual dependent-variable values and the predicted dependent-variable values is the smallest. We also discussed the fact that we want to square these differences, or *errors*, to eliminate the possible cancellation of positive and negative values, and also to give more weight to the larger errors, to try to avoid them. Finally, we added the squares of the errors, hence the name "sum of the squared errors," abbreviated SSE. In this section we use the optimization techniques described in this chapter to generalize how to find the equation of a model that best fits a set of points, the equation for which the sum of the squared errors is minimized. The method is called **least squares regression.**

Below are some sample problems to illustrate the variety of types of regression possible:

- You are studying for a quiz. Each week you have a similar quiz, consisting of ten problems. You have kept a record of the number of problems you got correct on the ten-question quiz and the number of hours spent studying before the quiz. You really have to get at least eight out of ten on this quiz. What is the minimum amount of time that you have to study to achieve this?
- The price of souvenir T-shirts has risen along with all other prices. You have been watching the prices over the past few years. Can you determine approximately how much the increase in the price of the T-shirts will be for next year?
- You frequently go with some friends to this little sandwich shop. Their prices seem to be quite reasonable, so you haven't been disturbed by the fact that they don't have an actual price list. They just write up what sandwiches they are offering each day. When you put your order in, they tell you the total cost. You would like to figure out how much they actually charge for a sandwich and a milkshake, the favorites of your group.

When you have finished this section you should know how to answer these questions *without using technology* to find the least squares regression model and:

- Know how to formulate a model of the sum of the squared errors in terms of the parameters of the model for single and multivariable linear and non-linear functions.

- Know how to minimize this SSE function using calculus by setting partial derivatives equal to 0.
- Know how to generalize the expressions for the partial derivatives to derive the general normal equations.
- Be able to use matrix theory to find the least squares regression model.

Finding the Linear Least Squares Regression

Sample Problem 1: You are preparing to study for a quiz. You have decided to let

y = the number of problems correct on a ten-question quiz, and
x = the total number of hours spent on the material covered in the quiz.

Table 1 shows your past experience on similar quizzes:

Table 1

x	y
2	4
5	6
6	7
9	8

Find the SSE for a general linear model for this data as a function of its parameters, m and b. Then set the partial derivatives equal to 0 and solve for m and b to derive the least-squares regression line.

Solution: The graph of this data is shown in Figure 8.5-1.

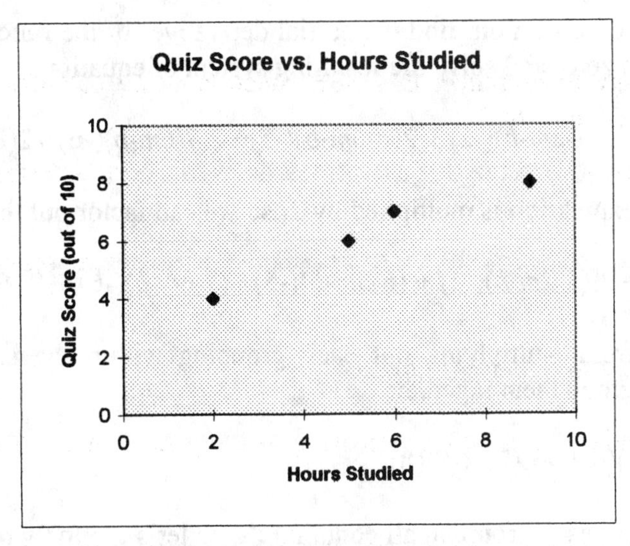

Figure 8.5-1

This data is clearly not perfectly linear, but a linear model would fit it fairly well, so is useful for rough approximation purposes. For example, you could use a linear model to get a quick rule-of-thumb estimate of the grade you would get for different amounts of studying.

The general equation for the linear model is $y = mx + b$. Table 2 shows the value of y as calculated from our model ($y = mx + b$), the distance (error) between the actual y value (from our data) and the predicted value of y as calculated from our model ($mx + b$), and finally, the square of this distance (the error squared), as we have already discussed in Section 5.5 .

Table 2

x	Actual y	Predicted $mx + b$	Error $y - (mx + b)$	$(\text{Error})^2$ $(y - mx - b)^2$
2	4	$2m + b$	$4 - (2m + b)$	$(4 - 2m - b)^2$
5	6	$5m + b$	$6 - (5m + b)$	$(6 - 5m - b)^2$
6	7	$6m + b$	$7 - (6m + b)$	$(7 - 6m - b)^2$
9	8	$9m + b$	$8 - (9m + b)$	$(8 - 9m - b)^2$

To find the best possible model, we want to minimize the total squared error. The sum of the squares of the errors is:

$$\text{SSE}(m,b) = (4\text{-}2m\text{-}b)^2 + (6\text{-}5m\text{-}b)^2 + (7\text{-}6m\text{-}b)^2 + (8\text{-}9m\text{-}b)^2$$

We wish to minimize the sum of the squares of the errors, so we will use our techniques from multivariable optimization: find the partial derivatives of the function, set the partial derivatives equal to zero, and solve the resulting system of equations.

$$SSE_m(m,b) = 2(4-2m-b)(-2) + 2(6-5m-b)(-5)+ 2(7-6m-b)(-6) +2(8-9m-b)(-9)$$

Each term in this expression is multiplied by 2, so we can factor out the 2 and we get:

$$SSE_m(m,b) = 2[(4-2m-b)(-2) + (6-5m-b)(-5)+ (7-6m-b)(-6) + (8-9m-b)(-9)]$$

The first entries from multiplying out each grouping within the square brackets are constants, so let's group them together:

$$\{(-2)(4)+(-5)(6)+(-6)(7)+(-9)(8)\}$$

The second entries in each grouping all contain m's, so let's group them together:

$$\{(-2)(-2m)+(-5)(-5m)+(-6)(-6m)+(-9)(-9m)\}$$

In each term, there are two negatives multiplied, so we can write this as:

$$\{(2)(2m)+(5)(5m)+(6)(6m)+(9)(9m)\}$$

The last entries in each term all contain b's, so let's group them together:

$$\{(2)(b)+(5)(b)+(6)(b)+(9)(b)\}$$

We now have:

$$SSE_m(m,b)=2[\{(-2)(4)+(-5)(6)+(-6)(7)+(-9)(8)\}+\{(2)(2m)+(5)(5m)+(6)(6m)+(9)(9m)\}+ \{(2)(b)+(5)(b)+(6)(b)+(9)(b)\}]$$

Simplifying each of these terms, we get:

$$SSE_m(m,b)=2[(-8+-30+-42+-72) + (4m+25m+36m+81m)+ (2b+5b+6b+9b)]$$

Further simplifying we have:

$$SSE_m(m,b)= 2(-152 + 146m + 22b)\ .$$

Going through the same process with the partial derivative with respect to b we get:

$$\begin{aligned} SSE_b(m,b) &= 2(4\text{-}2m\text{-}b)(-1) + 2(6\text{-}5m\text{-}b)(-1) + 2(9\text{-}6m\text{-}b)(-1) + 2(8\text{-}9m\text{-}b)(-1) \\ &= -2(4+6+7+8) + -2(-2m+-5m+-6m+-9m) + -2(-b+-b+-b+-b) \\ &= -2(25\text{-}22m\text{-}4b) \end{aligned}$$

Now. let's set the partial derivatives equal to zero:

$$SSE_m(m,b) = 2(-152 + 146m + 22b) = 0$$
$$SSE_b(m,b) = -2(25 - 22m - 4b) = 0$$

If we divide each equation by 2 and rewrite them in standard form, we get

$$\begin{aligned} 146m + 22b &= 152 \\ 22m + \ \ 4b &= \ \ 25 \end{aligned}$$

These equations are called the **normal equations** for this regression problem. **The normal equations are the *system* of equations involving the *parameters* of the model you are trying to formulate whose *solution* values are the *parameters* of the least-squares regression model.**

Solving this system of normal equations by any of the techniques discussed in Chapter 7, we get $m = 0.58$ and $b = 3.06$ as the only critical point. Therefore the equation of the line that best fits the data is:

$$y = f(x) = 0.58x + 3.06$$

This model would predict, for example, that if you studied for 10 hours ($x = 10$), then your score out of 10 (y value) would be given by

$$y = f(10) = 0.58(10) + 3.06 = 5.8 + 3.06 = 8.86 \ .$$

Translation: You should expect to get a score of about 9 out of 10 on the quiz if you study for 10 hours.

As a rough rule of thumb to use in your head, you could use a rounded form of the model, like $y = 0.6x + 3$. So, for example, if you studied 6 hours, you could expect on average about a score of $0.6(6) + 3 = 3.6 + 3 = 6.6$ (that is, probably somewhere around 6 or 7 out of 10). To get at least eight out of ten correct, you would have to study more than eight hours, since $0.6(8) + 3 = 7.8$.

Figure 8.5-2 shows a graph of the derived model overlaid over the data points. You can see that the fit is pretty good, but not perfect. The deviations (errors) correspond to the vertical distance from the model to each data point (so data points above the line have positive errors and those below have negative errors), and these are drawn in on the graph.

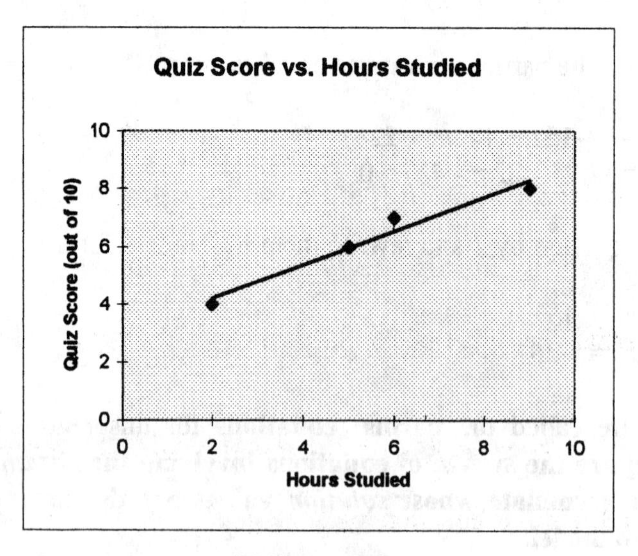

Figure 8.5-2

The regression line could also have been found using your calculator, or using the regression tool on a spreadsheet to verify our results. Try both methods on your own to convince yourself that this procedure is indeed what these technologies are doing behind the scenes. □

The Normal Equations for Linear Least Squares Regressions

In Sample Problem 1, our data was given by

x	y
2	4
5	6
6	7
9	8

and we derived the SSE to be given by

$$SSE(m,b) = (4-2m-b)^2 + (6-5m-b)^2 + (7-6m-b)^2 + (8-9m-b)^2 .$$

Suppose we were trying to fit a linear model $y = mx + b$ to a general data set of n points, given by:

Table 3

x	y
x_1	y_1
x_1	y_2
...	...
x_n	y_n

Analogous to Sample Problem 1, the model for SSE will be given by

$$SSE(m,b) = (y_1 - mx_1 - b)^2 + (y_2 - mx_2 - b)^2 + \ldots + (y_n - mx_n - b)^2$$

The first partial derivative of the SSE function, the partial with respect to m , set equal to 0, is:

$$SSE_m\,(m,b) = 2(y_1 - mx_1 - b)(-x_1) + 2(y_2 - mx_2 - b)(-x_2) + \ldots + 2(y_n - mx_n - b)(-x_n) = 0$$

If we divide both sides of this equation by 2, we get

$$(y_1 - mx_1 - b)(-x_1) + (y_2 - mx_2 - b)(-x_2) + \ldots + (y_n - mx_n - b)(-x_n) = 0 \ .$$

Now, multiplying out and simplifying, we get

$$-x_1 y_1 + mx_1^2 + bx_1 - x_2 y_2 + mx_2^2 + bx_2 - \ \ldots \ - x_n y_n + mx_n^2 + bx_n \ .$$

Remember that we are trying to solve for m and b , so let's combine like terms:

$$(x_1^2 + x_2^2 + \ \ldots \ + x_n^2)m + (x_1 + x_2 + \ \ldots \ + x_n)b = (x_1 y_1 + x_2 y_2 + \ \ldots \ + x_n y_n)$$

Recall the summation notation introduced in Section 4.2 of Volume 1 on sums and limits of sums and used in Section 5.5 to formulate models for SSE. The general form of the summation symbol is:

$$\sum_{i=k}^{n} f(i)$$

This reads: the sum of $f(i)$ as i is replaced by the integers from the lower limit k to the upper limit n, and it is shorthand for the sum $f(k) + f(k+1) + \ldots + f(n)$. Using this notation, we can rewrite our equation as

$$\left(\sum_{i=1}^{n} x_i^2\right)m + \left(\sum_{i=1}^{n} x_i\right)b = \left(\sum_{i=1}^{n} x_i y_i\right)$$

Now let's do the same for the partial of SSE with respect to b:

$$\text{SSE}_b\ (m,b) = 2(y_1 - mx_1 - b)(-1) + 2(y_2 - mx_2 - b)(-1) + \ldots + 2(y_n - mx_n - b)(-1) = 0$$

Again, we can divide both sides by 2, giving us

$$(y_1 - mx_1 - b)(-1) + (y_2 - mx_2 - b)(-1) + \ldots + (y_n - mx_n - b)(-1) = 0$$

Multiplying this out and combining like terms, we get

$$-y_1 + mx_1 + b - y_2 + mx_2 + b - y_2 + mx_2 + b = 0$$
$$(x_1 + x_2 + \ldots + x_n)m + (1 + 1 + \ldots + 1)b = (y_1 + y_2 + \ldots + y_n)$$

Using sigma notation, we can write this as

$$\left(\sum_{i=1}^{n} x_i\right)m + \left(\sum_{i=1}^{n} 1\right)b = \left(\sum_{i=1}^{n} y_i\right)$$

Now, the coefficient of b is just the sum of n 1's, so is in fact n, giving us the equation

$$\left(\sum_{i=1}^{n} x_i\right)m + (n)b = \left(\sum_{i=1}^{n} y_i\right)$$

We now have simplified both of our partial derivative equations, giving us the system. **For a set of n data points, $(x_1 y_1)$, $(x_2 y_2)$, ...$(x_n y_n)$ the coefficients of the least squares line $y = mx + b$ are the solutions of the normal equations:**

$$\left(\sum_{i=1}^{n} x_i^{\,2}\right)m + \left(\sum_{i=1}^{n} x_i\right)b = \left(\sum_{i=1}^{n} x_i y_i\right)$$
$$\left(\sum_{i=1}^{n} x_i\right)m + (n)b = \left(\sum_{i=1}^{n} y_i\right)$$

These equations are the general **normal equations** for single-variable linear regression. Remember, the sums are numbers that are the **coefficients** of m and b and so are the **constants** of the *equations*. Now we are left with a simple system of two equations in two unknowns, the unknowns being the *m* and the *b* that we are seeking.[1] We can then write our regression model as $y = mx + b$. The *m* is the slope of the line and the *b* is the intercept of the line.

Let's check that these equations are right by applying the data from Sample Problem 1 and calculating the various sums in the formulas (Table 4):

Table 4

	x_i	y_i	$x_i y_i$	x_i^2
	2	4	8	4
	5	6	30	25
	6	7	42	36
	9	8	72	81
Sums	22	25	152	146

Putting these sums and $n = 4$ into our normal equations above we have:

$$146m + 22b = 152$$
$$22m + 4b = 25$$

Notice that, indeed, these are the same normal equations we obtained directly earlier. We can now solve this set of equations for *m* and *b* using any of the methods that we learned in the previous section, and will get the same solution as before, corresponding to the model $y = 0.58x + 3.06$.

By recognizing the pattern in the terms of the partial derivatives we were able to devise a method of determining the normal equations that is considerably less tedious than doing it all by hand. This method also has the advantage that we can use either calculators or spreadsheets to do much of the work.

General Solution for Least Squares Regression Using Matrices

Suppose we define the matrix **X** to be the matrix whose columns are the data

[1] The general solutions of these equations are:

$$m = \frac{n\left(\sum_{i=1}^{n} x_i y_i\right) - \left(\sum_{i=1}^{n} x_i\right)\left(\sum_{i=1}^{n} y_i\right)}{n\left(\sum_{i=1}^{n} x_i^2\right) - \left(\sum_{i=1}^{n} x_i\right)^2} \quad \text{and} \quad b = \frac{\sum_{i=1}^{n} y_i - m\left(\sum_{i=1}^{n} x_i\right)}{n}$$

values for the independent variables, with a column of all 1's appended on the right. In the general case, it would look like:

$$\mathbf{X} = \begin{bmatrix} x_1 & 1 \\ x_2 & 1 \\ \vdots & \vdots \\ x_n & 1 \end{bmatrix}$$

Why would we want this matrix? Well, its transpose is

$$\mathbf{X}^{\mathrm{T}} = \begin{bmatrix} x_1 & x_2 & \cdots & x_n \\ 1 & 1 & \cdots & 1 \end{bmatrix},$$

and if we multiply the two we get

$$\mathbf{X}^{\mathrm{T}}\mathbf{X} = \begin{bmatrix} x_1 & x_2 & \cdots & x_n \\ 1 & 1 & \cdots & 1 \end{bmatrix} \begin{bmatrix} x_1 & 1 \\ x_2 & 1 \\ \vdots & \vdots \\ x_n & 1 \end{bmatrix}$$

$$= \begin{bmatrix} x_1^2 + x_2^2 + \cdots + x_n^2 & x_1 + x_2 + \cdots + x_n \\ x_1 + x_2 + \cdots + x_n & 1 + 1 + \cdots + 1 \end{bmatrix} = \begin{bmatrix} \sum_{i=1}^{n} x_i^2 & \sum_{i=1}^{n} x_i \\ \sum_{i=1}^{n} x_i & n \end{bmatrix}$$

This is the matrix of the coefficients from the left-hand sides (what we called the **A** matrix in Section 7.4) of the normal equations! In this situation, $\mathbf{X}^{\mathrm{T}}$ is $2 \times n$, $\mathbf{X}$ is $n \times 2$, and the product is 2×2. Check that the matrix multiplication does indeed give the results indicated above. Also, notice why the column of 1's is needed to make the result work. One way to think of these 1's is as the quantity that the constant parameter is being multiplied by, just like each variable is what its coefficient (the parameter for that term) is being multiplied by.

Now, to similarly get a matrix version of the right-hand sides of the normal equations, define **Y** to be the single column matrix (vector) of the data values of the dependent variable (so that the same row in **X** and **Y** corresponds to the same data point). In general, we would have

$$\mathbf{Y} = \begin{bmatrix} y_1 \\ y_2 \\ \vdots \\ y_n \end{bmatrix}$$

Now let's look at $\mathbf{X}^{\mathrm{T}}\mathbf{Y}$:

$$\mathbf{X}^{\mathrm{T}}\mathbf{Y} = \begin{bmatrix} x_1 & x_2 & \cdots & x_n \\ 1 & 1 & \cdots & 1 \end{bmatrix} \begin{bmatrix} y_1 \\ y_2 \\ \vdots \\ y_n \end{bmatrix} = \begin{bmatrix} x_1 y_1 + x_2 y_2 + \cdots + x_n y_n \\ y_1 + y_2 + \cdots + y_n \end{bmatrix} = \begin{bmatrix} \sum_{i=1}^{n} (x_i y_i) \\ \sum_{i=1}^{n} y_i \end{bmatrix}$$

Exactly what we need! Now all that remains is to define **M** to be the column matrix of the parameters of our model, with the constant (intercept) b at the end:

$$\mathbf{M} = \begin{bmatrix} m \\ b \end{bmatrix}$$

We can see that the normal equations have the form: $(\mathbf{X}^{\mathrm{T}}\mathbf{X})\mathbf{M} = \mathbf{X}^{\mathrm{T}}\mathbf{Y}$:

$$\begin{bmatrix} \sum_{i=1}^{n} x_i^{\,2} & \sum_{i=1}^{n} x_i \\ \sum_{i=1}^{n} x_i & n \end{bmatrix} \begin{bmatrix} m \\ b \end{bmatrix} = \begin{bmatrix} \sum_{i=1}^{n} (x_i y_i) \\ \sum_{i=1}^{n} y_i \end{bmatrix}$$

Let's go back to our first example to see that this works. Recall that the original data was:

x	y
2	4
5	6
6	7
9	8

The equations representing the two partial derivatives set equal to zero and rewritten in standard form were:

$$\begin{aligned} 146m + 22b &= 152 \\ 22m + 4b &= 25 \end{aligned}$$

We'll write our matrix **X** with the original x values and a column of 1's appended:

$$\mathbf{X} = \begin{bmatrix} 2 & 1 \\ 5 & 1 \\ 6 & 1 \\ 9 & 1 \end{bmatrix}, \text{ so } \mathbf{X}^{\mathrm{T}} = \begin{bmatrix} 2 & 5 & 6 & 9 \\ 1 & 1 & 1 & 1 \end{bmatrix}, \text{ so}$$

$$\mathbf{X}^{\mathrm{T}}\mathbf{X} = \begin{bmatrix} 2 & 5 & 6 & 9 \\ 1 & 1 & 1 & 1 \end{bmatrix}\begin{bmatrix} 2 & 1 \\ 5 & 1 \\ 6 & 1 \\ 9 & 1 \end{bmatrix} = \begin{bmatrix} 2(2)+5(5)+6(6)+9(9) & 2(1)+5(1)+6(1)+9(1) \\ 1(2)+1(5)+1(6)+1(9) & 1(1)+1(1)+1(1)+1(1) \end{bmatrix}$$

$$= \begin{bmatrix} 146 & 22 \\ 22 & 4 \end{bmatrix}$$

This matrix is the same as the coefficient matrix for the system of normal equations representing the partial derivatives being set equal to zero.

Now, we also know that $\mathbf{Y} = \begin{bmatrix} 4 \\ 6 \\ 7 \\ 8 \end{bmatrix}$, so

$$\mathbf{X}^{\mathrm{T}}\mathbf{Y} = \begin{bmatrix} 2 & 5 & 6 & 9 \\ 1 & 1 & 1 & 1 \end{bmatrix}\begin{bmatrix} 4 \\ 6 \\ 7 \\ 8 \end{bmatrix} = \begin{bmatrix} 2(4)+5(6)+6(7)+9(8) \\ 1(4)+1(6)+1(7)+1(8) \end{bmatrix} = \begin{bmatrix} 152 \\ 25 \end{bmatrix}.$$

This matrix is the same as the constant matrix for the system of normal equations representing the partial derivatives being set equal to zero.

So, we can write the matrix equation for the system of normal equations representing the partial derivatives being set equal to zero as:

$$(\mathbf{X}^{\mathrm{T}}\mathbf{X})\mathbf{M} = \mathbf{X}^{\mathrm{T}}\mathbf{Y}$$

As we learned in the last chapter, to solve a matrix equation like this, we simply need to find the inverse of the coefficient matrix on the left, and multiply that by the RHS matrix on the right Thus, the solution to *this* matrix equation is found by multiplying both sides of the equation by $(\mathbf{X}^T\mathbf{X})^{-1}$ on the left, yielding:

$$(\mathbf{X}^T\mathbf{X})^{-1}(\mathbf{X}^T\mathbf{X})\mathbf{M} = (\mathbf{X}^T\mathbf{X})^{-1}\mathbf{X}^T\mathbf{Y}\text{ , or}$$
$$\mathbf{IM} = (\mathbf{X}^T\mathbf{X})^{-1}\mathbf{X}^T\mathbf{Y}\text{ , or}$$
$$\mathbf{M} = (\mathbf{X}^T\mathbf{X})^{-1}\mathbf{X}^T\mathbf{Y}$$

Recall that for the least-squares problem, the parameters (the m and b, or **M**) are the decision variables, so the solution values for the parameters are given by

$$(\mathbf{X}^T\mathbf{X})^{-1}\mathbf{X}^T\mathbf{Y}\,.$$

In the example, $\mathbf{M} = \begin{bmatrix} m \\ b \end{bmatrix}$, so you can verify that the normal equations correspond to $(\mathbf{X}^T\mathbf{X})\mathbf{M} = \mathbf{X}^T\mathbf{Y}$. The inverse of the coefficient matrix is:

$$(\mathbf{X}^T\mathbf{X})^{-1} = \begin{bmatrix} .04 & -.22 \\ -.22 & 1.46 \end{bmatrix}.$$

Thus, the solution to the system of normal equations is given by

$$\mathbf{M} = \begin{bmatrix} m \\ b \end{bmatrix} = (\mathbf{X}^T\mathbf{X})^{-1}\mathbf{X}^T\mathbf{Y} = \begin{bmatrix} .04 & -.22 \\ -.22 & 1.46 \end{bmatrix}\begin{bmatrix} 152 \\ 25 \end{bmatrix} = \begin{bmatrix} .58 \\ 3.06 \end{bmatrix}.$$

In other words, $m = 0.58$ and $b = 3.06$, so the model is $y = 0.58x + 3.06$, as we found earlier. □

Least Squares Regressions using Matrices on a Spreadsheet

How could we do this on a spreadsheet? Easily! All we need to do is set up a spreadsheet with our dependent variable (usually y) column first (on the left), followed by the columns for the independent variable(s) (usually called x), followed by a column of 1's. These will be our **Y** (the first column, for the dependent variable) and **X** (the other columns: the independent variables and the 1's) matrices. In the spreadsheet shown in Table 5, we have entered the data and have added the column of 1's as discussed:

Table 5

	A	B	C	D	E	F	G
1	Linear Least-Squares Regression Model for Quiz Grade = f(Hours Studied)						
2							
3		Grade	Hrs. Studied				
4		y	x				
5	Week 1	4	2	1			
6	Week 2	6	5	1			
7	Week 3	7	6	1			
8	Week 4	8	9	1			
9							
10							
11	X^T =	2	5	6	9		
12		1	1	1	1		
13							
14	X^TX =	146	22		X^TY =	152	
15		22	4			25	
16							
17	$(X^TX)^{-1}$ =	0.04	-0.22	$(X^TX)^{-1}$	(X^TY) =	0.58	
18		-0.22	1.46			3.06	

Thus the matrix $\mathbf{X}$ is the block C5..D8, and $\mathbf{Y}$ is B5..B8 . We then used the transpose operation of the spreadsheet to write $\mathbf{X}^T$ in the block starting in cell B11 . Then we used the matrix multiplication operation to write $\mathbf{X}^T\mathbf{X}$ in the block starting in B14, and to write $\mathbf{X}^T\mathbf{Y}$ in the block starting in F14. Next, we used the matrix inverse operation on the spreadsheet to put $(\mathbf{X}^T\mathbf{X})^{-1}$ in the block starting in cell B17 . And finally, we used matrix multiplication again to put $(\mathbf{X}^T\mathbf{X})^{-1}(\mathbf{X}^T\mathbf{Y})$ in the block starting in cell F17. This is essentially what the automatic regression routine on the spreadsheet is doing. See your technology supplement for detailed directions.

We should also mention that this operation is even easier on a graphing calculator, since you can define the matrices for $\mathbf{X}$ and $\mathbf{Y}$ (you may need to call them $\mathbf{C}$ and $\mathbf{D}$, respectively), and write out the calculation $(\mathbf{X}^T\mathbf{X})^{-1}\mathbf{X}^T\mathbf{Y}$ as $(\mathbf{C}^T\mathbf{C})^{-1}\mathbf{C}^T\mathbf{D}$. When you do this, your answer will be the column vector $\mathbf{M}$, which will give the values of all of your parameters (the slope(s) first, then the intercept)!

Least Squares Multivariable Linear Regression

What if there are more than one independent variable? In Chapter 5 we studied several cases in which the output or dependent variable was related to more than one input. How can we handle such a situation?

To keep things as uncomplicated as possible, let's look at a very simple example.

Sample Problem 2: Table 6 shows the number of milkshakes and chicken sandwiches purchased and the total cost of the order at a diner where the bill is not broken down by item:

Table 6

Milkshakes	Chicken Sandwiches	Cost of Order
1	2	8
2	1	7
3	2	12

How much is the diner charging for the milkshakes and sandwiches and is there any fixed cost? To answer this question, find a linear cost function using the method of least squares.

Solution 2: To simplify the problem we will let x_1 = the number of milkshakes, x_2 = the number of chicken sandwiches and y = the total cost. Since y is the dependent variable and we are assuming a linear relationship, the model that we are seeking will be of the form:

$$y = m_1 x_1 + m_2 x_2 + b$$

We can now write the expressions for the predicted value of y, the errors, and the squares of the errors, as shown in Table 7.

Table 7

y	x_1	x_2	$m_1x_1 + m_2x_2 + b$	y-$(m_1x_1 + m_2x_2 + b)$	y-$(m_1x_1 + m_2x_2 + b)^2$
8	1	2	$m_1(1) + m_2(2) + b$	8-$(m_1 + 2m_2 + b)$	$(8 - m_1 - 2m_2 - b)^2$
7	2	1	$m_1(2) + m_2(1) + b$	7-$(2m_1 + m_2 + b)$	$(7 - 2m_1 - 2m_2 - b)^2$
12	3	2	$m_1(3) + m_2(2) + b$	12-$(3m_1 + 2m_2 + b)$	$(12 - 3m_1 - 2m_2 - b)^2$

We thus see that

$$SSE(m_1, m_2, b) = (8 - m_1 - 2m_2 - b)^2 + (7 - 2m_1 - m_2 - b)^2 + (12 - 3m_1 - 2m_2 - b)^2$$

We wish to find the coefficients m_1 and m_2 and the constant b. If we take the partial derivatives (**three** of them this time, since there are three variables: m_1, m_2 and b), we get

$$SSE_{m_1} = 2(8\text{-}m_1\text{-}2m_2\text{-b})(\text{-}1) + 2(7\text{-}2m_1\text{-}m_2\text{-b})(\text{-}2) + 2(12\text{-}3\text{-}m_1\text{-}2m_2\text{-b})(\text{-}3)$$

$$SSE_{m_2} = 2(8\text{-}m_1\text{-}2m_2\text{-b})(\text{-}2) + 2(7\text{-}2m_1\text{-}m_2\text{-b})(\text{-}1) + 2(12\text{-}3\text{-}m_1\text{-}2m_2\text{-b})(\text{-}2)$$
$$SSE_b = 2(8\text{-}m_1\text{-}2m_2\text{-b})(\text{-}1) + 2(7\text{-}2m_1\text{-}m_2\,\text{b})(\text{-}1) + 2(12\text{-}3m_1\text{-}2m_2\text{-b})(\text{-}1)$$

If we now set these equal to 0, divide through by 2 as before, and rewrite the resulting equations in standard form, we find that the normal equations are:

$$14m_1 + 10m_2 + 6\text{b} = 58$$
$$10m_1 + 9m_2 + 5\text{b} = 47$$
$$6m_1 + 5m_2 + 3\text{b} = 27$$

Now, let's see how the general matrix method described earlier obtains these same normal equations and solves them (Table 8).

Table 8

	A	B	C	D	E	F	G
1	Linear Regression Model for Cost = f(Shakes, Chicken Sandwiches)						
2							
3		Cost	Shakes	Sandwiches			
4		y	x_1	x_2			
5		8	1	2	1		
6		7	2	1	1		
7		12	3	2	1		
8							
9	X^T =	1	2	3			
10		2	1	2			
11		1	1	1			
12							
13	X^TX =	14	10	6		X^TY =	58
14		10	9	5			47
15		6	5	3			27
16							
17	$(X^TX)^{-1}$ =	0.5	-5E-16	-1		$(X^TX)^{-1}(X^TY)$ =	2
18		-4.78E-16	1.5	-2.5			3
19		-1	-2.5	6.5			-2.8E-14

Using any of the methods we learned in the last chapter, we get the solution m_1=2, m_2 =3, and b=0 (note that the matrix solution gives -2.8E-14, due to roundoff error), so the cost function can be written: $C(x,y) = 2x + 3y$. In other words, milkshakes cost \$2, chicken sandwiches cost \$3, and there is no fixed cost (cover charge). □

Least Squares Non-Linear Regression

Sample Problem 3: A student has a weekly test in both accounting and economics. There are only 10 hours to study for the two upcoming tests. Let t equal the number of hours spent studying accounting. Then 10- t is the number of hours spent studying for the economics test. The table below shows the hours spent studying for the accounting test and the average of the grades on the two tests each week, over an entire semester.[2]

t	0	1	2	3	4	5	6	7	8	9	10
Avg	87.5	88.5	90.5	91	90.5	90.5	88.5	87	86	86.4	82.5

This certainly does not look linear, because it first increases and then starts to decrease. Let's look at the data plot in Figure 8.5-3 .

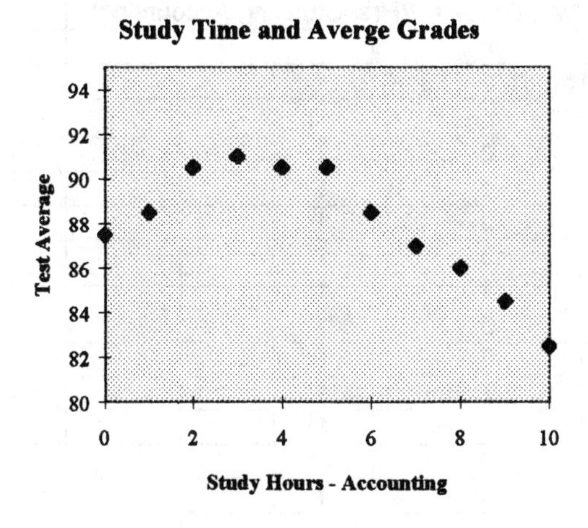

Figure 8.5-3

This data should be fit with a quadratic model. How do we do that, since the least squares regressions that we have seen fit only linear models? We want a model of the form: $y = ax^2 + bx + c$, or, in our case : $G(t) = at^2 + bt + c$.

Solution . We just saw how to fit a model of the form

$$y = m_1x_1 + m_2x_2 + b .$$

We can make our data fit this form. In order to do this, we have to **linearize** the data. In essence, we have a one-variable *nonlinear* function, but will be thinking of it as a two-variable *linear* function. If we take our variable t, in this case time spent on studying accounting, and square it, we can use this (t^2) in place of the x_1 in $y = m_1x_1 + m_2x_2 + b$

---

[2] Data from a Student Project.

and use the original data (t) in place of the x_2 . Note that b in the standard quadratic form is the coefficient of the linear term (x or t), while in our general regression form b is the intercept term.

Because the numbers in this problem are a little larger and we know that we will have to do a great deal of calculating, we will let technology do the work for us. The errors and their squares could be calculated as in the first two problems, and the partial derivatives could also be found in a similar way, but the results of these calculations are summarized in the spreadsheet in Table 9.

Table 9

	A	B	C	D	E	F	G	H	I	J	K	L
1	Quadratic Regression Model for Avg. Grade = f(Hrs. Studying Accounting)											
2												
3		Avg. Grade	t^2	Ac'g Study Hrs. (t)								
4		y	x_1	x_2								
5		87.5	0	0	1							
6		88.5	1	1	1							
7		90.5	4	2	1							
8		91	9	3	1							
9		90.5	16	4	1							
10		90.5	25	5	1							
11		88.5	36	6	1							
12		87	49	7	1							
13		86	64	8	1							
14		84.5	81	9	1							
15		82.5	100	10	1							
16												
17	$X^T =$	0	1	4	9	16	25	36	49	64	81	100
18		0	1	2	3	4	5	6	7	8	9	10
19		1	1	1	1	1	1	1	1	1	1	1
20												
21	$X^TX =$	25333	3025	385		$X^TY =$	33027.5					
22		3025	385	55			4770.5					
23		385	55	11			967					
24												
25	$(X^TX)^{-1} =$	0.0011655	-0.01	0.017482517	$(X^TX)^{-1}(X^TY) =$		-0.201					
26		-0.011655	0.126	-0.22027972			1.42413					
27		0.0174825	-0.22	0.58041958			87.8252					

Plugging in the values from our spreadsheet we see that the normal equations are:

$$
\begin{aligned}
25333\,m_1 + 3025\,m_2 + 385b &= 33027.5 \\
3025\,m_1 + 385\,m_2 + 55b &= 4770.5 \\
385\,m_1 + 55\,m_2 + 11b &= 967
\end{aligned}
$$

Using matrices to solve this, we get:

$$m_1 = -0.201048951,\ m_2 = 1.424125874, \text{ and } b = 87.82517483,$$

so our model would be:

$$G(t) = -0.201t^2 + 1.42t + 87.8$$

Translation: For example, if the student spends no time ($t = 0$) studying for accounting (and thus 10 hours on economics), she should get an average grade of $G(0) = 87.8$, or about 88, on the two tests. □

In this problem, we took a single-variable quadratic function and transformed it into a linear two-variable function (by thinking of the square of t as a separate variable). For single-variable cubic or quartic functions, we would simply extend the number of variables, analogous to the quadratic model we discussed here. For multivariable polynomials, we can also simply treat each *term* as a separate variable, and do a multivariable linear regression. We can also linearize exponential functions by taking the natural logarithm of the output values, fitting a least-squares line, and then finding the true value of the predicted output by raising e to the power of the value obtained from the linear model. This same principal can be applied to linearize some other nonlinear functions to be able to use the least-squares methodology and results.

Most graphing calculators will fit curves for functions of one variable, and most spreadsheets will calculate multivariable regressions, so we do not normally have to go through the process by hand, or even using the spreadsheet, to do many of the calculations. It is, however, an important application of minimization of multivariable functions, and it is important that we understand just what it is that the technology is doing. This also provides a means for *validating* a model obtained using a technological regression procedure.

Sample Problem 4: In Sample Problem 5 of Section 6.1, we performed a multivariable quadratic regression to find a model for the energy level of a gymnast as a function of the number of calories she ate, and the amount of exercise she got, in a day. The data, including the calculated values of the nonlinear terms, are repeated in Table 10.

Table 10

Energy	Calories	Exercise (min)			
y	c	x	c^2	cx	x^2
69	1800	0	3240000	0	0
77	1800	30	3240000	54000	900
74	1800	45	3240000	81000	2025
74	1800	75	3240000	135000	5625
69	1800	0	3240000	0	0
79	1750	60	3062500	105000	3600
88	1300	60	1690000	78000	3600
91	1300	45	1690000	58500	2025
54	2000	90	4000000	180000	8100
91	1300	75	1690000	97500	5625
85	1200	60	1440000	72000	3600
91	1500	60	2250000	90000	3600
66	1000	120	1000000	120000	14400
87	1250	60	1562500	75000	3600
88	1200	90	1440000	108000	8100
93	1400	60	1960000	84000	3600
82	1400	120	1960000	168000	14400
78	1000	30	1000000	30000	900
86	1300	105	1690000	136500	11025
57	2000	60	4000000	120000	3600

Use the techniques of this section to verify that the least-squares regression function is given by

$$E(c,x) = -93.1 + 0.260c + 0.146x - 0.0000944c^2 + 0.0000410cx - 0.00196x^2$$

Solution: The **Y** matrix will be the *first* column in Table 10. The **X** matrix will be the *rest* of the columns in Table 10, followed by a column of 1's. Doing the matrix calculation $(\mathbf{X}^T\mathbf{X})^{-1}\,\mathbf{X}^T\mathbf{Y}$ on a spreadsheet yields the **M** column matrix shown in Table 11.

Table 11

$(X^TX)^{-1}(X^TY) =$	
	0.2602388
	0.1459598
	-9.439E-05
	4.098E-05
	-0.0019584
	-93.142303

which does in fact verify the model obtained earlier. Notice that the constant term, -93.1, goes *last*, since the column of 1's was last, and that the other coefficients are also in the order in which they occurred in the **X** matrix.

Section Summary

Before you begin the exercises, be sure that you:

- Know how to set up a table and calculate the sum of the squared errors for linear, quadratic and multivariable models and an appropriate given set of data.
- Know that to get the least squares regression model, you can find all of the partial derivatives of the sum of the squared errors, set these partial derivatives equal to zero and solve the resulting system of equations, called the normal equation.
- For a single-variable linear regression, understand how you could set up a table of the independent and dependent variables, the square of the independent variable and the product of the independent variable and dependent variable, and use the sums of these columns to find the coefficients for the normal equations.
- Know how to set up the matrices **X**, **Y**, and **M** so that the solution to the normal equations can be found using the matrix equation: $(\mathbf{X}^T\mathbf{X})\mathbf{M} = \mathbf{X}^T\mathbf{Y}$, for single- or multivariable linear regression, or for single- or multivariable general polynomial (nonlinear) regression.
- Know that the solution to the matrix equation $(\mathbf{X}^T\mathbf{X})\mathbf{M} = \mathbf{X}^T\mathbf{Y}$ is given by $\mathbf{M} = (\mathbf{X}^T\mathbf{X})^{-1}(\mathbf{X}^T\mathbf{Y})$, and know how to calculate the answer by hand and using technology.

EXERCISES FOR SECTION 8.5:

Warm Up

1. For the following data:

x	y
1	9
2	6
3	4
4	1

a) Find $SSE(m,b)$ for the linear model $y = mx + b$.
b) Find the best-fit linear model by hand using calculus.

2. For the following data:

x	y
1	15
2	611
3	99
4	86

a) Find $SSE(m,b)$ for the linear model $y = mx + b$.
b) Find the best-fit linear model by hand using calculus.

3. For the following data:

x	y
0	2
1	5
2	6
3	9
4	8

a) Find $SSE(m,b)$ for the linear model $y = mx + b$.
b) Find the best-fit linear model by hand using calculus.

4. For the following data:

x	y
0	4
1	5
2	8
3	9
4	9

a) Find $SSE(m,b)$ for the linear model $y = mx + b$.
b) Find the best-fit linear mode by hand using calculus.

5. For the following data:

x	y
0	4
1	5
2	8
3	9
4	9

Use technology (calculating sums) and the normal equations for the least squares linear regression to find the best-fit linear model.

6. For the following data:

x	y
1	9
2	6
3	4
4	1

Use technology (calculating sums) and the normal equations for the least squares linear regression to find the best-fit linear model.

7. For the following data:

x	y
9	55
7	80
6	105
4	130
3	155

Use technology (calculating sums) and the normal equations for the least squares linear regression to find the best-fit linear model.

8. For the following data:

x	y
0	2
1	5
2	6
3	9
4	8

Use technology (calculating sums) and the normal equations for the least squares linear regression to find the best-fit linear model.

9. For the following data:

x	y
0	4
1	5
2	8
3	9
4	9

Use the matrix equation and technology to find the best-fit linear model.

10. For the following data:

x	y
55	9
80	7
105	6
130	4
155	3

Use the matrix equation and technology to find the best-fit linear model.

11. For the following data:

x	y	z
1	5	14
2	8	27
3	5	22
4	7	32
5	2	24
6	9	45

Use any of the methods discussed in this section to derive the best-fit linear model for z as a function of x and y. (See footnote [3])

---

[3] The normal equations for $z = ax + by + c$ are:

$$\sum_{i=1}^{n} x^2 a + \sum_{i=1}^{n} xyb + \sum_{i=1}^{n} xc = \sum_{i=1}^{n} xz$$

$$\sum_{i=1}^{n} xya + \sum_{i=1}^{n} y^2 b + \sum_{i=1}^{n} yc = \sum_{i=1}^{n} yz$$

$$\sum_{i=1}^{n} xa + \sum_{i=1}^{n} yb + nc = \sum_{i=1}^{n} z$$

12. For the following data:

x	y	z
2	7	16
25	5	11
36	8	15
48	4	7
54	1	0
63	6	14

Use any of the methods discussed in this section to derive the best-fit linear model for z as a function of x and y

13. For the following data:

x	y
6	385
8	302
10	231
12	175
15	123

Use any method discussed in this section to derive the best-fit quadratic model for y as a function of x.

14. For the following data:

x	y
0	4
1	5
2	8
3	9
4	9

Use any method discussed in this section to derive the best-fit quadratic model for y as a function of x.

Game Time

15. A small coffeehouse in Philadelphia has had the same act every year for the last few years. Each year they have tried charging a different price, with the effect on audience size that would be expected (fewer people come when the price is higher). The data from the last three years is as follows:

Price (dollars)	7	8	9
Paid Attendance	100	80	70

A look at the graph of this data suggests that it could be roughly modeled using a linear function. If we let x be the price and y be the paid attendance, the linear model would have the form $y = mx + b$.

a) What is the expression for the Sum of the Squares of the Errors (SSE) for this model and data?
b) Use this to derive the least squares linear regression model for the demand function.
c) Given that the act charges a fixed fee and that people attending the concert bring in an average of $1 in profit from refreshments each, use this demand function to determine what price should be charged for the concert to maximize profit.

16. Suppose you run a hand crafted furniture manufacturing business making chairs and tables. You have historical data about the costs associated with making different combinations of numbers of chairs and tables, and also about the relationships between the selling prices and sales of both types of furniture in a month, as given below:

DATA FOR CHAIRS AND TABLES
COST RECORDS

MONTH	TOTAL COSTS	NUMBER CHAIRS	NUMBER TABLES
JAN	$1,350	5	3
FEB	$1,660	8	4
MAR	$1,480	4	5
APR	$1,450	5	4
MAY	$1,460	8	2
JUN	$1,210	3	3
JUL	$1,600	10	2
AUG	$1,940	12	4
SEPT	$1,750	5	7
OCT	$1,920	6	8
NOV	$1,740	12	2
DEC	$1,080	4	1

Derive the cost function in terms of the quantities of chairs and tables made using the method of least squares.

17. A student on the golf team has noticed that he normally lands his tee shot in a spot that is 135 yards from the hole on the 8th hole on his home golf course. He normally uses a 9 iron to shoot for the hole from there, but has trouble with getting his distance right, and is trying to figure out the right amount of backswing to get the distance right. He goes out on a quiet day, and takes some trial shots from the same spot, measuring his backswing in degrees (no backswing is 0, 90 degrees is one quarter of a circle in arc, etc.). The table below gives the different backswing angles he tried, and the average distance from the hole for each. He looks at a graph of this data, and decides a quadratic model should fit quite well.

a) If the general form of the quadratic model is $f(x) = ax^2 + bx + c$, formulate an expression for the sum of the squares of the errors (SSE) for this model and this data.
b) Use calculus to find the parameter values that minimize the SSE.
c) Use that model to find the backswing angle that will minimize his distance from the hole.

Backswing Angle (degrees)	45	90	135	180
Distance from Hole (yards)	48	27	22	31

18. The table below shows the prices charged and the paid attendance for 3 years of concerts by 1 group at a local coffeehouse.[4]

Price $	7	8	9
Paid Attendance	100	80	70

a) We have already found the linear demand equation. This time, use the least squares method to find the best-fit quadratic model for these data.
b) Given that the act charges a fixed fee and that people attending the concert bring in an average of $1 in profit from refreshments each, use this demand function to determine what price should be charged for the concert to maximize profit.

19. Students took a poll and found the number of team T shirts that they could expect to sell at four different prices. The results of their poll is shown below:[5]

Price:	$8	$10	$12	$15
# Sold	302	231	175	123

a) Use any method to derive the best-fit quadratic model.
b) If the cost for the T shirts is $6.50 per shirt, plus a fixed cost of $20, how much should the students charge to make the largest profit?
c) What is that profit?

20. A group of students decided to sell "scrunchies". They estimated that they could sell 310 scrunchies at $2 each, 200 at $3.50 each and 100 at $5 each.
a) Use any method to derive the best-fit quadratic model.
b) If the cost for the scrunchies is $0.64 each for the materials plus $5.00 for scissors and needles, how much should the students charge to make the largest profit?
c) What is the maximum profit?

---

[4] Data from the Folk Factory Coffeehouse in Philadelphia.
[5] Data from a Student Project.

21. An athlete competes in the weight throw event for the Track and Field team. He has been trying to decide how many practice throws he should take and how many warm-up laps he should run before competing to achieve maximum distance on his throws. He recorded the following information:

Distance Feet	Warm-Up Throws	Laps
27	1	4
27.4	1	5
27.1	2	2
27.9	3	2
28.4	4	2
29.8	5	2
27.6	2	3
28	2	4
28.8	2	5
29	3	3
33	5	3
30.8	3	5
32	4	4
30.6	5	4
31.2	4	5
29.2	5	5

If D equals the distance in feet and t equals the number of warm-up throws and l equals the number of laps, then $D(t,l) = m_1 t + m_2 l + a_{11} t^2 + a_{12} tl + a_{22} l^2 + b$.

a) Use multivariable regression on a spreadsheet to model $D(t,l)$.

b) Use the matrix techniques of this section to verify your model.

22. A student wants to increase her energy level. She decided to split her exercising between cardiovascular workouts and weight lifting. She recorded the time she spent on each type of exercise and her energy level on a scale of 0 to 100. The results are shown below:

Energy Level	Cardio Minutes	Weight Minutes
70	40	20
62	25	15
67	30	15
72	30	20
63	30	15
77	35	20
79	35	25
73	25	20
81	45	20

If E equals the energy level and c equals the minutes spent on cardiovascular exercises and l equals the minutes spent on weight lifting, , then

$E(c,l) = m_1c + m_2l + m_3c^2 + m_4cl + m_5l^2 + b$.

a) Use multivariable regression on a spreadsheet to model $E(c,l)$.

b) Verify your model using the matrix techniques of this section.

Chapter 8 Summary

A partial derivative of a multivariable function with respect to one of the independent variables is the instantaneous rate of change of that function with respect to that *one* independent variable, holding all of the *other* independent variable values *constant.* To find the partial derivative of a multivariable function with respect to one of the independent variables, you treat all other independent variables as constants, and take the derivative using the same methods as are used for single-variable functions.

The partial derivative of a function, $F(x, y)$, with respect to one of its independent variables, say x, is denoted by $F_x(x, y)$ or $\frac{\partial F}{\partial x}$. The graph of a function of two variables such as $F(x, y)$ is a three-dimensional surface, and holding one of the variables (such as y) constant at a particular value (say, $y = k$), is like slicing the surface with the plane $y = k$ (perpendicular to the y axis, through the value k on the y axis), resulting in a two-dimensional curve. The partial derivative $\frac{\partial F}{\partial x}$ is then the slope of that two-dimensional curve at a given x value.

To optimize a smooth multivariable function, you can first find all of the partial derivatives of the multivariable function (one for *each* independent variable), then set the partial derivatives equal to zero and solve the resulting system of equations for the independent variables. If the original function was quadratic, the partial derivative equations will be linear, and matrices can be used to solve the system. Otherwise, substitution and elimination can be used.

If a point satisfies these conditions (all partial derivatives equal to 0), then the curves sliced by vertical planes perpendicular to each of the two independent variables' axes will have horizontal tangent lines, and that these two horizontal tangent *lines* will determine a **horizontal tangent *plane*** to the surface at the point. Any solutions of the system of equations resulting from setting the partial derivatives equal to 0 represent points (called **critical points**), each of which is a candidate to be a locally optimal solution (a local extremum), and therefore could also be a global optimum. If it is indeed the global optimum, then plugging in these values of the independent variables into the function will give the optimal value of the dependent variable.

A critical point is a *candidate* for an optimal solution, but is not necessarily optimal. It could be a saddle point, or the wrong kind of extremum (maximum versus minimum).

When solving an equation, if you *divide* both sides of the equation by an expression involving a variable, you need to check to see if that variable being 0 could be part of a solution, since otherwise you could lose solutions (you might not find *all* possible solutions). If you *multiply* both sides of the equation by an expression involving a variable, you need to *check* all of your solutions to make sure they satisfy the *original* equation, since otherwise you could end up with spurious roots.

Before you try to optimize a function using a spreadsheet, you must first *set up* the spreadsheet by designating a group of one or more adjacent cells as variable cells (inputs, independent variables, decision variables) and a *different* cell as the target cell (output, dependent variable, objective function to be optimized), with the formula (your objective function) entered into the target cell in terms of the variable cell(s). It is a good idea to label your variable and target cells and the problem you are solving, as well as writing out the *formula* for the target cell and any limit constraints you have imposed, so someone else (or you, later) can understand what is going on from a printout. You can test your formulas by entering some very simple starting values for the independent variable(s).

Optimization on a spreadsheet is a powerful tool, but must be used cautiously. The wrong choice of an initial solution could lead to a local optimum that is not the global optimum, or the procedure may not be able to find a solution at all, even if one exists.

For functions of 2-variables, to see if a critical point is probably a local extremum or not, you can test points adjacent to the critical point by finding points on a square grid around it, corresponding to all possible permutations of subtracting or adding a small increment to each variable, or leaving it unchanged (8 points in addition to the original critical point). If the function value at the critical point is better than the function values at the points on the grid around it, then the critical point is probably a local extremum. If not, it probably isn't. You can make the increment size the value of the smallest significant digit's place (for example, if your answer was rounded to 5.36 , use .01). This can be done on a spreadsheet, including a matrix of differences resulting from subtracting the function values at the grid points minus the function value at the critical point, which uses both relative and absolute addressing.

A critical point can also be tested for local and/or global optimality by checking random points with variable values within specified increment values of the critical point. This can also be done on a spreadsheet, and works for functions of *any* number of variables. As in the grid test, the output value of the test point can be subtracted off, to make it easier to see if the other points are above or below the test point. To test for local optimality, use relatively small increments (although for rounded test points, it may be best to use the value of the smallest significant digit times 10 or 100). To test for global optimality, choose a value for the increment so that the values of that variable that are generated will range over all of its reasonable values in the domain.

A critical point of a function of two variables can also be tested for local optimality using calculus. The Second Derivative Test says that for a critical point of $f(x, y)$ satisfying $f_{xx}f_{yy} - f_{xy}^2 > 0$: if $f_{xx} > 0$, then the critical point is a local minimum, and if $f_{xx} < 0$, then the critical point is a local maximum. Otherwise, the test is inconclusive.

If you are trying to optimize a quadratic function and if some of the signs of your squared term coefficients are strictly positive and some are strictly negative, your critical point (if one exists) *will not* be a local or global extremum. If you are trying to optimize a quadratic function, even if the squared term coefficient signs are all positive and you are minimizing, or all negative and you are maximizing, you still need to test further to see if a critical point is a local or global extremum.

To derive a least-squares regression model, you can find all of the partial derivatives of the sum of the squared errors (SSE), set these partial derivatives equal to zero and solve the resulting system of equations, called the normal equations. This system could also be expressed in the form $(\mathbf{X}^T\mathbf{X})\mathbf{M} = \mathbf{X}^T\mathbf{Y}$, where $\mathbf{X}$ is the matrix of the data values of all of the *terms* of the model (thought of as separate independent variables) followed by a column of 1's, $\mathbf{Y}$ is the single column of data output (dependent variable) values, and $\mathbf{M}$ is the column vector of the model parameters (in the same order as the columns of $\mathbf{X}$). The parameters of the least-squares model are then given by the matrix solution to this matrix equation: $\mathbf{M} = (\mathbf{X}^T\mathbf{X})^{-1} (\mathbf{X}^T\mathbf{Y})$.

Chapter 9: Constrained Optimization and Linear Programming

Introduction

We have seen how to optimize many kinds of functions of two or more variables, both by hand using calculus (setting partials equal to 0 and solving) and by using technology. We have seen how to find critical points and how to test them for local and global optimality. The calculus methods we learned were useful for many nonlinear functions, such as quadratic functions. But we also saw nonlinear examples where the calculus methods, and sometimes even technology, could not find a solution, or did not find a global optimum. Often this was true because there were no critical points, or because the critical points would not be feasible in a real-life situation, such as negative values. When our objective function is linear, the partial derivatives are constants, so we can't use our calculus concepts to maximize: there are no variables to solve for. When there are two decision variables, the graph of a linear function is a plane, usually slanted, so there *is* no maximum or minimum, and there are no critical points. This means we need some new concepts to understand and solve such problems.

In Section 9.1 , we discuss how multivariable optimization with a single constraint in the form of an *equation* (as opposed to an *inequality*) can be solved by finding critical points of a multivariable function that is formulated based on the objective function and any constraints you want to *force* to hold, from the original problem. The procedure is called the **method of Lagrange multipliers**. In addition to giving optimal solutions, the solution from this method also gives useful economic information about the cost of imposing the constraints.

One of the most common mathematical structures used to model situations in the real world is **linear programming** (often referred to as LP): maximizing or minimizing a linear objective function subject to linear inequality (or equation) constraints. This may not *sound* common, but it is used all the time in business and the management of organizations. In Section 9.2, we show how to formulate linear programming problems (LP's) and how to **solve LP's** that have only two decision variables **graphically**. In Section 9.3, we show how to solve LP's using matrices with a method called the **tabular simplex method**. Then in Section 9.4, we show how to solve LP's, as well as constrained nonlinear optimization problems, **using spreadsheet programs**.

Here are some examples of the kinds of problems the material in this section can help you solve:

- Consider again the example of the hand-crafted furniture business from previous chapters. You own this small business, and for now you make and sell only two products: chairs and tables (just a single style of each). Given restrictions based on the raw materials available to you and limited demand, how many chairs and tables should you make this month to maximize your profit?

- You are trying to decide what combination of foods you should eat, on average, for breakfast, to meet your nutritional goals at the lowest possible cost. What foods (and how much of each) should you plan for your breakfasts over the course of a week?

After studying this chapter, in addition to being able to solve problems like those above, you should:

- Know how to use Lagrange multipliers to optimize multivariable functions with one or more constraints in the form of *equations*, and how to use them to guide a search for an optimal solution with *inequality* constraints.
- Understand what a linear programming problem with two decision variables corresponds to graphically.
- Know how to solve *any* LP with two decision variables graphically.
- Know how to solve an LP in maximization form with any number of decision variables, each of which must be nonnegative, and with constraints that could be expressed in $\leq$ form with nonnegative right-hand sides (RHS's) using the tabular simplex method.
- Know how to interpret the final tableau of the simplex method to get information about the shadow prices for the constraints (the marginal effect of changing their RHS's), and how to recognize if there could be alternative optimal solutions.
- Know how to use technology to solve linear and nonlinear constrained optimization problems, and how to interpret the output reports.

Section 9.1: Optimization with Equality Constraints - Lagrange Multipliers

In Chapter 8, when we used calculus to find local extrema, it was almost exclusively for unconstrained objective functions. However, we were able to handle some problems with one major constraint if the constraint was in equality (equation) form, by solving the equation for one variable, then substituting the resulting expression for that variable in our objective function, and optimizing the objective function without additional constraints. In this section, we will learn about another way to handle problems with equality constraints using calculus, called the method of **Lagrange multipliers**. The basic idea of a Lagrange multiplier is that it is a variable that represents a *penalty* for *not* satisfying the constraint, and by *adjusting* the penalty, you can find just the *right* penalty to induce a solution that *perfectly* satisfies the constraint. This perfect penalty can be found using calculus. The resulting solution not only gives the *solution* to the original problem, but also gives you economic information about the *constraint*: the marginal effect of changing its constant term (right-hand side) on the optimal value of the objective function, sometimes called the **shadow price** or **dual value** of the constraint. Often, this means the marginal benefit of having one more unit of a limited resource, compared to the original limit value.

Here are some of the kinds of problems that the material in this section can help you solve:

- You use the United States Postal Service to send a great many packages. You want to order some shipping boxes that will have the maximum capacity and still stay within the limits prescribed by the Post Office. What should the dimensions of the box be?
- Most airlines restrict the size of luggage that you can carry on, usually by designating a limited overall measurement. What dimensions should the luggage be to maximize the volume and stay within airline restrictions?
- You make two kinds of hand-crafted shirts for a little extra cash, and have formulated a profit function based on the number of each you make in a month. But your day job and other activities put an upper limit on the total number of shirts you can make in a month. How many of each type of shirt should you make in a month? How much more profit could you make if you allowed yourself to make one *more* shirt a month?
- You are a food manufacturer. You know that you want your cans of peas to have a certain volume, and you want to find the dimensions that will minimize the cost of the metal.

- You manufacture special pellets, and need to construct boxes that will hold a particular volume of pellets in which to transport them. The bottom of the box is more expensive than the top and sides. What should be the dimensions of the box to minimize the cost?

After studying this section, in addition to being able to solve problems like those above, you should:

- Understand what kinds of problems can be solved using Lagrange multipliers.
- Know how to set up and solve problems using the method of Lagrange multipliers.
- Understand the basic concept behind Lagrange multipliers.
- Understand how to interpret the values of Lagrange multipliers in the solution to a problem.

Single-Variable Constrained Optimization

Let's start with a problem that you can solve easily, almost in your head, to illustrate the main ideas of this section.

Sample Problem 1: In your spare time, you make a certain special kind of hand-crafted embroidered shirt to make a little extra pocket money. You have a friend who sells them for you, to whom you give 10% of the profits as a commission. You have kept some records of different selling prices and demand in a month, and have determined that, to sell x shirts in a month, you should charge about $(36-x)$ dollars. Your cost works out to about $20 per shirt. Because of your day job and other activities, you have decided to make no more than 6 shirts in a month. How many shirts should you make to maximize your profit? How much more profit could you make if you allowed yourself to make one more shirt per month?

Solution: First, we can solve this problem using the methods covered in Volume 1. We will next talk about a different way to look at the problem, which we can then generalize to help solve constrained multivariable problems.

To start with, we need to formulate a model for the problem. As usual, we can start by formulating the demand function, then derive the revenue function and the cost function, then derive the profit function. In this problem, we can define *two* kinds of profit function: the *gross* profit (*before* paying your friend's commission) and the *net* profit (*after* paying your friend's commission). Of course, the *net* profit is what really goes in your pocket, but since it is just 90% of your *gross* profit, maximizing the gross profit will also maximize your net profit (and vice-versa). For simplicity, then, we will focus on the gross profit for the analysis, then just calculate the optimal net profit at the very end.

The demand function can be defined

$d(x)$ = the price you'd have to charge, in dollars, to sell exactly x shirts in a month

The description of the problem gives the symbol definition of $d(x)$ to be $36 - x$, but does not mention a domain. As usual, we will assume that negative selling prices don't make sense, so we will use a domain restricting x to be between 0 and 36 :

$$d(x) = 36 - x \ , \quad 0 \le x \le 36$$

As usual, the revenue is given by the selling price times the quantity, and the same domain makes sense, so if we define

$r(x)$ = the revenue in one month, in dollars, if you make x shirts and sell them for $(36\text{-}x)$ dollars each,

then the symbol definition will be

$$r(x) = (\text{price})(\text{quantity}) = [d(x)][x] = [36 - x][x] = 36x - x^2 \ , \quad 0 \le x \le 36$$

The cost is given as a straight \$20 per shirt, and the domain conceivably could be any nonnegative number of shirts, so we can define

$c(x)$ = the cost in one month, in dollars, if you make x shirts
$c(x) = 20x$, for $x \ge 0$

The gross profit, then, can be defined as

$p(x)$ = the gross profit in one month (before paying your friend's commission), in dollars, if you make x shirts and sell them for $(36\text{-}x)$ dollars each.
$p(x) = (\text{revenue}) - (\text{cost}) = (36x - x^2) - 20x = 16x - x^2$, for $0 \le x \le 36$

Notice that the revenue domain was a subset of the cost domain, so we have used the more restrictive domain, or in general the intersection of the two. Let's fully define a model for the gross profit for our problem. This time we will restrict the domain even further, since the problem says you want to put an upper limit of 6 on the number of shirts you make and sell in a month.

Verbal Definition: $p(x)$ = the gross profit in one month (before paying your friend's commission), in dollars, if you make x shirts and sell them for $(36\text{-}x)$ dollars each.

Symbol Definition: $p(x) = 16x - x^2$, for $0 \leq x \leq 6$

Assumptions: Certainty and divisibility. Divisibility implies fractions of shirts are possible. Certainty implies that the demand function is exact, and that there are no sales or discounts on the selling price or the materials comprising the cost.

As we have commented for similar problems before, the divisibility assumption can be made more realistic if we think of x as the *average* number of shirts made and sold in a month.

To solve this problem by hand using calculus, we can first take the derivative and set it equal to 0:

$p'(x) = D_x(16x - x^2) = 16 - 2x$ (taking the derivative)
$16 - 2x = 0$ (setting the derivative = 0)
$2x = 16$ (adding $2x$ to both sides, switching sides)
$x = 8$ (dividing both sides by 2)

Thus the unconstrained solution would be to make 8 shirts per month, for a profit of

$$p(8) = 16(8) - (8)^2 = 128 - 64 = 64$$

dollars. However, this solution is not in our domain. Since the gross profit function is continuous and smooth (a polynomial), this implies that the global maximum, if there is one, must occur at an endpoint of the domain. So we can simply evaluate $p(x)$ at 0 and 6 to find which is better:

$$p(0) = 16(0) - (0)^2 = 0 - 0 = 0$$
$$p(6) = 16(6) - (6)^2 = 96 - 36 = 60$$

This tells us that the global maximum over the given domain is $x = 6$. To validate this solution, let's look at a graph of the gross profit function, as shown in Figure 8.6-1 .

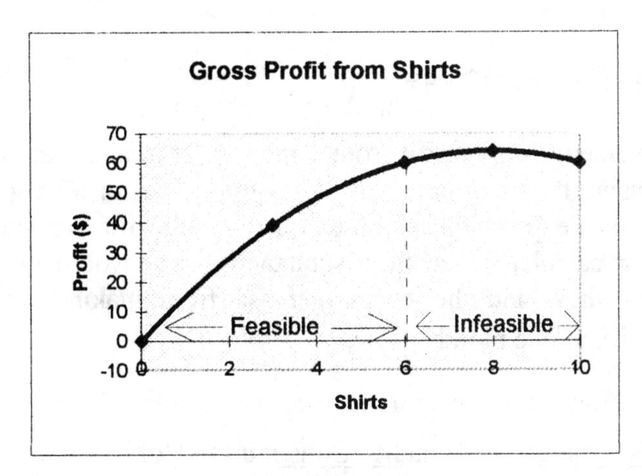

Figure 9.1-1

The graph in Figure 8.6-1 shows clearly that, indeed, over the domain between 0 and 6 for x, the global maximum occurs at $x = 6$, with a gross profit of 60.

In other words, you should make 6 shirts each month, and sell them for 36-6 = 30 dollars each, for a gross profit of $60. You will then pay your friend the commission of 10% of *that*, or $6, making your net profit $54 per month.

How can we answer the second part of the problem, and find out how much more profit you can earn if you allow yourself to make one more shirt? If we wanted to use calculus, we would recognize that the increased profit from one more shirt is exactly the idea of **marginal profit**, so we could find the instantaneous marginal profit by taking the derivative of the profit function and evaluating it at 6. We already know from before that the derivative is

$$p'(x) = 16 - 2x ,$$

so at $x = 6$, this will be

$$p'(6) = 16 - 2(6) = 16 - 12 = 4 .$$

In other words, we would expect to make *approximately* 4 more dollars in gross profit from each additional shirt, or the gross profit is increasing at a *rate* of $4 per shirt at $x = 6$. Recall that this is really a *linear approximation* from using the *tangent line* at $x = 6$, so will not be exact for a curve.

To get the exact increase in profit if we made one more shirt, we can simply evaluate the gross profit function at $x = 7$ (technically redefining $p(x)$ to include 7 in its domain):

$$p(7) = 16(7) - (7)^2 = 112 - 49 = 63 .$$

So the actual improvement in the profit from 1 more shirt in a month would be 3 dollars. Thus, our estimate using the marginal profit was quite good for a ballpark value, but not perfect. If we wanted to express these values for the *net* profit, we would just take 90% of the values we just calculated, so the instantaneous rate would be $0.9(4) = 3.60$, or \$3.60 per additional shirt, and the exact increase from making one shirt would be $0.9(3) = 2.70$, or \$2.70, in one month. □

Economic Incentives to Induce an Equality Constraint to Hold

As we mentioned earlier, we are now going to look at this simple problem from a different perspective.

Sample Problem 2: Suppose we are still in the situation described in Sample Problem 1, but that *you* have *not* set the upper limit of 6 shirts per month. At the same time, suppose your friend, who really cares about you, is concerned that you are working too hard on these shirts. Suppose you have been making 8 shirts per month recently, but have not been getting enough sleep because of it, which seems to be affecting your overall happiness significantly. Your friend decides that it would really be better if you made 6 shirts per month, but does not want to say this outright, or to try to convince you verbally. Your friend decides instead to use economic persuasion, by charging you an extra "fee" for every shirt *over* 6 that you make in a month, and offering you a *bonus* for every shirt *under* 6 that you make in a month. Suppose this fee, or penalty per shirt *over* 6 made in a month, is \$2 per shirt *over* 6, and that the bonus is also \$2 per shirt (*under* 6). The 10% commission will be charged *after* the fee has been subtracted from, or the bonus added to, your profit. How many shirts should you make to maximize your profit?

Solution: The situation is similar to before, but now we need to think about how to incorporate the fee and bonus into the problem. Since the fee is incurred *before* the 10% commission, it is really just like an additional cost. For the sake of language, let's continue to call the profit *just before* paying the commission (and *after* paying the fee or receiving the bonus) the "gross profit". Technically, the bonus could be thought of as a *negative* cost. This way, we could think of a \$2 penalty per shirt *over* 6, or a cost of \$2 times the number of shirts over 6, where if the number of shirts is *under* 6, we will consider the difference a *negative* number of shirts *over* 6. This means the cost function will simply become

$$c(x) = 20x + 2(x-6) .$$

Notice that if you made 9 shirts in a month, x would be 9, so the penalty term would be $2(9\text{-}6) = 2(3) = 6$ dollars (a *fee* of \$6, for 3 shirts *over* 6), while if you make 1 shirt, x would be 1, so the penalty term would be $2(1\text{-}6) = 2(\text{-}5) = \text{-}10$ dollars (a *bonus* of \$10, for 5 shirts *under* 6).

Thus, the gross profit can now be written as

$$
\begin{aligned}
p(x) &= [\text{revenue}] - [\text{cost}] \\
&= [36x - x^2] - [20x + 2(x-6)] \\
&= 36x - x^2 - 20x - 2(x-6) && \text{(removing brackets)} \\
&= 16x - x^2 - 2x + 12 && \text{(combining like terms, multiplying out)} \\
&= 12 + 14x - x^2 && \text{(combining like terms)}
\end{aligned}
$$

This time, the domain is limited only by the domain of the demand and revenue functions, so our model is given by:

Verbal Definition: $p(x)$ = the gross profit in one month (*after* the fee/bonus, and *before* paying your friend's commission), in dollars, if you make x shirts and sell them for $(36\text{-}x)$ dollars each.

Symbol Definition: $p(x) = 12 + 14x - x^2$, for $0 \le x \le 36$

Assumptions: Certainty and divisibility. Divisibility implies fractions of shirts are possible. Certainty implies that the demand function is exact, and that there are no sales or discounts on the selling price or the materials comprising the cost.

Once again, we can take the derivative and set it equal to 0 for the unconstrained solution (by inspection, we see again that the curve is a parabola opening down since the squared term has a negative coefficient):

$$
\begin{aligned}
p'(x) &= D_x(12 + 14x - x^2) = 14 - 2x = 0 \\
2x &= 14 && \text{(adding } 2x \text{ to both sides, switching sides)} \\
x &= 7 && \text{(dividing both sides by 2)}
\end{aligned}
$$

Since this value is in the domain, we know it is the global optimum. We can also see that the second derivative would be -2, which would show we have a local maximum and a curve that is always concave down, so the local maximum must also be a global maximum.

In other words, you will maximize your gross profit (as well as your net profit) if you make 7 shirts per month under this arrangement. □

This means that your friend's plan has been moderately successful: the economic incentives have induced you to cut down your production of shirts from 8 to 7 in a month. How could this get down to 6? What if the penalty (fee/bonus) was \$3 per shirt instead of \$2? Let's see what happens.

Sample Problem 3: Solve the problem in Sample Problem 2, changing the fee/bonus figure from \$2 per shirt to \$3 per shirt.

Solution: The discussion is the same; only the cost and gross profit functions will change. The new cost function will be

$$c(x) = 20x + 3(x\text{-}6)\ .$$

and the new gross profit function will be

$$\begin{aligned} p(x) &= [\text{revenue}] - [\text{cost}] = [36x - x^2]\text{-}[20x + 3(x\text{-}6)] \\ &= 16x - x^2 - 3(x\text{-}6) = 18 + 13x - x^2 \end{aligned}$$

so the new solution can be found by calculating

$$\begin{aligned} p'(x) &= D_x(18 + 13x - x^2) = 13 - 2x = 0 \\ 2x &= 13 \qquad \text{(adding } 2x \text{ to both sides, switching sides)} \\ x &= 6.5 \qquad \text{(dividing both sides by 2)} \end{aligned}$$

Thus the optimal number of shirts for you to make is an average of 6.5 each month (perhaps alternate: 6 one month and 7 the next). □

This is even closer to 6! How could we hit 6 exactly?

Sample Problem 4: Considering the problem defined in Sample Problem 2, find the fee/bonus rate per shirt that will induce you to make exactly 6 shirts per month to maximize your profit.

Solution: We have already done the steps several times for specific fee/bonus rates of \$2 and \$3. Now, we need to define a symbol for it. Let

λ = the fee/bonus rate, in dollars per shirt[1]

Then, if we think of λ as a constant, the cost function will be

---

[1] The Greek letter λ is the lower-case "lambda", which corresponds to "l" in English (the first letter in the name of the mathematician who developed this theory, Lagrange).

$$c(x) = 20x + \lambda(x\text{-}6) \ .$$

and the new gross profit function will be

$$\begin{aligned} p(x) &= [\text{revenue}] - [\text{cost}] = [36x - x^2]\text{-}[20x + \lambda(x\text{-}6)] \\ &= 16x - x^2 - \lambda(x\text{-}6) = 16x + 6\lambda - x^2 - \lambda x \end{aligned}$$

Since we think of λ as a constant, the new solution can be found by calculating

$$\begin{aligned} p'(x) &= D_x(16x + 6\lambda - x^2 - \lambda x) = 16 - 2x - \lambda = 0 \\ 2x &= 16 - \lambda && \text{(adding } 2x \text{ to both sides, switching sides)} \\ x &= \frac{16-\lambda}{2} = 8 - \frac{\lambda}{2} = 8 - 0.5\lambda && \text{(dividing both sides by 2)} \end{aligned}$$

Now if we want to induce the optimal value of x to be 6, we can just set the expression we got for x in terms of λ equal to 6 and solve for λ:

$$\begin{aligned} 8 - 0.5\lambda &= 6 \\ -0.5\lambda &= -2 && \text{(subtracting 8 from both sides)} \\ \lambda &= \frac{-2}{-0.5} = 4 && \text{(dividing both sides by -0.5)} \end{aligned}$$

In fact, from noticing the patterns of the solutions to Sample Problems 2 and 3, you may have already guessed or convinced yourself that $4 was in fact the answer to this problem.

Lagrange Multipliers and Their Interpretation as Shadow Prices

Thus the fee/bonus rate that will induce you to make exactly 6 shirts per month is $4 per shirt. □

Notice that, at this optimal fee/bonus rate, your friend does not actually collect any fee or pay you any bonus! In other words, using the $4 per shirt economic incentive structure will bring about exactly the goal your friend wanted, without helping or hurting either of you!

Furthermore, does the $4 per shirt sound familiar? If you recall from Sample Problem 1, it was exactly the instantaneous marginal profit at a production level of 6 shirts per month. In fact, the value λ is called a **Lagrange multiplier**, and in general it tells you **the marginal effect on your objective function per unit increase in the right-**

hand side (constant term) of your constraint. In our case, the constraint was $x = 6$, so the Lagrange multiplier, λ, told the marginal effect on the gross profit per unit increase in the number of shirts made per month (per additional shirt over 6).

Put differently, this is saying that the optimal fee/bonus rate (Lagrange multiplier value) is exactly the marginal benefit of increasing the right-hand side (RHS) of your constraint, or increasing the availability of a resource. Since we knew this marginal profit from Sample Problem 1, this property would mean we could have known right away the optimal fee/bonus rate.

Another way of looking at this relationship is to realize that your constraint is restricting your potential profit, so there is essentially a *price* you are paying for imposing the constraint. The Lagrange multiplier is the instantaneous *price* per unit of the resource represented by the constraint (your time, in our example), and so it is sometimes called a **shadow price**, or **dual value** (the problem of finding the values of the Lagrange multiplier(s) is sometimes called the *dual problem*).[2]

The Method of Lagrange Multipliers

You may have noticed that, in Sample Problem 4, we treated λ as a constant, even though it was really a decision variable. What would happen if we treated our expression for the gross profit as a function of *two* variables, and looked for a critical point? Let's try it. Define

$L(x,\lambda)$ = the gross profit in one month (*after* the fee/bonus with a rate of λ dollars per shirt, and *before* paying your friend's commission), in dollars, if you make x shirts and sell them for $(36\text{-}x)$ dollars each.

We have already found the expression for this function:

$$L(x,\lambda) = 16x - x^2 - \lambda(x-6) = 16x + 6\lambda - x^2 - \lambda x$$

Now let's try taking partial derivatives and setting them equal to 0:

$$L_x(x,\lambda) = 16 + 0 - 2x - \lambda = 16 - 2x - \lambda = 0$$
$$L_\lambda(x,\lambda) = 0 + 6 - 0 - x = 6 - x = 0$$

[2] For more details about duality and the dual of a problem, for which the theory is most fully developed in the case of a linear objective function with linear inequality constraints, see a book on linear programming or Operations Reserach.

Notice that the effect of the second partial derivative equation (after setting the partial equal to 0) is to force the constraint, $x = 6$, to hold, and that the first partial derivative equation is exactly what we used in our solution to Sample Problem 4 (we solved it for x, then set the result equal to 6 and solved for λ). In other words, **the solution** we obtained to Sample Problem 4 **could have been obtained by treating the Lagrange multiplier as a *variable* and finding the critical point for the function defined by $L(x,\lambda)$.**

Let's check the critical point we obtained to see if it is a local extremum of the $L(x,\lambda)$ function. Let's use the spreadsheet tests discussed in Section 8.4 . Sample results are shown in Tables 1 and 2.

Table 1

	A	B	C	D	E	F	G	H	I
10		1st var is:	x	& 2nd var is:	l		fn is:	L(x,l)	
11	Critical Point: 1st =		6	& 2nd =	4				
12	Increments ==>		1		1				
13									
14	B18=+16*B$21+6*$E18-B21^2-B21*$E18								
15									
16	fn(1st,2nd)				2nd var		fn(Other)-fn(Crit.Pt.)		
17	60	60	58		5		0	0	-2
18	59	60	59		4		-1	0	-1
19	58	60	60		3		-2	0	0
20									
21	5	6	7		<= 1st var =>		5	6	7

Table 2

	A	B	C	D	E	F	G	H
33	C38=+16*G38+6*H38-G38^2-G38*H38							
34								
35		fn name: L(x,l)			variable names ==>		x	l
36					Increments ==>		1	1
37		i	fn Value	fn(Other$_i$)-fn(Test)	Variables ==>		1st	2nd
38			60	0	<== Test Point ==>		6	4
39		1	60.00006	6.31724E-05			6.000084	3.250307
40		2	60.06772	0.0677156			5.688026	4.529029
41		3	59.98474	-0.015263506			6.018165	4.8221
42		4	59.78413	-0.21587064		Other	6.270248	4.528539
43		5	59.3481	-0.651902518		Points	5.204923	3.975152
44		6	60.19605	0.196052429			6.395142	3.108701
45		7	58.58959	-1.410409668			6.975262	4.470923
46		8	59.94329	-0.056709188			5.360974	4.550283

Table 1 suggests that our critical point *is* probably a local maximum, *but Table 2 proves that it is **not** a global maximum **nor** a global minimum*, and suggests that it is probably not a local maximum either. We could use the trial and error search procedure discussed in Section 8.4 to look for the global extrema, but that is not really the point. We were just curious to see if the critical point was a local or global extremum of the $L(x,\lambda)$ function. To prove it is a local extremum would require calculus.[3] However, Figure 9.1-2 shows quite convincingly that the critical point at $x = 6$ and $\lambda = 4$ is a saddle point, not an extremum, of the $L(x, \lambda)$ function. Notice that for slices holding λ constant, parabolas opening down are obtained so a solution for x is a maximum, while when cut at a different angle, the slices are parabolas opening up, so the solution is a minimum.

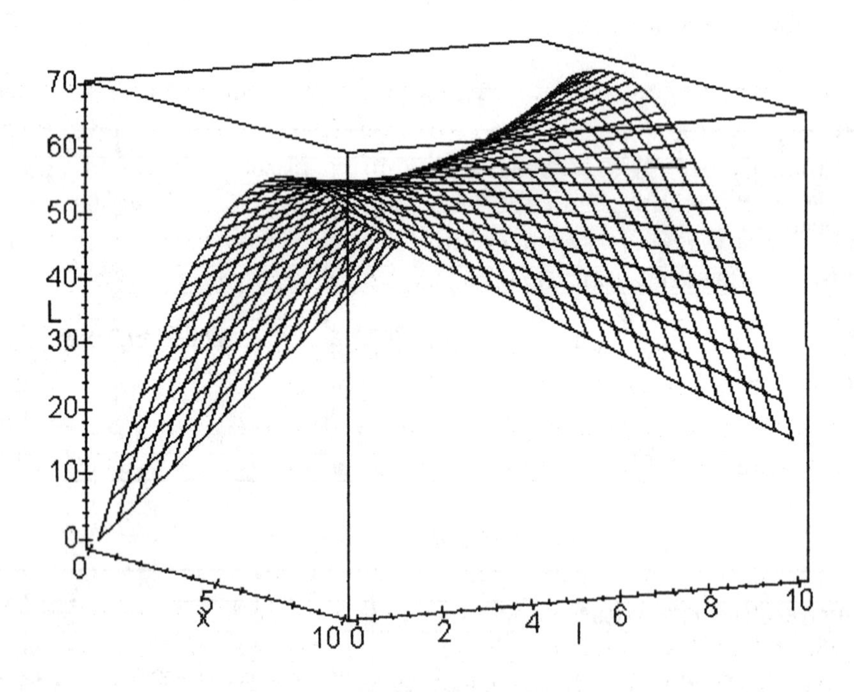

Figure 9.1-2

What we have done above as an alternative solution procedure to Sample Problem 4 (and to Sample Problem 1) is called the **method of Lagrange multipliers**. We were trying to optimize a function subject to an equality constraint,

Maximize $p(x)$, subject to $x = 6$,

which we could express more generally as

---

[3] $L_{xx} = -2$, $L_{x\lambda} = -1$, $L_{\lambda x} = -1$, and $L_{\lambda\lambda} = 0$, so $L_{xx}L_{\lambda\lambda}-L_{x\lambda}^2 = (-2)(0) - (-1)^2 = 0 - 1 < 0$, so the Second Derivative Test is inconclusive, and we cannot prove the critical point is a local extremum. A proof that the critical point is *not* a local extremum is beyond the scope of this book; see an advanced multivariable calculus book.

Optimize $z = f(x)$, subject to $g(x) = b$

We were able to find the solution to this constrained optimization problem by finding a critical point for the *unconstrained* function in the form

$$L(x,\lambda) = p(x) - (\lambda)[x-6]$$

which in the general form would be

$$L(x,\lambda) = f(x) - (\lambda)[g(x) - b]$$

Does this make sense in general? Recall that in our problem, the *unconstrained* global maximum of $p(x)$ occurred at $x = 8$, so we made λ a *positive* value and subtracted it times the amount *over* the constraint RHS value of 6. If our unconstrained solution had been at $x = 4$, we would have wanted to *penalize* values *below* 6 and give a *bonus* to values *above* 6. But this could be accomplished by simply changing the *sign* of λ ; the *form* of the $L(x,\lambda)$ function would still look exactly the same.

Exactly the same concept applies to the general case. **If the unconstrained global optimum to $f(x)$ is *feasible* (satisfies any and all constraints), then we are done, and there is no need for Lagrange multipliers because the constrained solution is the same as the unconstrained solution** and the constraint does not affect the optimal solution.

But if the unconstrained global optimum is *not* feasible, then we need to impose at least one constraint in equation form, something that the unconstrained solution *violates*. **To induce an equality constraint to hold, the Lagrange multiplier method subtracts a penalty variable (the Lagrange multiplier, λ) times the left-hand side minus the right-hand side (LHS - RHS) of the constraint. The idea is that by adjusting the penalty value (including the sign, as discussed above), a solution satisfying the constraint *exactly* can be induced.** And it turns out that **the perfect value of the Lagrange multiplier can be found by simply finding a critical point of the $L(x,\lambda)$ function,** sometimes called **the Lagrangean function.**

Multivariable Constrained Optimization with Lagrange Multipliers

If you think about it, the above logic would hold whether f and g were single-variable *or multivariable* functions! In other words, we could think of the variable x as standing for an ordered pair or a sequence of values of n variables in general. For example, in the case of functions of two variables, our original problem could be in the form

Optimize $z = f(x, y)$, subject to $g(x, y) = b$

and we would simply find a critical point for the Lagrangean function

$$L(x, y, \lambda) = f(x, y) - (\lambda)[g(x, y) - b]$$

or, in general, for functions of n variables, our original problem could be

Optimize $z = f(x_1, x_2, \ldots, x_n)$, subject to $g(x_1, x_2, \ldots, x_n) = b$

and we would find a critical point of the Lagrangean function

$$L(x_1, x_2, \ldots, x_n, \lambda) = f(x_1, x_2, \ldots, x_n) - (\lambda)[g(x_1, x_2, \ldots, x_n) - b]$$

Let's try an example with two decision variables.

Sample Problem 5: You are still making the same embroidered shirts as in Sample Problem 1, but you got a flash of insight and started making a different kind of shirt, with rhinestone decorations rather than embroidery. The cost for these shirts is the same as before, and initial sales suggest that to sell y rhinestone shirts you should charge about $(40 - y)$ dollars per shirt. Your friend indicates that the kinds of people buying the two types of shirts are completely different, and that sales of each should be completely independent of the other (they won't be affected by each other's prices, as we saw in some of the examples in Section 6.1). You still feel that you don't want to make more than a total of 6 shirts a month (they take about the same amount of time and effort to make). How many shirts of each kind should you make per month on average, how much should you charge for each, and what will be your optimal gross and net profit?

Solution: We already know the expression for the gross profit for the original embroidered shirts, $16x - x^2$. Using the same formulation steps for the new rhinestone shirts, the revenue will be given by

$$\text{rhinestone revenue} = (\text{price})(\text{quantity}) = (40 - y)(y) = 40y - y^2$$

and the cost will be $20y$, analogous to the original shirts, so the gross profit will be

$$\text{rhinestone gross profit} = (\text{revenue}) - (\text{cost}) = (40y - y^2) - (20y) = 20y - y^2 .$$

Thus the total gross profit will simply be the sum of the two individual gross profit functions:

$$\text{total gross profit} = f(x, y) = (16x - x^2) + (20y - y^2) = 16x + 20y - x^2 - y^2 .$$

The model for the problem we are trying to solve is thus given by:

Verbal Definition: $f(x, y)$ = the average total gross profit in one month, in dollars, from the two types of shirts, if you make an average of x embroidered shirts and y rhinestone shirts (and charge selling prices as determined by the demand functions)

Symbol Definition: $f(x, y) = (16x - x^2) + (20y - y^2) = 16x + 20y - x^2 - y^2$, s.t. $x+y \leq 6$, $x, y \geq 0$

Assumptions: Certainty and divisibility. Divisibility implies that fractions of shirts make sense. Certainty implies that the demand functions are exact, that there are no sales or discounts in the costs or selling prices, and that all conditions remain constant into the future (that demand will not tail off, for example).

As before, let's try to find the unconstrained solution first. If it is feasible, we don't have to do anything else. As always, we find the partial derivatives, and set them equal to 0 first:

$$f_x(x, y) = 16 + 0 - 2x - 0 = 16 - 2x = 0$$
$$f_y(x, y) = 0 + 20 - 0 - 2y = 20 - 2y = 0$$

As before, the solution to the first partial derivative equation is

$$x = 8$$

and the solution to the second is given by

$$2y = 20 , \text{ or } y = 10 .$$

Let's use our spreadsheet tests to check that this solution is locally and globally optimal. The results are shown in Tables 3 and 4.

Table 3

	A	B	C	D	E	F	G	H	I
10		1st var is:	x	& 2nd var is:	y		fn is:	**f(x,y)**	
11	**Test Point: 1st =**		8	**& 2nd =**	10				
12	**Increments ==>**		1		1				
13									
14	B18=+16*B$21+20*$E18-B21^2-E18^2								
15									
16	**fn(1st,2nd)**				**2nd var**		**fn(Other)-fn(Test Pt.)**		
17	162	163	162		11		-2	-1	-2
18	163	164	163		10		-1	0	-1
19	162	163	162		9		-2	-1	-2
20									
21	7	8	9		**<= 1st var**	**=>**	7	8	9

Table 4

	A	B	C	D	E	F	G	H
33	C38=+16*G38+20*H38-G38^2-H38^2							
34								
35		fn name:	**f(x,y)**		variable names ==>		x	y
36					**Increments ==>**		8	10
37		**i**	**fn Value**	**fn(Other$_i$)-fn(Test)**	**Variables ==>**		**1st**	**2nd**
38			164	0	**<== Test Point ==>**		8	10
39		1	30.54106	-133.4589419			1.723703	0.301184
40		2	162.3576	-1.642414282			7.05721	9.131921
41		3	131.4867	-32.51334892			10.19081	4.735619
42		4	110.4388	-53.56124657	**Other**		7.675024	2.688664
43		5	142.6019	-21.39813929	**Points**		11.99865	12.32571
44		6	106.8852	-57.11476607			1.890488	14.44844
45		7	64.56185	-99.4381471			15.29541	16.79818
46		8	87.53191	-76.46809125			0.660835	5.245555

Table 3 shows very strong evidence that our test point is a local maximum, and Table 4 shows very strong evidence that it is also a global maximum. Notice that we have chosen increments to span the constrained feasible region, although in Table 4, only *two* of the randomized points are feasible (Other Point number 1, (1.72,0.301), and Other Point number 8, (0.661,5.25)). Repeated recalculations gave similar results, and revealed no other points with a higher profit than (8,10).

Thus our unconstrained solution is (8,10): make 8 embroidered shirts and 10 rhinestone shirts. Both values satisfy the nonnegativity constraints, but, unfortunately, do not meet the constraint of making no more than 6 shirts total on average each month.

This is exactly the situation in which the method of Lagrange Multipliers is useful! We can impose the number-of-shirts constraint as an *equation* (force you to make a total of *exactly* 6 shirts, not just *at most* 6) and see what we get. Our revised problem, then is to

$$\text{Maximize } f(x, y) = 16x + 20y - x^2 - y^2 \ , \quad \text{subject to} \quad x + y = 6$$

The Lagrangean function for this problem will then be

$$\begin{aligned} L(x, y, \lambda) &= f(x, y) - (\lambda)[g(x, y) - b] = (16x + 20y - x^2 - y^2) - (\lambda)[(x + y) - (6) \\ &= 16x + 20y + 6\lambda - x^2 - x\lambda - y^2 - y\lambda \end{aligned}$$

We now want to find a critical point for this function (notice that it is quadratic), and will have *three* partial derivatives to set equal to 0 and solve:

$$\begin{aligned} L_x(x, y, \lambda) &= 16 + 0 + 0 - 2x - \lambda - 0 - 0 = 16 - 2x - \lambda = 0 \\ Ly(x, y, \lambda) &= 0 + 20 + 0 - 0 - 0 - 2y - \lambda = 20 - 2y - \lambda = 0 \\ L_\lambda(x, y, \lambda) &= 0 + 0 + 6 - 0 - x - 0 - y = 6 - x - y = 0 \end{aligned}$$

Since the Lagrangean function was quadratic, the partial derivative equations are all linear, so we can solve this system using matrices! Let's write it in standard form first (subtracting the constant from both sides and then reversing the signs, or multiplying both sides by -1, of each equation):

$$\begin{aligned} 2x \quad\quad + \lambda &= 16 \\ 2y + \lambda &= 20 \\ x + y \quad\quad &= 6 \end{aligned}$$

Our solution will then be given by

$$\mathbf{X} = \begin{bmatrix} x \\ y \\ \lambda \end{bmatrix} = \mathbf{A}^{-1}\mathbf{B} = \begin{bmatrix} 2 & 0 & 1 \\ 0 & 2 & 1 \\ 1 & 1 & 0 \end{bmatrix}^{-1} \begin{bmatrix} 16 \\ 20 \\ 6 \end{bmatrix} = \begin{bmatrix} 2 \\ 4 \\ 12 \end{bmatrix}$$

In other words, $x = 2$, $y = 4$, and $\lambda = 12$. The gross profit for this solution is given by

$$\begin{aligned} f(x, y) &= 16x + 20y - x^2 - y^2 \\ f(2,4) &= 16(2) + 20(4) - (2)^2 - (4)^2 = 32 + 80 - 4 - 16 = 92 \ . \end{aligned}$$

How could we check to try to have confidence that this is indeed the global maximum to the constrained problem? Let's try the testing methods discussed in Section 8.4 . The results are shown in Tables 5 and 6.

Table 5

	A	B	C	D	E	F	G	H	I
10		1st var is:	x	& 2nd var is:	y		fn is:	f(x,y)	
11	Test Point: 1st =		2	& 2nd =	4				
12	Increments ==>		1		1				
13									
14	B18=+16*B$21+20*$E18-B21^2-E18^2								
15									
16	fn(1st,2nd)				2nd var		fn(Other)-fn(Test Pt.)		
17	90	103	114		5		-2	11	22
18	79	92	103		4		-13	0	11
19	66	79	90		3		-26	-13	-2
20									
21	1	2	3		<= 1st var =>		1	2	3

Table 6

	A	B	C	D	E	F	G	H
33	C38=+16*G38+20*H38-G38^2-H38^2							
34								
35		fn name: f(x,y)			variable names ==>		x	y
36					Increments ==>		4	4
37		i	fn Value	fn(Other i)-fn(Test)	Variables ==>		1st	2nd
38			92	0	<== Test Point ==>		2	4
39		1	10.14497	-81.85502548			-1.17261	1.650253
40		2	66.63403	-25.36596971			0.239279	3.905972
41		3	68.35809	-23.64190943			2.041362	2.245225
42		4	130.1968	38.19683186	Other		3.892735	5.884961
43		5	89.1301	-2.869898898	Points		0.121596	6.4222
44		6	92.15779	0.157790528			1.517028	4.539846
45		7	152.5914	60.59142575			5.460223	7.773319
46		8	68.39846	-23.60153706			3.477047	1.331411

In this case, Table 5 gives the best picture of what is going on. Notice that the increments used are both 1, which means that the constraint is just satisfied along the main diagonal, from the upper left (point (1,5)) to the lower right (point (3,3)). Along this main diagonal, you can see that there seems to be a local maximum at our solution point (test point), (2,4). The feasible side of the constraint is below and to the left of it, and in this direction the profit drops off. There are grid points with better profits above and to the right of our critical point, but none of them are feasible.

In Table 6, we have done a test for global optimality by using larger increments. In this case, since the domain was all nonnegative values of x and y whose sum is no more than 6, and since our test point was (2,4), increments of 4 for both variables would span the entire **feasible region** of feasible solutions. Notice that this did include some negative values of x and a good number of other infeasible solutions. Notice also that there are plenty of solutions with larger profits than our test point (positive differences in the "fn(Other $_i$)-fn(Test)" column), but if you check carefully, *all* of them are infeasible.

Repeated recalculations of other random points (hitting the F9 key) shows this to be consistently true. Notice that there is one *infeasible* point (Other Point number 5, or $Other_5$, (0.121,6.42)) which still has a *lower* profit than the Test Point (negative difference).

From these tests, we conclude with great confidence that (2,4) is the global maximum for the constrained problem.

Translation: You should make 2 embroidered shirts and 4 of the new rhinestone shirts, for an average total gross profit per month of $92. The commission on this will be $0.10(92) = 9.20$ dollars, so your average net profit per month will be $92 - 9.20 = 82.60$ dollars. The selling price for the embroidered shirts should be $36 - 2 = 34$ dollars and the selling price for the new rhinestone shirts should be $40 - 4 = 36$ dollars. In other words, this new product line in your business has increased your average gross and net profit per month by more than 50% (from $60 to $92 for the gross profit, for example)! Start thinking about other possibilities! Also, notice that the shadow price of your constraint limiting you to 6 shirts per month has gone up from $4 per shirt in the one-shirt version of the problem to $12 per shirt (the optimal value of the Lagrange multiplier) here in the two-shirt version! You may need to re-think whether the time and effort to make one more shirt per month is worth the additional $12 (approximately) in gross profit, or $0.9(12) = 10.80$ dollars in net profit. □

<u>Graphical Interpretation of the Lagrange Multiplier Method with 2-Variable Problems</u>

Tables 3 and 5 gives us a rough picture of what is going on graphically, but let's do this more completely. Our gross profit function is a quadratic, and from the tables seems to have the shape of a rounded hilltop. Let's confirm this by graphing it in three dimensions, as shown in Figure 9.1-3

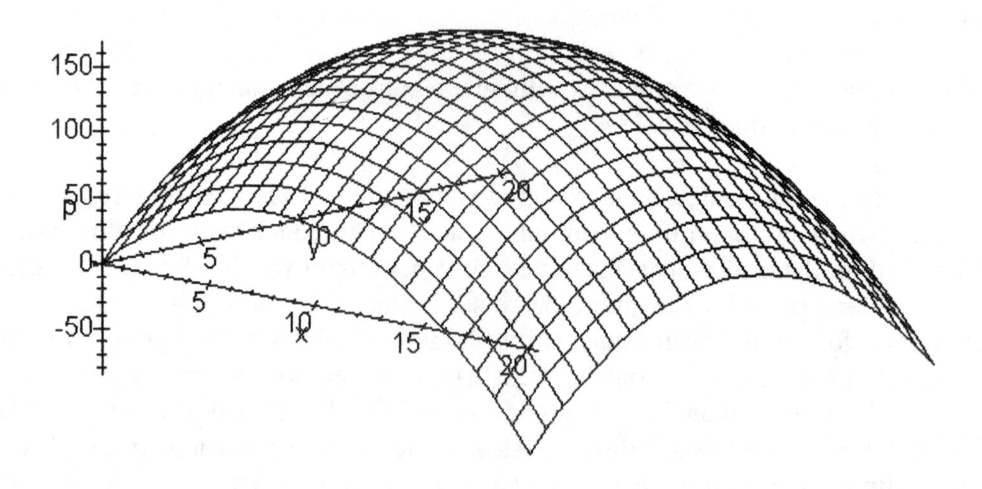

Figure 9.1-3

Figure 9.1-3 confirms that our function is indeed in the shape of a rounded hilltop. Let's look at a different way to graph this same surface, which will be helpful to us in understanding what the method of Lagrange multipliers is doing graphically, shown in Figure 9.1-4 .

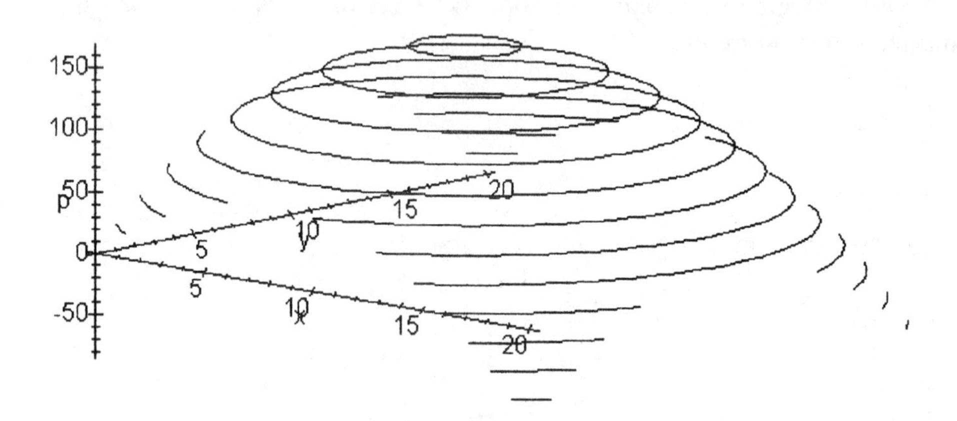

Figure 9.1-4

Figure 9.1-4 is graphing what are usually called the **level curves** of the surface. These are all of the points on the surface that are at the same vertical height (have the same function value), so if the surface here was a real-world hilltop, each level curve would be all the points at a particular altitude. In Figure 9.1-4 , there are level curves for profit values of 20, 40, 60, …, 160 . Recall that the global maximum occurred at (8,10) and had a profit of 164.

Now, if you imagine shifting your viewing perspective in Figure 9.1-4 so that your eye is directly *above* the maximum point, you could see a graph looking something like that in Figure 9.1-5 .

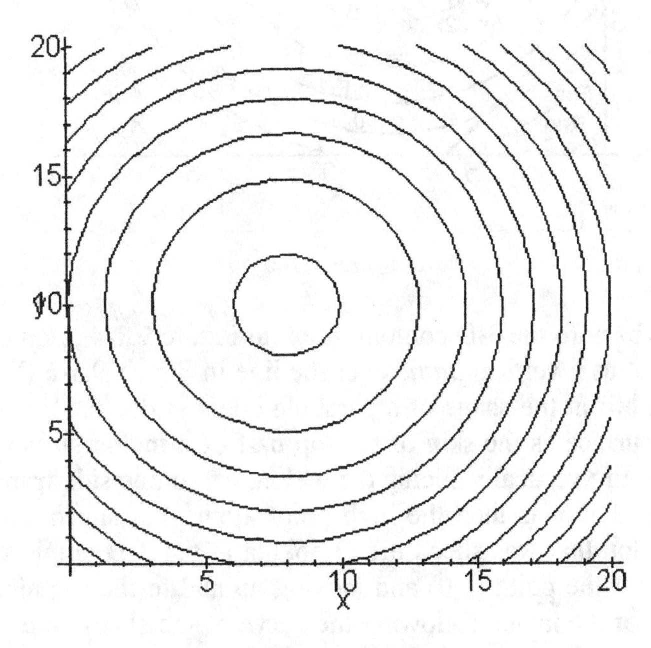

Figure 9.1-5

This is exactly the idea of a **contour map**, or a topographical map you might use to plan a hike in the mountains, where the **contour lines** are simply the level curves we saw in Figure 9.1-4 . Now let's draw in our constraint corresponding to $x + y = 6$ onto this contour graph, as shown in Figure 9.1-6 .

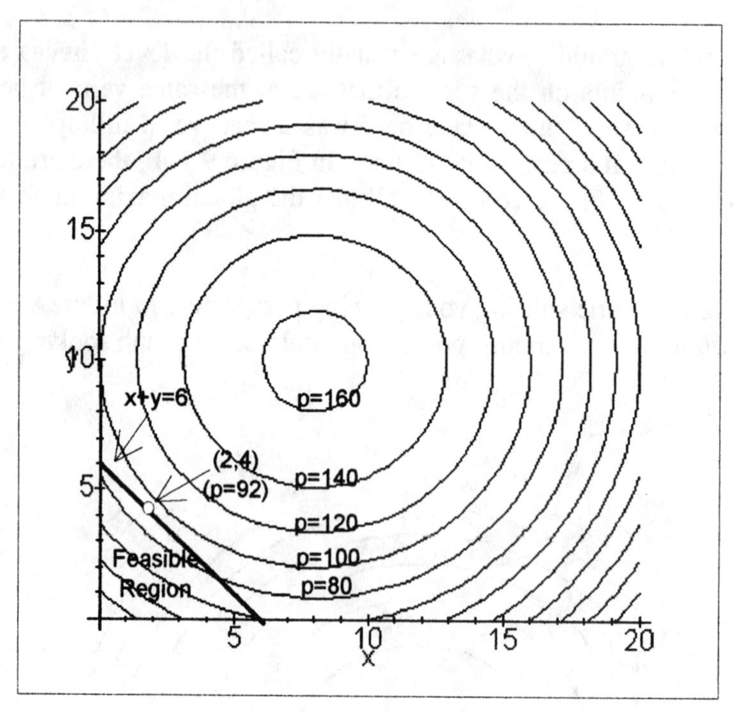

Figure 9.1-6

If you move back to the 3-D contour graph in Figure 9.1-4 , you could think of the constraint $x + y = 6$ as a vertical *plane* over the line in Figure 9.1-6 (like a wall), which would intersect the hill in the shape of a parabola opening down. It might be helpful to think of the profit surface as the skin of the top half of a melon sitting on a table. The constraint is like a knife vertically slicing the melon, off to one side from the center. Our optimal solution at (2,4) was then the high point along that sliced curve, or along the smaller piece of melon that was sliced off. Looking at the 2-D graph in Figure 9.1-6, if you picture starting at the point (6,0) and moving up and to the left along the constraint line (along the skin of the melon following the curve of the slice), you are cutting across higher contour lines and moving to higher altitudes until you get to (2,4), at which point you are instantaneously level, and then as you move further up and left, you start moving back *down* and cutting across lower contour lines.

Figure 9.1-7 illustrates the intersection of the $x + y = 6$ plane and the profit surface.

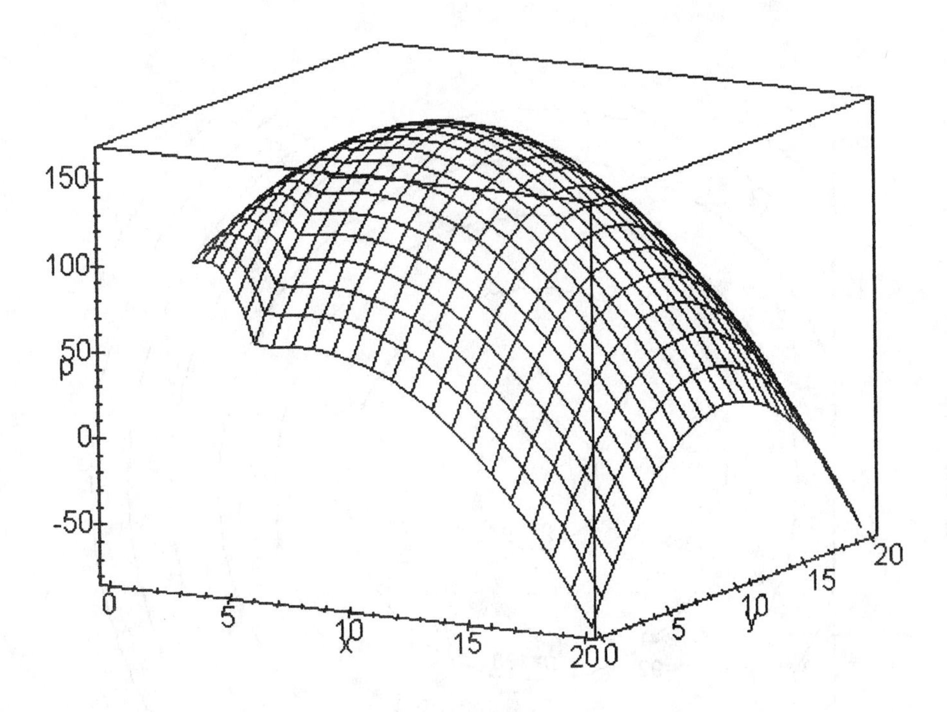

Figure 9.1-7

Notice how the leftmost corner seems to have been sliced off the surface, and the edge of the slice is the leftmost small arc. We are looking for the high point on this small arc.

If you picture more and more contour curves to fill in the other profit values on Figure 9.1-6 (like 70 and 90, or even all the even numbers), you can picture that **the optimal solution is exactly at the point along the constraint line where it is perfectly tangent to some contour curve**. Even on Figure 9.1-6 , you can see that the tangent at the point on the $p = 100$ contour curve nearest to (2,4) would be very nearly parallel to the constraint line $x+y = 6$. Let's see this even more clearly by looking at Figure 9.1-8 .

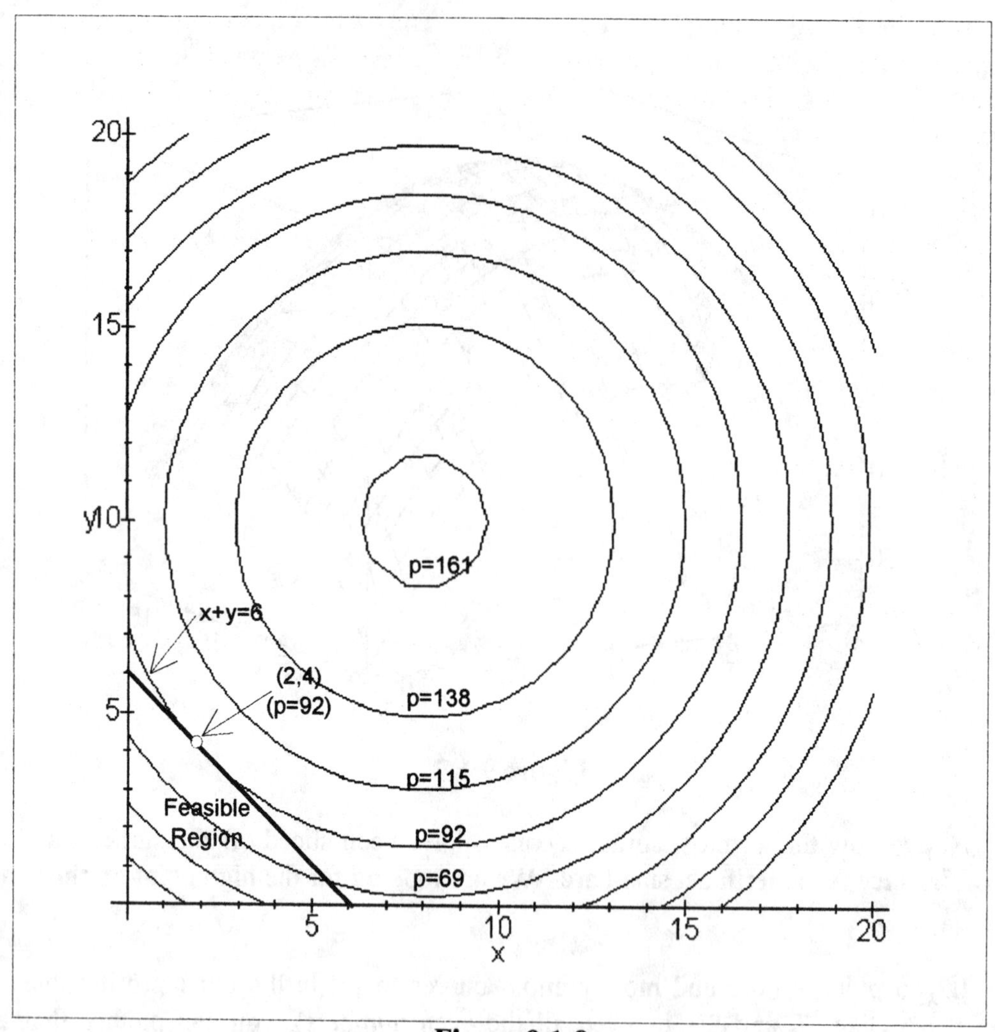

Figure 9.1-8

In Figure 9.1-8 , it is quite clear that the constraint line $x+y=6$ is *tangent* to the contour curve corresponding to $p=92$ at the global constrained optimal solution (2,4). Again, this makes sense, since it is saying that as you move along the constraint line, at that point you are instantaneously traveling at a level height on the surface (*along* a level or contour curve), in between going up and going down along the surface.

Lagrange Multipliers for Functions of 3 or More Variables

Let's now see how we could apply this method to some of the other constrained multivariable problems we have already seen.

Sample Problem 6: The Post Office restricts the dimensions of parcels by requiring that the length plus the girth (the distance all the way around the package in the direction of the address) be no more that 108 inches. What dimensions will maximize the volume you can send?

Solution: We have already formulated this problem in Section 5.2 . Right now, we want to use the *initial* formulation we did, *before* eliminating one variable. The model was given by

Verbal Definition: $V(l,w,h)$ = the volume of a box, in cubic inches, if the length is l inches, the width is w inches, and the height is h inches

Symbol Definition: $V(l,w,h) = lwh$ for $l + 2w + 2h \leq 108$ and $l,w,h \geq 0$

Assumptions: Certainty and divisibility.

Clearly, the unconstrained solution is unbounded: for example, if you make l, w, and h arbitrarily large, you will get an arbitrarily large volume. It makes sense that, since you are limited to 108 inches for the length plus the girth, you will want to use all that you are allowed. If you didn't, you could increase one of the measurements a little and improve your volume, so your initial solution couldn't have been the maximum.

Thus we will try imposing the single constraint

$$l + 2w + 2h = 108$$

and so our problem will be to

Maximize $V(l,w,h) = lwh$ subject to $l + 2w + 2h = 108$

Let's first form the Lagrangean function:

$$L(l,w,h,\lambda) = lwh - (\lambda)(l + 2w + 2h - 108) = lwh - \lambda l - 2\lambda w - 2\lambda h + 108\lambda$$

Now let's find the partial derivatives and set them equal to 0:

$$L_l(l,w,h,\lambda) = wh - \lambda = 0$$
$$L_w(l,w,h,\lambda) = lh - 2\lambda = 0$$
$$L_h(l,w,h,\lambda) = lw - 2\lambda = 0$$
$$L_\lambda(l,w,h,\lambda) = -l - 2w - 2h + 108 = 0$$

How can we solve this system of equations? Well, it is not linear, so unfortunately we can't solve it easily with matrices. That leaves us substitution and elimination.

Notice that in the second and third equations it would be easy to eliminate the λ by simply changing the signs of both sides of one of the equations, and them adding them together. Let's do it:

$$(-1)(lh - 2\lambda) = (-1)(0)$$
$$lw - 2\lambda = 0$$

$$\begin{aligned} -lh + 2\lambda &= 0 \\ \underline{lw - 2\lambda} &\underline{= 0} \\ -lh + lw &= 0 \end{aligned}$$

If we add lh to both sides, we get

$$lw = lh .$$

Now, your first instinct might be to cancel the l's , but as we discussed in Section 8.2 , you shouldn't just do that blindly. It is better to go back to the form that had 0 on one side and factor:

$$\begin{aligned} -lh + lw &= 0 \\ (l)(-h + w) &= 0 && \text{(factoring out } l\text{)} \\ l = 0 \ \text{ or } \ -h + w &= 0 && (uv = 0 \text{ means } u = 0 \text{ or } v = 0) \\ l = 0 \ \text{ or } \ w &= h && \text{(adding } h \text{ to both sides of the 2}^{\text{nd}} \text{ equation)} \end{aligned}$$

In other words, before canceling out the l's , we need to check what would happen if l were 0. In this problem, a length of 0 inches would mean a volume of 0, which is certainly not optimal, so the only solution we need to consider is

$$w = h$$

This means that we have found a solution for w in terms of h , so we can go back to our equations and make this substitution, and see where it gets us:

$$\begin{aligned} wh - \lambda &= 0 \\ lh - 2\lambda &= 0 \\ lw - 2\lambda &= 0 \\ -l - 2w - 2h + 108 &= 0 \end{aligned} \qquad \begin{aligned} (h)h - \lambda &= 0 \\ lh - 2\lambda &= 0 \\ l(h) - 2\lambda &= 0 \\ -l - 2(h) - 2h + 108 &= 0 \end{aligned} \qquad \begin{aligned} h^2 - \lambda &= 0 \\ lh - 2\lambda &= 0 \\ lh - 2\lambda &= 0 \\ -l - 4h &= -108 \end{aligned}$$

Notice that the second and third equations are now the same, so we only need one of them. Also, we can change the signs of both sides of the last equation to get the reduced system

$$\begin{aligned} h^2 - \lambda &= 0 \\ lh - 2\lambda &= 0 \\ l + 4h &= 108 \end{aligned}$$

Notice that, in the *first* equation, λ is easy to solve for in terms of h, and in the *last* equation, l is easy to solve for in terms of h. We could then take these results and substitute into the middle equation for λ and l to get one equation that only involves h. Let's try it:

$$h^2 - \lambda = 0 \qquad \lambda = h^2$$
$$l + 4h = 108 \qquad l = 108 - 4h$$

$$lh - 2\lambda = 0 \qquad (108 - 4h)h - 2(h^2) = 0$$
$$108h - 4h^2 - 2h^2 = 0$$
$$108h - 6h^2 = 0$$
$$6h(18 - h) = 0$$
$$6h = 0 \quad \text{or} \quad 18 - h = 0$$
$$h = 0 \quad \text{or} \quad h = 18$$

Thus the solution for h is $h = 0$ or $h = 18$. As we discussed for l, the 0 solution could not be optimal, so we are left with

$$h = 18$$

and now we can substitute back to get the values of the other variables:

$$h = 18$$
$$\lambda = h^2 = (18)^2 = 324$$
$$l = 108 - 4h = 108 - 4(18) = 108 - 72 = 36$$
$$w = h = 18$$

This would suggest a solution of $l = 36$, $w = 18$, $h = 18$, and $\lambda = 324$. The fact that λ is positive indicates, that if the sum of the length and girth could be a *larger* number, we could increase the volume, so suggests that making that constraint **binding** (forcing it to hold as an *equation*) was the right thing to do. Notice also that all of our original decision variables are nonnegative, as they must be to be feasible.

Let's check this solution for local and global optimality. Since our original objective function (the volume function) is a function of *three* variables, we cannot use the grid method of testing, but we can use the randomized method with small and large increments to check local and global optimality, respectively. A sample result for the test for local optimality is shown in Table 7.

Table 7

	A	B	C	D	E	F	G	H	I
33	C38=+G38*H38*I38								
34									
35		fn name:	V(l,w,h)		variable names ==>		l	w	h
36					Increments ==>		1	1	1
37		i	fn Value	fn(Other$_i$)-fn(Test)	Variables ==>		1st	2nd	3rd
38			11664	0	<== Test Point ==>		36	18	18
39		1	12116.45	452.4516766			36.36236	17.57132	18.96352
40		2	11744.32	80.31553373			35.04441	18.40878	18.20472
41		3	11846.75	182.7545295			35.92322	18.05758	18.26268
42		4	11651.51	-12.49375512	Other		36.48769	17.53488	18.21096
43		5	12133.31	469.3096354	Points		35.7144	18.07935	18.79114
44		6	11289.53	-374.4703682			36.66466	17.64731	17.44816
45		7	11663.13	-0.867539759			35.52909	18.17973	18.05692
46		8	10857.93	-806.0656263			36.80234	17.19727	17.15585

From looking at Table 7, it doesn't appear that our solution is locally optimal, since there are points with both higher and lower volumes. But remember that, like in Sample Problem 5, we need to check that "better" solutions (higher volumes) are also *feasible*. In this problem, checking feasibility can't easily be done by just looking at the solution, so let's add a column to keep track of the value of the length plus the girth, which needs to be less than or equal to 108. The results are shown in Table 8.

Table 8

	A	B	C	D	E	F	G	H	I
33	C38=+G38*H38*I38								
34									
35		fn name:	V(l,w,h)		variable names ==>		l	w	h
36					Increments ==>		1	1	1
37	LHS	i	fn Value	fn(Other$_i$)-fn(Test)	Variables ==>		1st	2nd	3rd
38	108		11664	0	<== Test Point ==>		36	18	18
39	109.432	1	12116.45	452.4516766			36.36236	17.57132	18.96352
40	108.2714	2	11744.32	80.31553373			35.04441	18.40878	18.20472
41	108.5637	3	11846.75	182.7545295			35.92322	18.05758	18.26268
42	107.9794	4	11651.51	-12.49375512	Other		36.48769	17.53488	18.21096
43	109.4554	5	12133.31	469.3096354	Points		35.7144	18.07935	18.79114
44	106.8556	6	11289.53	-374.4703682			36.66466	17.64731	17.44816
45	108.0024	7	11663.13	-0.867539759			35.52909	18.17973	18.05692
46	105.5086	8	10857.93	-806.0656263			36.80234	17.19727	17.15585

From Table 8, we can see that the only *better* solutions were in fact infeasible. Notice also that there *is* one other point with a length plus girth very *close* to 108, but it has a *lower* volume. This helps validate that our solution is the maximum *along* the constraint line as well as for general feasible points nearby. Having checked many other points with recalculated random points, we can be quite confident that our solution is locally optimal.

To check global optimality, we can now use larger increments to span the entire feasible region. In this case, we can just use the values of the variables for the increments. Table 9 shows some sample results.

Table 9

	A	B	C	D	E	F	G	H	I
33	C38=+G38*H38*I38								
34									
35		fn name:	V(l,w,h)		variable names ==>		l	w	h
36					Increments ==>		36	18	18
37	LHS	i	fn Value	fn(Other $_i$)-fn(Test)	Variables ==>		1st	2nd	3rd
38	108		11664	0	<== Test Point ==>		36	18	18
39	86.19926	1	3240.828	-8423.17197			41.26136	18.13879	4.330163
40	169.7696	2	42515.69	30851.6902			70.07855	21.1204	28.72513
41	103.731	3	4566.494	-7097.505665			9.520236	32.2168	14.88857
42	81.06442	4	2307.672	-9356.328244	Other		38.87899	3.344245	17.74847
43	168.9156	5	43561.51	31897.50527	Points		59.18708	31.4972	23.36706
44	107.8993	6	5654.631	-6009.369497			9.43117	22.08137	27.15268
45	111.2025	7	11345.75	-318.2524391			37.98708	24.33351	12.27418
46	132.3512	8	19124.59	7460.590591			36.4005	31.05983	16.91553

Once again, the only points that are better are infeasible, and this was also true for many recalculations. Notice again that the one other point with a volume close to 108 (Other Point number 6) has a much lower volume, validating that our solution is globally optimal *both* along the constraint line and over the entire feasible region.

Translation: The best dimensions for the box are a length of 36 inches, a width of 18 inches, and a height of 18 inches (with the address written parallel to the width or the height), for a total volume of 11,664 cubic inches. If the length plus the girth were allowed to be a larger number, the volume would increase at an instantaneous rate of 324 cubic inches per additional inch. □

One of the most important problems in Finance is the problem of finding an optimal combination of investments from a given group of investments, sometimes referred to a **modern portfolio theory**. This builds on our discussion of risk and return in Section 6.4 . There we mentioned that investors want to maximize return and minimize risk (as measured by the *variance*), but that higher returns tend to be associated with investments that also have higher risks.

Using some statistical concepts beyond the scope of this book, it is possible to get an expression for the *overall return* and the *overall risk* of a potential *combination* of investments (**portfolio**). It is then possible to formulate the problem by holding *one* of the quantities (overall return or overall risk) fixed at some target value, and optimizing the *other* subject to that constraint. For instance, you could fix a desired rate of return, and then try to minimize the risk. This would then involve minimizing a quadratic risk function (since variance involves squaring) subject to a linear constraint. The Lagrange multiplier would then tell you the "cost" (in terms of added risk) per unit improvement in the expected return, so the sign should be positive.

For example, you could fix your overall return to be 11% and try to minimize your overall risk. In this case, the value of the Lagrange multiplier would tell you the effect of raising your return target on the optimal risk, or how much increased risk you'd have to take on per percentage point of increased return. In this case, the objective function turns out to be a quadratic function.

Alternatively, you could fix your risk at a certain value for the variance (perhaps corresponding to a standard deviation of 4 percentage points in overall return), and try to maximize your return. In this case, the Lagrange multiplier would tell you the effect of increasing the target risk on the optimal return, or how much of an increase in return you could expect per unit increase in the risk.

Lagrange Multipliers with More Than One Equality Constraint

What if you want to force *more than one* constraint to be binding? Let's look at a problem that has that property.

Sample Problem 7: For the Post Office box problem of Sample Problem 6, suppose that, for personal reasons, since you have to carry your package for some distance, you don't want the shape to be too oblong, so you want the length to be no more than 2 feet. What would be the new solution?

Solution: Since we have already solved Sample Problem 6, we have a great head start on this problem. Unfortunately, our optimal solution to that problem had a length of 36 inches, which violates our new constraint. This suggests that we want to *keep* the 108 inches constraint binding, but now we would *also* like the new length constraint to be binding. In other words, we want to impose the equality constraint

$$l = 24$$

in addition to the first constraint.

How can we deal with *multiple* equality constraints using Lagrange multipliers? The most obvious strategy would be to have a *different* Lagrange multiplier for each equality constraint. Let's try it. We are solving the problem

$$\text{Maximize } V = lwh \text{ subject to } l + 2w + 2h = 108 \text{ and } l = 24 .$$

We can no longer just use a single λ variable, since we might need *different* penalties on the two different constraints to get them *both* to hold perfectly. So let's call them λ_1 and λ_2 for the first and second constraints, respectively. Then our Lagrangean function would become

$$\begin{aligned} L(l,w,h,\lambda_1,\lambda_2) &= lwh - (\lambda_1)(l + 2w + 2h - 108) - (\lambda_2)(l - 24) \\ &= lwh - \lambda_1 l - 2\lambda_1 w - 2\lambda_1 h + 108\lambda_1 - \lambda_2 l + 24\lambda_2 \end{aligned}$$

and the partial derivatives, set equal to 0, are

$$\begin{aligned} L_l(l,w,h,\lambda_1,\lambda_2) &= wh - \lambda_1 - \lambda_2 = 0 \\ L_w(l,w,h,\lambda_1,\lambda_2) &= lh - 2\lambda_1 = 0 \\ L_h(l,w,h,\lambda_1,\lambda_2) &= lw - 2\lambda_1 = 0 \\ L_{\lambda_1}(l,w,h,\lambda_1,\lambda_2) &= -l - 2w - 2h + 108 = 0 \\ L_{\lambda_2}(l,w,h,\lambda_1,\lambda_2) &= -l + 24 = 0 \ . \end{aligned}$$

As before, the second and third constraints can be solved simultaneously to get

$$w = h \ ,$$

and the new equation from L_{λ_2} tells us that

$$l = 24 \ .$$

If we substitute h in for w and 24 in for l into the L_{λ_1} equation, we get

$$\begin{aligned} -(24) - 2(h) - 2h + 108 &= 0 \\ -4h &= -84 && \text{(combining like terms and subtracting 84)} \\ h &= \frac{-84}{-4} = 21 && \text{(dividing by -4 and simplifying)} \end{aligned}$$

This will mean that

$$w = 21$$

also, since we had $w = h$, and so now we only need to find λ_1 and λ_2. From the second equation we get

$$\begin{aligned} lh - 2\lambda_1 &= 0 \\ (24)(21) - 2\lambda_1 &= 0 && \text{(substituting 24 for } l \text{ and 21 for } h\text{)} \end{aligned}$$

$504 = 2\lambda_1$ (multiplying out and adding $2\lambda_1$ to both sides)

$\lambda_1 = \frac{504}{2} = 252$ (switching sides, dividing by 2, and simplifying) .

From the first equation we get

$wh - \lambda_1 - \lambda_2 = 0$

$(21)(21) - 252 - \lambda_2 = 0$ (substituting 21 for both w and h, and 252 for λ_1)

$441 - 252 = \lambda_2$ (multiplying out and adding λ_2 to both sides)

$\lambda_2 = 189$ (switching sides and combining like terms)

Let's check this for local optimality. Sample results are shown in Table 10.

Table 10

	A	B	C	D	E	F	G	H	I
33	C38=+G38*H38*I38								
34									
35		fn name:	V(l,w,h)		variable names ==>		l	w	h
36					Increments ==>		1	1	1
37	LHS	i	fn Value	fn(Other$_i$)-fn(Test)	Variables ==>		1st	2nd	3rd
38	108		10584	0	<== Test Point ==>		24	21	21
39	108.4124	1	10757.35	173.3533778			24.42415	21.65839	20.33573
40	106.1433	2	10064.19	-519.8058834			23.71967	21.15877	20.05302
41	107.008	3	10190.53	-393.4658774			23.23151	21.01271	20.87553
42	106.1622	4	10041.75	-542.2484452	Other		23.55131	21.0494	20.25605
43	108.216	5	10643.79	59.79417518	Points		24.03409	21.26959	20.82139
44	107.5031	6	10610.22	26.21991635			24.85749	20.4335	20.88932
45	108.9111	7	10657.23	73.22562792			23.23872	21.04733	21.78888
46	107.9039	8	10560.97	-23.02856147			24.01002	20.8012	21.14575

From Table 10, we can see that the only better solutions violate *only* the total measurement constraint (like Other Point 7), *only* the length constraint (like Other Point 6), or *both* constraints (like Other Point 1). Notice also that Other Point 8 has a length of almost exactly 24 and a total measurement of just a little less than 108, but has a lower volume. From repeated recalculations with similar results, we have strong evidence that this solution is locally optimal.

To check global optimality, we can again simply make the increments equal to the values of the variables in the solution. A sample result is shown in Table 11.

Table 11

	A	B	C	D	E	F	G	H	I
33	C38=+G38*H38*I38								
34									
35		fn name:	**V(l,w,h)**		variable names ==>		l	w	h
36						**Increments ==>**	24	21	21
37	**LHS**	**i**	**fn Value**	**fn(Other$_i$)-fn(Test)**		**Variables ==>**	**1st**	**2nd**	**3rd**
38	108		10584	0	**<== Test Point ==>**		24	21	21
39	189.7447	1	58127.03	47543.02516			43.51766	37.3843	35.7292
40	78.50506	2	174.1573	-10409.84274			0.457661	19.09393	19.92976
41	119.1723	3	7956.696	-2627.304409			16.5161	12.36407	38.964
42	128.5527	4	4678.277	-5905.723122	**Other**		5.154912	24.20483	37.49408
43	51.04403	5	775.4607	-9808.539323	**Points**		14.1526	14.72452	3.721194
44	170.6566	6	28654.95	18070.95054			20.40465	40.14061	34.98536
45	121.8005	7	5506.973	-5077.026522			8.14559	16.95634	39.8711
46	174.7677	8	42027.55	31443.54847			35.38412	39.99227	29.69953

From Table 11 and repeated recalculations, we again see that the only better solutions are infeasible, giving strong evidence that our solution is globally optimal.

Translation: The best dimensions for the box are a length of 24 inches, a width of 21 inches, and a height of 21 inches (with the address written parallel to the width or the height), for a total volume of 10,584 cubic inches. If the length plus the girth were allowed to be a larger number, the volume would increase at an instantaneous rate of 252 cubic inches per additional inch beyond 108. If the length restriction were relaxed, the volume would increase at an instantaneous rate of 189 cubic inches per additional inch beyond 24. □

The General Method of Lagrange Multipliers

To solve the problem

Optimize $z = f(x_1,x_2,\ldots,x_n)$, subject to

$$g_1(x_1,x_2,\ldots,x_n) = b_1$$
$$g_2(x_1,x_2,\ldots,x_n) = b_2$$
$$\ldots$$
$$g_m(x_1,x_2,\ldots,x_n) = b_m ,$$

the optimal solution will be a critical point of the Lagrangean function

$$L(x_1,x_2,\ldots,x_n, \lambda_1, \lambda_2,\ldots, \lambda_m)$$
$$= f(x_1,x_2,\ldots,x_n) - (\lambda_1)[g_1(x_1,x_2,\ldots,x_n) - b_1] - (\lambda_2)[g_2(x_1,x_2,\ldots,x_n) - b_2] - \ldots - (\lambda_m)[g_m(x_1,x_2,\ldots,x_n) - b_m]$$

= (original objective function) - (1st Lagrange multiplier)(LHS - RHS of 1st constraint)
- (2nd Lagrange multiplier)(LHS - RHS of 2nd constraint)
- ... - (last Lagrange multiplier)(LHS - RHS of last constraint)

As usual, to find critical points, take the partial derivative with respect to each of the variables (original decision variables *and* Lagrange multipliers), set each equal to 0, and solve for all of the variables. Use the techniques of Section 8.4 to test your solution(s) for local and global optimality.

For problems with *inequality* constraints, first try solving the unconstrained problem. If that global optimum is feasible, then it is also the global optimum for the *constrained* problem. If it is *not* feasible, determine which constraints are violated, and try forcing them to be binding (hold as *equations* rather than *inequalities*) using Lagrange multipliers. Again, use the techniques of Section 8.4 to test your answer, and also check to see if any *better* solutions are *feasible*. If so, you may need to try a different combination of constraints to be binding to find the global constrained optimal solution.

Section Summary

Before trying the exercises, be sure that you:

- Understand that Lagrange multipliers can be used directly to optimize an objective function subject to one or more *equality* constraints, but can also be used indirectly to solve problems with *inequality* constraints.
- Understand that the basic idea behind the method of Lagrange multipliers is to impose a *penalty* (the Lagrange multiplier) per unit *violation* of a constraint (its

LHS minus its RHS), which is analogous to charging a *fee* for violating the constraint on *one* side and offering a *bonus* for violating it on the *other* side. By adjusting this penalty (possibly switching the sign if necessary, to *reverse* which side you *reward*), solutions that come closer and closer to satisfying the constraint can be obtained, and a *perfect* value can be determined so that the equality constraint is satisfied *exactly*.

- Understand that the perfect value of the Lagrange multiplier(s) can be found by finding critical points for the Lagrangean function, which is formed by starting with the original objective function and subtracting a Lagrange multiplier for each equality constraint *times* (its LHS minus its RHS).
- Understand that a solution can be checked for optimality using the techniques of Section 8.4 .
- Understand that the optimal value of a Lagrange multiplier for a constraint corresponds to its **shadow price**, the instantaneous rate of change of the *optimal* objective function value per unit increase in the RHS of that constraint.

EXERCISES FOR SECTION 9.1:

Warm Up

1. Use the method of Lagrange multipliers to solve the problem
   Minimize $f(x, y) = 4x^2 + 2y^2$, subject to $x + 3y = 5$

2. Use the method of Lagrange multipliers to solve the problem
   Maximize $f(x, y) = 8x + 6y - 4x^2 - 2y^2$, subject to $x + y = 5$

3. Use the method of Lagrange multipliers, if necessary, to solve the problem
   Minimize $f(x, y) = 4x^2 + 2y^2$, subject to $x + 3y \geq 5$

4. Use the method of Lagrange multipliers, if necessary, to solve the problem
   Maximize $f(x, y) = 8x + 6y - 4x^2 - 2y^2$, subject to $x + y \leq 5$

5. Use the method of Lagrange multipliers to solve the problem
   Minimize $f(x, y, z) = 4x^2 + 2y^2 + z^2$, subject to $x + 3y + 2z = 5$

6. Use the method of Lagrange multipliers to solve the problem
   Maximize $f(x, y, z) = 8x + 6y - 4x^2 - 2y^2 - z^2$, subject to $x + y + z = 5$

7. Use the method of Lagrange multipliers to solve the problem
   Minimize $f(l, w, h) = l + w + h$, subject to $lwh = 8$

8. Use the method of Lagrange multipliers to solve the problem
   Maximize $f(x, y) = 4x^2 + y^2$, subject to $8x + 2y = 20$

9. Use the method of Lagrange multipliers to solve the problem
   Minimize $f(x, y, z) = x^2 + 2y^2 + 4z^2$, subject to $x + y + z = 8$ and $z = 2x$

10. Use the method of Lagrange multipliers to solve the problem
   Maximize $f(x, y) = 4x^2 + y^2$, subject to $8x + 2y = 20$ and $x, y \geq 0$.

Game Time

11. You need to decide how to allocate your available investment dollars right now. You are considering three different investments: X, Y, and Z. You have looked at the

available data on how these investments have performed in the past, computing percent return for each year, then estimating your expected overall return and risk. You decide to let x, y, and z represent the *fraction* of your money you will invest in X, Y, and Z, respectively, right now. From the data, you estimate that the overall risk of such a mix of these investments is given by

$$V(x,y,z) = 4.2x^2 + 2.1y^2 + 0.6z^2 - 0.088xz + 1.4yz$$

You estimate that your overall rate of return, in percentage points, is given by

$$M(x,y,z) = 18x + 12y + 7.5z$$

You decide that you want to achieve an expected overall return of 13 percentage points, and otherwise you want to minimize your overall risk.

a) If you do not put any restrictions on the magnitudes of x, y, and z other than achieving your expected overall return goal exactly, formulate your problem and solve it using the method of Lagrange multipliers.

b) Interpret the value of your Lagrange multiplier in your answer to (a).

c) Now suppose you add the restriction that your variables must sum to 1 (you want to allocate *all* of your available money, and only to these three investments. Formulate and solve the new problem using the method of Lagrange multipliers.

12. Suppose you want to ship some loose material from a European country. You want to maximize the volume of your package, but the length plus the girth must be no more than 3 meters.

a) Formulate your problem (don't eliminate any variables).

b) Use Lagrange multipliers to solve your problem.

c) Interpret the value of your Lagrange multiplier in your answer to (b). Explain what this tells you about the desirability of having the length plus the girth be *less* than 3 meters.

d) Try solving your problem by eliminating a variable. Do you get the same answer? What *don't* you get, and why is it significant?

13. If a box costs $2 per square foot for sides and top and $3 per square foot for the bottom and it must have a volume of 6 cubic feet,

a) Formulate the problem of minimizing the cost without eliminating variables.

b) Use Lagrange multipliers to find the optimal dimensions and cost.

c) Interpret the value you got for your Lagrange multiplier in (b).

14. If the cost of producing each chair in the furniture problem can be reduced to $60 while the cost of producing the tables remains at $100 (that is, if the coefficient of c in the profit function increases by 10, from 140 to 150), so that the profit function is now:

$$P(c,t) = 150c + 200t + 2ct - 4c^2 - 12t^2 - 700,$$

and if your experience tells you that the number of chairs you make should be *at least* 3 times the number of tables,

a) Formulate a model for your problem.
b) Use Lagrange multipliers to determine how many chairs and tables should be made to maximize the profit. What is the maximum profit?

15. If the cost of producing each table can be reduced to \$80 (so the coefficient of t in the profit function increases by 20, from 200 to 220) and the cost of producing the chairs remains at \$60 , so that the profit function is now

$$P(c,t) = 150c + 220t + 2ct - 4c^2 - 12t^2 - 700,$$

and if your experience tells you that the number of chairs you make should be *at least* 3 times the number of tables,

a) Formulate a model for your problem.
b) Use Lagrange multipliers to determine how many chairs and tables should be made to maximize the profit. What is the maximum profit?

16. You are a soup manufacturer, and your marketing research has indicated that the ideal volume for a soup can is 100 cubic inches. Your costs for soup cans are directly proportional to the amount of metal used, which is essentially the total surface area of the can.

a) Formulate a model for this problem.
b) Use Lagrange multipliers to find the optimal dimensions for your can.
c) Interpret the value of your Lagrange multiplier in (b).

Section 9.2: Solving Linear Programming Problems Graphically

In Section 5.2 we had some practice in formulating models from verbal descriptions. Some of the examples we worked with turned out to involve multivariable expressions that were *all linear*, both in the main function and in the domain inequalities and equations, called the **constraints**. When you are trying to maximize or minimize the main function in such a case, you are trying to solve what is called a **linear programming** (LP) problem. The main function is then called the **objective function**, whatever you are trying to maximize or minimize, like profit or cost. The constraints are divided into two groups: **structural constraints**, the major restrictions on possible solutions, and **nonnegativity constraints**, which force variables to be nonnegative. In this section, we see the standard format for defining a linear programming problem. We will show how to solve programming problems **graphically** when they have with only two independent variables called the **decision variables**.

Here are some examples of linear programming problems you should be able to solve graphically by the end of this section:

- Consider again the example of the hand-crafted furniture business from previous chapters. You own this small business, and for now you make and sell only two products: chairs and tables (just a single style of each). Given known restrictions on available raw materials and demand, how many chairs and tables should you make this month to maximize your profit?
- You are trying to decide what combination of foods you should eat, on average, for breakfast, to meet your nutritional goals at the lowest possible cost. What foods, and how much of each, should you plan for your breakfasts over the course of a week?

In addition to being able to solve the above kinds of problems, by the end of this section you should:

- Recognize a linear programming problem from a verbal description.
- Be able to formulate a linear programming problem into standard LP format.
- Know how to graph inequality and equation constraints for a linear program with two decision variables.
- Know how to graph the contour lines for the objective function of a linear program.
- Know how to determine the feasible region (if there is one) and the optimal solution(s) for a linear program graphically.

Sample Problem 1: Consider again the example of the hand crafted furniture business from previous chapters. You own this small business, and for now you make and sell only two products: chairs and tables (just a single style of each). The chairs each use 3 units of wood and 8 units of hardware, and the tables each use 6 units of wood and 4 units of hardware. You have on hand 60 units of wood and 100 units of hardware. You will make a profit of $140 on each chair and $200 on each table, and that at the selling price you have chosen, you expect the demand for chairs to be at most 11 for the coming month. The demand for tables should be sufficiently large that you can sell all the tables you can make. You are planning for the next month, and you will not have time to order and receive any new resources before the end of the month. Formulate the problem of how many chairs and tables to make, and solve it graphically.

Solution 1: This seems as if it could be a linear programming problem because you want to optimize (maximize) your profit from the sales of the chairs and tables, and there are some constraints on the numbers of chairs and tables that you can make. To be sure, we will have to check our objective function and constraints after we have formulated them, and make sure everything is linear. The question in this problem (the main decision to be made) is how many chairs and tables to make this month. This suggests that the variables (often called the **decision variables**) we need to define are:

c = the number of chairs to make this month, and
t = the number of tables to make this month.

We could start by testing different combinations of chairs and tables to see if they satisfy our needs. In Table 1, each row gives a different possible combination of chairs and tables, the profit from each type of furniture individually and the total profit from both together, and, similarly, the amount of wood and hardware used by each type of furniture individually and the total amount of each resource used.

For example, the first row corresponds to making 3 chairs and 8 tables. The profit from the 3 chairs is $(\$140)(3) = \$420$ ($140 profit for each of the 3 chairs). Similarly, the profit from the tables is $(\$200)(8) = \$1600$. Thus the total profit is $\$420 + \$1600 = \$2020$. In an analogous way, the 3 chairs use $(3)(3) = 9$ units of wood (3 units for each chair) and the tables use $(6)(8) = 48$ units of wood, for a total of $9 + 48 = 57$ units of wood. Similarly, the 3 chairs use $(8)(3) = 24$ units of hardware and the 8 tables use $(4)(8) = 32$ units of hardware, for a total of $24 + 32 = 56$ units of hardware. The other cases are all done in the same way.

Table 1

ORDER NUM	QUANT CHAIRS	QUANT TABLES	PROFIT CHAIRS	PROFIT TABLES	TOTAL PROFIT	WOOD CHAIRS	WOOD TABLES	<=60 TOTAL WOOD	HRDWR CHAIRS	HRDWR TABLES	<=100 TOTAL HRDWR
1	3	8	$420	$1,600	$2,020	9	48	57	24	32	56
2	12	4	$1,680	$800	$2,480	36	24	60	96	16	112
3	15	5	$2,100	$1,000	$3,100	45	30	75	120	20	140
4	12	3	$1,680	$600	$2,280	36	18	54	96	12	108
5	10	8	$1,400	$1,600	$3,000	30	48	78	80	32	112
6	2	21	$280	$4,200	$4,480	6	126	132	16	84	100
7	0	10	$0	$2,000	$2,000	0	60	60	0	40	40
8	9	9	$1,260	$1,800	$3,060	27	54	81	72	36	108
9	11	3	$1,540	$600	$2,140	33	18	51	88	12	100
10	3	15	$420	$3,000	$3,420	9	90	99	24	60	84
11	14	4	$1,960	$800	$2,760	42	24	66	112	16	128
12	10	4	$1,400	$800	$2,200	30	24	54	80	16	96
13	3	8	$420	$1,600	$2,020	9	48	57	24	32	56
14	9	7	$1,260	$1,400	$2,660	27	42	69	72	28	100
15	11	5	$1,540	$1,000	$2,540	33	30	63	88	20	108
16	8	7	$1,120	$1,400	$2,520	24	42	66	64	28	92
17	6	6	$840	$1,200	$2,040	18	36	54	48	24	72
18	7	5	$980	$1,000	$1,980	21	30	51	56	20	76
19	14	3	$1,960	$600	$2,560	42	18	60	112	12	124
20	2	15	$280	$3,000	$3,280	6	90	96	16	60	76
21	10	5	$1,400	$1,000	$2,400	30	30	60	80	20	100
22	8	8	$1,120	$1,600	$2,720	24	48	72	64	32	96
23	7	10	$980	$2,000	$2,980	21	60	81	56	40	96
24	2	9	$280	$1,800	$2,080	6	54	60	16	36	52
25	12	0	$1,680	$0	$1,680	36	0	36	96	0	96
26	18	3	$2,520	$600	$3,120	54	18	72	144	12	156
27	10	6	$1,400	$1,200	$2,600	30	36	66	80	24	104
28	13	3	$1,820	$600	$2,420	39	18	57	104	12	116
29	8	9	$1,120	$1,800	$2,920	24	54	78	64	36	100
30	9	5	$1,260	$1,000	$2,260	27	30	57	72	20	92
31	17	2	$2,380	$400	$2,780	51	12	63	136	8	144
32	20	0	$2,800	$0	$2,800	60	0	60	160	0	160
33	0	25	$0	$5,000	$5,000	0	150	150	0	100	100
34	11	0	$1,540	$0	$1,540	33	0	33	88	0	88
35	2	4	$280	$800	$1,080	6	24	30	16	16	32
36	4	8	$560	$1,600	$2,160	12	48	60	32	32	64
37	11	8	$1,540	$1,600	$3,140	33	48	81	88	32	120
38	7	5	$980	$1,000	$1,980	21	30	51	56	20	76
39	6	7	$840	$1,400	$2,240	18	42	60	48	28	76
40	10	2	$1,400	$400	$1,800	30	12	42	80	8	88
41	9	1	$1,260	$200	$1,460	27	6	33	72	4	76
42	12	7	$1,680	$1,400	$3,080	36	42	78	96	28	124
43	12	4	$1,680	$800	$2,480	36	24	60	96	16	112
44	9	3	$1,260	$600	$1,860	27	18	45	72	12	84
45	7	6	$980	$1,200	$2,180	21	36	57	56	24	80
46	3	8	$420	$1,600	$2,020	9	48	57	24	32	56
47	5	12	$700	$2,400	$3,100	15	72	87	40	48	88
48	7	4	$980	$800	$1,780	21	24	45	56	16	72
49	8	1	$1,120	$200	$1,320	24	6	30	64	4	68
50	10	1	$1,400	$200	$1,600	30	6	36	80	4	84
51	11	2	$1,540	$400	$1,940	33	12	45	88	8	96
52	7	11	$980	$2,200	$3,180	21	66	87	56	44	100
53	3	5	$420	$1,000	$1,420	9	30	39	24	20	44
54	10	21	$1,400	$4,200	$5,600	30	126	156	80	84	164
55	4	17	$560	$3,400	$3,960	12	102	114	32	68	100
56	5	20	$700	$4,000	$4,700	15	120	135	40	80	120
57	4	22	$560	$4,400	$4,960	12	132	144	32	88	120
58	8	15	$1,120	$3,000	$4,120	24	90	114	64	60	124
59	10	20	$1,400	$4,000	$5,400	30	120	150	80	80	160
60	10	18	$1,400	$3,600	$5,000	30	108	138	80	72	152
61	5	3	$700	$600	$1,300	15	18	33	40	12	52
62	2	7	$280	$1,400	$1,680	6	42	48	16	28	44
63	5	6	$700	$1,200	$1,900	15	36	51	40	24	64
64	1	8	$140	$1,600	$1,740	3	48	51	8	32	40
65	6	2	$840	$400	$1,240	18	12	30	48	8	56

Figure 1 is a graph of the 65 data points (possible solution values) listed in Table 1, with the number of chairs plotted on the horizontal axis and the number of tables plotted on the vertical axis. For example, the first case is graphed as the point (3,8). Graphing the data gives us a little better picture of the problem, but it is still far from clear what the best solution is.

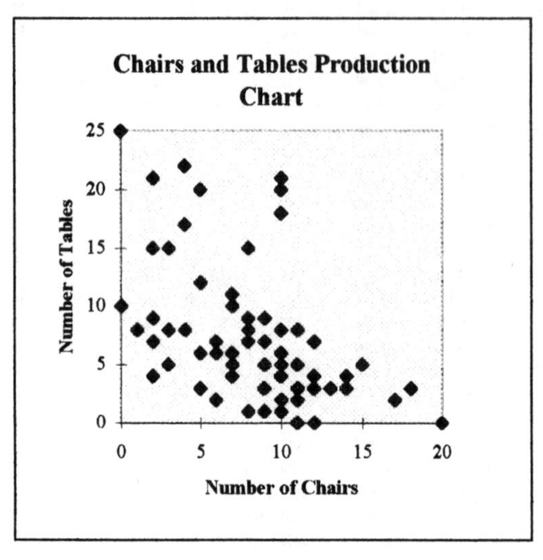

Figure 1

The graphs in Figure 2 show the different constraints, or restrictions, on what solutions are *possible,* or **feasible**. For example, one constraint is the restriction on the demand for chairs this month, which we are told is no more than 11 chairs. The upper left chart shows the combinations (solution points) from Table 1 that satisfy this constraint, those for which the number of chairs is 11 or less. Notice that graphically, the **feasible** points, the ones satisfying the constraint condition, are all along or to the left of the vertical line passing through the number 11 on the horizontal (c) axis. The upper right graph shows the solution points that satisfy the hardware constraint, where the total amount of hardware needed by that solution, as listed in the table, is not more than 100 units. Notice that these also have the property of being on or to one side of a line, this time a slanted line. The lower left graph is similar to the upper right one, and corresponds to the wood constraint, all the solutions using no more than 60 units of wood. Again, these points all lie on or to one side of a slanted line, but the line is a little different than the one for the hardware constraint. The lower right graph is then all of those solution points from the table that satisfy *all* of the constraints. You can see that they correspond to a region that seems to be a polygon, formed by the edge lines of the various constraints.

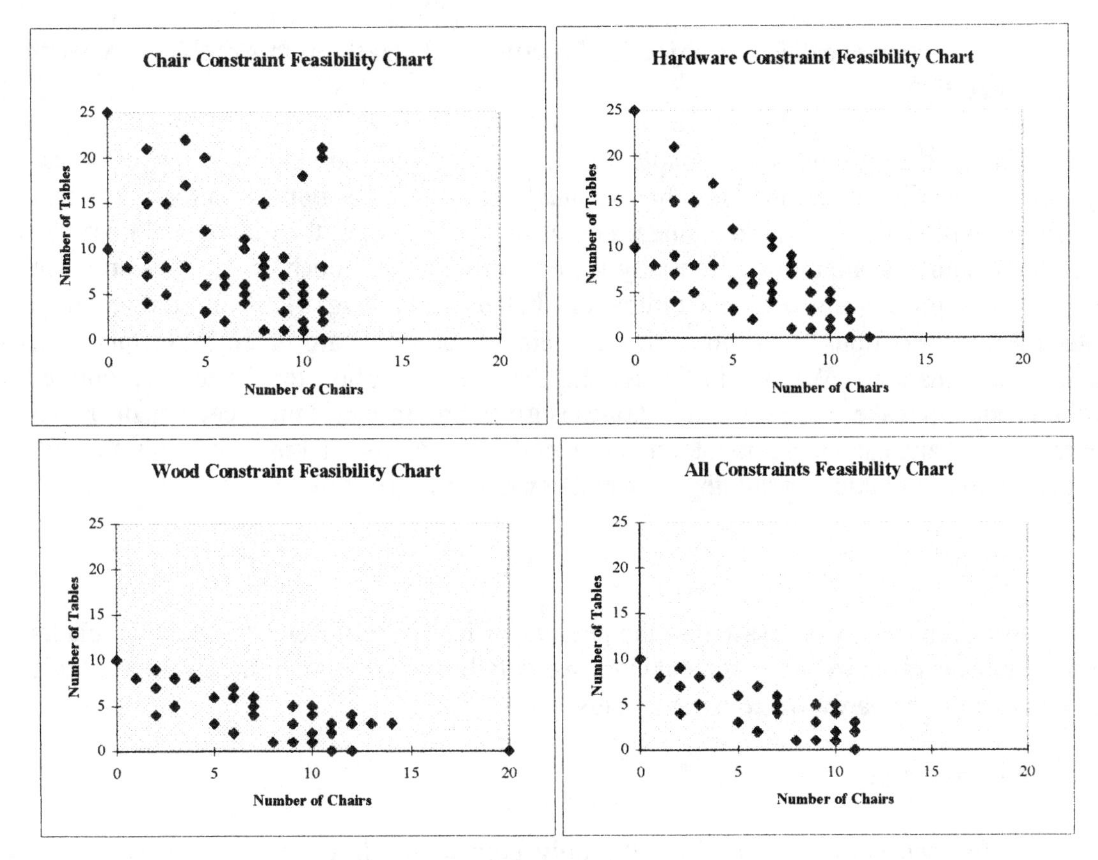

Figure 2

We can see some patterns, but it would be very difficult to make a good decision from these graphs. It is obvious that there are many possible solutions, but what is the optimum solution? There has to be a better way!

Standard Format for Linear Programming Problems

Our goal, or criteria for choosing the best of the possible solutions, is the total profit. Recall the calculations shown earlier for the first line of the table. We showed that since the profit from each chair is \$140, if we make 3 chairs, our profit will be $(\$140)(3) = \$420$. In general, if we make c chairs, the profit from them will be $140c$: the profit from 1 chair would be \$140, from 2 chairs \$280, etc. Similarly, since tables yield a profit of \$200 each, if we make t tables, the profit from them will be $200t$: the profit from 1 table would be \$200, from 2 tables \$400, etc. Thus the total profit, in dollars, for the month will be:

$$P = P(c,t) = 140c + 200t \ .$$

This function is called our **objective function**, and clearly in this problem we want to **maximize** P.

Unlike the problems in Chapter 7, when we were solving systems of linear *equations*, we will not assume here that we *must* use *all* units of both resources, although clearly we will probably want to come pretty close to that, to get the most profit possible. Still, the resource availabilities are really *upper limits* on how much of each resource we can use this month (*allowing* full utilization, but *not requiring* it). Thus the resource **constraints** (restrictions, conditions on the decision variables) are given by inequalities, rather than equations. We saw in Section 5.2 that the expression for the total amount of wood needed to make c chairs and t tables is given by $3c + 6t$ (since each chair uses 3 units of wood and each table uses 6 units of wood). Since this total must be within the 60 units of wood available this month, the wood constraint is given by

$$3c + 6t \leq 60 .$$

In a similar way, we saw that the amount of hardware needed to make c chairs and t tables is given by $8c + 4t$, and since we only have 100 units of hardware available for this month, the hardware constraint is given by

$$8c + 4t \leq 100 .$$

These two constraints are not the only constraints, however. We also have a **demand constraint** on the number of chairs we can sell (no more than 11), also an upper limit, which in symbols is simply:

$$c \leq 11 .$$

These three constraints are called the **structural constraints**. They form the upper structure of the feasible area.

In addition, the number of chairs and tables we make cannot be negative. This is a lower limit, giving us the **nonnegativity constraints**:

$$c \geq 0$$
$$t \geq 0$$

These constraints form the base, or foundation, of the feasible area. We could conceivably think of negative values for c or t as *refunds* (someone returning a piece of furniture and getting their money back), but that doesn't really make sense in the context of the decision we have to make here.

We have one more restriction to consider: logically, the number of chairs and tables we make this month should also be whole numbers (integers), with no fractions or decimals. On the other hand, we could *interpret* a fractional answer to mean that we build *part* of a chair or table this month, to be completed the following month. In such a case, the profit may not be *received* this month, but it would be reasonable to consider the *fractional* profit "*earned*" this month to be included in P (for example, if the optimal solution came out to be 7.5 chairs, the profit from the 0.5 of a chair would be (0.5)($140) = $70 for half a chair) and could be considered to have been earned this month.

Just as we did in Section 5.2, we can now formulate our mathematical model.

Verbal Definition: $P(c,t)$ = the profit, in dollars, from making and selling c chairs and t tables

Symbol Definition: $P(c,t) = 140c + 200t$, for $3c + 6t \le 60$
$8c + 4t \le 100$
$c \le 11$
$c \ge 0$
$t \ge 0$

Assumptions: Certainty and divisibility. Certainty implies the profits per chair and per table, and the demand for chairs, are known exactly, and that the demand for tables is essentially unlimited. Divisibility implies fractions of chairs and tables can be made and sold.

In linear programming form we can write:

P = profit in dollars
c = number of chairs
t = number of tables

Maximize $P = 140c + 200t$,

subject to the constraints:
(1) $3c + 6t \le 60$ — structural constraint
(2) $8c + 4t \le 100$ — structural constraint
(3) $c \le 11$ — structural constraint
(4) $c \ge 0$ — nonnegativity constraint
(5) $t \ge 0$. — nonnegativity constraint

This is usually written in even more compact form:

$$\begin{aligned} \text{Max. } P &= 140c + 200t \\ \text{s.t.} \quad 3c + 6t &\le 60 \\ 8c + 4t &\le 100 \\ c &\le 11 \\ c,t &\ge 0 \end{aligned}$$

The above format is what we will call the **standard format for a linear programming problem** (more precisely, for the symbol definition of the model). We use the term **linear** because, in the objective function and in the left-hand sides of all of the constraints (considered as functions), any variables that appear have an exponent of 1, and there are no mixed terms with products of variables. This means that the graph is a **line** for functions of one variable (equations involving two decision variables) or a **plane** for functions of two variables (equations involving three decision variables). An example of a **nonlinear** function is:

$$Q(c,t) = 140c + 200t - 4c^2 - 12t^2 + 2ct - 700$$

Both c and t appear with exponents different from 1 ($-4c^2$ and $-12t^2$) and there is a mixed term ($2ct$).

Another way to understand this distinction is that all partial derivatives of a linear function are constants (for example, $P_c(c,t)=140$ and $P_t(c,t)=200$). If a function has any partial derivative that contains one of the variables, then the function must be nonlinear. For example, $Q_c(c,t)=140-8c+2t$, which involves both c and t, so $Q(c,t)$ is nonlinear. Furthermore, the graph of P is a plane, while the graph of Q is a curved (nonlinear) surface, as shown in Figure 3.

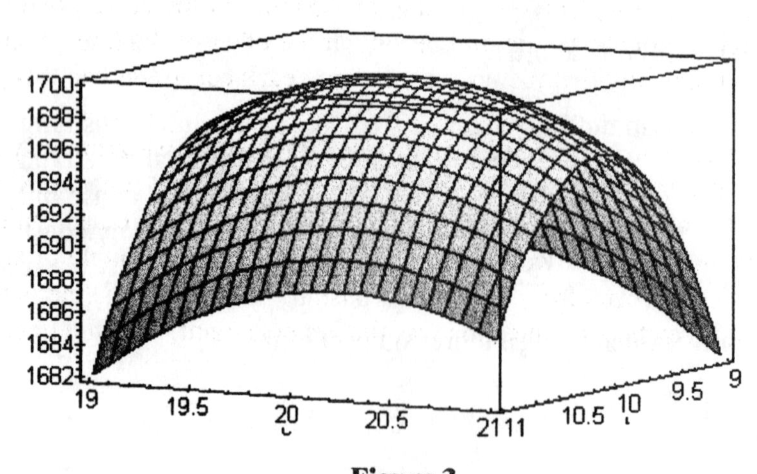

Figure 3

As we have discussed many times, when we use a mathematical model to help us solve a problem, we should include any assumptions that are being made. Most linear programming problems make several assumptions, including the **certainty assumption** and the **divisibility assumption**, like virtually every problem we have studied in this course. Notice **that the standard format for a linear programming problem** does not list the assumptions, since **it is really just a format for the symbol definition of the model. If you are solving a real problem, always be sure to specify your assumptions when presenting a model!** Remember that we discussed this issue in the chair and table problem, indicating that there is a reasonable way to interpret fractional solutions (part of a chair means we start it this month and finish it next month), so we don't need to restrict our solutions to be integers. When a variable does have an integer restriction, if the solution value is large, rounding off (up or down, depending on the constraints) usually gives a reasonable solution. However, if the solution value is small, special integer programming techniques are needed, which are beyond the scope of this book.

Graphing the Constraints

Now that we know what is meant by linear programming, how do we solve these problems? Since our sample problem has only two variables in the constraints, we can **graph** the constraints to determine the **feasible region**, the set of all possible feasible solutions to the problem, satisfying *all* of the constraints. We can also represent different values of the objective function graphically, and from examining these and the feasible region, it is possible to find the optimal solution(s).

Since all of the constraints are linear (given that there are two decision variables), they are all determined by straight lines. Inequality constraints (as in our example) will always correspond to a half-plane on one side or the other of the corresponding line; that is, the feasible region will always lie on one side or the other of the line. Remember that Figure 2 showed graphically that the points satisfying each constraint had this property.

Some constraints in an LP problem are easy to graph. If we arbitrarily make c the horizontal axis and t the vertical axis, the graph of the demand constraint $c \leq 11$ is simply all points to the *left* of (or *on*) the vertical line $c = 11$. We can indicate this on our graph by drawing in the vertical line $c = 11$ and drawing in a couple of arrows to the left of it (Figure 4). We can also identify the constraint of which this line is the **edge** by writing in a “3” in parentheses (since this was the 3rd constraint) near the line.

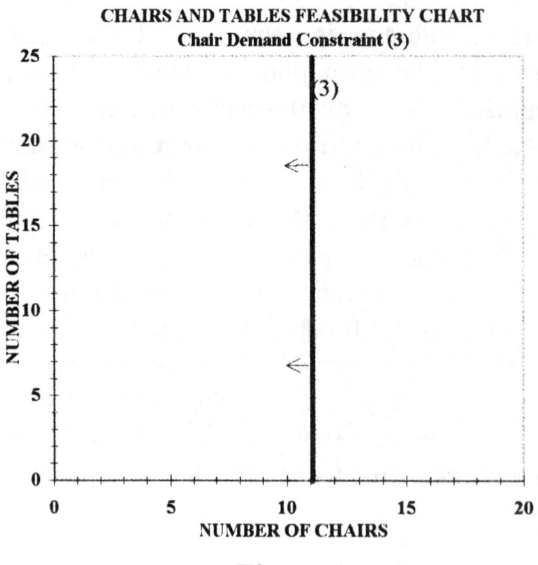

Figure 4

Similarly, the nonnegativity constraint $c \geq 0$ (constraint 4) is simply all points to the right of the t axis (whose equation is $c = 0$), and $t \geq 0$ (constraint 5) is all points above the c axis (whose equation is $t = 0$). We can draw in appropriate arrows for these as well to show which side of the line is feasible, and label them 4 and 5 in parentheses to identify the lines on the graph (Figure 5). Note that these two nonnegativity constraints together restrict us to solutions in the **1st quadrant** (upper right) of the graph. Most LP problems have such nonnegativity constraints on all of the variables.

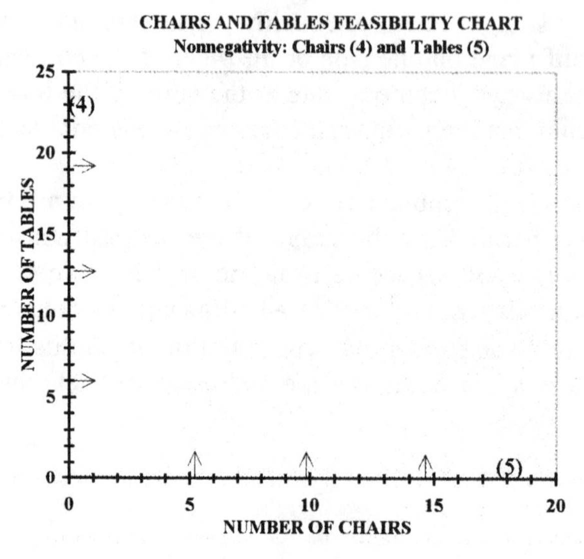

Figure 5

To graph the more complicated constraints, such as our resource constraints, we start by graphing the edge of the constraint. For the wood constraint $3c + 6t \leq 60$ we begin by graphing the edge *line*: $3c + 6t = 60$. An easy way to graph a line is to find the **intercepts** on both axes. To find the intercept for one variable, set the **other** variable equal to 0 and solve for the first variable.

We find the c intercept of this line by setting $t = 0$:

$$\begin{aligned} 3c + 6t &= 60 \\ 3c + 6(0) &= 60 \\ 3c &= 60 \\ c &= 20 \end{aligned}$$

Thus $c = 20$ is the c intercept.

Similarly, to get the t intercept, we set $c = 0$:

$$\begin{aligned} 3c + 6t &= 60 \\ 3(0) + 6t &= 60 \\ 6t &= 60 \\ t &= 10 \end{aligned}$$

Thus the t intercept is $t = 10$.

$t = 60/6 = 10$.[1]

Now, since the intercepts are different points, the graph is simply the line connecting the two intercepts. If both intercepts turn out to be 0 you can simply set one variable equal to some other convenient value (like 1 or 10) and solve for the other variable to get a second point through which the line must go.

How do we know which side of the edge line to shade for our wood constraint? The easiest way is to choose a test point that is not on the line, and see if it satisfies the inequality constraint (plug in the two values and see if they make the inequality true). If your intercepts were not both 0, the easiest test point to use is the origin, (0,0), as is possible for our wood constraint. Plugging in these values, we get

---

[1] If you think about what ended up being the final calculation, you can greatly simplify the process just by looking at the equation. If you want the c intercept, simply cover over the t term (with your finger, or just in your head), and you get $3c = 60$, which you can solve in your head to be $c = 60/3 = 20$. In other words, you get the right-hand-side constant divided by the coefficient of the variable whose intercept you want. For t, that would simply be

$$3(0) + 6(0) \overset{?}{\leq} 60 .$$

Since $0 \leq 60$ is true, that means our test point, the origin, should be included in the shaded region. Therefore, in our example, we shade below (to the left of) the edge line. Often, instead of shading, we simply draw in arrows to indicate the direction that would be shaded, as we did for constraints 3, 4, and 5, and label the constraint with a 1 in parentheses (Figure 6). If the intercepts are both 0, simply choose a different test point that is not on the line, see if it satisfies the inequality, and indicate the appropriate direction *toward* the test point if it satisfied the inequality and in the *other* direction if not.

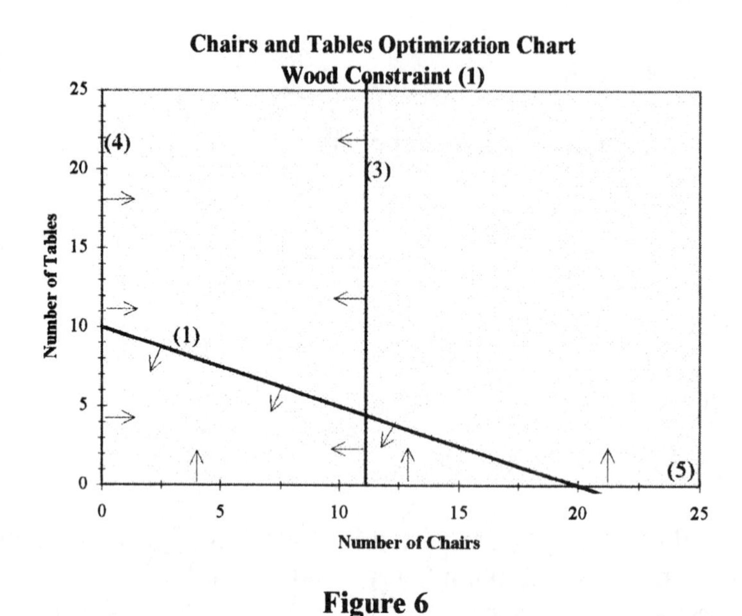

Figure 6

For the hardware constraint, what are the intercepts? The constraint is

$$8c + 4t \leq 100 ,$$

so the edge line is $8c + 4t = 100$. Setting $t = 0$, we get

$$\begin{aligned} 8c + 4(0) &= 100 \\ 8c &= 100 \\ c &= 12.5 , \end{aligned}$$

so the c intercept is $= 12.5$. Similarly, setting $c = 0$ we get:

$$\begin{aligned} 8(0) + 4t &= 100 \\ 4t &= 100 \\ t &= 25\ , \end{aligned}$$

so the t intercept is $t = 25$.

We now draw in the edge line and label it (2). Using the test point (0,0) again, we get $8(0) + 4(0) \leq? \ 100$, which is true, so we draw in arrows toward the origin, down and to the left (Figure 7).

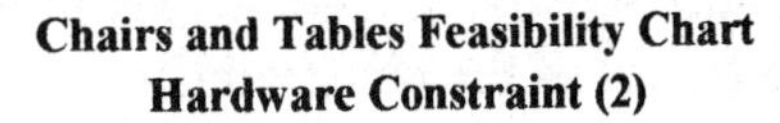

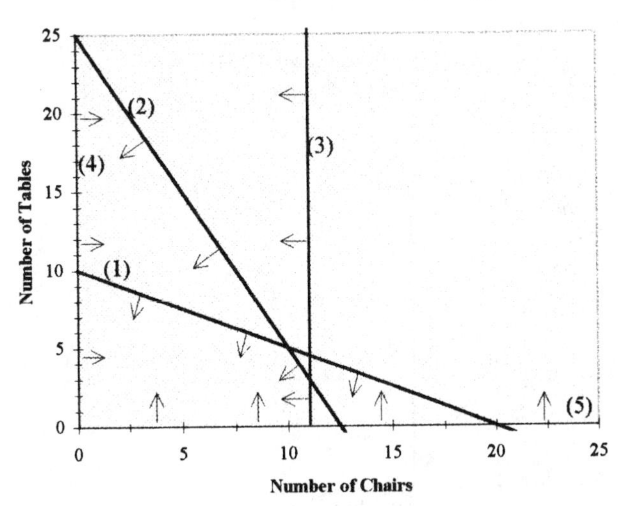

Figure 7

Graphing the Feasible Region

We have now graphed all of the constraints for our problem. The feasible region is the **intersection** of all of these graphs. These are the points that are feasible for *all* of the constraints, where the separate feasible areas overlap. This region is normally a single polygon (an irregular pentagon in our example) and is always **convex**, meaning that it has no dents in it; that is, the line segment joining any two points in the region stays completely within the region. (Figure 8)

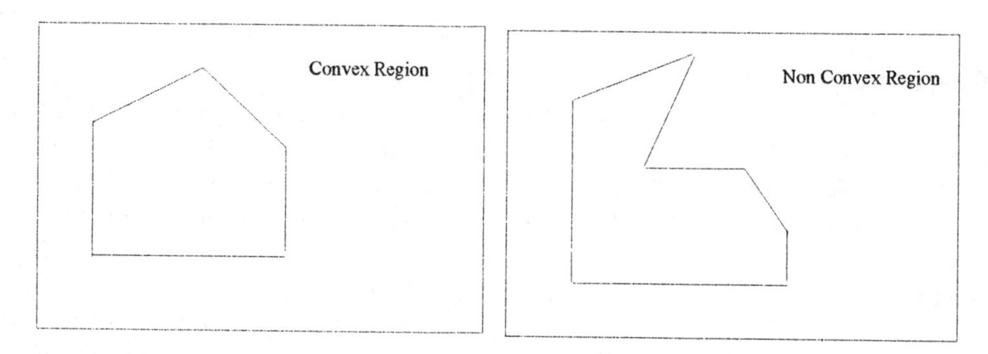

Figure 8

Usually, by looking at all the constraint edges and arrows, you can see what region this will be (Figure 9). If that is difficult, you can shade each inequality in a different way (vertical stripes, horizontal stripes, angled stripes in either direction, etc.), and look for the region that is shaded for *all* of the constraints.

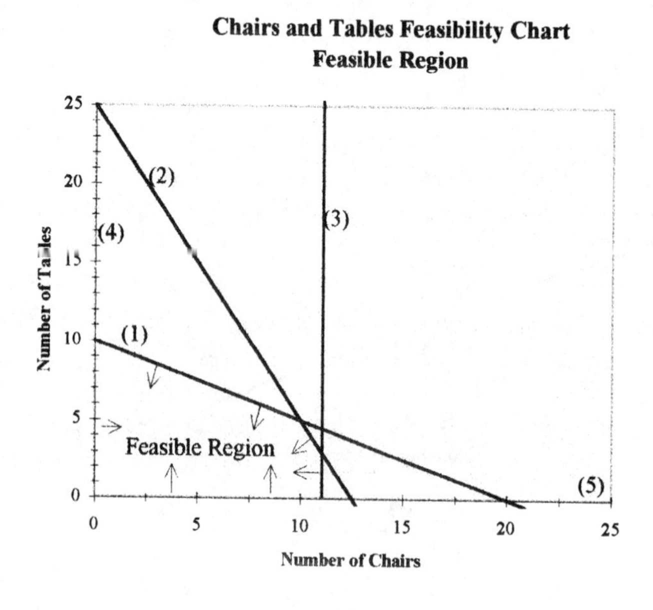

Figure 9

Compare the graph in Figure 9 to the graph of individual points in Figure 2 to see how they correspond to each other.

Graphing the Objective Function

Now we have a graphical representation of all the feasible solutions to our problem. How do we decide which one(s) are optimal? If we had a small number of solutions, we could just plug their values into the objective function, and choose the best (highest or lowest, depending on whether we were maximizing or minimizing). But a polygon region has an infinite number of solutions. We got a feel for this in Figure 2, which showed possible solution points that satisfied all of the constraints. The main reason there are so many possible solutions is the divisibility assumption; because all real numbers are possible, this would even be true for a line segment! We need another strategy. Note that our objective function is a function of two variables, but its value (**P** in our example) is a third variable. To fully graph this, we would need a third axis (coming straight up out of the page, from the origin). Normally in linear programming, we do not include a constant term in the objective function. If there is one, we can just add it to our final objective function value. Since the objective function is linear, it corresponds to a plane passing through the origin (if both decision variables are 0 and there is no constant term, the objective function value must be 0 also). Normally this plane will be slanted, and will lie above the feasible region.

For any particular solution in the feasible region, if we extend a line vertically (in the direction of the objective function variable) to the objective function plane, the height of the point where it hits the plane is the value of the objective function at that point. If we do this for the whole feasible region, the resulting points on the slanted plane will be a distorted image of the feasible region, as if we put a light source under the feasible region and **projected** the shape onto the slanted plane (like a screen for showing slides, but tilted at an angle). We can then think of our problem as finding the highest (or lowest) point of the projected feasible region on the slanted plane.

Figure 10 shows what this looks like for our furniture problem. The vertical axis is the P axis. The lower axis shown (in the foreground) is the c axis. The axis in between (in the background) is the t axis.

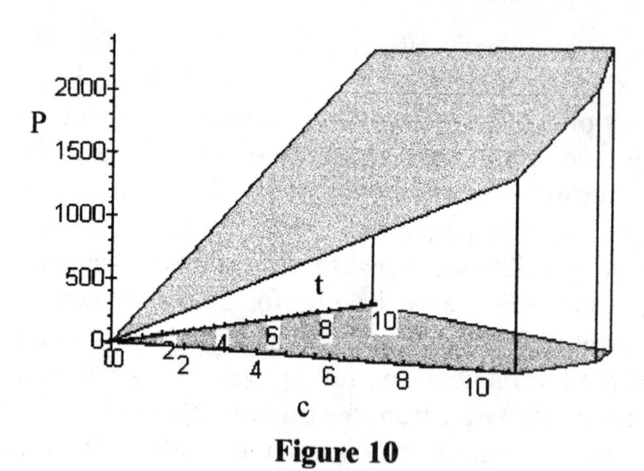

Figure 10

Another way to think of this problem is to picture a field that is flat but slanted, with a fenced-in pasture on it (the projected feasible region). We want to find the point within the fenced-in pasture that is highest (or lowest).

One way to represent this third dimension on a two-dimensional graph is to do what topographical maps do: use contour lines connecting all points at the same altitude. Since the objective function is a plane, these contour lines will simply be parallel straight lines. Each one will correspond to the graph of the line given by a specific objective function value. In our P example, we have $P = 140c + 200t$. If we choose $P = 1400$, we get the equation $1400 = 140c + 200t$. We can graph this by finding the intercepts as discussed earlier ($c = 10$ and $t = 7$). We will draw it on the graph with a dotted line, which we can label with the objective function value, $P = 1400$ (Figure 11). We chose 1400 for the value of P because it is the smallest whole number that both 140 and 200 divide into evenly (the lowest common denominator), but any number could be used.

Once we have one contour line, the others are all parallel. This is because we can think of the objective function as a line of the form $y = mx + b$. In our case the objective function is: $P = 140c + 200t$. Since we have been graphing the lines with the chairs on the horizontal (x) axis, we can rewrite the objective function in the general form of a linear function: $200t = -140c + P$. Solving for t we get $t = (-7/10)c + P/200$. No matter what value P takes, the slope of this line is always -7/10. Any line that represents any value of the profit P will have this same slope. An easy one to draw in is the one through the origin, which will have an objective function value of 0 (Figure 11).

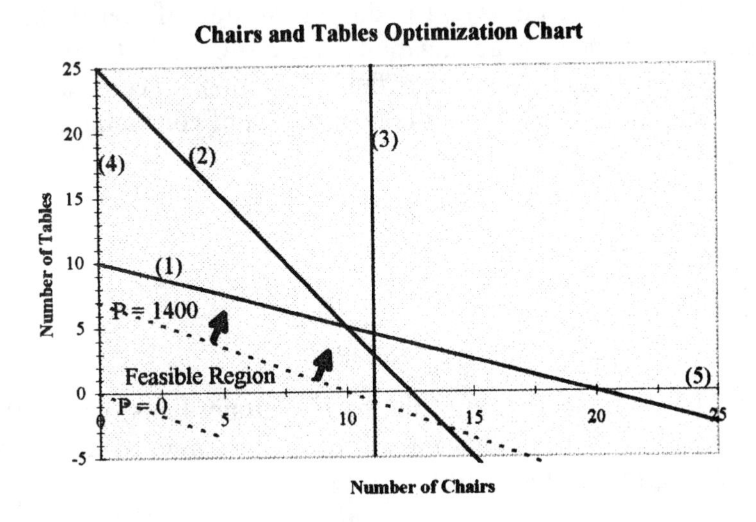

Figure 11

Once we have graphed two contour lines, we can see which direction gives us the better solutions: in the direction of the higher value if we are maximizing, and the other direction if we are minimizing. We will indicate this direction on the better contour line (away from the other) with a *thick* arrow (to distinguish it from a constraint direction), and can even mark "Better direction" beside it for clarity. Now, if we picture *all* of the possible contour lines that pass over the feasible region, we want to pick **the one that is best**, if there is one. As can be seen in Figure 11, this **will always occur at a point (or points) on the *edge* of the feasible region**. This also means that **there will always be at least one *corner point* that is optimal. If more than one point is optimal, it would have to be an entire edge of the feasible region. This will occur if the contour lines are *parallel* to a constraint that is on the good side of the feasible region. In this case there would usually be two optimal corner points.**

The above observation gives us another way to determine the optimal solution(s). If we don't trust our graphing ability or eyesight, we could simply test all the corner points by plugging them into the objective function, and seeing which is best. As an in-between strategy, we could only do this for the corner points that visually seem possibly optimal.

Corner Points

How do we find corner points? Notice that every corner point is the intersection of at least two constraint edges. Thus, to find a particular corner point, we can simply choose two constraints whose edges pass through it and solve the two edge equations simultaneously. For example, the origin is the intersection of constraints 4 and 5, the nonnegativity constraints whose edge equations are $c = 0$ and $t = 0$. As can be seen from the graph, the other corner points are at the intersections of constraint edges 1 and 4, 1 and 2, 2 and 3, and 3 and 5. To find the intersection of constraint edges 1 and 4, the edge equations are

$$3c + 6t = 60 \text{ and } c = 0:$$
$$3(0) + 6t = 60$$
$$6t = 60$$
$$t = 10$$

Thus the corner point (the intersection of the two constraint lines) is $c = 0$ and $t = 10$. Similarly, the intersection of constraints 3 and 5 yields the corner point $t = 0$ and $c = 11$. For constraints 2 and 3, the equations are

$$8c + 4t = 100 \text{ and } c = 11,$$
$$8(11) + 4t = 100$$
$$4t = 100 - 8$$
$$4t = 12$$
$$t = 3$$

so the solution is $c = 11$ and $t = 3$.

The last intersection, of constraints 1 and 2, yields the equations

$$3c + 6t = 60 \text{ and } 8c + 4t = 100.$$

We already studied this system in Section 7.3 (Sample Problem 4), and it could be solved by any of the methods we discussed earlier (substitution, elimination, Gaussian elimination using augmented matrices, $\mathbf{X} = \mathbf{A}^{-1}\mathbf{B}$, etc.). As we saw then, the solution is $c = 10$ and $t = 5$.

Now we know our corner points are (0,0), (0,10), (10,5), (11,3), and (11,0). If we plot $P = 1400$ and $P = 0$ as discussed earlier (Figure 11), we see that the contour lines that are higher and to the right are better. Looking at the graph, it appears that (10,5) may be the optimal solution, but it is difficult to be sure. To check this, let's calculate the P values for all of the corner points:

Corner (c,t)	$P(c,t)$ $140c+200t$	P
(0,0)	140(0) + 200(0)	0
(0,10)	140(0) + 200(10)	2000
(10,5)	140(10)+ 200(5)	2400
(11,3)	140(11) + 200(3)	2140
(11,0)	140(11) + 200(0)	1540

We were right! In fact, the unique optimal solution is (10,5).

Translation: Given the resource and demand constraints for the coming month, we should make 10 chairs and 5 tables to maximize our profit at $2400. □

Sample Problem 2: You have two breakfast cereals that you like, Special K™ and Raisin Bran. You have just set up a dietary plan for your consumption of fat, fiber, protein, and calories. You want to choose your breakfast cereal so you minimize the cost while satisfying your dietary goals. You check the prices and determine that one ounce of Special K costs 40 cents and an ounce of Raisin Bran costs 30 cents. Suppose that one ounce of Special K contains 0 grams of fat, 1 gram of dietary fiber, 6 grams of protein and 150 calories. (This includes ½ cup of skim milk. Eating dry cereal is not very appealing.) One ounce of Raisin Bran contains 1 gram of fat, 7 grams of dietary fiber, 4 grams of protein and 210 calories (this also includes ½ cup of skim milk). You have decided that you want to get at most 2 grams of fat, at least 5 grams of fiber, at most 350 calories, and exactly 9 grams of protein at breakfast. What combination of the two cereals meets your specifications at the lowest cost?

Solution: Let's display the information in a more organized form, as in Table 2.

Table 2

	Special K	**Raisin Bran**	**Comments**
Cost (cents per ounce)	40	30	Minimize
Fat (grams per ounce)	0	1	at most 2 grams
Fiber (grams per ounce)	1	7	at least 5 grams
Calories (kCal per ounce)	150	210	at most 350 kCal
Protein (grams per ounce)	6	4	exactly 9 grams

As we did for similar problems in Section 5.2 , we can formulate a model for this problem as follows:

Verbal Definition: $C(K,R)$ = the cost, in cents, for K ounces of Special K and R ounces of Raisin Bran

Symbol Definition: $C(K,R) = 40K + 30R$, for $R \leq 2$

$$K + 7R \geq 5$$
$$150K + 210R \leq 350$$
$$6K + 4R = 9$$
$$K \geq 0$$
$$R \geq 0$$

Assumptions: Certainty and divisibility. Certainty implies prices stay stable and nutritional goals stay the same. Divisibility implies fractions of ounces are possible.

Putting this into linear programming form, we get:

K = oz. of Special K
R = oz. of Raisin Bran
C = cost, in cents

Minimize $C = 40K + 30R$
subject to
(1) $R \leq 2$ (fat constraint)
(2) $K + 7R \geq 5$ (fiber constraint)
(3) $150K + 210R \leq 350$ (calorie constraint)
(4) $6K + 4R = 9$ (protein constraint)
(5) $K \geq 0$ (nonnegativity constraint)
(6) $R \geq 0$ (nonnegativity constraint)

In streamlined standard LP format, this would be:

Min. $C = 40K + 30R$
s. t.
$R \leq 2$
$K + 7R \geq 5$
$150K + 210R \leq 350$
$6K + 4R = 9$
$K,R \geq 0$

Figure 12 shows constraint 1, the fat constraint, $R \leq 2$. This is quite easy to draw because its edge is the horizontal line R = 2. The point (0,0) satisfies the constraint, $0 \leq 2$, so the arrows point toward (0,0).

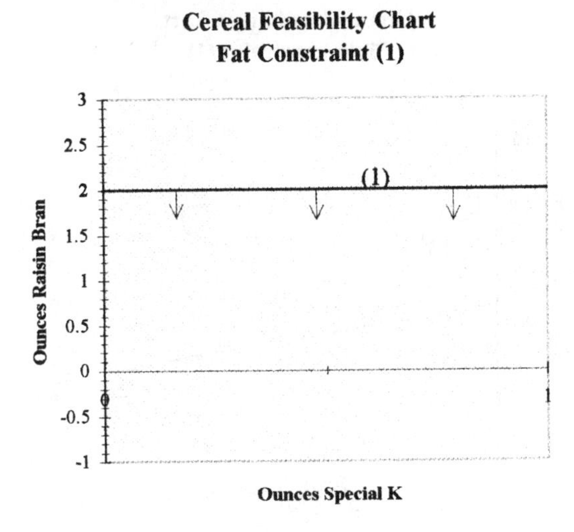

Figure 12

Figure 13 shows constraint 2, the fiber constraint, $K + 7R \geq 5$. Again we find the intercepts so that we can draw the line that represents the fiber constraint:

$$\begin{aligned} K + 7R &= 5 \\ 0 + 7R &= 5 \\ 7R &= 5 \\ R &= 5/7 \end{aligned}$$

$$\begin{aligned} K + 7(0) &= 5 \\ K + 0 &= 5 \\ K &= 5 \end{aligned}$$

Since the point (0,0) does not satisfy the constraint $(0 + 7(0) < 5)$, we draw our arrows pointing *away* from (0,0).

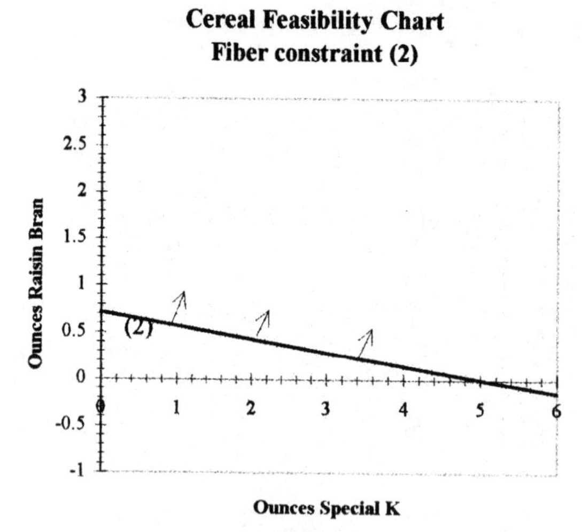

Figure 13

Figure 14 shows constraint 3, the calorie constraint: $150K + 210R \leq 350$. The intercepts are given by

$$\begin{aligned}150K + 210R &= 350\\ 150(0) + 210R &= 350\\ 210R &= 350\\ R &= 350/210 = 5/3\end{aligned}$$

$$\begin{aligned}150K + 210(0) &= 350\\ 150K &= 350\\ K &= 350/150 = 7/3\end{aligned}$$

The point (0,0) satisfies the constraint, $150(0) + 210(0) \leq 350$, so the arrows point toward (0,0)

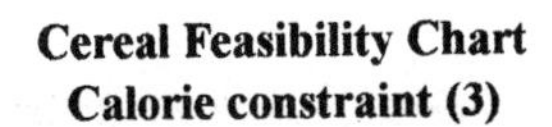

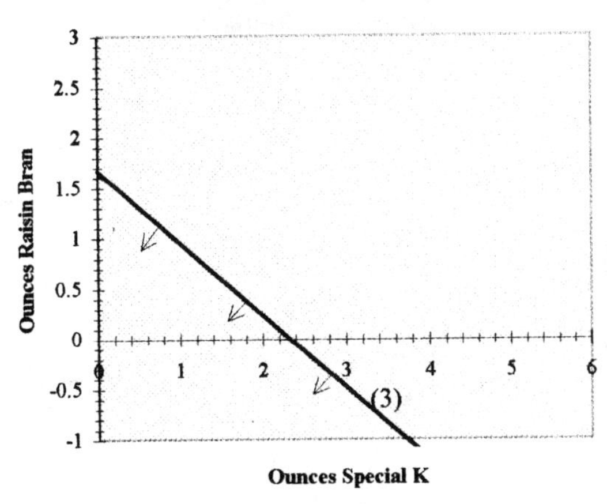

Figure 14

Figure 15 shows constraint 4, the protein constraint: $6K + 4R = 9$. The intercepts are:

$$6K + 4R = 9$$
$$6(0) + 4K = 9$$
$$4K = 9$$
$$K = 9/4$$

$$6K + 4(0) = 9$$
$$6K = 9$$
$$K = 9/6 = 1\ 1/2$$

This is an equality constraint, so only points on the line itself satisfy this constraint, and we do not add any arrows on the graph.

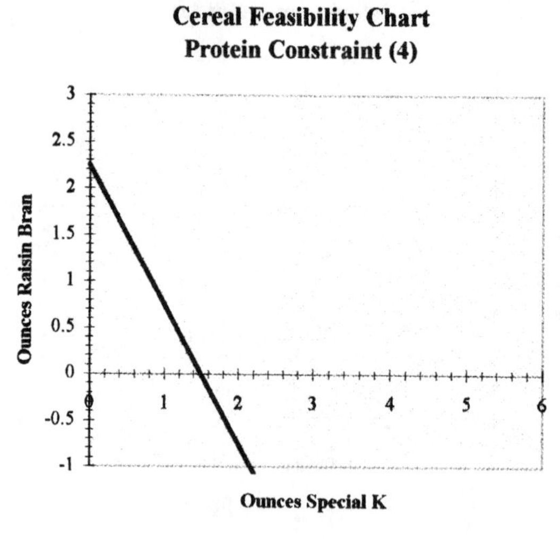

Figure 15

Figure 16 shows all of the constraints, including the nonnegativity constraints. We followed the procedures described above to graph this problem, remembering that the equality constraint (constraint 4, the protein constraint) means that our feasible region will be a line segment only.

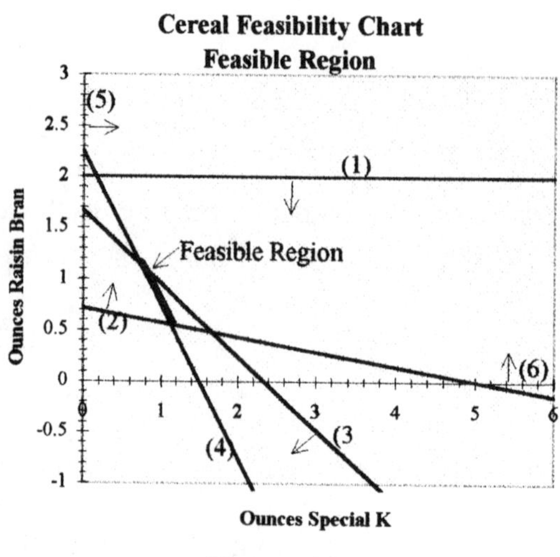

Figure 16

In Figure 17 we add lines that represent the objective function, $C = 40K + 30R$, the cost of the cereals, for the cost values of 90 cents and 45 cents. At 90 cents, the R intercept is 3 and the K intercept is $90/40 = 2.25$. At 45 cents, the intercepts are simply cut in half ($K = 1.125$ and $R = 1.5$). Note that we are minimizing this time, so **smaller** values of C are preferable.

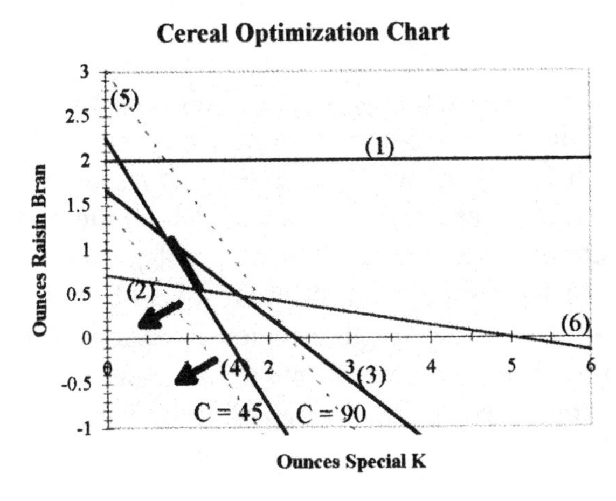

Figure 17

From looking at the graph, it appears that the optimal solution is at the intersection of constraints 2 and 4,

$$(2)\ K + 7R \geq 5\ \text{(fiber constraint) and (4)}\ 6K + 4R = 9\ \text{(protein constraint).}$$

Solving the system of linear equations simultaneously, we find that $K = 43/38 \approx 1.13$, and $R = 21/38 \approx 0.55$, at a cost of $C = 2350/38 \approx 61.84$. However, there is only one other corner point, so we might as well check it, too. It is at the intersection of constraints 3 and 4,

$$(3)\ 150K + 210R \leq 350\ \text{(calorie constraint) and (4)}\ 6K + 4R = 9\ \text{(protein constraint).}$$

Solving this system of linear equations gives $R = 25/22 \approx 1.14$ and $K = 49/66 \approx 0.76$, at a cost of $C = 745/11 \approx 67.73$.

The first solution we found was indeed the unique optimal solution.

Translation: We should eat, on average, about 1.13 oz. of Special K and 0.55 oz. of Raisin Bran every morning, for a minimum cost of about 62¢. If a bowl holds approximately one ounce of cereal, we could mix a little more than a bowl of Special K with a little more than half of a bowl of Raisin Bran, if we like the taste of them together. Or we could eat the two parts of the solution in separate bowls every morning. Or we could work out some system of alternating between the two cereals, as long as the averages per day come close to the above solution. For example, we could eat about 2¼ bowls of Special K every other day, alternating with a little more than 1 bowl of Raisin Bran on the other days. □

If, when we have drawn all of our constraint lines, there is no region that satisfies all of the constraints, then there are no feasible solutions to the problem. If any of your constraints are equalities, then the graph of the constraint is the edge line itself, so you would not draw any direction arrows, and the feasible region would have to be a subset (a part) of that line. In such a case, you can simply darken the part of that line that satisfies all of the other constraints (if there are any such points). Similarly, if two constraints had the same edge line but were shaded on different sides of the line, the intersection would be the line itself. Sometimes the feasible region is **unbounded** (extends to infinity in at least one direction), such as if we were maximizing and all of our example **structural constraints** (all constraints except the nonnegativity constraints) had all been $\geq$ instead of $\leq$ constraints.

Section Summary

Before you proceed to the exercises, make sure that you:

- Know how to recognize when a problem is a linear program, and know how to express any linear programming problem in the standard LP format.
- Understand that a convex set is one with "no dents" in it, where any line segment joining two points in the set is contained completely within the set, and that an LP always has a convex feasible region.
- Know that a maximization LP with two decision variables corresponds graphically to looking for the highest point on the plane of the objective function within the image of the convex feasible region vertically projected onto that plane. This is analogous to looking for the high point of a fenced-in pasture on a flat slanted field.
- Know how to graph both inequality and equation constraints in two variables, including finding intercepts (plug in 0 for the other variable) and knowing which side of an inequality to shade (plug in the origin or another point not on the edge).
- Know how to graph contour lines for the objective function of an LP with two decision variables (pick a sample value with easy intercepts, then plot a parallel line through the origin).

- Know how to determine which direction corresponds to better values of the objective function (depending on whether you are maximizing or minimizing) and how to mark it with dark arrows.
- Know how to determine visually the optimal corner point of an LP with two decision variables (or at least determine a small number of candidates).
- Know how to find the corner point at the intersection of the edges of any two constraints of an LP with two decision variables (solve the system of two linear equations corresponding to the edges by elimination, substitution, matrices, etc.).
- Know how to evaluate the objective function value of an LP at a corner point, and to decide which candidate solution(s) are in fact optimal.

EXERCISES FOR SECTION 9.2:

Warm Up

Solve the following linear programs graphically:

1. Maximize $z = 3x + 4y$
   subject to $x + y \leq 20$
   $5x + 8y \leq 120$
   $x, y \geq 0$

2. Max. $P = 5x_1 + 4x_2$
   s.t. $3x_1 + 4x_2 \leq 24$
   $10x_1 + 8x_2 \leq 62$
   $x_1, x_2 \geq 0$

3. Max. $z = 5x + 7y$
   s.t. $3x + 5y \leq 60$
   $6x + 4y \leq 90$
   $y \leq 5$
   $x \geq 0$

4. Min. $z = 2x + 3y$
   s.t. $3x - y \geq 12$
   $x + y \geq 8$
   $y \leq 6$
   $x, y \geq 0$

5. Min. $z = 2x + 3y$
   s.t. $3x - y \geq 12$
   $x + y \geq 8$
   $y \leq 6$
   $x \leq 10$

6. Min. $z = 2x + 3y$
   s.t. $3x - y = 12$
   $x + y \geq 8$
   $y \leq 6$
   $x, y \geq 0$

7. Max. $z = 5x + 7y$

$$\begin{aligned} \text{s.t.} \quad & x - 4y \leq 0 \\ & x - 3y \geq 12 \\ & x, y \geq 0 \end{aligned}$$

8. Max. $z = 6x + 10y$

$$\begin{aligned} \text{s.t.} \quad & 3x + 5y \leq 60 \\ & 6x + 4y \leq 90 \\ & y \leq 5 \\ & x, y \geq 0 \end{aligned}$$

Game Time

9. Graphically solve a modified version of the furniture problem, if the profit from each chair is \$100 rather than \$140:

$$\begin{aligned} \text{Max } P &= 100c + 200t \\ \text{s.t.} \quad 3c + 6t &\leq 60 \\ 8c + 4t &\leq 100 \\ c &\leq 11 \\ c, t &\geq 0 \end{aligned}$$

10. Graphically solve a different modified version of the furniture problem, if the demand limit on the number of chairs that can be sold drops from 11 to 10 :

$$\begin{aligned} \text{Max } P &= 140c + 200t \\ \text{s.t.} \quad 3c + 6t &\leq 60 \\ 8c + 4t &\leq 100 \\ c &\leq 10 \\ c, t &\geq 0 \end{aligned}$$

11. Graphically solve the furniture problem if both of the modifications from exercises 9 and 10 are made at the same time:

$$\begin{aligned} \text{Max } P &= 100c + 200t \\ \text{s.t.} \quad 3c + 6t &\leq 60 \\ 8c + 4t &\leq 100 \\ c &\leq 10 \\ c, t &\geq 0 \end{aligned}$$

12. Suppose you are trying to raise money for your sports team by selling T-shirts and hats. You can order both items from the same company, which will charge you \$4 for each hat and \$6 for each shirt, plus there is a fixed cost of \$30 for any order. Your team

has $100 to spend for everything, and has decided to sell the shirts for $12 and the hats for $7, at which prices you think you can sell no more than 15 hats and 10 shirts. Assume that both items are "one size fits all" and that the demands for the two products are independent for this exercise. Graphically find the optimal solution to this problem. What is your team's optimal profit? What would you do (how many of each item would you order) in this situation?

13. Suppose you run a small candle-making business, making regular and deluxe candles. You can sell the regular candles for a profit of $4 each and the deluxe candles for a profit of $7 each. Regular candles use 6 oz. of wax and 5 inches of wick material, while deluxe candles use 9 oz. of wax and 6 inches of wick material. You have 120 oz. of wax and 110 inches of wick material in your studio at the moment, and will not get any more supplies until tomorrow. Graphically determine how many candles of each type you should make today. What assumption is violated here? How could you deal with it?

14. You are trying to decide between two dog foods for your Golden Retriever, Ralph. Purina™ costs $32 for a 10-pound bag, and each ounce of food has 4 grams of fat, 8 grams of protein, and 200 calories. Iams™ costs $20 for a 5-pound bag, and each ounce of food has 1 gram of fat, 6 grams of protein, and 220 calories. You want Ralph to get at least 1000 calories per day, but not more than 1400 calories. You also want him to get no more than 10 grams of fat and at least 40 grams of protein. What combination of the two dog foods should you give Ralph every day?

Section 9.3: The Simplex Method

We have seen how we can solve a linear programming problem with two decision variables graphically. Technically, we could also solve problems with three decision variables graphically, but in most cases it is not practical. For problems with four or more decision variables, graphing is impossible, so we need another solution method. In this section we discuss a method called the **simplex method** that is extremely efficient and widely used. We start off our discussion by looking again at our hand-crafted furniture example from Chapters 7 and 8, taking advantage of being able to represent the problem graphically to help understand what the simplex method is all about.

Here are some examples of the kind of linear programming problems you should be able to solve using the Simplex Method by the end of this section:

- Consider again the example of the hand-crafted furniture business from previous chapters. You own this small business, and for now you make and sell only two products: chairs and tables (just a single style of each). Given restrictions on the raw materials available and on demand, how many chairs and tables should you make this month?
- You are trying to decide what combination of foods you should normally eat for breakfast, to meet your nutritional goals at the lowest possible cost. What foods (and how much of each) should you plan for your breakfasts over the course of a week?

In addition to being able to solve the above kinds of problems, by the end of this section you should:

- Understand the basic idea of the simplex method: when it is used and its underlying geometric (graphical) logic.
- Understand how to set up and use the tabular (matrix) simplex method to solve linear programming problems (LP's) that are written in "Maximize" form, and whose constraints are nonnegativity constraints, possibly with constraints written in "≤" form that only have nonnegative right-hand-side constants (RHS's).
- Understand what the major steps of the simplex method correspond to graphically, algebraically, and with respect to matrix theory.
- Understand how to interpret a simplex method tableau (matrix), including the values of all the variables at the corner point it denotes.

- Understand what a corner point corresponds to with regard to the values of the variables in an LP.

Example 1: Consider again the example of the hand-crafted furniture business from previous chapters and from Sample Problem 1 of Section 8.1 . You own this small business, and for now you make and sell only two products: chairs and tables (just a single style of each). Chairs each use 3 units of wood and 8 units of hardware, tables each use 6 units of wood and 4 units of hardware. You have on hand 60 units of wood and 100 units of hardware. You will make a profit of $140 on each chair and $200 on each table, and that at the selling price you have chosen, you expect the demand for chairs to be at most 11, but that the demand for tables should be sufficiently large that it will not impose any additional restrictions on the variables (the resources will be more restrictive). You are planning for the next week, and you will not have time to order and receive any new resources before the end of the week. How many chairs and tables should you make this week?

To give a simple analogy of how the simplex method works, remember the metaphor we gave earlier for two-variable linear programming: it can be thought of as trying to find the highest point within a fenced-in pasture on a flat but slanted field (a slanted plane). To understand the Simplex Method, let's assume also that we are trying to find the highest point in the pasture in the midst of a heavy fog. The simplex method uses the insight we saw in the last section: if an optimal solution exists, then there must be at least one corner point of the feasible region that is optimal. Therefore, we can restrict our search for an optimal solution to just corner points. In our pasture analogy, that means we can choose some corner point (place where the fence forms an angle) to start, see if it is optimal, and if not, move along the edge of the pasture (along the fence) to a better corner point. We then repeat this process until we find an optimal corner point.

How do we know if a particular corner is optimal in this fenced-in pasture maximization? We simply check the slant as we move in the two fence directions coming out from the corner: if both directions are slanting *down* in elevation (or if neither is slanting strictly up; that is to say, it could be level in either or both directions), then the corner point is optimal, and we can stop. In other words, we are optimal if, as we move along the fence in *any* direction away from the corner, our altitude does *not* strictly increase, such as at (10,5) in Figure 1.

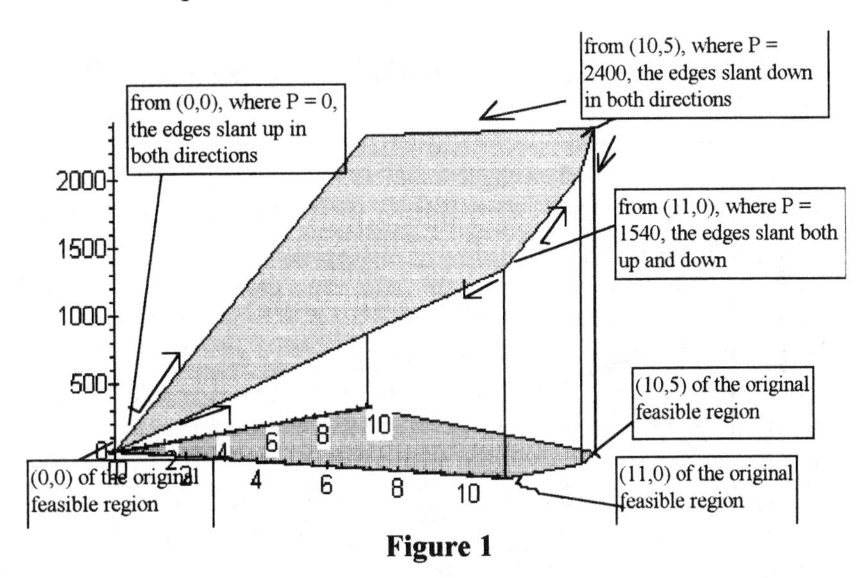

Figure 1

If either direction does slant *up* (strictly), then we can find a better (higher) corner point, so the current corner is not optimal, such as at (0,0). In fact, at (0,0), moving in either direction slants up. How do we decide which direction to follow? If only *one* direction goes up, such as at (11,0), clearly that is the one we want. If *both* directions go up (such as from (0,0)), we choose the one that slants up at the *steepest* angle (improves at the fastest rate). This rule makes even more sense if you remember the fog, since we can't see *where* each direction will take us.

The simplex method makes intuitive sense in the above example. It is guaranteed to find an optimal solution in a finite number of steps, partly because of our assumptions that the field is flat (the objective function is linear) and the feasible region is **convex** (that is, has *no dents*, as discussed in Section 8.1, with reference to Figure 7). Otherwise, the fence could snake up and down the hill several times, so we wouldn't know for sure if we were at an optimal corner. Another comment before we leave this example is to recognize that there will be another optimal corner if (and only if) one fence direction from an optimal corner is level (goes neither up nor down); the next corner that *that* direction leads to is another (alternative) optimal corner. Of course, *all* of the points along that fence (edge) are also optimal then. The example shown below is the same as our original chairs and tables example, except that the profit (objective function) is given by $P_2 = 100c + 200t$ (parallel to constraint 1), so the two corner points (0,10) and (10,5) are both optimal, as are all the points on the line segment between them (Figure 2).

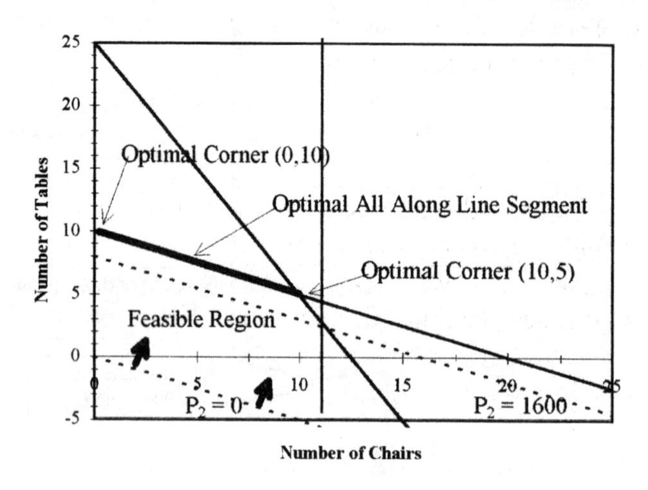

Figure 2

How can we solve our problem in Example 1 by the simplex method? We will learn a matrix form of the simplex method, called the **tabular simplex method**. To use it, we need to rewrite our problem in a slightly different form. First, we rewrite the objective function so that all the *variables* are on the *left*, and the *constant term* (number) is on the *right hand side* (RHS) of the equation. In our example the objective function was

$$P = 140c + 200t \ .$$

Since our profit will vary with the number of chairs and tables, P, c, and t are *all* variables and must appear on the left hand side of the equation, so we rewrite it in the form

$$P - 140c - 200t = 0 \ .$$

Next, we need to rewrite each constraint as an *equation* rather than an *inequality*. Let's look first at our demand constraint, $c \leq 11$. If we make only 5 chairs, we would have $5 \leq 11$, which is, of course, true. But we want the constraint written as an equation, and $5 \neq 11$ $(5 < 11)$, so we must *add* a variable to take up the slack. This variable is called a **slack variable**. Since this is the third constraint, we will denote the slack variable as s_3. We define

$$s_3 = 11 - c$$

(the right-hand side minus the left-hand side of the constraint). So when $c = 5$,

$$s_3 = 11 - c = 11 - 5 = 6$$

so the $c \leq 11$ constraint is satisfied and $s_3 \geq 0$, like all of the original decision variables. In general, whenever $c \leq 11$, s_3 will be nonnegative, and vice-versa. Figure 3 shows how $s_3 \geq 0$ corresponds to being on the feasible side of constraint (3). Rewriting $s_3 = 11 - c$ with the *variables* on the *left* and the *constants* on the *right*, we get

$$c + s_3 = 11 \ .$$

Similarly we add a different **slack variable** to the LHS of each **structural constraint** (the constraints that are not nonnegativity constraints), which we call s_i for the *i*'th constraint. We can define s_1 and s_2 in the same way, so they will also be nonnegative when the corresponding constraint is satisfied. The final form for the resulting equations is

$$3c + 6t + s_1 = 60 \text{ and } 8c + 4t + s_2 = 100 \ .$$

Notice that when any slack variable is equal to 0 at a point, its corresponding constraint holds as an equation (we say the constraint is **binding**) at the point (for example, $c \leq 11$ at (11,0)). This is because the left-hand side (LHS) is equal to the right-hand side (RHS), which graphically means that the point is on the **edge** of the constraint.

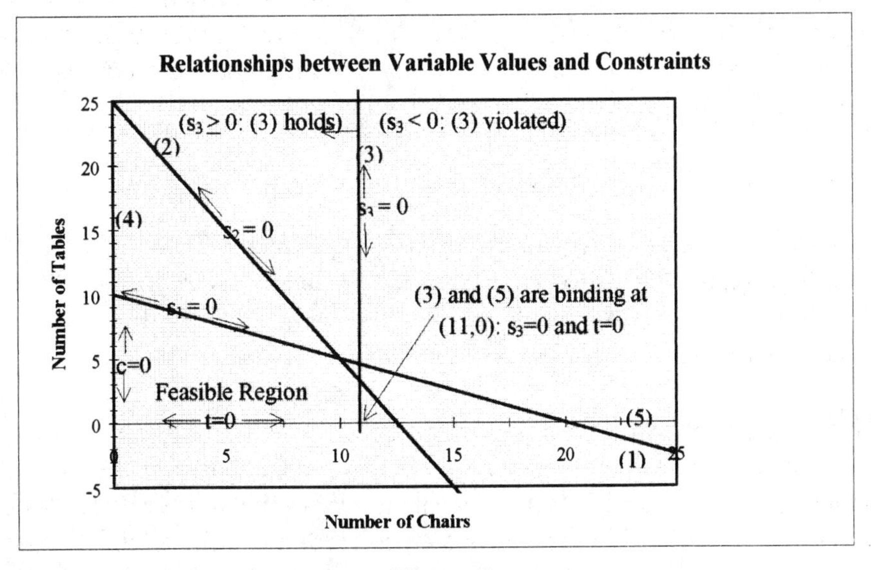

Figure 3

Now we can rewrite our problem in the following form:

Maximize *P*, where $P - 140c - 200t = 0$

subject to
$$3c + 6t + s_1 = 60$$
$$8c + 4t + s_2 = 100$$

$$c \qquad + s_3 = 11$$
$$c, t, s_1, s_2, s_3 \geq 0$$

P = profit in dollars
c = number of chairs
t = number of table
s_1 = wood slack
s_2= hardware slack
s_3 = chair slack

In the simplex method, we always assume that all of the decision variables are nonnegative, so in the matrices we will not mention those constraints. Furthermore, there is no standard way to indicate whether the problem is being maximized or minimized in the matrices, so we just need to keep that in mind as we solve the problem.

In this book, we will only study the tabular simplex method for one special case of LP problems: maximization problems, with all constraints of the $\leq$ type, all having right-hand side values that are ≥ 0, and all variables being nonnegative. It is possible to convert any problem into the form of a maximization problem with all nonnegative variables and all structural constraints in $\leq$ form [1]. However, even after doing all of that, there is no way to guarantee that the right-hand sides (RHS's) of all the $\leq$ constraints will be nonnegative. Techniques for handling other cases are discussed in more advanced books on Operations Management and Operations Research. The basic concepts of the simplex method are all contained in this case, however.

As we said in the pasture analogy, **the simplex method starts at a corner, tests it for optimality, and if it is not optimal, moves to a better adjacent corner along an edge of the feasible region, then repeats these steps until an optimal solution is found**.

The Initial Corner-Point Solution

Because of the special case we are considering, the origin (all decision variables equal to 0) will always be a feasible corner point (the intersection of the original nonnegativity constraints), since the RHS's are all nonnegative. As that is an easy feasible corner for a starting point, it is where we will always start. If we write the equations in their rewritten versions in matrix form (as we did in Chapter 6), we get Table 1.

[1] To convert from a minimization problem to a maximization problem, just negate the objective function; to change $\geq$ constraints to $\leq$, just multiply through both sides by -1; to change = constraints to $\leq$ form, rewrite each as both a $\leq$ and $\geq$ constraint, then negate the $\geq$ one; and to change variables that can be negative to nonnegative form, rewrite them as a difference of two nonnegative variables.

Table 1

Basic Vars	P	c	t	s_1	s_2	s_3	RHS
	1	-140	-200	0	0	0	0
	0	3	6	1	0	0	60
	0	8	4	0	1	0	100
	0	1	0	0	0	1	11

In this case, the origin means the corner point (0,0) ($c = 0$ and $t = 0$), which is the intersection of constraints 4 and 5 (the nonnegativity constraints). Notice that since $c = 0$ and $t = 0$, the columns for c and t can be temporarily ignored, and the resulting matrix is an identity matrix augmented by the RHS's. This is the final form we worked toward in Chapter 7 because the solution can be immediately read from the matrix (just cover the c and t columns with your finger). In this case, that means $P = 0$, $s_1 = 60$, $s_2 = 100$, and $s_3 = 11$ (the same solution you'd get from the equations if you plugged in $c = 0$ and $t = 0$).

The Simplex Method: Initial Solution

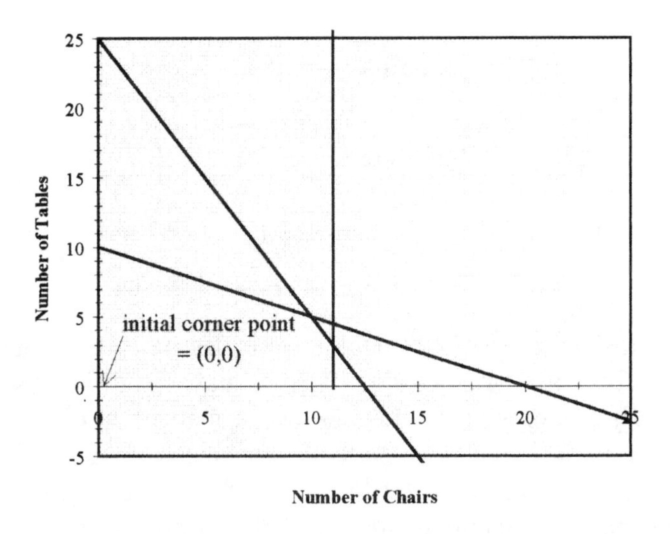

Figure 4

Let's now take a minute to explore the concept of a corner point solution. In a problem with two original decision variables (a two-dimensional problem), a corner point is the intersection of the edges of two constraints. Actually, it is the intersection of the original form of those constraints, but with any inequality signs replaced by equal signs. This means that two constraints are **binding** at such a point (for example, constraints one and two are binding at (10,5)). We said earlier that **a structural constraint being binding corresponds**

to a slack variable being 0. A nonnegativity constraint, on the other hand, **is binding when the corresponding decision variable equals 0** (for example $c \geq 0$ being binding means $c = 0$). In both cases, a constraint being binding corresponds to a variable being 0, and a variable equaling 0 corresponds to a binding constraint, for all of the decision and slack variables (not including the objective function variable). Thus a corner point in a problem with two decision variables corresponds to a solution in which two variables equal 0 (for example, at (0,10), constraints 1 and 4 are binding, so s_1 and c are both 0).

For problems with n decision variables, it similarly works out that a corner point corresponds to a solution in which n variables are 0 (the number of equations needed to solve for n variables). The variables that are equal to 0 are called the **nonbasic variables**. All of the other variables are called **basic variables**, and collectively as a set (excluding the objective function variable) they are sometimes referred to as **the basis**. In our example, at the corner point (0,10), s_1 and c are both 0, so they are the nonbasic variables. The other variables (t, s_2, and s_3) are the basic variables, and so the basis is $\{t, s_2, s_3\}$. The objective function variable *acts like* a basic variable, although it is sometimes treated in a different way, as we shall see. We list it under the heading "Basic Variables," but put it in parentheses to reflect this difference. If you noticed in Table 1, the column under "Basic vars" was not filled in. Now we can fill that in (Table 2).

Table 2

Basic Vars	P	c	t	s_1	s_2	s_3	RHS
(P)	1	-140	-200	0	0	0	0
s_1	0	3	6	1	0	0	60
s_2	0	8	4	0	1	0	100
s_3	0	1	0	0	0	1	11

For our initial solution, the origin, the nonbasic variables are c and t, and their values are both 0 by definition. The basic variables are (P), s_1, s_2, and s_3, and the basis is $\{s_1, s_2, s_3\}$. Each basic variable (and P) occurs in one, and only one, equation, and these labels show which variable corresponds to which equation. Note that the values of the basic variables can then be read off quickly from the tableau, by looking at the corresponding RHS values. Notice also that the columns corresponding to P and the basic variables correspond to columns of an identity matrix. This will always be true of a simplex tableau in proper form. Furthermore, the nonbasic variables will be all of the *other* variables, whose columns are *not* identity matrix columns.

This same structure will exist for every simplex tableau (matrix) at every step of the algorithm, and each will correspond to a corner point solution. You can read off the values of the basic variables and the objective function directly from the corresponding RHS's, and the values of all the other variables are 0. As we create new tableaus, we will *update* the basic variables in the "Basic vars" column, so we can read the values of the objective function and the basic variables from each tableau.

The Optimality Test

Once we have a corner point, how do we test it for optimality? In the pasture analogy, we said to look along the two fence directions from the corner, and if neither slants strictly up, then we are optimal. What does this correspond to in the tableau? Remember that a corner point corresponds to a point with two binding constraints, or two variables that equal 0 (the nonbasic variables). When we move along an edge toward another corner point, one of the constraints is no longer binding (we'll be moving along the other constraint, so it stays binding). Therefore, one of the nonbasic variables will become positive. For example, if we move from our initial solution (0,0) to (0,10), constraint 4 stays binding (c stays 0), but constraint 5 is no longer binding (t becomes positive) (Figure 5).

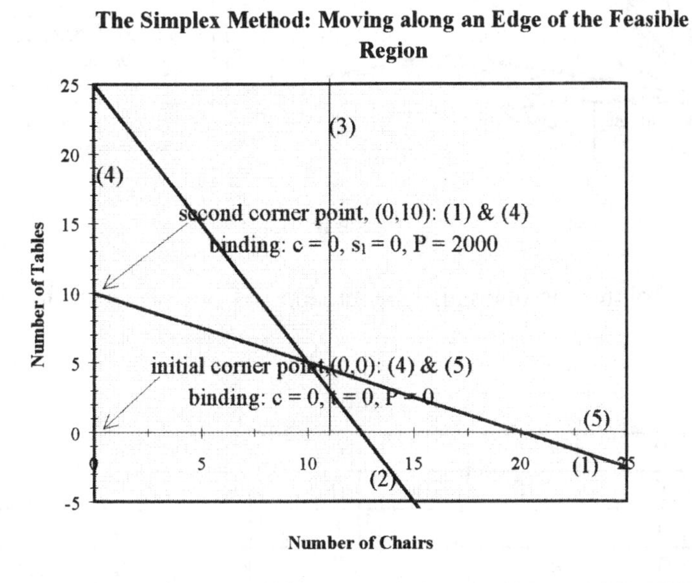

Figure 5

Remember that the original form of our objective function was $P = 140c + 200t$. So if c becomes positive (increases from 0), P will also increase, since the coefficient of c is positive in this original form of the P equation. Similarly, if t increases, P will increase as well. Another way to understand this relationship is to notice that the partial derivatives ($P_c(c,t) = 140$ and $P_t(c,t) = 200$) are both positive, so the rate of change of P with respect to each variable is positive (increases as they do). Since we are trying to maximize P, increasing either nonbasic variable will improve P, so moving in either direction slants up, and therefore our initial solution (the origin) is not optimal (Figure 6). This was true because the coefficients of both c and t in the original form of P were both positive, which corresponds to both being negative in the tableau when we moved the variable terms to the left-hand side of the equation (Table 2). On the other hand, if both coefficients had been ≥ 0 in the tableau, it would mean that neither direction slanted strictly up, and so we would be optimal. In general, **if all of the variable coefficients (not necessarily the RHS) in the objective function row of the tableau are ≥ 0, then the current corner point solution is optimal for a maximization problem.**

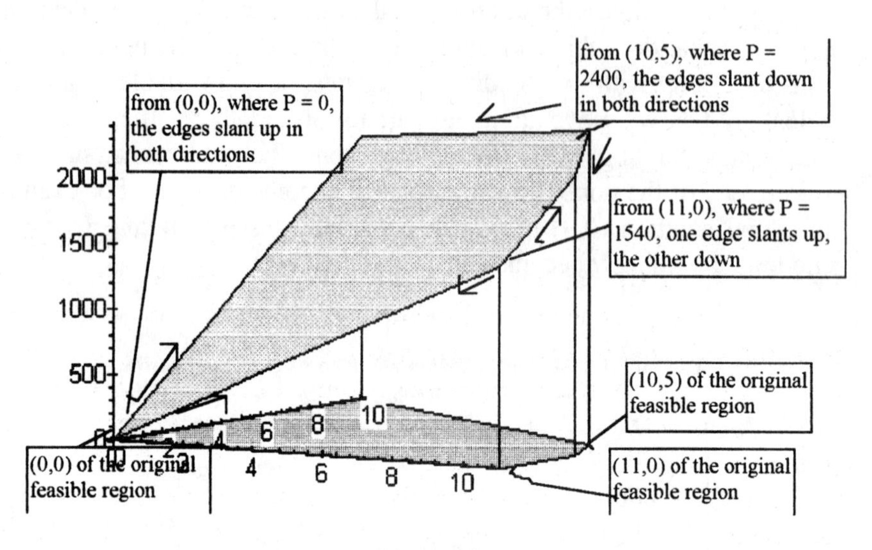

Figure 6

If the current solution is optimal, we can stop. If not, we need to move to a better corner point.

Table 2

Basic Vars	P	c	t	s_1	s_2	s_3	RHS
(P)	1	-140	-200	0	0	0	0
s_1	0	3	6	1	0	0	60
s_2	0	8	4	0	1	0	100
s_3	0	1	0	0	0	1	11

Moving to an Adjacent (Better) Corner Point

Recall that we rewrote the objective function $P = 140c + 200t$ as $P - 140c - 200t = 0$. Thus **the negative of the coefficient of each nonbasic variable in the (*P*) row of the tableau** ($-(-140) = 140$ for c and $-(-200) = 200$ for t) corresponded to the effect on P of increasing that variable by one unit from 0 to 1. This **is the marginal effect of that variable on *P*, or the**

partial derivative of P with respect to that variable, which is also the steepness of the slope in that direction. In the pasture analogy, we said the simplex method always chooses the steeper direction slanting up, so in the tabular simplex method, that means we choose to increase **the variable with the most negative coefficient in P**. Since this variable will no longer be 0, it will no longer be a nonbasic variable. We say it will be *entering* the basis, and so it **is called the entering variable**. The column for that variable in the tableau is called the **pivot column**. In our example, you can see that t increases P at the fastest rate (a partial derivative, or steepness, of 200 versus 140), so it will be the entering variable. On the 3D graph (Figure 6), you can see that this is because moving up in that direction (along the t axis, further in the background) from the origin is steeper than moving up in the other direction (along the c axis, further in the foreground). If you consider just the c axis and the vertical P axis, the slope of the edge line is, in fact, 140.

In general, **we choose the variable with the most negative coefficient in the objective function row of the tableau to be the entering variable** (ties can be broken arbitrarily). This choice is quick and easy, but **does not *guarantee* that we will get the most actual *improvement*, only that we are improving at the fastest *rate***. For instance, we might not be able to go very far in the steeper direction, but might be able to go a long distance in the less steep direction, with the latter resulting in a larger improvement. For a specific example, suppose we could only move 2 units in the t direction (up the t axis), for a total improvement of (200)(2) = 400, but 10 units in the c direction (to the right along the c axis), for a total improvement of (140)(10) = 1400 (both are starting from (0,0)). The problem is that determining the distance you can go in a direction takes time. In the foggy pasture analogy, you'd have to go in one direction until you reached the next corner, then run back and go along the *other* fence direction until you reached the next corner in *that* direction.

The choice of an **entering variable** corresponds to choosing a fence direction in the pasture metaphor. How far do we move in that direction? Until we hit the next corner point! That corner point will have its own tableau, its own basis, and its own nonbasic variables. In fact, the basis will be almost the same as the prior basis, but since the entering variable will be coming into the basis, another variable has to *leave* the basis (which we will call, of course, the **leaving variable**).

In our example, t has the most negative coefficient in the P row of our initial tableau, so t is the **entering variable**. This means graphically that we are increasing t and keeping c at 0, so we are moving up the t axis. What is the next corner point we will hit? From Figure 7 we can see it is (0,10).

The Simplex Method: Choice of a Leaving Variable

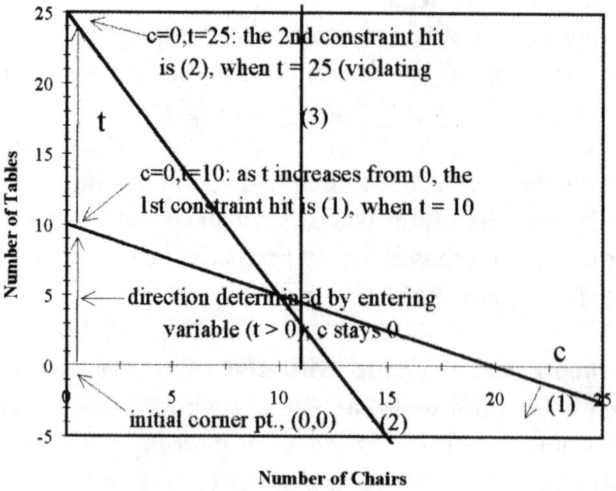

Figure 7

But for higher-dimensional problems, we can't just look at the graph, so we need a more general method. Another graphical way to think of the problem is that we are trying to determine what constraint we will hit first in the direction we have chosen. Remember that **hitting a constraint corresponds to a variable becoming 0**, so the question is equivalent to finding **what variable will become 0 first as the entering variable increases**.

Consider the first constraint:

$$3c + 6t + s_1 = 60 .$$

In the current situation (moving to the next corner point), $c = 0$ and will stay 0, so we can rewrite the equation

$$6t + s_1 = 60.$$

If we solve it for s_1, we get

$$s_1 = 60 - 6t .$$

When will s_1=0 ? s_1 will hit 0 when

$$60 - 6t = 0 ,$$

or when

$$t = 60/6 = 10 .$$

Note that, if t had had a negative (or 0) coefficient in the tableau, it would have had a positive (or 0) coefficient when we solved for s_1, so t could have increased forever (to infinity) without making $s_1 = 0$. For example, the coefficient of t in the s_3 row is 0, and we could move up the t axis forever without ever hitting the 3rd constraint edge ($c = 11$), since it is parallel to the t axis. Thus the **only candidates for the leaving variable are those basic variables for which the entering variable has a *positive* coefficient in the tableau** (positive elements of the pivot column, ignoring the P row). As in the example above, when the maximum increase was 60/6 = 10, **the maximum increase in the entering variable before the basic variable hits 0 for a given basic variable row is the RHS/(pivot column element)**. For the rows where the coefficient is 0 or negative, we will call the maximum increase infinity. Since we want to determine the *first* basic variable to hit 0, we choose **the smallest of these maximum increase ratios to determine the leaving variable** (breaking ties arbitrarily). We call the corresponding row the **pivot row** of the tableau. The value in both the pivot row and the pivot column is called the **pivot element**.

In Table 3 we show the calculations for the leaving variable:

Table 3

Basic Vars	P	c	t	s_1	s_2	s_3	RHS	Max. Inc.
(P)	1	-140	**-200**	0	0	0	0	-------
s_1	0	3	**6**	1	0	0	60	60/6=**10**
s_2	0	8	4	0	1	0	100	100/4=25
s_3	0	1	0	0	0	1	11	∞

Thus, t is the entering variable since t has the most negative coefficient in the P row, and the pivot column is the t column. The leaving variable is s_1 (since 10 is the smallest maximum increase) and its row is the pivot row. The pivot element is 6. Notice that the coefficient of s_3 in the t column is not positive, so we call the maximum increase infinity.

Now we need to find the tableau for the new corner point. Since t will be entering the basis and replacing s_1 as the basic variable in the second row, we now want the column of t to look like an identity matrix column, with a 1 in the second row and 0's elsewhere. To do this, we do exactly what we did for Gaussian elimination in Chapter 7. First we get the 1 in the desired place by dividing the old pivot row through by the pivot element, which is equivalent to multiplying by the reciprocal. In our example, we take the original row 2 (the row labeled s_1) and divide every element of it by 6 (which is the same as multiplying by 1/6). We call the result the "new pivot row", or R_2' (the new Row 2), and put it in row 2 in the next tableau ($(1/6)R_2 \rightarrow R_2'$). Notationally, R_i means row *i* of the last tableau and R_i' means row i of the *next* tableau, which is being calculated.

$R_2/6 \rightarrow R_2'$	0/6	3/6	**6/6**	1/6	0/6	0/6	60/6

R_2'	0	1/2	1	1/6	0	0	10

We replace the old row 2 by the new row 2 in our tableau.

	Basic Vars	*P*	*c*	*t*	s_1	s_2	s_3	RHS
	(*P*)							
R_2'	*t*	0	1/2	**1**	1/6	0	0	10
	s_2							
	s_3							

Note that we have changed the "Basic Vars" column to show that "*t*" is now in the basis.

Now we need to make *t* have the proper form for a basic variable in the **other** rows as well, so we have to have the other entries in the t column equal 0. To do this, we take multiples of this new pivot row and add them to each of the other rows to get the desired 0's in the other rows. For each row, we can simply multiply the new pivot row by the opposite of the value in the pivot column. For example, we can multiply the new pivot row by 200 and add it to the old *P* row to get the new *P* row to cancel out the -200 that is there ($200R_2'+R_1 \rightarrow R_1'$).

$200R_2'$	0	100	200	100/3	0	0	2000

R_1=(*P*)	1	-140	**-200**	0	0	0	0

So $200R_2'+R_1 \rightarrow R_1'$ gives us the following:

R_1'	(*P*)	1	-40	0	100/3	0	0	2000

Next we multiply the new pivot row by -4 and add it to the old s_2 row to get the new s_2 row ($-4R_2'+R_3 \rightarrow R_3'$).

$-4R_2'$		0	-2	-4	-2/3	0	0	-40

R_3	s_2	0	8	4	0	1	0	100

$-4R_2'+R_3 \to R_3'$	s_2	0	6	0	-2/3	1	0	60

Note that the s_3 row already has a 0 where we need it, so the row doesn't have to change at all.

The result of these row operations can be given in one step (find the new pivot row first, then the others), resulting in Table 4.

Table 4

Basic Vars	P	c	t	s_1	s_2	s_3	RHS
(P)	1	-40	0	100/3	0	0	2000
t	0	1/2	1	1/6	0	0	10
s_2	0	6	0	-2/3	1	0	60
s_3	0	1	0	0	0	1	11

Here, the nonbasic variables, c and s_1, are both 0, $P = 2000$, $t = 10$, $s_2 = 60$, and $s_3 = 11$, so we are at the corner point (0,10), as expected (Figure 8). Now try on your own to find the entering variable and the leaving variable, and do the iteration to get the next tableau.

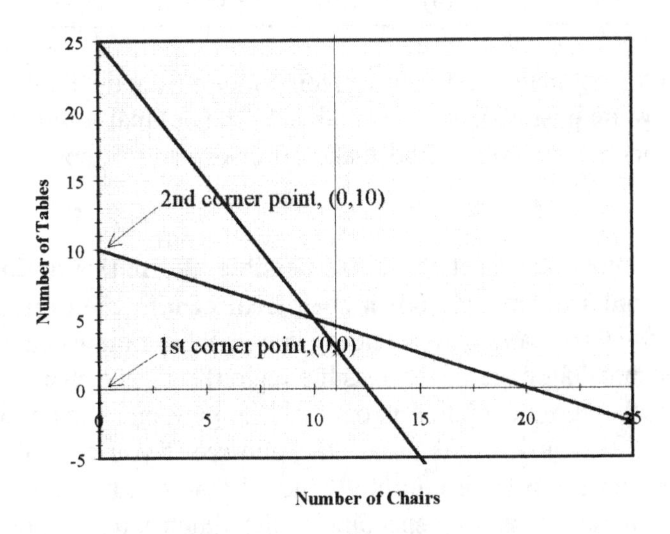

Figure 8

The new entering variable will be c (the only coefficient in the P row that is negative), so its column is the pivot column. The maximum increase calculations are given in Table 5.

Table 5

Basic Vars	P	c	t	s_1	s_2	s_3	RHS	Max Inc.
(P)	1	**-40**	0	100/3	0	0	2000	------
t	0	1/2	1	1/6	0	0	10	10/.5=20
s_2	0	**6**	0	-2/3	1	0	60	**60/6=10**
s_3	0	1	0	0	0	1	11	11/1=11

This time the leaving variable is s_2 (the smallest maximum increase is 10), so the pivot element is (again, by pure coincidence) 6. So we divide the pivot row by 6 (multiply by 1/6) to get the new pivot row, then do row operations to get the 0's where we need them, resulting in Table 6. Notice that R_1 now refers to Row 1 from Table 5, which we had called R_1' for the purpose of the last iteration. In general, **think of R_i as Row *i* from what is at the moment the *old* (or current) tableau, and R_i' as Row *i* of the *new* tableau.** From here on, we won't label the rows, but we will use this convention consistently.

Table 6

Row Operation	Basic Vars	P	c	t	s_1	s_2	s_3	RHS
$40R_3'+R_1 \to R_1'$	(P)	1	0	0	260/9	20/3	0	2400
$-.5R_3'+R_2 \to R_2'$	t	0	0	1	2/9	-1/12	0	5
$(1/6)R_3 \to R_3'$	c	0	1	0	-1/9	1/6	0	10
$-R_3'+R_4 \to R_4'$	s_3	0	0	0	1/9	-1/6	1	1

Now we see that the optimality test indicates this solution is optimal (all coefficients in the objective function row are positive), and we are done! The optimal solution is $c = 10$, $t = 5$, and $P = 2400$. So, once again, we should make 10 chairs and 5 tables, for a profit of \$2400. □

To see from an optimal tableau if **there are possible alternative optimal solutions**, simply look to see **if a nonbasic variable has a coefficient of 0 in the objective function.** This makes sense, since the 0 indicates that increasing that variable from its current value of 0 (moving along the corresponding edge of the feasible region) will not change the objective function value. Thus, if the current solution is optimal, the new one should be, too. If so, **make that variable enter the basis** in the usual way, and see if you get a different corner point. If so, **this new corner point is also optimal** (and all points in between). The process could even be repeated a number of times, depending on the dimension of the problem.

•

Basic Steps for the Simplex Method (for Maximization Problems):

- Given a feasible corner point solution, identify the entering variable: the column with the *most negative* coefficient (furthest left on a number line), called the pivot column.

- Identify the leaving variable: the row with the smallest maximum increase (the smallest ratio of RHS to positive pivot column coefficient), called the pivot row.
- The value in both the pivot row and pivot column is called the pivot element.
- Use Gaussian elimination (elementary row operations) to get a 1 in place of the pivot element and 0's in all of the other entries of the pivot column.
- Repeat until there are no more negative coefficients in the objective function row.

In the first two steps, ties can be broken arbitrarily. If there are no positive coefficients in the pivot column, the optimal solution is unbounded (a variable can go to infinity and make the objective function value arbitrarily large).

Sample Problem 2: Suppose in our original furniture problem, the profit from each chair was \$100 instead of \$140. Then the final tableau turns out to be that given in Table 7.

Table 7

Basic Vars	P	c	t	s_1	s_2	s_3	RHS	Max Inc.
(P)	1	0	0	300/9	**0**	0	2000	-----
t	0	0	1	2/9	-1/12	0	5	∞
c	0	1	0	-1/9	**1/6**	0	10	**60**
s_3	0	0	0	1/9	-1/6	1	1	∞

Are there any alternative optimal solutions? If so, find them.

Solution: Notice that s_2 is nonbasic in the final tableau, and has a coefficient of 0 in the (P) row. If we make s_2 enter, then c will be the leaving variable, and the new tableau becomes Table 8.

Table 8

Basic Vars	P	c	t	s_1	s_2	s_3	RHS
(P)	1	**0**	0	50/3	0	0	2000
t	0	1/2	1	1/6	0	0	10
s_2	0	6	0	-2/3	1	0	60
s_3	0	1	0	0	0	1	11

Notice that now c is nonbasic and has a coefficient of 0 in the P row. However, if we make it enter, we will get back to the first optimal tableau shown above (try this for yourself to confirm this fact). This means that we have exactly two different corner point solutions, (10,5) and (0,10), and that all points in between them are also optimal. Since these fall along constraint 1, they could be described by $\{(c,t): 3c + 6t = 60, 5 \leq t \leq 10\}$ or $\{(c,t): 5 \leq t \leq 10, c = 20-2t\}$. Figure 9 illustrates this problem graphically. Notice that **in a problem with two**

decision variables like this, you get alternative optimal solutions when a constraint is parallel to the objective function (on the optimal side of the feasible region), and **you get an entire line segment that is optimal**. This means that there are in fact an *infinite* number of optimal solutions!

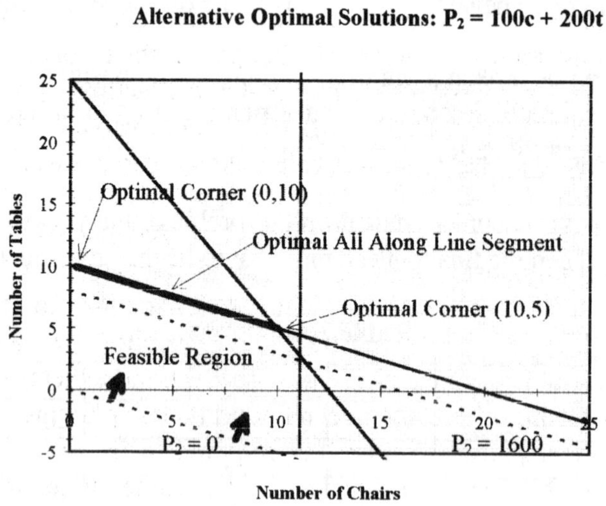

Alternative Optimal Solutions: $P_2 = 100c + 200t$

Figure 9

<u>Section Summary</u>

Before you proceed to the exercises, be sure that you:

- Understand that for a linear programming problem (an LP) in "Maximize" form with inequality constraints and two decision variables, the goal of the Simplex Method is analogous to finding the highest point(s) within a fenced-in pasture on a flat but slanted field (plane) in the midst of dense fog.

- Understand that the basic idea of the Simplex Method for a maximization problem using the fenced-in pasture analogy is to start at a corner point and check the slope of the fence in both directions from that corner to test optimality (if neither edge slopes strictly *up*, then the current corner is optimal). If the current corner is not optimal, then move up along the steeper edge coming out of it until you hit another corner, then repeat the process until you reach an optimal point.

- Understand how to add in slack variables to "≤" structural constraints that have nonnegative right-hand-side constants (RHS's), and how to interpret their values (graphically for problems with two decision variables): if the slack variable is 0 at a point, then the point is on the *edge* of the constraint; if the slack variable is *positive* at a point, then the point is *strictly within* the constraint, on the *feasible* side; and if the slack variable is *negative* at a point, then the point is *strictly beyond* the constraint, on the *infeasible* side.

- Understand that a constraint is **binding** at a given point if its left-hand-side value (with the values of the point plugged in) is *equal* to its RHS value (if it holds as an *equation*). If the constraint is binding at the point, then *some* variable (for a nonnegativity constraint, its decision variable; or for a structural constraint, its slack variable) is equal to zero. At a feasible corner point, those variables that are zero and are needed to determine the corner point are called the *nonbasic variables*.

- Understand that if an LP is in "Maximize" form with n nonnegative original decision variables and m structural constraints in "$\le$" form with nonnegative RHS's, a feasible corner point will correspond to the intersection of n constraints. In other words, at a corner point, there will be n binding constraints, and therefore n variables required to be 0 (n nonbasic variables). Since there are m constraints, there will be m slack variables, in addition to the original n decision variables, plus the variable for the objective function. The variables that are not required to be zero to determine the corner point are called the *basic variables*. So at a corner point there will be n nonbasic variables, m basic variables, and the objective function variable, which acts similarly to a basic variable but is treated in a special way (it can *never leave* the basis).

- Understand how to enter a linear programming problem into an initial tableau (matrix) with row and column headings, when the problem is written in "Maximize" form, with nonnegativity constraints and structural constraints that are only of the "$\le$" type with nonnegative RHS's.

- Understand how to look at a simplex tableau and be able to determine the values of the basic variables and the objective function (all given by the RHS's) and the nonbasic variables (all 0).

- Understand that the initial simplex tableau corresponds to the origin (all of the original decision variables are 0).

- Understand how to test whether a feasible corner point corresponding to a simplex tableau is optimal, and why this is the case algebraically (the only feasible changes in the nonbasic variables will make things worse). For a maximization problem, this will be when the coefficients of all of the variables in the top (objective function) row of the tableau are nonnegative.

- If a corner point is not optimal, understand how to select an **entering variable** (to enter the basis) by finding the variable with the *most negative* coefficient in the top row (for a problem in "Maximize" form), and what this corresponds to graphically (moving up the steepest edge) and algebraically (the partial derivative indicating the fastest rate of improvement).

- Given an entering variable (pivot column), understand how to compute the Maximum Increase in that entering variable that will keep all of the basic variables feasible (nonnegative) by computing the value of the ratio $\frac{RHS}{\text{value in the pivot column}}$ for each basic variable that has a *positive* coefficient in the pivot column. The Maximum Increase will be ∞ for basic variables (rows), where the coefficient in the pivot column is 0 or negative. It corresponds to how far you can move in the chosen direction (along an edge) before hitting that constraint.

- Understand that the **leaving variable** (the variable the entering variable *replaces* in the basis) will be the basic variable with the *smallest Maximum Increase ratio* (the *first* constraint hit when moving in the chosen direction along the chosen edge), and that row will be the *pivot row*.

- Understand that given an entering variable and a leaving variable, a simplex iteration corresponds to doing the row operations (as in Gaussian elimination) needed to make the pivot column look like a column of an identity matrix, with a 1 in the pivot row, and 0's everywhere else. This can usually be done by dividing the pivot row by the pivot element (multiplying by its reciprocal), then adding a multiple of the *new* pivot row to each old row, where the multiple is simply the *negative* (opposite sign) of the value in the old pivot column for that row. Graphically, this is moving from one corner to an adjacent corner. Algebraically, it is solving for the new basic variables in terms of the new nonbasic variables.

- Understand when alternative optimal solutions exist and how to find them.

EXERCISES FOR SECTION 9.3:

Warm Up

Solve the following linear programs by hand by the tabular simplex method. If there are any multiple solutions, indicate all possible optimal solutions.

1. Maximize $z = 3x + 4y$
   subject to $x + y \leq 20$
   $5x + 8y \leq 120$
   $x, y \geq 0$

2. Max. $P = 5x_1 + 4x_2$
   s.t. $3x_1 + 4x_2 \leq 24$
   $10x_1 + 8x_2 \leq 62$
   $x_1, x_2 \geq 0$

3. Max. $z = 5x + 7y$
   s.t. $3x + 5y \leq 60$
   $6x + 4y \leq 90$
   $y \leq 5$
   $x, y \geq 0$

4. Min. $z = 2x + 3y$
   s.t. $3x - y \leq 12$
   $x + y \leq 8$
   $y \leq 6$
   $x, y \geq 0$

5. Min. $z = 2x + 3y$
   s.t. $3x - y \leq 12$
   $x + y \leq 8$
   $y \leq 6$
   $x \leq 10$
   $x \geq 0$

6. Min. $z = -2x + -3y$
   s.t. $3x - y \geq -12$
   $x + y \leq 8$
   $y \leq 6$
   $x, y \geq 0$

7. Max. $z = 5x + 7y$

   s.t. $x - 4y \leq 0$

   $x - 3y \leq 12$

   $x, y \geq 0$

8. Max. $z = 6x + 10y$

   s.t. $3x + 5y \leq 60$

   $6x + 4y \leq 90$

   $y \leq 5$

   $x, y \geq 0$

Game Time

9. Use the tabular simplex method to solve a modified version of the furniture problem: the profit from each chair is \$100 rather than \$140. Thus the problem is:

$$
\begin{aligned}
\text{Max } P &= 100c + 200t \\
\text{s.t.} \quad 3c + 6t &\leq 60 \\
8c + 4t &\leq 100 \\
c &\leq 11 \\
c, t &\geq 0
\end{aligned}
$$

10. Use the tabular simplex method to solve a different modified version of the furniture problem: the demand limit on the number of chairs that can be sold drops from 11 to 10.

$$
\begin{aligned}
\text{Max } P &= 140c + 200t \\
\text{s.t.} \quad 3c + 6t &\leq 60 \\
8c + 4t &\leq 100 \\
c &\leq 10 \\
c, t &\geq 0
\end{aligned}
$$

11. Use the tabular simplex method to solve the furniture problem if both of the modifications from exercises 9 and 10 are made at the same time.

$$
\begin{aligned}
\text{Max } P &= 100c + 200t \\
\text{s.t.} \quad 3c + 6t &\leq 60 \\
8c + 4t &\leq 100 \\
c &\leq 10 \\
c, t &\geq 0
\end{aligned}
$$

12. You are trying to raise money for your sports team by selling T-shirts and hats. You can order both items from the same company, which will charge you \$4 for each hat and

$6 for each shirt, plus there is a fixed cost of $30 for any order. Your team has $100 to spend for everything, and has decided to sell the shirts for $12 and the hats for $7, at which price you think you can sell no more than 15 hats and 10 shirts. Assume that both items are "one size fits all" and that the demands for the two products are independent for this exercise. Use the tabular simplex method to find the optimal solution to this problem. What is your team's optimal profit? What would you do (how many of each item would you order) in this situation?

13. You run a small candle-making business, making regular and deluxe candles. You can sell the regular candles for a profit of $4 each and the deluxe candles for a profit of $7 each. Regular candles use 6 oz. of wax and 5 inches of wick material, while deluxe candles use 9 oz. of wax and 6 inches of wick material. You have 120 oz. of wax and 110 inches of wick material in your studio at the moment, and will not get any more supplies until tomorrow. Use the tabular simplex method to determine how many candles of each type you should make today. What assumption causes a problem here? How could you deal with it?

Section 9.4 - Linear and Non-Linear Optimization on Spreadsheets

In the previous two chapters we have seen how spreadsheets can be used to help us solve problems that involve matrices and multivariable regression. In Section 9.2 we saw how we can graphically solve linear programming problems in which there were only two decision variables. In Section 9.3 we saw how we could use the simplex method to solve linear programming problems in which there were more than two decision variables. In Section 8.3 , we learned how to optimize nonlinear functions using a spreadsheet optimizing tool. In this section we focus on using spreadsheets to solve problems similar to those in Section 8.3 for both linear and nonlinear optimization problems, except that we have added more complicated limitations or **constraints** on the variables or changing cells, that must be met. Using a spreadsheet optimizing tool allows us to solve both linear and nonlinear optimization problems with any number of complex constraints, as well as problems that involve more than two decision variables. There are also separate packages for linear and nonlinear programming (constrained optimization) available at many institutions.

Here are some examples of the kind of problems you might want to solve using a spreadsheet optimizing function:

- Consider again the example of the hand-crafted furniture business from previous chapters. You own this small business and for now you make and sell only two products: chairs and tables (just a single style of each). Given restrictions on the availability of raw materials and on the demand for furniture, how many chairs and tables should you make this month?
- You are trying to decide what combination of foods you should eat normally for breakfast, to meet your nutritional goals at the lowest possible cost. What foods (and how much of each) should you plan for your breakfasts over the course of a week?
- Your airline has size restrictions on carry-on luggage, based on the dimensions. What dimensions will give you the most volume so you can take the most possible stuff with you?

In addition to being able to solve the above kinds of problems, by the end of this section you should:

- Know how to enter a linear or nonlinear programming problem with constraints into a spreadsheet so that you can use its optimizing capabilities.
- Know how to enter the constraints in the optimizing tool.
- Know how to interpret the report given by the optimizing tool.

A popular reference on the use of a spreadsheet program[1] states:

> "Use 'Solver' to find the best solution to a problem. Solver is normally helpful for the following types of problems:
> - *Product Mix.* Maximizing the return on products given limited resources to build those products.
> - *Staff Scheduling.* Meeting staffing levels at a minimum cost with specified employee satisfaction levels.
> - *Optimal Routing.* Minimizing transportation costs between a manufacturing site and points of sale.
> - *Blending.* Blending materials to achieve a certain quality level at a minimum cost."

The hand crafted furniture problem that we have been studying fits the first category. The cereal problem is related to the blending type. Both examples are very simple, with only two variables and a small number of constraints. For this reason we were able to solve them without too much difficulty using either the graphical method or the simplex method. The graphical method cannot be used for problems involving more than three variables, and the simplex method, when done by hand, becomes tedious when a large number of variables are involved. It is a great advantage to have technology help us in the solution of these problems, particularly when that technology is readily available in most academic and business settings.

When using a spreadsheet program to solve this type of problem, it is helpful to set up your problem carefully before you enter it into the spreadsheet. This will help you avoid mistakes and the necessity of redoing the spreadsheet..

Linear Programming Using a Spreadsheet

Sample Problem 1: We have already formulated the hand crafted furniture problem:

$$\begin{aligned} \text{Max. } P &= 140c + 200t \\ \text{s.t. } 3c + 6t &\le 60 \\ 8c + 4t &\le 100 \\ c &\le 11 \\ c,t &\le 0. \end{aligned}$$

Solve this problem using an optimizing tool in a spreadsheet.

---

[1] Special Edition Using Excel for Windows 95, Ron Pearson, Que Corporation, Indianapolis, IN, p. 839. "Solver" is the Excel equivalent of Quattro Pro's Optimizer.

Solution: As we saw in Section 8.3, the way spreadsheet programs set up an optimization problem is to use the contents of **a cell to correspond to each variable**, and **these need to be adjacent in a rectangular block exclusively for this purpose** (a row is recommended, for simplicity). Therefore, we need to have a cell in our spreadsheet that represents each variable in the problem; in our case, that means the number of chairs (c) and the number of tables (t). Functions that represent the left hand side (LHS) of a constraint, such as $3c + 6t$ or $8c + 4t$, and the objective function itself, such as the $140c+200t$ that determines the value of P, are each stored as a formula within a cell. The variables in the formula are replaced by the cell addresses that correspond to each of them. To lay out the problem formulation, we can label a column for each variable, one for the total or LHS of each row, including the formula for the objective function P, and one for the constant or RHS of each constraint. We have set up rows for the quantities of chairs and tables; the coefficients and totals of the objective function (e.g., "Profit"); and each of the constraints (e.g., "Wood", "Demand", etc.). Not all of this is absolutely necessary; all that you really need are cells for each of the variables, and a cell for each total formula (for the objective function and the LHS's of each constraint). But we recommend the way we have laid things out here for any linear programming problem, so that anyone looking at a printout of your problem will know exactly what it means.

	A	B	C	D	E	F	G
1	HAND CRAFTED FURNITURE COMPANY						
2					LHS/	RHS	
3		CHAIRS	TABLES		FORMULA	CONSTANT	
4	QUANTITY						
5							
6	PROFIT EACH	140	200				
7	PROFIT TOTAL						
8							
9	WOOD EACH	3	6				
10	WOOD TOTAL					60	
11							
12	HARDWARE EACH	8	4				
13	HARDWARE TOTAL					100	
14							
15	DEMAND - CHAIRS					11	

For useful tracking information when solving an LP using a spreadsheet optimizing tool, the contribution per unit (coefficient) of each variable to each appropriate function can be placed in a separate row. These are the coefficients of the variables in our objective functions and constraints. To enter the total profit from the chairs (the contribution of the chairs to the profit), we want to multiply the quantity of chairs (in B4) by the profit from each chair (B6), so we enter the formula (+B4*B6) in B7. Similarly, we put (+C4*C6) in C7. We continue with the total wood used by the chairs by entering the quantity of chairs times the units of wood used per chair: (+B4*B9) in cell B10. The total wood used by the tables (+C4*C9) is entered in C10. Next we fill in the hardware used by the chairs and tables: (+B4*B12) is entered in B13 and (+C4*C12) is entered in C13. Finally, we fill in the Total column by entering the total profit from the chairs plus the total profit from the tables in E7: (+B7+C7); total wood is entered in E10: (+B10+C10); total hardware is entered in E13: (+B13+C13); and the number of chairs is entered in E15: (+B4).

	A	B	C	D	E	F	G
1	HAND CRAFTED FURNITURE COMPANY						
2					LHS/	RHS	
3		CHAIRS	TABLES		FORMULA	CONSTANT	
4	QUANTITY						
5							
6	PROFIT EACH	140	200				
7	PROFIT TOTAL	0	0		0		
8							
9	WOOD EACH	3	6				
10	WOOD TOTAL	0	0		0	60	
11							
12	HARDWARE EACH	8	4				
13	HARDWARE TOTAL	0	0		0	100	
14							
15	DEMAND - CHAIRS				0	11	

Note that all of the total rows and the grand total column have 0's in them. That is because we have no "values" entered in our variable row. We don't know the number of chairs and tables to make to optimize our profit. If you put in 1's for the values of all the variables, you will see all the individual products, making it easier to spot errors. To further check, it is a good idea to change the quantities from 1 to 2 and check to be sure that the totals change. Here we have changed the quantity of chairs to 1 and the quantity of tables to 2.

	A	B	C	D	E	F	G
1	HAND CRAFTED FURNITURE COMPANY						
2					LHS/	RHS	
3		CHAIRS	TABLES		FORMULA	CONSTANT	
4	QUANTITY	10	5				
5							
6	PROFIT EACH	140	200				
7	PROFIT TOTAL	1400	1000		2400		
8							
9	WOOD EACH	3	6				
10	WOOD TOTAL	30	30		60	60	
11							
12	HARDWARE EACH	8	4				
13	HARDWARE TOTAL	80	20		100	100	
14							
15	DEMAND - CHAIRS				10	11	

We should mention that, with this setup, you can do trial and error experimenting yourself, by plugging in different values for the variables and seeing whether the constraints are satisfied and checking the profit value. You can also see the effect of changing the parameters of the problem (the coefficients and constants), To change the profit on the chairs from 140 to 100, you could simply change cell B6 from 140 to 100, and all the other calculations would be adjusted automatically. Putting +B7+C7 into E7 also helps understand the expression for total profit by breaking it up into pieces. If you just want to do the minimum needed to run a spreadsheet optimizing tool, however, you would just have cells B4 and C4 for the variables, and the formulas for the objective function and constraints LHS's: 140*B4+200*C4 in E7; 3*B4+6*C4 in E10; and 8*B4+4*C4 in E16. Whichever method you choose, **you must do this setup of variable cells and formulas in your spreadsheet *before* going into the spreadsheet optimizing tool**.

At this point, before solving the problem, consider whether you want to specify an initial solution for the problem. If you do not, the program will take whatever values are in the specified cells for the variables. For any that do not have entries, it will assume a value of 0. If you want something else, simply enter the desired values into the variable cells of the spreadsheet at any point in the process. The initial solution does *not* have to be feasible. We have entered (1,2) as a solution just to be sure our formulas are entered correctly.

Before solving the problem, and *after completing the above steps*, if you want, you can also *show* in more detail on the spreadsheet what the original problem formulation is. To fill in the details of your formulation (as we have done on the printout), you can go back to the spread sheet and type it in, using the cell addresses for the variables rather than the variable names.

	A	B	C	D	E	F	G	H	I	J
1	HAND CRAFTED FURNITURE COMPANY									
2					LHS/	RHS				
3		CHAIRS	TABLES		FORMULA	CONSTANT				
4	QUANTITY	10	5							
5										
6	PROFIT EACH	140	200					Max. E7 = 140*B4+200*C4		
7	PROFIT TOTAL	1400	1000		2400			s.t.	3*B4+6*C4 <= 60	
8									8*B4+4*C4 <= 100	
9	WOOD EACH	3	6						B4 <= 11	
10	WOOD TOTAL	30	30		60	60			B4 >= 0	
11									C4 >= 0	
12	HARDWARE EACH	8	4							
13	HARDWARE TOTAL	80	20		100	100				
14										
15	DEMAND - CHAIRS				10	11				

In our example, we have typed in **Max. E7 = 140*B4+200*C4** into cell H6, in cell H7 we typed **s.t.**, and in cells I7 to I11, we have typed in the constraints:

3*B4+6*C4 <= 60
8*B4+4*C4 <= 100
B4 <= 11
B4 >= 0
C4 >= 0

If you think you might ever want to go back to work with this problem after working on one or more other problems, it is a good idea to save the model and problem formulation. The specific directions for this vary with the spreadsheet being used. See your technology supplement for details.

We are now ready to ask the program to solve our problem: to optimize our profit while staying within the constraints. The steps for optimizing are similar in most spreadsheet programs.

Directions for Solving Linear and Non-Linear Optimizations using a Spreadsheet

1. Select the spreadsheet optimizing function.
2. Identify the target or solution cell.
3. Identify the variable cells.
4. Enter your constraints. Constraints must be entered one at a time. To be safe, be sure to enter nonnegativity constraints as well (unless you know that your spreadsheet assumes them - see your technology supplement).

In most spreadsheets, you can indicate whether your optimization is a linear program or not. This does not need to be specified, but if you indicate that your problem is linear, the package can usually find the solution faster, and give your a more detailed solution (or even display Simplex Method tableaus). When you are ready, tell the spreadsheet optimization tool to solve the problem. The solutions will appear in the variable cells and in the target or solution cell.

Some spreadsheet programs ask you if you want to display and save an answer report and/or a sensitivity report after the solution is found. In this case, just select the items you want on the menu. These reports will sometimes be placed on different worksheets in the workbook. For some spreadsheet programs you can select an area on your spreadsheet for your answer report before you run the optimizer. As usual, see your technology supplement for details.

Interpretation of the Spreadsheet Analysis of Linear Optimization Problems

For our example problem with initial solution (1,2), a sample Answer Report is shown below. Notice that the report gives a starting (initial) and final (hopefully optimal) value for the Target Cell (objective function) and for the Adjustable Cells (the original decision variables).

Microsoft Excel 7.0a Answer Report
Worksheet: [LPCHAI~1.XLS]SolverchairSheet3
Report Created: 11/13/97 0:15

Target Cell (Max)

Cell	Name	Original Value	Final Value
$E$7	PROFIT TOTAL FORMULA	540	2400

Adjustable Cells

Cell	Name	Original Value	Final Value
$B$4	QUANTITY CHAIRS	1	10
$C$4	QUANTITY TABLES	2	5

Constraints

Cell	Name	Cell Value	Formula	Status	Slack
$E$10	WOOD TOTAL FORMULA	60	$E$10<=$F$10	Binding	0
$E$13	HARDWARE TOTAL FORMULA	100	$E$13<=$F$13	Binding	0
$E$15	DEMAND - CHAIRS FORMULA	10	$E$15<=$F$15	Not Binding	1

For the Constraints, the sample Answer Report gives even more useful information. First it lists the cell containing the formula for the left-hand side of the constraint, then a name associated with the LHS, then the final value of that quantity, and then the form of the constraint (the LHS formula cell address, the relation type: $\leq$ or $\geq$ or = , and the RHS, which could be a numerical value or a cell address). Next, it indicates whether the constraint is **binding** (whether the LHS = the RHS), as we discussed in Section 9.2, at the final solution point. The next column then gives the **slack**, which **corresponds to the absolute value of the difference between the RHS and the LHS of the constraint**. For a $\leq$ constraint, this is simply the value of the slack variable. In any case, if the constraint is binding, the slack should be 0, and if the slack is not 0, the constraint is probably not binding. The only possible exception would be when there is a **very** small value, such as 1.4E-15, which could be due to round off error. Recall that 1.4E-15 means $1.4\times10^{-15} = 0.0000000000000014$.

In our example, the optimal solution is (10,5), which is the intersection of the first two constraints, so both constraints are shown as binding. On the other hand, constraint 3 ($c\leq11$) is not binding, and its slack is 1 (= the value of $s_3 = 11 - c = 11 - 10$ in the final tableau, which is |RHS - LHS|). Also, constraints 4 and 5 are not binding, and have slacks of 10 and 5. For example, for the fourth constraint,

$$\text{slack} = |\text{RHS - LHS}| = |0 - \text{B4}| = |0 - c| = |0 - 10| = |-10| = 10$$

at the optimal corner point, corresponding to the values of the decision variables.

For your convenience, the final simplex tableau that we obtained for this problem in Section 9.2 is shown again in Table 1.

Table 1

Basic Vars	P	c	t	s_1	s_2	s_3	RHS
(P)	1	0	0	260/9	20/3	0	2400
t	0	0	1	2/9	-1/12	0	5
c	0	1	0	-1/9	1/6	0	10
s_3	0	0	0	1/9	-1/6	1	1

A sample Sensitivity Report is shown below, which gives even more detailed information about the solution and the effects of potential changes in the original parameters (objective function coefficients and constraint RHS constants). The sample Sensitivity Report has more useful information concerning the decision variables and the constraints.

Microsoft Excel 7.0a Sensitivity Report
Worksheet: [LPCHAI~1.XLS]Solverchairsolutionsheet4
Report Created: 10/7/97 4:42

Changing Cells

Cell	Name	Final Value	Reduced Cost	Objective Coefficient	Allowable Increase	Allowable Decrease
$B$4	QUANTITY CHAIRS	10	0	140	260	40
$C$4	QUANTITY TABLES	5	0	200	80	130

Constraints

Cell	Name	Final Value	Shadow Price	Constraint R.H. Side	Allowable Increase	Allowable Decrease
$E$10	WOOD TOTAL FORMULA	60	28.88888889	60	90	9
$E$13	HARDWARE TOTAL FORMULA	100	6.666666667	100	6	60

For the decision variables there is a column called "Reduced Cost" in the Sensitivity Report. These values (both are 0 above) tell us that changing the non-negativity constraints from >= 0 to >= 1 would have no effect on the optimal solution. Another way to interpret these, to help understand the name, is that they tell you how much the *cost* of each item or decision variable (chairs and tables, in the example) would have to be *reduced* (in other words, how much the *profit* would have to be *increased*, in the example) in order for that decision variable to enter the basis. In the example, both decision variables are already in the basis in the optimal solution, so the cost doesn't have to be reduced (the profit doesn't have to be increased) at all, so the reduced cost values are both 0. This calculation is done individually and independently for each decision variable (seeing how much its coefficient would have to change to bring *it* into the basis, holding all of the *other* coefficients constant at their *original* values).

If our optimal solution didn't have any chairs in it, it would mean that the profit from the chairs wasn't high enough to make producing them worthwhile compared to the tables. But if the profit from the chairs increased in that case, eventually it would become worth making them, and that decision variable would enter the basis. The Reduced Cost of the chairs would then be the *minimum* increase in the profit necessary to make c enter the basis in an optimal solution.

The Reduced Cost values correspond to the coefficients in the objective function of each decision variable in the tabular Simplex Method in general (except possibly for the sign, depending on the type of optimization and the specific package you are using), which you can confirm in our example by referring back to Table 1.

The Allowable Increase and Allowable Decrease values for each decision variable are the maximum increase and decrease in the objective function coefficient for that decision variable for which the optimal basis would remain the same. In our example, this means the coefficient values for which *both* types of furniture are worth producing. These ranges are determined individually and independently (changing only one coefficient at a time, while keeping the others at the original value).

As you can see from the sample Sensitivity Report for our example, the allowable increase for the chairs (decision variable c) is 260. Since the original objective function coefficient (profit per chair) was 140, this means that the profit per chair could go as high as $140+260=\$400$ before the chairs would be so much *more profitable* than the tables that we would stop making the tables (so t would become 0 and drop out of the basis). Similarly, the allowable decrease for the profit per chair is 40, which means the profit could decrease by 40 from 140 to $140-40=\$100$ before the chairs were so *unprofitable* compared to the tables that we would stop making them (so c would become 0 and drop out of the basis). In either case, we assume that the profit per table remains at $200 per table.

Similarly, the allowable increase and decrease for tables are 80 and 130, respectively, so the profit per table could be anywhere between $200-130=\$70$ and $200+80=\$280$ to keep tables profitable enough to be produced (keep t in the basis) in the optimal solution, holding the profit per chair fixed at $140 per chair. Some optimization packages may specify the actual range of values (like 70 to 280) rather than the allowable increase and decrease values; see your technology supplement to check this. Usually the labeling on the report will make it clear.

To understand the allowable increase and decrease values for a decision variable, you could try changing that coefficient repeatedly and re-solve the problem to see when that variable switches between being basic and non-basic (leaves or enters the basis). Try this with the example on your own.

From our definitions, notice that, if a variable is *not* in the basis of an optimal solution and has a positive reduced cost, then the reduced cost will correspond to the allowable increase or decrease (depending on the type of optimization), since it indicates how much *better* the coefficient must be before the variable will enter the optimal basis. In our example, if the original objective function coefficient of c had been 80 instead of 140, then c would not have been in the optimal basis and would have a reduced cost of 20. This would mean that the profit per chair would have to increase by \$20 to \$100 to make chairs worth producing, so this would be the allowable increase for c. The allowable decrease for c would technically be infinity, since the profit per chair would already be so low that chairs would not be produced, so lowering the profit further would not change anything. Instead of "infinity," some packages will simply give a huge number (see Sample Problem 3 below, where an allowable increase is given as "1E+30", which means 10^{30}, which is a 1 followed by 30 zeroes!).

For the constraints, there is a column called Shadow Price. The **shadow prices** of the constraints, reflect the marginal effect on the objective function value of changing the RHS of the constraint. You can think of it as the **change** in the objective function (whichever direction that might be) if the constraint RHS were **increased** by 1 unit. More precisely, the value is simply the **partial derivative of the optimal objective function value with respect to the RHS of the constraint** (or the **Lagrange multiplier** for that constraint).

From the Sensitivity Report for our example, you can see that the shadow price of the wood constraint is approximately 28.89. This means that if we had one more unit of wood (61 board-feet, or 1 more board-foot than the original 60 board-feet), then our profit would go up by about $\$28.89$. Since the original optimal profit was \$2400, this would make the new profit about $\$2428.89$. In a sense, then, you can think of this as the *value* of each additional board-foot of wood, or the maximum *price* you would pay for an additional board-foot (hence the name shadow *price*), over and above whatever cost of wood is already included in the profits for each kind of furniture.

Similarly, the shadow price of the hardware constraint is about 6.67, which means that each additional package of hardware (at least initially) over the original 100 would bring in an additional $\$6.67$ in profit.

In some spreadsheet packages, the Reduced Cost and Shadow Price columns are both labeled "Dual" or "Dual Value". These refer to some advanced concepts in linear programming. To learn more about this, see a book on linear programming or Operations Research.

The columns for Allowable Increase and Allowable Decrease for the constraints are analogous to those for the variables, but refer in this case to the RHS's of the constraints. For example, the original RHS for the wood constraint was 60 (you had 60 board-feet of wood available). The Sensitivity Report says that the Allowable Increase is 90 and the allowable Decrease is 9, so the original optimal basis (chairs and tables both being produced) will remain optimal (although the values of each can change) as long as the RHS of the wood constraint is between 60-9=51 and 60+90=150 board-feet. This is exactly the range over which the interpretation of the shadow price applies (the change in the optimal profit per unit increase in the RHS). As before, it assumes we hold all of the other parameters of the problem (objective function and constraint coefficients and constraint RHS's) fixed at their original values.

Similarly, for the hardware constraint, the allowable increase is 6 and the allowable decrease is 60 (from the original RHS value of 100 packages of hardware), so the original optimal basis stays optimal (*both* types of furniture are worth producing) as long as the amount of hardware available is between 100-60=40 and 100+6=106 packages, assuming that all of the *other* parameters stay fixed at their original values.

Let's look back again at the tableau in Table 1, reprinted here for your convenience:

Basic Vars	P	c	t	s_1	s_2	s_3	RHS
(P)	1	0	0	260/9	20/3	0	2400
t	0	0	1	2/9	-1/12	0	5
c	0	1	0	-1/9	1/6	0	10
s_3	0	0	0	1/9	-1/6	1	1

Note how the Shadow Price value in the Sensitivity Report corresponds to the coefficient of the slack variables for each constraint in the objective function row of the optimal Simplex Method tableau (for example, $260/9 = 28.89$ and $20/3 = 6.67$). The shadow price of the wood constraint is approximately 28.89, which appears as 260/9 in the simplex tableau in the objective function row in the s_1 column. This corresponds to the constraint

$$3c + 6t \le 60 \ ,$$

since s_1 was the slack variable added to that constraint. Increasing the RHS would mean we had *more* wood on hand, which would *increase* the magnitude of both intercepts and the size of the feasible region, making more solutions possible, and therefore making a larger profit possible. This would also be true for *any* $\leq$ constraint.

As we said before, if we had 61 units of wood available instead of 60 (everything else in the problem staying the same as in the original), our optimal profit would increase from $2400 to approximately $2428.89 . Because the constraints and the objective function still all have the same slopes as before, the optimal solution is still at the intersection of the first two constraints, now $(1')$ and (2) instead of (1) and (2) . See Figure 9.4-1 to see what this corresponds to graphically.

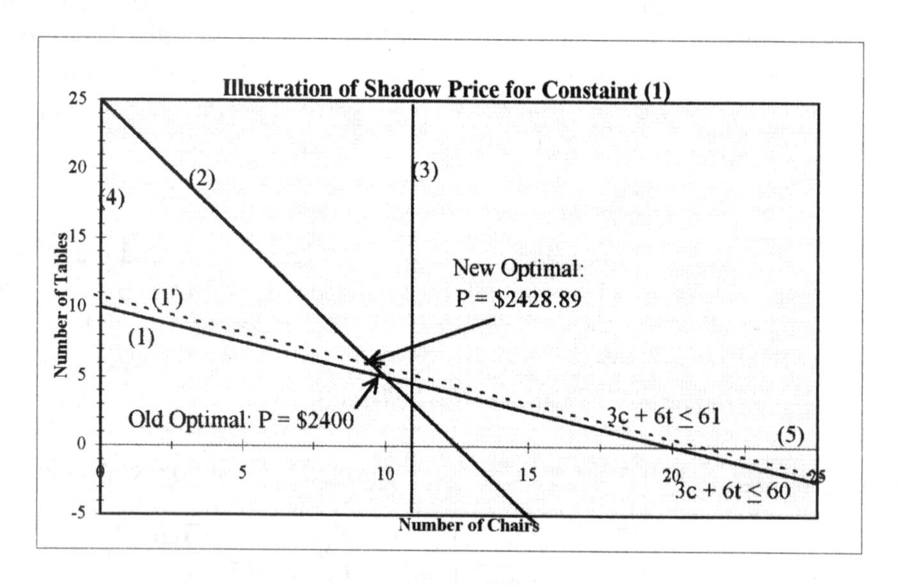

Figure 9.4-1

The only time this interpretation of the shadow price as the change in the optimal objective function if the RHS of the corresponding constraint is increased by 1 unit is not exactly right is when a basic variable equals 0 (called **degeneracy**) or the variable of interest can't increase by 1 unit because of the other constraints. Even so, it still gives the magnitude of the **instantaneous rate of change**, or **partial derivative** of the objective function value with respect to the RHS of each constraint *if* no such barriers existed (that is, *ignoring* such barriers), *assuming* that the constraints that *were* binding *continue* to be binding.

By similar reasoning, since the shadow price for the second constraint is approximately 6.67 , if we had 101 units of hardware and everything else was as in the original problem, then our profit would increase to approximately $2406.67 , as shown in Figure 9.4-2.

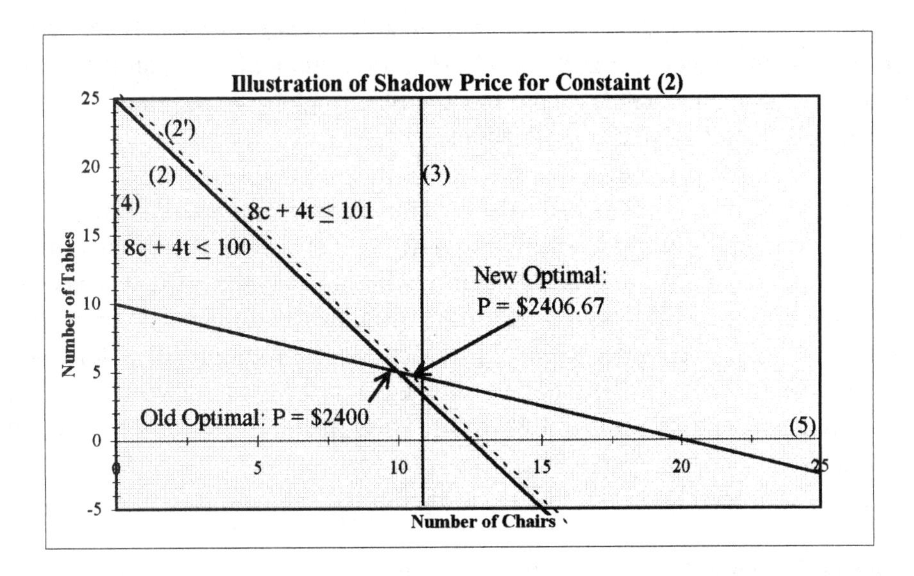

Figure 9.4-2

On the other hand, notice that the shadow price of the demand constraint $c \leq 11$ is not shown. This is because this constraint is not binding at the optimal solution, as we saw when we solved the problem graphically. Thus, if we relax the constraint (increase the RHS from 11 to 12, since this is again a $\leq$ constraint), the value of P will not change, because that constraint was not involved in determining the optimal corner point, so changing it won't affect our optimal solution. This is true in general: **any constraint that is not binding will have a shadow price of 0, and any constraint with a non-zero shadow price must be binding**. This property is referred to as **complementary slackness**. It *is* possible for a binding constraint to have a dual value of 0, however, if there are alternative optimal solutions (such as when the objective function is parallel to one of the *other* binding constraints at the optimal solution being presented). For an example, see Sample Problem 2 below.

Some spreadsheet packages may include a row for constraints like the nonnegativity constraints and our demand constraint $c \leq 11$, but the sample given (Excel) did not. In this example, the shadow prices would have all been 0 (as discussed above) and the allowable increases and decreases would have reflected the solution $c = 10$ and $t = 5$. For example, for the $c \leq 11$ constraint, the allowable decrease would have been 1 and the allowable increase would have been infinity (or 1E+30).

If the solution in the Answer Report does not seem right, you can change the initial solution by changing the initial values you put into the Variable Cell(s) (at the spreadsheet level). After this is done, simply go back to the spreadsheet optimizing tool menu, select **Solve** again, and proceed as before. This should not be necessary in linear programming, except possibly to try to find an alternative optimal solution (for example, if a shadow price of a binding constraint or the reduced cost of a nonbasic variable is 0), but even that could be complicated. On the other hand, it could be very important in optimizing a nonlinear problem, since local optimal solutions are not necessarily globally optimal. For an example, see Sample Problem 5 below, the cubic function.

Again, for the special case of LP that we considered in Section 9.2 , the reduced costs for the decision variables correspond to their coefficients in the objective function row of the simplex tableau. Thus, if a variable is *nonbasic* and has a dual value of 0, it suggests there are likely to be alternative optima. It is also true more generally that if a constraint is binding and the shadow price is 0, then there are likely to be alternative optimal solutions. Further discussion of these concepts can be found in more advanced texts in Operations Management, Management Science, and Operations Research. Thus, again, if the constraint is binding (so the slack variable is nonbasic, with a value of 0) and the dual value or shadow price is also 0, there is likely to be at least one more optimal corner point, as shown in Sample Problem 2.

Sample Problem 2: Let's look at a slight modification of our example problem, where we only change the profit coefficient of chairs, from \$140 to \$100. Find the new solution using a spreadsheet optimizing tool.

Solution : To clarify the difference, let's call the profit function $P_2 = 100c + 200t$. If we make this change in our problem formulation:

	A	B	C	D	E	F	G	H	I	J
1	HAND CRAFTED FURNITURE COMPANY									
2					LHS/	RHS				
3		CHAIRS	TABLES		FORMULA	CONSTANT				
4	QUANTITY	1	2							
5										
6	PROFIT EACH	100	200							
7	PROFIT TOTAL	100	400		500			Max. E7 = 100*B4+200*C4		
8								s.t.	3*B4+6*C4 <= 60	
9	WOOD EACH	3	6						8*B4+4*C4 <= 100	
10	WOOD TOTAL	3	12		15	60			B4 <= 11	
11									B4 >= 0	
12	HARDWARE EACH	8	4						C4 >= 0	
13	HARDWARE TOTAL	8	8		16	100				
14										
15	DEMAND - CHAIRS				2	11				

and start at the same initial solution as before, (1,2), we get the following Answer Report:

Target Cell (Max)

Cell	Name	Original Value	Final Value
$E$7	PROFIT TOTAL FORMULA	500	2000

Adjustable Cells

Cell	Name	Original Value	Final Value
$B$4	QUANTITY CHAIRS	1	0
$C$4	QUANTITY TABLES	2	10

Constraints

Cell	Name	Cell Value	Formula	Status	Slack
$E$10	WOOD TOTAL FORMULA	60	$E$10<=$F$10	Binding	0
$E$13	HARDWARE TOTAL FORMULA	40	$E$13<=$F$13	Not Binding	60
$E$15	DEMAND - CHAIRS FORMULA	10	$E$15<=$F$15	Not Binding	1
$B$4	QUANTITY CHAIRS	0	$B$4>=0	Binding	0
$C$4	QUANTITY TABLES	10	$C$4>=0	Not Binding	10

and the following Sensitivity Report:

Changing Cells

Cell	Name	Final Value	Reduced Cost	Objective Coefficient	Allowable Increase	Allowable Decrease
$B$4	QUANTITY CHAIRS	0	0	100	0	1E+30
$C$4	QUANTITY TABLES	10	0	200	1E+30	0

Constraints

Cell	Name	Final Value	Shadow Price	Constraint R.H. Side	Allowable Increase	Allowable Decrease
$E$10	WOOD TOTAL FORMULA	60	33.33333333	60	6	60
$E$13	HARDWARE TOTAL FORMULA	40	0	100	1E+30	60
$E$15	DEMAND - CHAIRS FORMULA	10	0	11	1E+30	1

Figure 9.4-3 shows the graph for this revised problem:

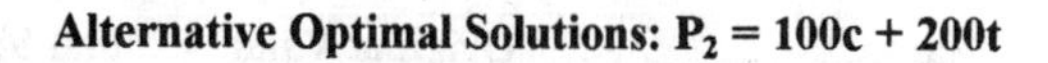

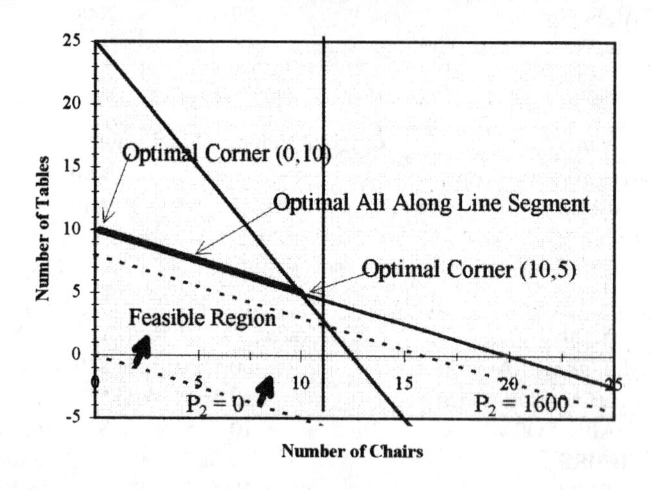

Figure 9.4-3

Notice from Figure 9.4-3 that the old optimal solution (10,5) is again optimal, so we can make 10 chairs and 5 tables, this time for a profit of $2000 (compared to $2400 before). Notice also from Table 2 that, although constraint 2 is binding (its slack variable, s_2, is 0, since it is not in the basis), its shadow price (the *coefficient* of its slack variable, s_2, in the objective function) is 0 (usually a binding constraint has a nonzero dual value).

Table 2

Basic Vars	P	c	t	s_1	s_2	s_3	RHS
(P)	1	0	0	300/9	**0**	0	2000
t	0	0	1	2/9	-1/12	0	5
c	0	1	0	-1/9	1/6	0	10
s_3	0	0	0	1/9	-1/6	1	1

As we saw in the preceding section, this means that increasing s_2 while keeping $s_1 = 0$ (moving to the left along constraint 1 from (10,5)) will not affect our profit, so all feasible solutions along the ray in that direction are also optimal (Figure 9.4-3). In other words, this means we have alternative optimal solutions. If we make s_2 the entering variable in a simplex iteration, we find that (0,10) is another optimal corner point. Table 3 shows this optimal solution.

Table 3

Basic Vars	P	c	t	s_1	s_2	s_3	RHS
(P)	1	**0**	0	50/3	0	0	2000
t	0	1/2	1	1/6	0	0	10
s_2	0	6	0	-2/3	1	0	60
s_3	0	1	0	0	0	1	11

Notice that the reduced cost of c, which corresponds to the coefficient of c in the objective function row of the tableau, is now 0, even though c is not a basic variable. It is a nonbasic variable, so has a value of 0. In other words, we can increase c (move to the right along the first constraint from $(0,10)$) and still get an optimal solution; in fact, this will get us right back to $(10,5)$, as noted earlier.

To recognize that there is an alternative corner point just from the spreadsheet output is difficult, but possible. The key is to look for a decision variable with a value of 0 whose reduced cost is also 0, *and* whose allowable increase is positive. This means that the variable can already be part of an optimal basis *and* that it can increase by some positive amount, thus arriving at another optimal solution. In the example above, notice that this is the case for the decision variable for the chairs, c. Another possible situation indicating alternative optimal solutions would be a binding constraint whose shadow price is 0 and whose allowable increase (in its RHS) is positive. This means that increasing the RHS will not change the objective function value (so it stays optimal) *and* that a positive change is possible. This does not occur in the solution reports for this example.

The above analysis is an example of "**What-if**" exploration (or **sensitivity analysis**), which means exactly what it says: we can explore *what* would happen *if* we made certain changes to our original problem. The above change would certainly be something under our control (we could lower the selling price of chairs and make less profit per chair). This exploration suggests that such a change would not be a good idea, since our optimal profit goes down. This is the kind of information that is summarized in the Sensitivity Report, with the reduced costs, shadow prices, and allowable increases and decreases. In a production context, the allowable increases and decreases show how much the problem parameters can change without changing which *combination* of products that should be produced (and which products should not be produced).

Thus the solution to the problem is to produce no chairs and 10 tables, or 10 chairs and 5 tables, or any solution in between. In fact, Table 3 gives us an easy way to describe all such solutions in between these 2 corner points. The row for the variable t corresponds to the equation

$$\tfrac{1}{2}c + t + \tfrac{1}{6}s_1 = 10$$

But since both solutions lie on the edge of constraint 1, for all optimal solutions, s_1 will be 0, and so we can plug in 0 for s_1 and solve for t, yielding:

$$t = 10 - \frac{1}{2}c = 10 - 0.5c \ , \ \text{for } 0 \le c \le 10 .$$

For example, if $c = 4$, then $t = 10 - 0.5(4) = 10 - 2 = 8$, so one optimal solution would be to make 4 chairs and 8 tables. □

Sample Problem 3: Let's try another "what if" exploration with our **original** problem (so the profit function is back to P=140c+200t). Suppose our demand constraint had been $c \le 10$ instead of $c \le 11$. Use a spreadsheet optimizing tool to find the new solution.

Solution: The formulation is given by:

	CHAIRS	TABLES	FORMULA	CONSTANT		
QUANTITY	1	2				
PROFIT EACH	140	200				
PROFIT TOTAL	140	400	540		Max. E7 =	140*B4+200*C4
					s.t.	3*B4+6*C4 <= 6
WOOD EACH	3	6				8*B4+4*C4 <= 1
WOOD TOTAL	3	12	15	60		B4 <= 10
						B4 >= 0
HARDWARE EACH	8	4				C4 >= 0
HARDWARE TOTAL	8	8	16	100		
DEMAND - CHAIRS			1	10		

Note that graphically, this means we have **three** constraints passing through the optimal point (10,5), so one of the basic variables is going to have a value of 0 (a situation called **degeneracy**, although not the kind that most of us think of!). This is because we will have an extra binding constraint at the optimal solution, so an extra variable will be zero. The graph of the solution is shown in Figure 9.4-4.

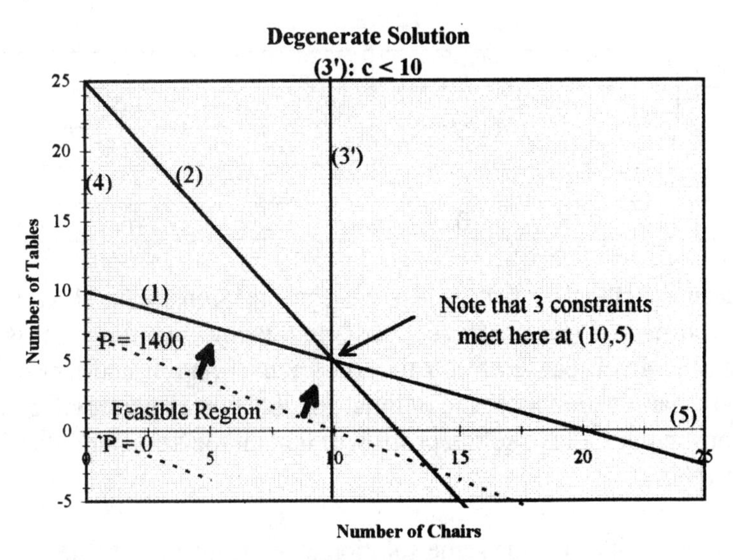

Figure 9.4-4

One Simplex Method tableau corresponding to the optimal solution is given in Table 4:

Table 4

Basic Vars	P	c	t	s_1	s_2	s_3	RHS
(P)	1	0	0	260/9	20/3	0	2400
t	0	0	1	2/9	-1/12	0	5
c	0	1	0	-1/9	1/6	0	10
s_3	0	0	0	1/9	-1/6	1	0

Looking at the tableau in Table 4, we see that our original solution point is still optimal, as we know it should be from the graph. The shadow price of the demand constraint (the coefficient of the third slack variable, s_3, in the objective function row) is 0, which should mean that if we increase the demand restriction from 10 to 11, our profit should not go up. In fact, we know that this is indeed the case from the solution to our original problem. The shadow price of the second constraint is still 6.67, as before. The problem is that the extra binding constraint (the demand constraint $c \leq 10$) keeps us from increasing c, so the *actual* effect of increasing the hardware limit would be *no change* in the optimal profit. The shadow price of 6.67 tells us that *if* that extra constraint were not there, the effect would be to increase the profit by 6.67 .

Another possible optimal simplex tableau is given in Table 5:

Table 5

Basic Vars	P	c	t	s_1	s_2	s_3	RHS
(P)	1	0	0	100/3	0	40	2400
t	0	0	1	1/6	0	-1/2	5
c	0	1	0	0	0	1	10
s_2	0	0	0	-2/3	1	-6	0

This tableau is telling us that, when we consider constraints 1 and 3 binding, the shadow price of constraint 2 (the coefficient of s_2 in the objective function) is 0. Looking at Figure 4, you can see this because, if you fix constraints 1 and 3, and move constraint 2 up and to the right in a parallel direction, the optimal feasible corner point will remain (10,5), the intersection of constraints 1 and 3, so the profit will remain the same, and the shadow price of constraint 2 *should* be 0.

A third possible tableau at the same solution again is given in Table 6:

Table 6

Basic Vars	P	c	t	s_1	s_2	s_3	RHS
(P)	1	0	0	0	50	-260	2400
t	0	0	1	0	¼	-2	5
c	0	1	0	0	0	1	10
s_1	0	0	0	1	-3/2	9	0

Notice that this time, the corner point is in fact optimal (as we know from the graphical solution in Figure 4), but the simplex tableau does not satisfy the optimality conditions. This is because, when we consider constraints 2 and 3 to be the binding constraints, the tableau indicates that making s_3 enter the basis and replacing s_1 (since its maximum increase ratio is 0, which is the smallest) would be the next step. Graphically, this corresponds to moving along constraint 2 up and to the left. If we did not have constraint 1 to worry about, this would in fact improve the profit (the shadow price would be 260), but the fact is that constraint 1 keeps us from moving in that direction *at all*, and doing the iteration would just end us up with the tableau in Table 4, where we would stop. Another way to think of this is that the maximum increase in the entering variable is 0, so we can't move from the corner point at all. But we don't *know* that the corner point is optimal until we think of it in the right way, by finding the combination of constraints to consider binding that *shows* the corner is optimal.

Thus, degenerate solutions can be tricky when solving a problem using the simplex method, because you may be at a corner point that is optimal without knowing it. They are also special because the shadow price values can be misleading. They may indicate changing the RHS of a constraint will improve the objective function, but another constraint may block that improvement from being possible. You can recognize this by looking at the allowable increase and decrease values. At a degenerate solution, one of them will be 0 for such a blocked improvement.

For example, let's look at a sample Answer Report for this problem:

Target Cell (Max)

Cell	Name	Original Value	Final Value
$E$7	PROFIT TOTAL FORMULA	540	2400

Adjustable Cells

Cell	Name	Original Value	Final Value
$B$4	QUANTITY CHAIRS	1	10
$C$4	QUANTITY TABLES	2	5

Constraints

Cell	Name	Cell Value	Formula	Status	Slack
$E$10	WOOD TOTAL FORMULA	60	$E$10<=$F$10	Binding	0
$E$13	HARDWARE TOTAL FORMULA	100	$E$13<=$F$13	Binding	0
$E$15	DEMAND - CHAIRS FORMULA	10	$E$15<=$F$15	Binding	0
$B$4	QUANTITY CHAIRS	10	$B$4>=0	Not Binding	10
$C$4	QUANTITY TABLES	5	$C$4>=0	Not Binding	5

and the accompanying Sensitivity Report:

Changing Cells

Cell	Name	Final Value	Reduced Cost	Objective Coefficient	Allowable Increase	Allowable Decrease
$B$4	QUANTITY CHAIRS	10	0	140	1E+30	40
$C$4	QUANTITY TABLES	5	0	200	80	200

Constraints

Cell	Name	Final Value	Shadow Price	Constraint R.H. Side	Allowable Increase	Allowable Decrease
$E$10	WOOD TOTAL FORMULA	60	33.33333333	60	0	30
$E$13	HARDWARE TOTAL FORMULA	100	0	100	1E+30	0

Notice that this solution corresponds to the Table 5 tableau (easiest to see from the shadow prices). These shadow prices imply that constraints 1 and 3 are considered the binding constraints (s_1 and s_3 are the nonbasic variables). Constraint 1 has a shadow price of approximately 33.33, but the allowable increase is 0, so in fact increasing the availability of wood would not improve the objective function.

Thus the solution to this problem is again to make 10 chairs and 5 tables, for a profit of \$2400. The only difference from the original analysis is in the interpretation of some of the shadow prices. □

Nonlinear Programming Using a Spreadsheet

Sample Problem 4: An airline allows one carry-on bag, the sum of whose three dimensions must be no more than 50 inches. Furthermore, the length can be no more than 26 inches, the width no more than 24 inches, and the height no more than 10 inches, so that it can fit under the seat. What dimensions would maximize the volume you can carry on?

Solution: This problem can be formulated as follows:

Let $l =$ the length of the bag (in inches),
$w =$ the width of the bag (in inches), and
$h =$ the height of the bag (in inches).

We want to find values of l, w, and h that will

Maximize $V = lwh$

subject to

$$l + w + h \leq 50$$
$$l \leq 26$$
$$w \leq 24$$
$$h \leq 10$$

Using an initial solution of (26,24,10) (note that this is not a feasible solution, since the sum of the values is more than 50), the formulation and Answer Report are as follows:

Luggage Problem				LHS/	RHS		
	Length	Width	Height	Formula	Constant		
Inches	26	24	10				
Volume				6240		Max. E5 = B3*C3*D3	
Dimen. Sum				60	50	s.t.	B3+C3+D3<=50
Max. Length				26	26		B3<=26
Max. Width				24	24		C3<=24
Max. Height				10	10		D3<=10

Target Cell (Max)

Cell	Name	Original Value	Final Value
$E$5	Volume Formula	6240	3999.999997

Adjustable Cells

Cell	Name	Original Value	Final Value
$B$3	Inches Length	26	20.00000005
$C$3	Inches Width	24	19.99999993
$D$3	Inches Height	10	10

Constraints

Cell	Name	Cell Value	Formula	Status	Slack
$E$7	Dimen. Sum Formula	49.99999998	$E$7<=$F$7	Binding	0
$E$8	Max. Length Formula	20.00000005	$E$8<=$F$8	Not Binding	5.999999951
$E$9	Max. Width Formula	19.99999993	$E$9<=$F$9	Not Binding	4.000000065
$E$10	Max. Height Formula	10	$E$10<=$F$10	Binding	0
$B$3	Inches Length	20.00000005	$B$3>=0	Not Binding	20.00000005
$C$3	Inches Width	19.99999993	$C$3>=0	Not Binding	19.99999993
$D$3	Inches Height	10	$D$3>=0	Not Binding	10

Changing Cells

Cell	Name	Final Value	Reduced Gradient
$B$3	Inches Length	20.00000005	0
$C$3	Inches Width	19.99999993	0
$D$3	Inches Height	10	0

Constraints

Cell	Name	Final Value	Lagrange Multiplier
$E$7	Dimen. Sum Formula	49.99999998	199.9999934
$E$8	Max. Length Formula	20.00000005	0

The printout indicates that the optimal solution (rounded off) is $l = 20$, $w = 20$, and $h = 10$, with $V = 4000$. In words, our carry-on bag should be 20"x20"x10", for a maximum volume of 4000 cubic inches. As we have seen in some earlier problems, there is some roundoff error often when you use a computer to solve a problem like this. If you round off your answers here, even to 4 or 5 places (more accurate than most real-world situations), however, the answer is correct.

Notice that the shadow price (now called a **Lagrange Multiplier**, or again a Dual value in some spreadsheets, for general optimization problems that are not specified to be linear) of the first constraint, Dimen. Sum Formula, is 200. This means that if the allowed sum of the dimensions was 51 inches instead of 50 inches, the maximum volume would increase approximately 200 cubic inches (the instantaneous rate of change, or partial derivative, would be exactly 200 cubic inches of volume per inch increase in the allowed sum). In fact, the new optimal solution (use a spreadsheet optimizing tool yourself to check this!) would be 20.5"x20.5"x10", with a volume of 4202.5 cubic inches (an increase of 202.5, which is close to 200, but not the same). In linear programming problems, the dual values give us the actual changes (since **derivatives are linear *approximations* in general, but give *exact* answers for linear functions**), but this is not usually true in nonlinear programming.

A spreadsheet optimizing tool can also be used for single-variable functions, even with no constraints (or with only upper and/or lower bounds on the variable), like the maximum and minimum functions on many graphing calculators. Let's use an example of this type to show how the initial values can affect the final solution.

Sample Problem 5: Consider again the problem of maximizing the function $f(x) = x^3 - 9x^2 + 24x$, for $0 \le x \le 6$, Sample Problem 7 of Section 8.3 . Use a spreadsheet optimizing tool to find the solution.

Solution: The following printout shows the layout, formulation, and Answer Report for the initial value $x = 3$:

Maximize f(x) = x^3-9x^2+24x for 0<=x<=6

x	f(x)
2	20

Target Cell (Max)

Cell	Name	Original Value	Final Value
$C$3	f(x)	18	20

Adjustable Cells

Cell	Name	Original Value	Final Value
$B$3	x	3	2.000000015

Constraints

Cell	Name	Cell Value	Formula	Status	Slack
$B$3	x	2.000000015	$B$3>=0	Not Binding	2.000000015
$B$3	x	2.000000015	$B$3<=6	Not Binding	3.999999985

Notice that the final value of x comes out to be 2 (when rounded off), with an objective function value of 20. Now, suppose we change the initial guess to 5, and select **Solve** again. This time, the printout is as follows:

Maximize f(x) = x^3-9x^2+24x for 0<=x<=6

x	f(x)
6	36

Target Cell (Max)

Cell	Name	Original Value	Final Value
$C$3	f(x)	20	36

Adjustable Cells

Cell	Name	Original Value	Final Value
$B$3	x	5	6

Constraints

Cell	Name	Cell Value	Formula	Status	Slack
$B$3	x	6	$B$3>=0	Not Binding	6
$B$3	x	6	$B$3<=6	Binding	0

This time the final value of x is 6, with an objective function value of 36, which is clearly better than the first solution we got. To see what happened, let's graph the function over its domain (Figure 9.4-5).

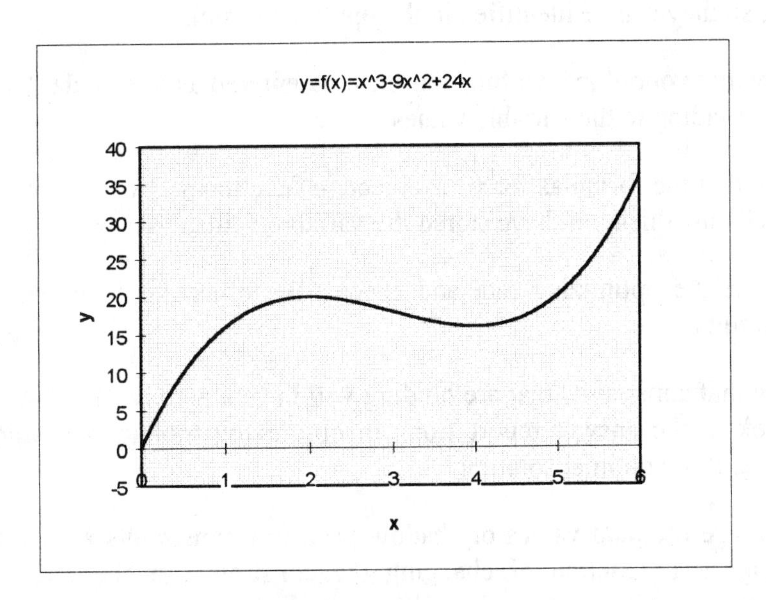

Figure 9.4-5

Now we can see that there is a local (relative) maximum at $x = 2$, which is what the first search obtained, but the global maximum over the domain is actually $x = 6$, as the second search found. The point of this exercise is to show that **nonlinear optimization can be very tricky**, and in general uses many advanced techniques and concepts to try to find global optima. However, for realistic problems, you will usually have some idea of where an optimal solution might lie, and can enter those values for your initial values of the variables. Even if you get a reasonable-looking solution, it might be worth your while to try a few other randomly located initial solutions, just in case you might find something better. Also, **be aware that the software might not be able to find an optimal solution for some nonlinear problems**.

One option in the optimization routine of a spreadsheet is to specify a value that you want your objective function (target cell) to *equal*, rather than maximizing or minimizing it. This is essentially a way to use the routine as a way to solve one or more equations. For example, if you have a system of equations, you could specify the LHS of one of them as your target cell and make the RHS (constant) of that constraint the target value. Then you can simply fill in the other equations as constraints.

Section Summary

Before you begin the exercises for this section, be sure that you:

- Know that your variable values should correspond to, and be entered into, adjacent cells, so they can be identified in the optimizing tool.
- Know that your objective function must be entered as a formula that refers to the cells corresponding to the variable values.
- Know that the formulas for your objective function and your constraints must refer to the cells in which you have stored the variable values.
- Can use the optimizing tool and specify the objective function, variable cells and constraints.
- Know that constraints that are binding will have a zero slack value, and so know how to look at the answer report from an optimizing tool to see which constraints are binding at the optimal solution.
- Recognize the dual values or shadow prices of constraints as the marginal effect on the objective function of changing the constraint, or the effect on the objective function per unit change in the RHS constant of that constraint.
- Understand intuitively why a constraint that is nonbinding at a solution point should have a dual value (shadow price) of 0, and why constraints with a nonzero dual value must be binding, at that point.

- Know that a **degenerate** solution to an LP means a solution in which a basic variable has a value of 0 (an extra constraint, beyond those *needed* to determine the corner point, is binding). In a problem with two decision variables, this means three constraint lines going through a corner point.

- Realize that the starting values that you select may affect the "optimal" solution.

- Are aware of both the power and the limitations of spreadsheet optimizing tools.

EXERCISES FOR SECTION 9.4:

Warm Up

1. Solve the following linear program using a spreadsheet program.

$$\begin{aligned} \text{Maximize } & z = 3x + 4y \\ \text{subject to } & x + y \leq 20 \\ & 5x + 8y \leq 120 \\ & x, y \geq 0 \end{aligned}$$

2. Solve the following linear program using a spreadsheet program

$$\begin{aligned} \text{Max. } & P = 5x_1 + 4x_2 \\ \text{s.t. } & 3x_1 + 4x_2 \leq 24 \\ & 10x_1 + 8x_2 \leq 62 \\ & x_1, x_2 \geq 0 \end{aligned}$$

3. Solve the following linear program using a spreadsheet program.

$$\begin{aligned} \text{Max. } & z = 5x + 7y \\ \text{s.t. } & 3x + 5y \leq 60 \\ & 6x + 4y \leq 90 \\ & y \leq 5 \\ & x, y \geq 0 \end{aligned}$$

4. Solve the following linear program using a spreadsheet program

$$\begin{aligned} \text{Min. } & z = 2x + 3y \\ \text{s.t. } & 3x - y \geq 12 \\ & x + y \geq 8 \\ & y \leq 6 \\ & x, y \geq 0 \end{aligned}$$

5. Solve the following linear program using a spreadsheet program

$$\begin{aligned} \text{Min. } & z = 2x + 3y \\ \text{s.t. } & 3x - y \geq 12 \\ & x + y \geq 8 \\ & y \leq 6 \\ & x \leq 10 \end{aligned}$$

6. Solve the following linear program using a spreadsheet program

$$\begin{aligned} \text{Min. } & z = 2x + 3y \\ \text{s.t. } & 3x - y = 12 \\ & x + y \geq 8 \\ & y \leq 6 \\ & x, y \geq 0 \end{aligned}$$

7. Solve the following linear program using a spreadsheet program

$$\begin{aligned} \text{Max. } & z = 5x + 7y \\ \text{s.t. } & x - 4y \leq 0 \\ & x + 3y \leq 12 \\ & x, y \geq 0 \end{aligned}$$

8. Solve the following linear program using a spreadsheet program

$$\begin{aligned} \text{Max. } & z = 6x + 10y \\ \text{s.t. } & 3x + 5y \leq 60 \\ & 6x + 4y \leq 90 \\ & y \leq 5 \\ & x, y \geq 0 \end{aligned}$$

Game Time

9. Use a spreadsheet program to solve a modified version of the furniture problem: the profit from each chair is $100 rather than $140 (the profit from each table is still $200); the chairs use 8 units of hardware and the tables use 4 units of hardware; the chairs use 3 units of wood and the tables use 6 units of wood; we have on hand 100 units of hardware and 60 units of wood; and no more than 11 chairs can be sold.

10. Use a spreadsheet program to solve a different modified version of the furniture problem, if the demand limit on the number of chairs that can be sold drops from 11 to 10:

$$\begin{aligned} \text{Max } P = \; & 140c + 200t \\ \text{s.t. } & 3c + 6t \leq 60 \\ & 8c + 4t \leq 100 \\ & c \leq 10 \\ & c, t \geq 0 \end{aligned}$$

11. Use a spreadsheet program to solve the furniture problem if both of the modifications from exercises 9 and 10 are made at the same time:

$$\begin{aligned} \text{Max } P &= 100c + 200t \\ \text{s.t.} \quad 3c + 6t &\le 60 \\ 8c + 4t &\le 100 \\ c &\le 10 \\ c,t &\ge 0 \end{aligned}$$

12. You are trying to raise money for your sports team by selling T-shirts and hats. You can order both items from the same company, which will charge you $4 for each hat and $6 for each shirt, plus there is a fixed cost of $30 for any order. Your team has $100 to spend for everything, and has decided to sell the shirts for $12 and the hats for $7, at which price you think you can sell no more than 15 hats and 10 shirts. Assume that both items are "one size fits all" and that the demands for the two products are independent for this exercise. Use a spreadsheet program to decide how many of each item you should order in this situation. What is your team's optimal profit? What assumption is violated here?

13. Suppose you run a small candle-making business, making regular and deluxe candles. You can sell the regular candles for a profit of $4 each and the deluxe candles for a profit of $7 each. Regular candles use 6 oz. of wax and 5 inches of wick material, while deluxe candles use 9 oz. of wax and 6 inches of wick material. You have 120 oz. of wax and 110 inches of wick material in your studio at the moment, and will not get any more supplies until tomorrow. Use a spreadsheet program to determine how many candles of each type you should make today. What assumption causes a problem here? How could you deal with it?

14. You are trying to decide between two dog foods for your Golden Retriever, Ralph. Purina costs $32 for a 10-pound bag, and each ounce of food has 4 grams of fat, 8 grams of protein, and 200 calories. Iams costs $20 for a 5-pound bag, and each ounce of food has 1 gram of fat, 6 grams of protein, and 220 calories. You want Ralph to get at least 1000 calories per day, but not more than 1400 calories. You also want him to get no more than 10 grams of fat and at least 40 grams of protein. Use a spreadsheet program to determine what combination of the two dog foods you should give Ralph every day to meet your requirements and minimize cost.

15. (Problem adapted from Introduction to Operations and Production Management , Nydick et al.) Six Star Refinery blends four petroleum components into three grades of gasoline: regular, premium and ultra-premium. The component specifications are given:

GRADE	COMPONENT SPECIFICATIONS
REGULAR	40% component A 20% component B 10% component C remainder component D
PREMIUM	50% component C remainder component D
ULTRA-PREMIUM	10% component A 60% component C remainder component D

Component A costs $9 per barrel, B costs $7 per barrel, C costs $12 per barrel and D costs $6 per barrel. Regular gasoline sells for $12 a barrel, premium for $15 per barrel and ultra premium for $18 per barrel. Write a function expressing the total cost of the components if you produce r barrels of regular, p barrels of premium, and u barrels of ultra-premium. If there are 100 barrels of each component available, fully define a model for the profit, and then use a spreadsheet program to determine how many barrels of each grade of gasoline should be made to maximize profit.

16. If the material for the four sides (including the front and back) and the top of a box costs $2 per square foot and the material for the base of the box (which must be stronger) costs $3 per square foot, and the box must have a volume of 6 cubic feet, use a spreadsheet program to determine the dimensions of the box that will minimize the cost.

17. In one day, the cost, in hundreds of dollars, to produce a serving of Caramel Cream Fudge and Chocolate Caramels can be modeled by: $C(f,c) = 21.71f + 3.54c$ where f = price, in dollars, of a serving of fudge and c = the price, in dollars, of a serving of caramels and the revenue, in hundreds of dollars, from the sale of the fudge and caramels is given by: $R(f,c) = -0.43(f)^2 + 19.56f + 1.45fc - 1.93(c)^2 + 13.97c$. Use a spreadsheet program to find the price to charge for each candy to maximize the profit, and find the maximum profit as well.

18. A cup in the shape of a cone is to be made from a circular piece of material with a diameter of 10 inches. Use a spreadsheet program to determine the measurements (height and radius) of the cone that will maximize the volume. ($V = \frac{\pi r^2 h}{3}$)

19. You want to choose a breakfast cereal that will minimize your costs while satisfying some dietary goals. One ounce of Special K costs $0.40 and contains no fat, 1 gram of dietary fiber, 150 calories and 6 grams of protein. One ounce of Raisin Bran cost $0.30 and contains 1 gram of fat, 7 grams of dietary fiber, 210 calories and 4 grams of protein. You have decided that you want to get at most 2 grams of fat, at least 5 grams of fiber, at most 350 calories, and exactly 9 grams of protein at breakfast. Use a computer spreadsheet program to determine what combinations of the two cereals meets your specifications at the lowest cost. What would be the effect of raising the fiber requirement by 1 gram? What would be the effect of raising the protein requirement 1 gram? What would be the effect of raising the calorie limit 1 calorie?

Chapter 9 Summary

Lagrange multipliers can be used directly to optimize an objective function subject to one or more *equality* constraints, but can also be used indirectly to solve problems with *inequality* constraints. The basic idea behind the method of Lagrange multipliers is to impose a *penalty* (the Lagrange multiplier) per unit *violation* of a constraint (its LHS minus its RHS), which is analogous to charging a *fee* for violating the constraint on *one* side and offering a *bonus* for violating it on the *other* side. By adjusting this penalty (possibly switching the sign if necessary, to *reverse* which side you *reward*), solutions that come closer and closer to satisfying the constraint can be obtained, and a *perfect* value can be determined so that the equality constraint is satisfied *exactly*. The perfect value of the Lagrange multiplier(s) can be found by finding critical points for the Lagrangean function, which is formed by starting with the original objective function and subtracting a Lagrange multiplier for each equality constraint *times* (its LHS minus its RHS). The optimal value of a Lagrange multiplier for a constraint corresponds to its **shadow price**, the instantaneous rate of change of the *optimal* objective function value per unit increase in the RHS of that constraint

Linear programming (LP) problems are problems that involve optimizing a linear objective function subject to linear inequality or equality constraints. A convex set is one with "no dents" in it, where any line segment joining two points in the set is contained completely within the set, and an LP always has a convex feasible region. A maximization LP with two decision variables corresponds graphically to looking for the highest point on the plane of the objective function within the image of the convex feasible region vertically projected onto that plane. This is analogous to looking for the high point of a fenced-in pasture on a flat slanted field.

To solve a two-variable LP graphically, first graph the constraints and the feasible region. Then pick a sample objective function value, and draw in a contour line for that value, plus a second line through the origin, parallel to it. Indicate which direction is better, and visually try to determine the corner point in the feasible region that achieves the best objective function value.

For a linear programming problem (an LP) in "Maximize" form with inequality constraints and two decision variables, the goal of the Simplex Method is analogous to finding the highest point(s) within a fenced-in pasture on a flat but slanted field (plane) in the midst of dense fog. The basic idea of the Simplex Method for a maximization problem using the fenced-in pasture analogy is to start at a corner point and check the slope of the fence in both directions from that corner to test optimality (if neither edge slopes strictly *up*, then the current corner is optimal). If the current corner is not optimal, then move up along the steeper edge coming out of it until you hit another corner, then repeat the process until you reach an optimal point.

To solve a problem with the Simplex method, first add in slack variables to "≤" structural constraints that have nonnegative right-hand-side constants (RHS's). To interpret their values (graphically for problems with two decision variables): if the slack variable is 0 at a point, then the point is on the *edge* of the constraint; if the slack variable is *positive* at a point, then the point is *strictly within* the constraint, on the *feasible* side; and if the slack variable is *negative* at a point, then the point is *strictly beyond* the constraint, on the *infeasible* side.

A constraint is **binding** at a given point if its left-hand-side value (with the values of the point plugged in) is *equal* to its RHS value (if it holds as an *equation*). If the constraint is binding at the point, then *some* variable (for a nonnegativity constraint, its decision variable; or for a structural constraint, its slack variable) is equal to zero. At a feasible corner point, those variables that are zero and are needed to determine the corner point are called the *nonbasic variables*.

If an LP is in "Maximize" form with n nonnegative original decision variables and m structural constraints in "≤" form with nonnegative RHS's, a feasible corner point will correspond to the intersection of n constraints. In other words, at a corner point, there will be n binding constraints, and therefore n variables required to be 0 (n nonbasic variables). Since there are m constraints, there will be m slack variables, in addition to the original n decision variables, plus the variable for the objective function. The variables that are not required to be zero to determine the corner point are called the *basic variables*. So at a corner point there will be n nonbasic variables, m basic variables, and the objective function variable, which acts similarly to a basic variable but is treated in a special way (it can *never leave* the basis).

The Simplex method uses tableaus, or augmented matrices to solve an LP. Each tableau corresponds to a corner point solution. The basic variables can be recognized because their columns look like columns of an identity matrix; their values can be read off from the RHS values in the tableau. The other variables are nonbasic, and all have a value of 0. The Simplex method starts at the origin. Given a Simplex tableau, you can check it for optimality by looking at the signs of the coefficients in the objective function row. If you are maximizing, these should all be non-negative for a solution to be optimal (you can only go down from that point). If it is not optimal, select the variable with the most negative coefficient (steepest improvement) in the objective function row to be your entering variable (pivot column). Of all the positive entries in the pivot column, take the ratio of the RHS over the pivot column coefficient. This is the maximum increase in the entering variable to keep all constraints feasible. The Maximum Increase will be ∞ for basic variables (rows), where the coefficient in the pivot column is 0 or negative. It corresponds to how far you can move in the chosen direction (along an edge) before hitting that constraint. Choose the *smallest* of these to determine the leaving variable (the first constraint hit in that direction), and the pivot row.

Given an entering variable and a leaving variable, a simplex iteration corresponds to doing the row operations (as in Gaussian elimination) needed to make the pivot column look like a column of an identity matrix, with a 1 in the pivot row, and 0's everywhere else. This can usually be done by dividing the pivot row by the pivot element (multiplying by its reciprocal), then adding a multiple of the *new* pivot row to each old row, where the multiple is simply the *negative* (opposite sign) of the value in the old pivot column for that row. Graphically, this is moving from one corner to an adjacent corner. Algebraically, it is solving for the new basic variables in terms of the new nonbasic variables.

Once a solution is obtained with all the coefficients in the objective function row having the correct signs, an optimal solution has been found. If a nonbasic variable has a coefficient of 0, there could be alternative optimal solutions, which could be found by making that variable enter the basis.

The coefficients of the slack variables in the final objective function row give the shadow prices (Lagrange multiplier values) for the corresponding constraints, which is the marginal effect on the optimal objective function value of increasing the RHS of that constraint.

Linear or nonlinear optimization with constraints can be performed with spreadsheets. Output reports give the values of the variables, which constraints are binding (0 slack), the shadow prices, and sometimes the range of parameter values that would keep the basis optimal. For nonlinear problems, different starting values can yield different solutions, so testing procedures should be used. Nonbinding constraints will always have shadow prices of 0, since changing them slightly will not affect the optimal solution.